Springer Tracts in Mechanical Engineering

Springer Tracts in Mechanical Engineering (STME) publishes the latest developments in Mechanical Engineering - quickly, informally and with high quality. The intent is to cover all the main branches of mechanical engineering, both theoretical and applied, including:

- Engineering Design
- Machinery and Machine Elements
- Mechanical Structures and Stress Analysis
- Automotive Engineering
- Engine Technology
- Aerospace Technology and Astronautics
- Nanotechnology and Microengineering
- Control, Robotics, Mechatronics
- MEMS
- Theoretical and Applied Mechanics
- Dynamical Systems, Control
- Fluids Mechanics
- Engineering Thermodynamics, Heat and Mass Transfer
- Manufacturing
- Precision Engineering, Instrumentation, Measurement
- Materials Engineering
- Tribology and Surface Technology

Within the scope of the series are monographs, professional books or graduate textbooks, edited volumes as well as outstanding PhD theses and books purposely devoted to support education in mechanical engineering at graduate and post-graduate levels.

Indexed by SCOPUS. The books of the series are submitted for indexing to Web of Science.

Please check our Lecture Notes in Mechanical Engineering at http://www.springer.com/series/11236 if you are interested in conference proceedings.

To submit a proposal or for further inquiries, please contact the Springer Editor **in your country**:

Dr. Mengchu Huang (China)
Email: mengchu.Huang@springer.com
Priya Vyas (India)
Email: priya.vyas@springer.com
Dr. Leontina Di Cecco (All other countries)
Email: leontina.dicecco@springer.com

More information about this series at http://www.springer.com/series/11693

Yong Chen

Automotive Transmissions

Design, Theory and Applications

机械工业出版社
CHINA MACHINE PRESS

Springer

Yong Chen
Hebei University of Technology
Tianjin, China

ISSN 2195-9862 ISSN 2195-9870 (electronic)
Springer Tracts in Mechanical Engineering
ISBN 978-981-15-6702-5 ISBN 978-981-15-6703-2 (eBook)
https://doi.org/10.1007/978-981-15-6703-2

Jointly published with China Machine Press
The print edition is not for sale in China (Mainland). Customers from China (Mainland) please order the print book from: China Machine Press.
ISBN of the Co-Publisher's edition: 978-7-111-59945-6

This Springer imprint is published by the registered company Springer Nature Singapore Pte Ltd.
The registered company address is: 152 Beach Road, #21-01/04 Gateway East, Singapore 189721, Singapore

Foreword

Professor Chen Yong has been engaged in the research and development of automotive transmissions for a long time. He worked in the world famous automatic transmission company (JATCO) of Nissan Motor for 19 years, and completed the development and mass production of many automatic transmission products; after that, he worked in domestic automobile enterprises for nine years, leading the independent research and development of manual and automatic transmissions of Geely Automobile, and achieved fruitful results; now, he is devoted to academic study in colleges and universities and committed to sorting out and systematically refining his practical experience and technical accumulation of transmission research over 30 years, and sharing it with others so as to benefit the industry and help the development and progress of China's automotive transmission business. The book *Theory, Design and Application of Automotive Transmissions* of the leading coauthor Chen Yong accumulates the profound knowledge and valuable experience of first-class transmission experts and scholars at home and abroad, elaborates the international advanced technology and development trend of automotive transmissions and emphatically explains the mature design theory in the automotive transmission field and the latest international research and application progress. It can be used as an engineering technical book for automotive and transmission R&D engineers and a reference book for graduates and undergraduates to learn basic theories and carry out academic research. It is a rare masterpiece of both theory and practice.

Zhao Fuquan
Rotating Chairman of FISITA (2018–2020)

President of Automotive Industry
and Technology Strategy Institute
Tsinghua University, China

Preface

In order to change the output torque and speed of the engine and other power systems, it is usually necessary to install a transmission in the transport machinery. This book deals mainly with the transmissions of road vehicles. Under different driving conditions, such as standing start, climbing, turning and acceleration, the torque and speed required by the driving wheel of the vehicle are constantly changing, while the range of torque and speed change that the engine can provide is limited. The transmission is to adjust the performance of the engine by changing the gear ratio and transfer the engine power to the wheels smoothly, reliably and economically, so as to adapt well to the demands of external load and road conditions and achieve the best match between the characteristic field provided by the engine and that required by the vehicle.

The automotive transmission is a high technology and process level of typical products in mass production, and its development and design shall be oriented to market demands while meeting environmental and regulatory requirements. Regulatory requirements (such as energy conservation and emissions policies) and user requirements must be fully considered. The main design objective of the automotive transmission is to achieve the optimal transformation from engine or motor power to the vehicle driving force within the wide speed range of the vehicle, so as to ensure the dynamic property, acceleration and fuel economy of the vehicle. Meanwhile, the application reliability and service life of new technologies and processes shall be considered. The transmission design will be increasingly challenging given the increasing demand for fuel consumption, emissions and driveability, especially shift comfort and response speed. The current types of transmission for passenger vehicles, including Manual Transmission (MT), Automatic Transmission (AT), Dual Clutch Transmission (DCT), Continuously Variable Transmission (CVT) and Automated Mechanical Transmission (AMT) will remain largely unchanged for a long time. However, the application of hybrid transmissions will grow substantially. The first author of this book worked as a senior technician in the AT&CVT R&D in the R&D center of JATCO in Japan for 19 years and was in charge of the R&D of AT and MT in Geely Automobile for a long time after returning from abroad. He was deeply aware of the rapid progress of transmission technology and accelerated R&D cycle brought by computer simulation technology, and also deeply aware of the urgent demand for systematic explanation of the transmission development technology in the research and development.

Therefore, the main purpose of this book is to systematically summarize the main technical areas of the MT and AT development technology based on practical development experience, comprehensively elaborate the theory and development characteristics of the transmissions and display the main development processes of the transmissions. In particular, many development processes are based on the experience of trial and error. As a reference book providing main technical information for the research and development of transmission technology, this book is suitable for engineers working in the field of automobiles and related power transmission machines and graduate students at school. I would also like to thank Dr. Guo Lishu, senior technical expert of Geely Commercial Vehicle Research Institute, Dr. Gao Bingzhao of the College of Automotive Engineering, Jilin University, for their hard work in this book. I would like to thank Dr. Tian Hua from SGM Powersoon Research Institute, my doctoral student Zang Libin, Qiu Zizhen, Li Guangxin, Wei Changyin, my assistant Cao Zhan and other graduate students from the NEV Research Center of the Hebei University of Technology, as well as others who have worked hard for this book. I would like to thank the Tianjin Science and Technology Association for subsidizing this book.

Tianjin, China

Yong Chen

Contents

1 **Introduction** 1
- 1.1 Transmission Functions and Requirements 2
- 1.2 Types, Advantages and Disadvantages of Transmissions 3
- 1.3 Basic Structure of Transmission 12
- 1.4 Development Status and Trend of Transmission 14
- Bibliography 21

2 **Manual Transmission** 23
- 2.1 Overview 23
- 2.2 Transmission Drive Mechanism 23
- 2.3 Synchronizer 26
- 2.4 Transmission Operating Mechanism 38
- 2.5 Determination of Gear Ratio 42
- 2.6 Joint Working of Engine and Transmission 49
- 2.7 Shift Performance Evaluation 54
- 2.8 New Technologies of MT 54
- Bibliography 60

3 **Automatic Transmission** 61
- 3.1 Overview 61
- 3.2 Composition and Control Principle of AT 63
- 3.3 Mechanical Structure of AT 66
- 3.4 AT Speed Change Process Analysis 78
- 3.5 Hydraulic Control System of AT 84
- 3.6 Electronic Control System of aT 98
- 3.7 AT Performance Tests 99
- 3.8 Development Direction of aT 103
- Bibliography 108

4 Continuously Variable Transmission 109
4.1 Overview 109
4.2 Composition of CVT 111
4.3 Composition and Drive Theory of VDT Belt 117
4.4 Composition and Principle of Hydraulic Control System 129
4.5 CVT Electronic Control System 140
4.6 Control of CVT 145
4.7 Main Performance Tests of Metal Belt CVT 145
4.8 CVT Upgrade 148
Bibliography 153

5 Dual Clutch Transmission 155
5.1 Overview 155
5.2 System Composition and Working Principle of DCT 157
5.3 Typical Design Scheme of DCT 159
5.4 Dual Clutch 162
5.5 Select-Shift Actuator 177
5.6 Hydraulic Control System 180
5.7 Control System Hardware Design 190
5.8 Control System Software Design 195
Bibliography 199

6 Automated Mechanical Transmission 201
6.1 Overview 201
6.2 Composition and Working Principle of AMT Control System 202
6.3 Shifter 207
6.4 Clutch 210
6.5 Select-Shift Actuator 214
6.6 Hydraulic Control System 217
6.7 AMT Control Strategy 222
6.8 AMT Performance Evaluation Indexes 227
Bibliography 232

7 Transmission for New Energy Vehicle 233
7.1 Overview 233
7.2 Power and Economy Performance of HEV 235
7.3 AMT Hybrid Transmission 240
7.4 AT Hybrid Transmission 243
7.5 CVT Hybrid Transmission 248
7.6 DCT Hybrid Transmission 254
7.7 Planetary Gear Hybrid Transmission 257
7.8 Electric Vehicle Transmission 261
7.9 Other Hybrid Power Plants 267
Bibliography 271

8 Transmission Design . . . 273
8.1 Gear Design . . . 273
8.2 Shaft Design . . . 292
8.3 Bearing Selection and Design . . . 297
8.4 Case Design . . . 300
8.5 Parking Mechanism Design . . . 308
8.6 Synchronizer Design . . . 328
8.7 Selection of Seals . . . 345
8.8 Transmission Ventilation Design . . . 350
8.9 Transmission Tests . . . 351
Bibliography . . . 356

9 Transmission Fluid . . . 357
9.1 MTF . . . 357
9.2 ATF . . . 358
9.3 CVTF . . . 358
9.4 DCTF . . . 360
9.5 Performance Requirements and Tests for Transmission Fluid . . . 362
9.6 Selection of Transmission Fluid . . . 366
Bibliography . . . 369

10 Design of Hydraulic Torque Converter . . . 371
10.1 Working Principle and Characteristics of Hydraulic Torque Converter . . . 371
10.2 Pre-design of Hydraulic Torque Converter . . . 383
10.3 Blade Shape Design . . . 391
10.4 Numerical Simulation and Analysis of Internal Flow Field of Hydraulic Torque Converter . . . 409
10.5 Parameter Adjustment of Hydraulic Torque Converter . . . 428
10.6 Hydraulic Torque Converter Matching with Engine . . . 432
Bibliography . . . 435

11 Planetary Gear Drive . . . 437
11.1 Theoretical Calculation of Transmission Efficiency of Planetary Gear Train . . . 439
11.2 Transmission Efficiency Test of Planetary Gear Train . . . 454
11.3 Theoretical Calculation of Vibration and Noise of Planetary Gear Train . . . 464
11.4 Vibration and Noise Test of Planetary Gear Train . . . 470
Bibliography . . . 483

12 Electronic Control System of Automatic Transmission . . . 485
12.1 Introduction to AT Electronic Control System . . . 485
12.2 AT Control System Development . . . 488

12.3 Validation of System Validity 503
12.4 Control Strategies of AT 506
12.5 Modeling of Hydraulic Torque Converter with Lockup Clutch 507
12.6 Lockup Clutch Oil Circuit 508
12.7 Study on Control Strategies for Lockup Process of Lockup Clutch 510
Bibliography 532

13 Automobile and Transmission Vibration and Noise 533
13.1 Vibration and Noise Foundation 533
13.2 Automobile Vibration and Noise 537
13.3 Typical Automobile Vibration and Noise 545
13.4 Analysis of Transmission Vibration and Noise 553
13.5 Typical Transmission Vibration and Noise Control 563
Bibliography 569

Chapter 1
Introduction

Since the small torque range of the automobile engine cannot adapt to the requirements of automobile driving under various road conditions, the transmission that can change the speed ratio and drive torque is adopted in the automotive drivetrain to make the torque and speed output by the engine and other power systems vary in a considerable range. This book deals mainly with the transmissions of road vehicles. Under different driving conditions, such as standing start, climbing, turning and acceleration, the torque and speed required by the driving wheel of the vehicle are constantly changing, while the range of torque and speed change that the engine can provide is limited. The transmission is to adjust the performance of the engine by changing the gear ratio and transfer the engine power to the wheels smoothly, reliably and economically, so as to adapt well to the demands of external load and road conditions and achieve the best match between the characteristic field provided by the engine and that required by the vehicle.

The automotive transmission is a high technology and process level of typical products in mass production, and its development and design shall be oriented to market demands and fully consider the user requirements while meeting environmental and regulatory requirements (such as energy conservation and emissions policies), as shown in Fig. 1.1. The main design objective of the automotive transmission is to achieve the optimal transformation from engine or motor power to the vehicle driving force within the wide speed range of the vehicle, so as to ensure the dynamic property, acceleration and fuel economy of the vehicle. Meanwhile, the application reliability and service life of new technologies and processes shall be considered. The transmission design will be increasingly challenging given the increasing demand for fuel consumption, emissions and driveability, especially shift comfort and response speed. The current types of transmission for passenger vehicles, including manual transmission (MT), automatic transmission (AT), continuously variable transmission (CVT), dual clutch transmission (DCT) and automated mechanical transmission (AMT) will remain largely unchanged for a long time. However, the hybrid transmissions will be more widely used and their number will increase dramatically. The author of this book worked as a senior technician in the

Y. Chen, *Automotive Transmissions*, Springer Tracts in Mechanical Engineering,
https://doi.org/10.1007/978-981-15-6703-2_1

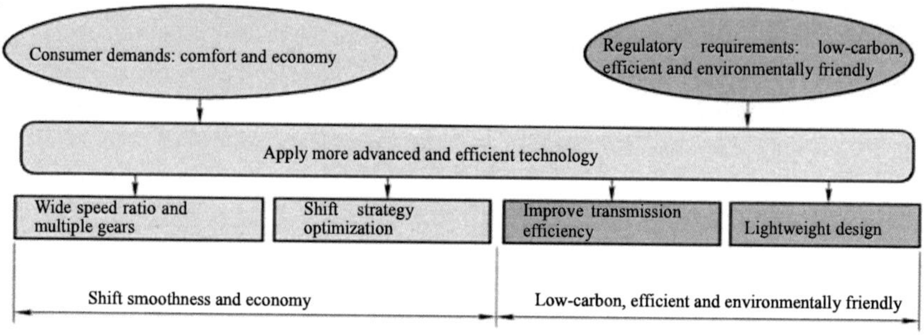

Fig. 1.1 Transmission design requirements

AT&CVT R&D in the R&D center of JATCO in Japan for 19 years and was in charge of the R&D of AT and MT in Geely Automobile for a long time after returning from abroad. He was deeply aware of the rapid progress of transmission technology and accelerated R&D cycle brought by computer simulation technology, and also deeply aware of the urgent demand of the R&D personnel for systematic explanation of the transmission development technology.

Based on the author's over 30 years of practical experience in the development of automotive transmissions at home and abroad, this book systematically summarizes the main technical areas of the MT and AT development technology, comprehensively elaborates the theory and development characteristics of the transmissions and displays the main development processes of the transmissions. Many development processes are based on the experience of trial and error. As a reference book providing main technical information for the research and development of transmission technology, this book is suitable for engineers in the field of automobiles and related power transmission machines and graduate students at school.

1.1 Transmission Functions and Requirements

The function of the transmission is to change the torque and speed of the engine according to the requirements of the vehicle in different driving conditions, so that the vehicle has the right traction and speed, and keeps the engine working in the most favorable working conditions. To ensure the vehicle reversing and the powertrain separation, the transmission must have forward as well as reverse and neutral. When power output is required, power output devices shall also be provided.

Main requirements for the transmission:

(1) Guarantee good dynamic and economic indicators of the vehicle. This requirement is met by choosing the appropriate transmission gear number and gear ratio according to the vehicle dead weight capacity, engine performance parameters, tire performance parameters and vehicle use requirements in the overall vehicle design.

(2) Reliable work and easy control. Automatic gear dropping, gear mixing and shift impact are not allowed in the transmission during the vehicle driving.
(3) High safety. Guarantee the safe and reliable vehicle driving in any working conditions.
(4) Small in size and light in weight. This requirement is met by reasonable use of engineering plastics and other non-metallic materials, and the use of advanced material forming technology and heat treatment technology.
(5) Low cost. Minimize the cost under the premise of meeting the vehicle requirements for the transmission.
(6) High transmission efficiency. In order to reduce the gear engagement loss and bearing friction loss, it is necessary to improve the manufacture and assembly quality of components and reduce the gear churning loss. For example, the proper lubricating oil and installed capacity can be selected to reduce the churning loss and friction loss, thus improving the transmission efficiency.
(7) Low noise. This requirement can be met by adopting helical gears, selecting a reasonable modification coefficient, making the axial modification and profile modification and improving the manufacturing accuracy and assembly stiffness.
(8) Meet the maximum input torque requirements.
(9) Meet the drive mode requirements.
(10) Meet the vehicle layout and installation requirements.
(11) Meet the reliable ramp parking requirements except for the MT.
(12) Meet the limp home requirements.
(13) With respect to the electrical controlled transmission, the hardware of the control system shall meet the requirements of electromagnetic compatibility and anti-interference, and the software shall meet the requirements of ISO26262, ASPICE, AUTOSAR and other standards.
(14) Consider factors such as driving pleasure and shift comfort and minimize the shift impact.

1.2 Types, Advantages and Disadvantages of Transmissions

I. Types of automotive transmissions

Depending on the fixation of the gear ratio, the transmission may be classified into stepped transmission and continuously variable transmission (CVT). The stepped transmission, with fixed gear ratio, including ordinary transmission and planetary transmission, is gear-driven and mostly widely used; the CVT, with the gear ratio changing in a certain range, includes electric and hydraulic types. The variable speed drive component of the electric CVT is DC series motor and the drive component of the hydraulic CVT is hydraulic torque converter. The CVT can overcome the sudden shift, slow throttle response, high fuel consumption and other shortcomings of the AT and is mostly used in trolley buses and heavy-duty vehicles.

The transmissions used in the passenger vehicles currently include MT, AT, CVT, DCT, AMT and VIT, as well as the BEV transmission and HEV transmission evolved on the basis of the above transmissions, as shown in Fig. 1.2.

1. **Manual transmission**

Manual transmission (MT) is also known as mechanical transmission. The engaging position of the gears in the transmission is changed by moving the gear shift lever by hand, thus changing the gear ratio to achieve variable speed. It take a place in the transmission family because of its high transmission efficiency, high torque transfer, easy manufacture, low cost, simple structure, low fuel consumption, short acceleration time and short development cycle. The current MT is mainly 5 speed or 6 speed MT, covering SUV, middle-sized vehicles and compact vehicles. In order to reduce fuel consumption and improve driving pleasure, the transfer torque increases

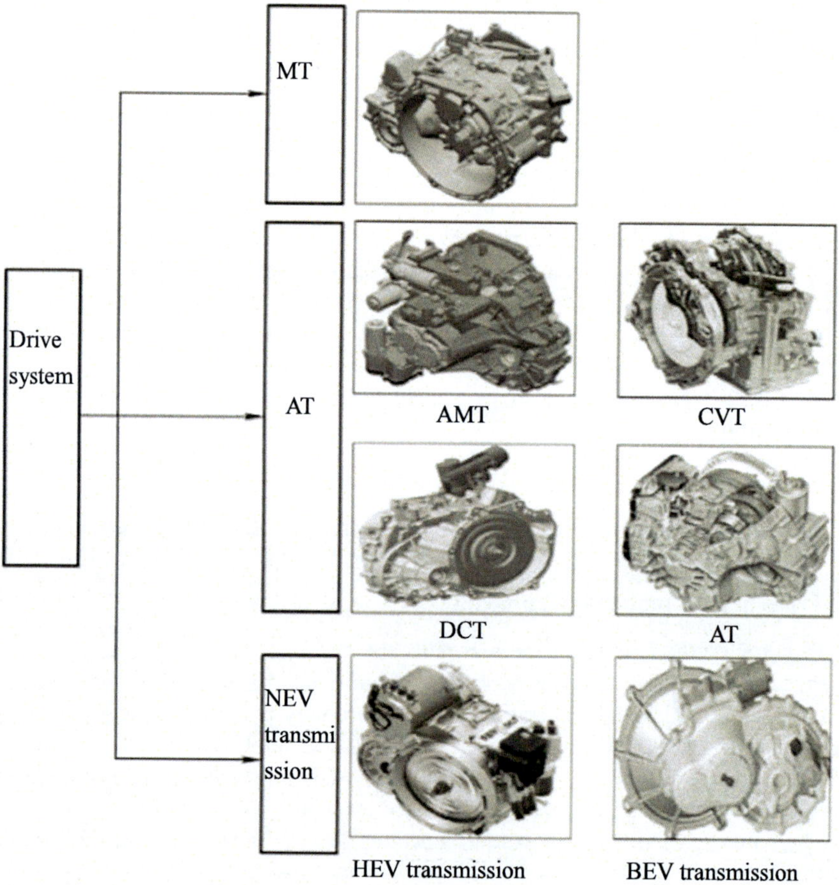

Fig. 1.2 Main types of transmissions for passenger vehicles

step by step and the 6 speed transmission in the MT will become the mainstream. ZF and BMW are developing the 7 speed MT currently.

2. **Automatic transmission**

The automatic transmission (AT) is also known as stepped AT. According to the different arrangement modes of shafts, AT can be divided into fixed shaft AT and rotating shaft AT. Due to the large space occupied by the fixed shaft (parallel shaft), it is impossible to achieve more gears (limited to 5-speed and 6-speed), which is currently the main technology of Honda. The vast majority of transmission companies use the rotating shaft technology solutions. AT may be divided into 4AT, 5AT, 6AT, 7AT, 8AT and 9AT by gear. GM, Volkswagen and other automotive companies have begun to develop 10AT. With mature technology and small size, 4AT is still the main AT carried by small cars, although its comfort and economy are poor; 5AT is slightly better than 4AT in terms of comfort and economy, but has still obvious abrupt shift and limited late development space; 6AT has relatively superior economy and comfort and has become a mainstream trend for small, compact and middle-sized vehicles; 7AT and 8AT have been adopted by Lexus, BMW, Audi, Benz and other high-end imported models. 8AT has been successfully developed by domestic Shengrui Transmission Co., Ltd. and installed in Landwind; 9AT represents the innovation of science and technology has been developed successfully by ZF and Aisin Seiki.

3. **Continuously variable transmission**

The continuously variable transmission (CVT) becomes an important branch of automotive automatic transmission field since it can produce continuously changing gear ratio, achieve the best match between the engine and powertrain, simplify the operation, reduce the driver's labor intensity, increase driving safety, driving smoothness and comfort and improve emission. CVT is classified into metal pushing V-belt type CVT, chain type CVT and KRG. With light weight, small volume, simple structure, smooth shift, relatively high cost, inconvenient maintenance and limited carrying capacity, the metal pushing V-belt type CVT is used in small and compact cars and has more room for improvement because of its unparalleled comfort. The chain type CVT is the main technology of Schaeffler in terms of CVT and has higher transmission efficiency and more compact structure than the metal pushing V-belt type CVT, but its disadvantage is that the protruding pin on the side of the chain will produce noise when it comes into contact with the pulley point. At present, the KRG may be unfamiliar to most people, but this kind of transmission may become the mainstream transmission for low-emission vehicles in the future. Its low cost, high efficiency, simple structure and multiple advantages in function and comfort deserve people's attention.

4. **Dual clutch transmission**

The dual clutch transmission (DCT), as a new AT, arranges the transmission gears by odevity respectively on the two input shafts connected with two clutches, completes the shift process and achieves power shift through the alternate switching of clutches. DCT has the advantages of AT and AMT. With high transmission efficiency, simple structure and low production cost, it guarantees the dynamic and economy performance of the vehicle and greatly improves the vehicle operating comfort. DCT is divided into DDCT and WDCT. DDCT is also divided into electro-hydraulic DDCT and electric DDCT. The former is complicated in structure, needs to be improved in reliability and is used in Volkswagen vehicles; the latter is simple and reliable in structure and will become the mainstream trend in the development of small and medium torque AT in the future. With large carrying capacity, WDCT will be used greatly in the intermediate class and above vehicles.

5. **Automated mechanical transmission**

The automated mechanical transmission (AMT), based on the traditional fixed shaft transmission, controls the hydraulic or electric actuating system through the electronic control unit by use of the electronic technology and automatic transmission theory to achieve the clutch disengagement and engagement, gear selecting and shifting, so as to achieve the automatic control of start and shift. AMT has a certain development space in mini and small cars due to its high transmission efficiency, easy manufacturing and low cost, but it has not much development space in models above the compact level due to power failure in the shift process.

The AMT usually consists of electrically controlled hydraulic AMT and electrically driven AMT. The AMT with the core of electro-hydraulic actuator in the electrically controlled hydraulic AMT has been widely applied in Chery QQ3, Riich M1, Chevrolet Sail, SAIC MG3 and other models; compared with the electrically controlled hydraulic AMT, the electrically driven AMT has simpler structure, better reliability and lower cost and is the mainstream trend of AMT development.

6. **Vit**

VIT is a new concept of high-power and high-efficiency mechanical CVT successfully developed on the basis of the meshing principle of sliding vane CVT movable teeth. The working surface of the sliding vane CVT movable teeth is overlapped by a series of sliding vanes and any shape of meshed tooth profile can be formed by free stepless slip of the sliding vanes. This design idea is equivalent to the "differentiation and reintegration" of the gears. That is, any required tooth profile, i.e. the sliding vane CVT movable teeth can be formed by organic combination of multiple elements. Since the slip direction of the sliding vanes is different from the force direction, the sliding vanes are free to deform with the current meshed tooth profile, while the force direction is perpendicular to the free slip direction or the angle between them is self-locking in the equivalent friction angle during the power transmission. Therefore, the sliding vanes will not change the shape of the tooth profile when bearing force. With the "rigid and flexible fusion, and movable teeth solidification" effect, high carrying capacity and transmission efficiency, VIT is the true sense of "movable teeth meshing CVT" and is applied in saloon cars, passenger cars, trucks and other high-power and high-torque vehicles.

7. **BEV transmission**

The BEV transmission mainly includes single reduction gear, multi-speed transmission and the wheel-side drive motor integrating reducer and motor. At present, the single speed reducer with fixed speed ratio is mostly used in the small electric vehicles. This drive mode has simple structure and low manufacturing cost, but it puts forward higher requirements for the traction motor that the traction motor shall provide higher instantaneous torque in the constant torque area and higher running speed in the constant power area, so as to meet the vehicle acceleration performance requirements and maximum speed design requirements. Meanwhile, the single speed reducer with fixed speed ratio has the problem of low motor utilization efficiency. In order to ensure the maximum vehicle speed, the speed ratio of the reducer is often relatively small, which makes the traction motor in a long-term high torque and high current working condition, and relatively low motor efficiency, thus wasting the battery energy and reducing the driving range. The electric vehicle drivetrain tends to be multi-speed to make the electric vehicles better meet their dynamic performance and reduce their requirements for traction motors and batteries. Oerlikon Graziano developed a two-speed transmission to match small electric vehicles. Antonov designed a new efficient 3 speed AT for BEV that optimizes the powertrain size, weight and development costs while improving the energy efficiency and guaranteeing the dynamic performance.

8. **HEV transmission**

(1) Hybrid drive type: before introducing the HEV transmission, it is important to know the hybrid drive type. As shown in Fig. 1.3, the hybrid drive mainly includes tandem hybrid drive, parallel hybrid drive, power-split hybrid drive and other hybrid drive.

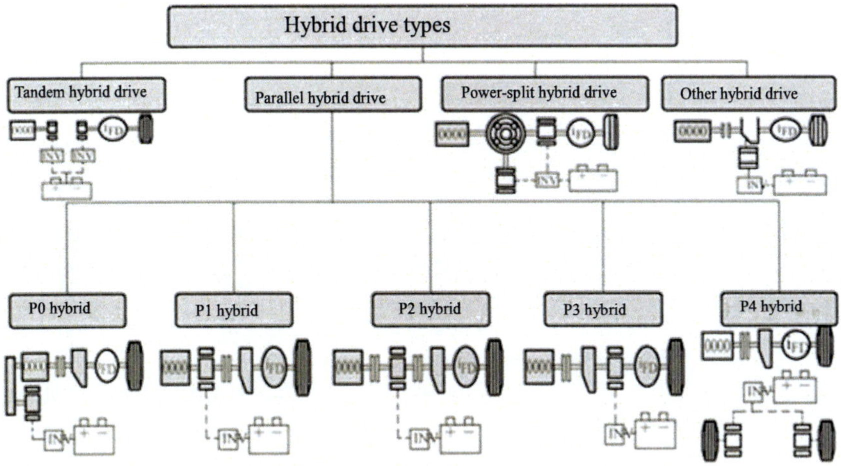

Fig. 1.3 Hybrid drive types

(1) Tandem hybrid drive. The engine is completely decoupled from the driving wheel and simply drives the generator to charge the power battery, which drives another motor through the motor controller to keep the vehicle moving. This scheme has low overall efficiency due to many power drive links, but it is comfortable because the engine is completely decoupled from the driving wheel.
(2) Parallel hybrid drive. Depending on the motor arrangement on the whole vehicle, it is divided into P0 (BSG), P1 (ISG), P2, P3 and P4 structural hybrid drives. In P0 structural hybrid drive, the motor is integrated in the position of the generator of the traditional engine, playing the role of power generation, assistance and starting, and acting as the generator in energy recovery to recover the braking energy; in P1 structural hybrid drive, the motor is integrated at the output end of the engine crankshaft, playing the role of power generation, assistance and starting, and acting as the generator in energy recovery to recover the braking energy; in P2 structural hybrid drive, the motor is also integrated between the engine and the transmission. The difference is that the motor is connected to the engine and transmission separately through the clutch, playing the role of power generation, assistance and starting, and acting as the generator in energy recovery to recover the braking energy. The motor drives the vehicle alone, either fully hybrid or plug-in hybrid, which is one of the main forms of hybrid drive; in P3 structural hybrid drive, the motor is integrated at the output end of the transmission, playing the role of assistance and power generation, and acting as the generator in energy recovery to recover the braking energy. The motor drives the vehicle alone, either fully hybrid or plug-in hybrid; in P4 structural hybrid drive, the engine drives a drive axle and the motor drives the other drive axle. The motor plays the role of assistance and power generation and acts as the generator in energy recovery to recover the braking energy. The motor can also drive the vehicle alone, either fully hybrid or plug-in hybrid.
(3) Power-split hybrid drive. It integrates the engine and two motors by means of the planetary gear train to achieve various functions of the hybrid drive. A typical example is Toyota Prius HEV, also known as the eCVT because of the use of planetary gear train and motors to shift the output end of the engine and planetary gear train.
(4) Other hybrid drive. A typical example is Honda Fit 7DCTH hybrid drive, in which, the motor is connected to an input shaft of 7DCT through the drive mechanism, forming a new and unique drive form. In this scheme, the motor needs to be synchronously tracked in shift of each gear connected with the motor input shaft, so as to reduce the difference between the active and passive speeds of the synchronizer and prolong the service life of the response synchronizer.

(2) Common transmissions on HEV: the traditional automotive transmissions are used in HEVs, among which AT is more widely used, as follows:

The AMT is the ideal choice for HEV transmissions. The electrically driven transmission based on AMT is characterized by coupling the motor used in the NEV with AMT through high-intensity silent chain drive, which solves the power failure problem during shift. This new electrically driven transmission fully combines the advantages of the motor and the AMT to significantly reduce fuel consumption. The 7H-AMT hybrid transmission developed by FEM based on AMT is characterized by that the drive motor transfers power through other gears and outputs a certain torque in the upshift to eliminate the impact caused by power failure during shift.

The application schemes of AT in HEV include coupling the motor at the input end of the transmission and coupling the motor at the output end of the transmission. The recent new scheme is to replace the hydraulic torque converter of the AT with motor. In these schemes, the lubrication system shall be improved and the mechanical fuel pump of the traditional AT is replaced with an electronic fuel pump or a new electronic fuel pump is added; otherwise, the EV mode will be difficult to meet the system lubrication requirements and fast start-stop requirements.

CVT is most frequently used in the mass produced HEVs. The HEV AT is characterized by powertrain integration. That is, the motor is integrated with the transmission, making the system structure more compact, power drive more stable and control performance better.

The applications of DCT in HEV mainly include that the drive motor is connected to the input shaft 1 of the transmission through the reducing gear, that the drive motor is connected to the input shaft 2 of the transmission through the reducing gear and that the drive motor is connected to the input shaft of the transmission through the reducing gear. The advantage of the first two schemes is that the motor drive can change the speed, but the disadvantage is that the motor needs to be synchronously tracked in the synchronous engagement of the gear of the input shaft connected to the motor; the advantage of the latter scheme is that the motor may not be synchronously tracked in the shift, but the disadvantage is that the motor drive cannot change speed.

In addition, the PRIUS hybrid power system is a typical example of PSHEV. The biggest feature of this system is to use a planetary gear train to couple two motors and an engine together, so that a single planetary gear train can realize the functions of CVT. See Table 1.1 for typical hybrid transmission applications.

II. Advantages and disadvantages of transmissions

The transmission, as an important part of the vehicle powertrain system, determines the power output of the vehicle and has a direct impact on the fuel economy, comfort and reliability of the vehicle. Different types of transmissions have different characteristics. The advantages and disadvantages of mainstream transmissions in today's market are shown in Table 1.2.

Table 1.1 Typical hybrid transmission applications

Structure	Manufacturer	Model	Transmission
Start-stop-BSG	Buick	LaCrosse	6AT
	Chevrolet	Malibu	6AT
	Chery	A5 BSG	Original transmission
ISG (E-M-C-T)	Honda	Fit Hybrid	MT/LVT
	Honda	CIVIC Hybrid (II)	CVT
	Honda	CIVIC Hybrid (III)	CVT
	Honda	Insight	CVT
	Honda	CR-Z	MT/LVT
	BMW	BMW 7 Hybrid	8AT
	Benz	Benz S400 Hybrid	7 AT
ISG (E-C-M-T)	Hyundai	Hyundai Sonata hybrid power	6AT
	Nissan	Nissan Fuga	7 AT
	Audi	Audi A6 Hybrid	8AT
	Audi	Q5 Hybrid quattro	8AT
	Audi	Q7 Hybrid (EOL)	
	Porsche	Porsche Panamero S Hybrid	8AT
	Volkswagen	Volkswagen Touareg	8AT
Parallel rear axle drive	Peugeot	Peugeot 3008	6 AMT
Series-parallel/dual-motor single planetary gear train	Toyota	Toyota Prius	Single reduction gear
	Nissan	Nissan Altimn Hybrid	Single reduction gear
	Ford	Fusion Hybrid	–
	Ford	Escape Hybrid	
	Ford	C-MAX Hybrid	
Series-parallel/dual-motor dual planetary gear train	Lexus	RX400h/HighlanSer Hybrid	Single reduction gear
Series-parallel/dual-motor three-planetary gear train	Lexus	GS450h/LS600h	Single reduction gear
	BMW	BMW X6	7 AT
	Benz	Benz ML450	7 AT

Table 1.2 Advantages and disadvantages of transmissions

	Transmission form	Torque	Efficiency	Comfort	Shift smoothness	Reliability	Fuel economy	Emission performance	Life	Cost	Scope of application
MT	Stepped manual	Large	High	Poor	Power failure	Good	Good	Relatively good	Good	Low	Unlimited power Unlimited cost
AT	Stepped automatic	Large	Low	Good	Approximately uninterrupted power	Good	Bad	Relatively good	Relatively good	High	Unlimited power Limited cost
CVT	Stepless automatic	Limited	Low	Best	Power without interruption	Relatively good	Relatively good	Good	Bad	High	Limited power Limited cost
DCT	Stepped automatic	Large	High	Good	Power approximately without interruption	Relatively good	Relatively good	Good	Relatively good	High	Unlimited power Limited cost
AMT	Stepped automatic	Large	High	Relatively good	Power failure	Bad	Good	Relatively good	Relatively good	Relatively low	Unlimited power Unlimited cost

1.3 Basic Structure of Transmission

The transmission consists of a case, a drive part and a shift control device.

1. **Case**

As the basic part, the case is used to mount and support all parts of the transmission and to store the lubricating oil, above which, there is a precise bore for mounting the bearing. The transmission bears variable load, so the case shall be rigid enough, with complex ribs on the inner, most of which are castings (made of gray cast iron, commonly HT200), as shown in Fig. 1.4.

For the convenience of installation, the transmission part and the shift control device are often made into split type, and the transmission cover is bolted to the case and positioned reliably. The case is provided with refueling and fuel drain hole and fuel level inspection ruler hole, and the heat dissipation should also be considered.

2. **Drive part**

The drive part consists of the transmission gears, shaft, bearing and other driving media. The geometric dimensions of the shaft are determined by the checking calculation of the strength and stiffness; the material is selected mainly depending on whether its stiffness meets the requirements. The carbon steel has nearly equal elasticity modulus with the alloy steel, so the shaft is generally made of carbon steel (usually steel 45) and the alloy steel is used only when the gear and shaft are integrated or when the bearing is under heavy load. The gears are usually made of low carbon alloy steel (e.g. 20CrMnTi and 20MnCrS). The shaft is mostly splined with the gears and has the advantages of good centering, reliable transmission of power and small extrusion stress. The spline part and the bearing mounting site of the shaft are surface hardened. The shaft is mainly supported by a rolling bearing, with simple lubrication, high efficiency, small radial clearance and reliable axial positioning and

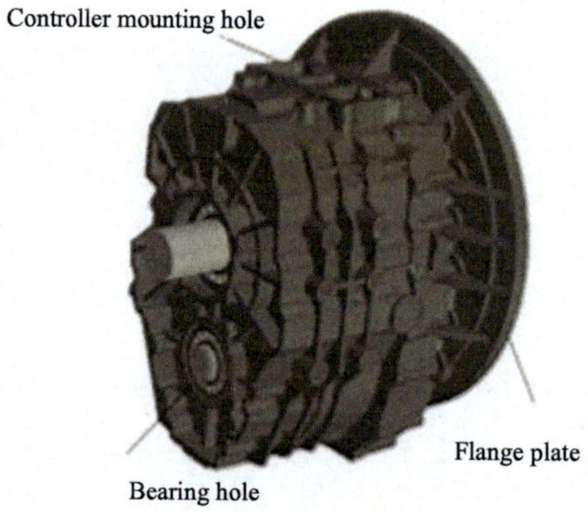

Fig. 1.4 Transmission case

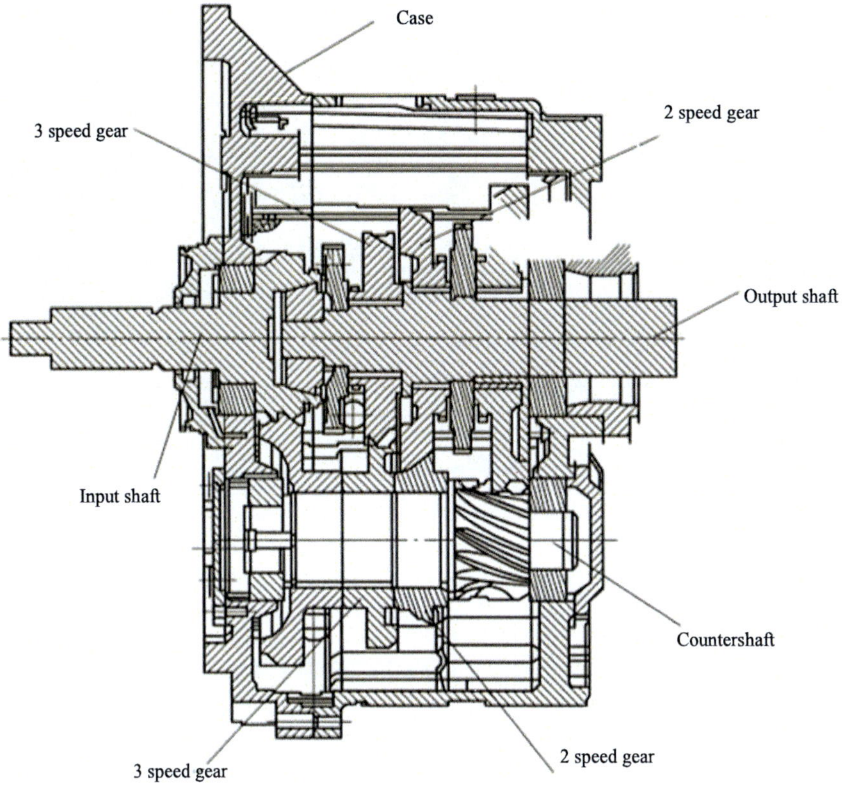

Fig. 1.5 Transmission profile

is mainly lubricated by splash lubrication (v>25 m/s, thrown to the wall as long as the viscosity is appropriate). The transmission profile is shown in Fig. 1.5.

3. **Shift control device**

In MT, the driver controls the shift, while in AT, the electronic actuating system completes shift partly or entirely relying on a lot of automation technology. The neutral, reverse and park are still completed by the driver by controlling the shift control device. The elements in the shift control device are selected according to the transmission type and vehicle type. The engaging elements of the transmission for passenger vehicles mainly include:

(1) Internal engaging elements: shift fork (Fig. 1.6a), shift synchronizer (Fig. 1.6b), locking device, multi-disk clutch and brake.
(2) External engaging elements: shift level system, inhaul cable and gear shift lever

Fig. 1.6 (b) Shift synchronizer

1.4 Development Status and Trend of Transmission

I. Development status of transmissions

The demand for transmissions varies greatly from region to region. The CVT has benefited to a certain extent from the continued demand among Japanese users for small cars that change speeds automatically, making it the best-selling AT type in Japan. Figure 1.7 shows the demand for transmissions in Japan. The main transmission enterprises in Japan develop to the high gears (8 and 9) AT. Moreover, the CVT is developing fast and gradually developing towards high torque.

AT possesses absolute advantage in the US, mainly because the consumers require simple control and comfortable driving of vehicles but are not sensitive to fuel consumption, thus forming the AT-dominated AT market in the US. Figure 1.8 shows the demand for transmissions in the US. The main transmission enterprises in the US are currently targeting the 6AT. With the continuous development of the wet clutch technology of BorgWarner, the WDCT will also grow rapidly in the US. The

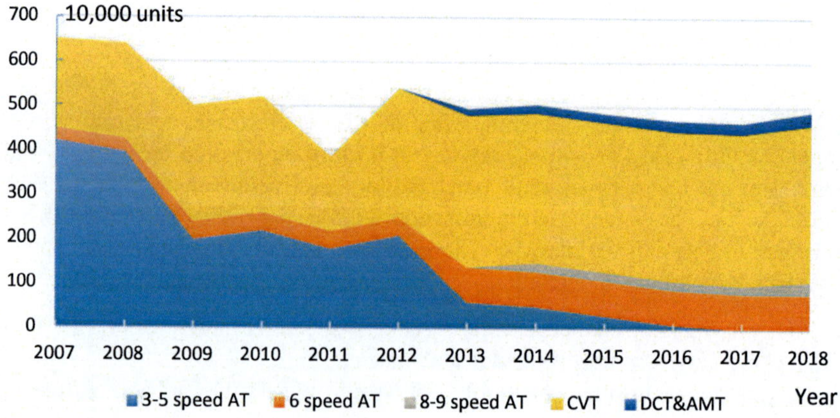

Fig. 1.7 Demand for transmission in Japan

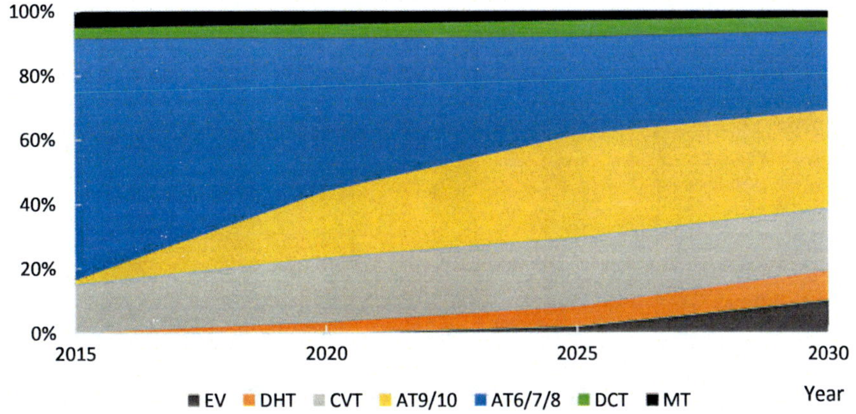

Fig. 1.8 Demand for transmission in the US

higher gear AT produced by the European transmission company is adopted for the high-grade vehicles.

The European consumers pay attention to driving experience and driving pleasure, like the feeling brought by manual control of the machinery, and think a great deal of fuel consumption. Therefore, after emergence, the DCT with energy saving and kinetic characteristics immediately became the darling of the European market. Figure 1.9 shows the demand for transmissions in Europe. The main transmission enterprises in Europe have developed towards the high speed (8 and 9) AT and DCT, and the engineering companies are also pushing the hybrid power technology vigorously.

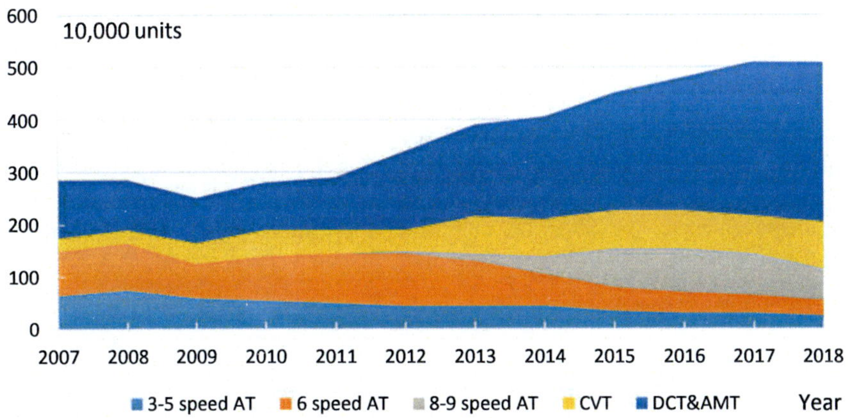

Fig. 1.9 Demand for transmission in Europe

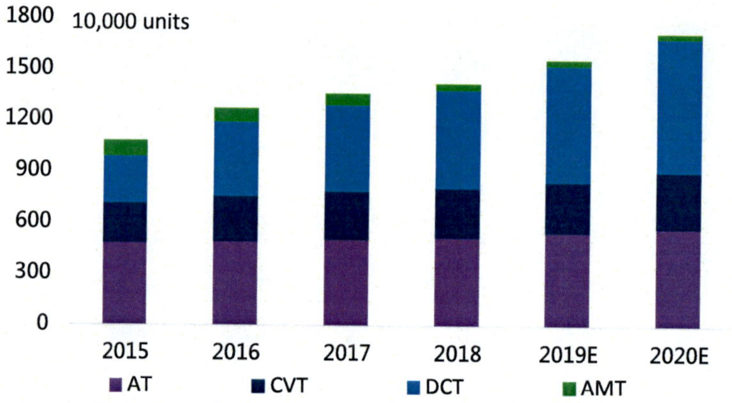

Fig. 1.10 Demand for transmission in China

In China, an emerging auto market, the traditional AT was the main choice for previously automatic transmission vehicles. However, in recent years, the biggest factors influencing consumers' decision to buy cars are "price" factors (vehicle fuel consumption and vehicle price). The number of vehicles assembled with DCT has increased significantly, and the share of vehicles assembled with CVT has also increased. Figure 1.10 shows the demand for transmissions in China. Due to accumulated experience in MT development and good process inheritance of AMT and DCT, the transmission technology develops rapidly in recent years with the involvement of the engineering companies with outstanding transmission development capability, and the companies are also trying to make breakthroughs in the AT and CVT fields.

As AT continues to be multi-speed, DCT grows most and the demand for the AT is increasing year by year. Figure 1.11 shows the demand for AT.

The carrying status of the AT in the passenger vehicles in China is shown in Fig. 1.12. The 4 speed and 5- speed ATs are mainly assembled in 1.3–1.6 L small cars and compact cars; the 6-7-8 speed ATs are mainly assembled in the vehicles with the displacement of 2.0 L and above; the CVT (without hydraulic torque converter) is mainly assembled in 1.5–1.8 L compact cars; the CVT (hydraulic torque converter) is mainly assembled in 1.5–2.5 L vehicles; the DDCT is mainly assembled in the vehicles with the displacement of 2.0 L and below and the WDCT in the vehicles with the displacement above 2.0 L; the number of applications of turbocharged engine and DCT combination will increase rapidly.

II. Development trend of transmissions

Energy conservation, environmental protection, safety and high efficiency is the theme of the development of automotive transmissions, with the pursuit of comfort, economy and safety from the perspective of consumers and the pursuit of high efficiency, low carbon and environmental protection from the perspective of regulations. Thus, the transmission companies are required to apply advanced and efficient design, manufacturing and control technologies and advanced materials to achieve wide gear

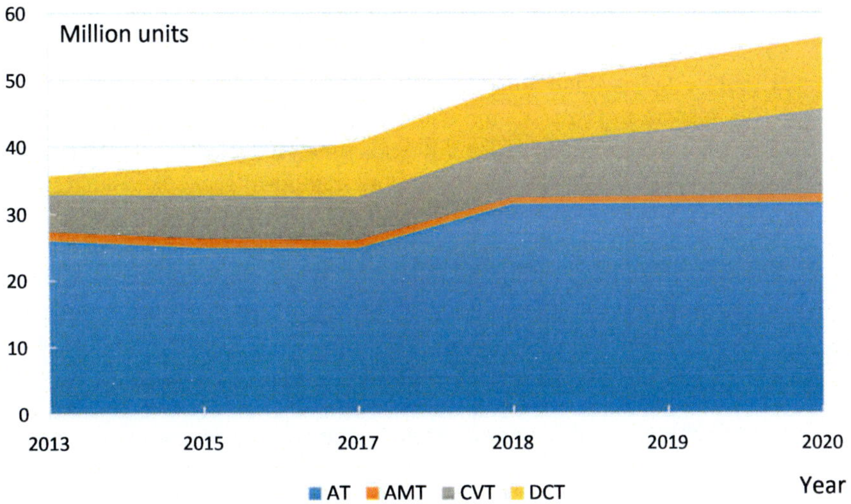

Fig. 1.11 Demand for AT

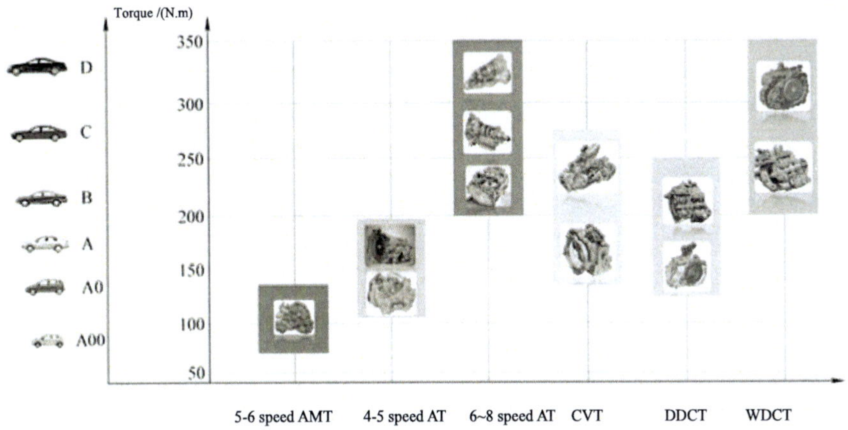

Fig. 1.12 Carrying status of the AT in the passenger vehicles in China

ratio and multi-speed of transmission, optimization of the shift strategy, improvement of the transmission efficiency and light weight of transmissions.

1. **Development trend of MT**

   (1) Multi-speed and large gear ratio of transmission: the three-axis 6 speed transmission was gradually replacing the dual-axis 5 speed transmission under the dominance of some European, American and Japanese companies from two or three years ago. The 7 speed MT supplied by ZF for Porsche 911 Carrera and Carrera S was used in the passenger vehicles for the first time.

With increased low speed gear ratio and more reasonable gear, it achieves the comprehensive optimization of dynamic and economy performance.

(2) High efficiency and reduced NVH: the following measures can be taken to improve the transmission efficiency: multi-speed transmission; reduce the immersion height of the differential mechanism in the lubricating oil and change the lubricating mode from traditional gear splash lubrication to combination of gear splash lubrication and oil guide lubrication, so as to reduce the churning loss of the lubricating oil; replace the traditional conical bearing with ball bearing and roller bearing to reduce the bearing friction loss; use efficient transmission lubricating oil. To meet the requirement of noise reduction, the best match of the clutch and transmission shaft with the transmission shall be considered in addition to the corresponding measures taken for the transmission, such as precise control of the backlash in circular tooth of the transmission matching gear, selection of appropriate gear material, application of low noise bearing and full consideration to the impact of the gear engagement on the transmission structure in the design phase.

(3) Light weight and low cost: the light weight and low cost can be achieved by use of advanced forming technology, reasonable plastics and other non-metallic materials. The lightweight design of the transmission case based on CAE can shorten the development cycle, reduce the development cost and improve the product competitiveness.

(4) Further reduce the space occupied by the transmission and optimize the space and size of all parts. Due to the application of the start-stop system, a highly integrated solution with low cost and high reliability that can identify the neutral and reverse is applied through the detection of reliable neutral position signals.

2. **Development trend of AT**

(1) Development trend of hydraulic torque converter: flattened hydraulic torque converter, increasing torque ratio, extended locking range and sliding friction range and improved hydraulic torque converter range. With the application of the engine supercharging technique, the engine torque is increasing and higher requirements are put forward for the vibration reduction of the transmissions. The torque converter with centrifugal pendulum vibration absorber and the torque converter with turbine vibration absorber are presented, significantly improving the vibration damping performance of the torque converter. The hydraulic part is optimized to improve the hydraulic torque converter capacity and optimize the axial space.

(2) Multi-speed AT. The 6~9 speed AT is gradually replacing the 4 or 5 speed transmission. With the increased gears, the transmission may have a larger gear ratio range and a reasonable gear ratio distribution. ZF has successfully developed 9AT and some companies are already working on 10AT.

(3) AT modular design. The modular design is very obvious in AT, including modularization of hydraulic torque converter, modularization of hydraulic valve body,

modularization of clutch, modularization of brake, etc. According to different user requirements, various schemes can be realized through different module combinations, reducing the design changes, shortening the development cycle and improving the product competitiveness.

(4) Application of multiple solenoid valves. Multiple solenoid valves are adopted to control the shift, which can significantly improve the shift quality. To control the system pressure and achieve shift, six PWM solenoid valves with high flow capacity are set in the 6AT of ZF, simplifying the valve body structure and improving the comprehensive performance of the transmission. The transmission efficiency can be further improved if the leakage-free solenoid valve or the near-leakage-free solenoid valve is used.

(5) Component integration to reduce the mass. For example, the ZF6H26 employs a gear called Lepetler, which reduces the mass of the gear train by 11 kg; in A750E/A750F of Toyota, 3 clutches are integrated into the same clutch hub; the dog clutch used for ZF9HP is longer but smaller than the ordinary clutch and looks like a spline; some transmission companies use magnesium alloy transmissions, further reducing the weight of the transmission.

(6) Use of new materials and new processes: the application of new lightweight and high-strength materials and the application of the stamping forming technology make great contributions to reducing cost and weight.

3. **Development trend of CVT**

(1) Improve the CVT efficiency

  (1) Reduce the cone disk pressing loss. The applied pressing force and the adjustment and pressing force are optimized by using a torque sensor to effectively reduce the flow of the hydraulic system in the adjustment process, so that a smaller hydraulic pump can be used, which not only reduces the fuel consumption of the pump, but also reduces the related energy loss.
  (2) Reduce the chain loss. The connecting piece and oscillation pin are used, characterized by high efficiency, small size, reliable work and low noise.
  (3) Reduce the hydraulic pump loss and use the low-energy electric hydraulic pump.
  (4) Reduce the CVT bearing loss and replace the hydraulic torque converter with the clutch.

(2) Reduce the CVT cost. With respect to the processing technology, the cone disk group stamped by the steel plates can be used as the main part to reduce the cost and weight and the shaft machined by cold extrusion can also be used to reduce the cost; according to different market requirements, the dry clutch or wet clutch can be used instead of the hydraulic torque converter to reduce the cost; the CVT case may be optimized to reduce the weight and cost.

(3) Comprehensive optimization of the engine and CVT precise control. The torque sensor has been widely used, especially for small batch supercharged engines, to optimize the pressing force and achieve precise control of CVT; it can be

integrated with the engine for integrated control to further reduce the fuel consumption and emission.

(4) Increase the drive torque. Increasing CVT transfer torque has been the focus of research and development. The new structure of CVT drivetrain, such as the steel belt designed by BOSCH using new flexible ring materials, with better dynamic performance, lower cost, higher efficiency and smaller center distance, can transfer higher torque. The modular design can meet the special customer requirements in the rapidly developing market, shorten the R&D cycle, and improve the market competitiveness. The CVT is one of the coupling mechanisms between the HEV engine and the drive motor. An HEV with a CVT drivetrain can reduce the fuel consumption by 30% and emissions by 50%.

4. **Development trend of DCT**

(1) Wide gear ratio, multi-speed and light weight: achieve the wide gear ratio, multi-speed and light weight on the premise of ensuring transmission efficiency.

(2) Modular design: e.g. electro-hydraulic actuator of the motor-controlled DCT, gear selecting and shifting actuator and wet dual clutch.

(3) Integrated control: the DCT control is combined with engine, ABS, ESP, EPS and ACC control to realize the integrated control of the powertrain, improve the performance of the powertrain and optimize the control effect.

(4) Hybrid powertrain: the DCT is combined with the motor/generator to form the hybrid powertrain, enabling separate engine drive, separate motor drive, and combined engine and motor drive. When braking, the motor/generator is in the generating state, and the kinetic energy of the vehicle is converted into electric energy and stored. This scheme provides the HEV with a dynamic coupling device that is easy to implement.

(5) Integration and intelligence: all functions of the transmission are integrated into one unit to reduce the fault sources and improve reliability. In terms of control strategy, adaptive control, fuzzy control and other intelligent control methods are adopted to improve the adaptive ability of DCT.

5. **Development trend of AMT**

(1) Powertrain integration: the AMT control is combined with engine, ABS, ESP, EPS and ACC control through integrated control to realize the integrated control of the powertrain, improve the performance of the powertrain and optimize the control effect.

(2) Use of new structure: for example, ZEROSHIFT has developed a new AMT technology that uses a series of sliding claws and tooth sockets to allow both gears to engage at the same time when shifting gears. It has all the advantages of a dual clutch, with a simple design and low cost.

(3) Hybrid powertrain: the AMT is combined with the wheel-side motor/generator to form the hybrid powertrain, enabling separate engine drive, separate motor drive, and combined engine and motor drive. During the shift, the motor drives the vehicle to improve the dynamic performance of the vehicle; when braking, the motor/generator is in the generating state, and the kinetic energy of the

vehicle is converted into electric energy and stored. This scheme provides the HEV with a dynamic coupling device that is easy to implement.

(4) Integration and intelligence: all functions of the transmission are integrated into one unit to reduce the fault sources and improve reliability; in terms of control strategy, adaptive control, fuzzy control and other intelligent control methods are adopted to improve the adaptive ability of AMT.

In conclusion, due to the increasing traffic congestion and the increasing number of female drivers, more and more users choose automatic transmission for the convenience of driving; the AT technology continues to be mature, the quality and efficiency are improving and the coverage of models is increasing; in recent years and in the next few years, most of the AT in mass production is DCT, and the output of Chinese and European models based on DCT will exceed that of Japanese models based on CVT; 6AT has entered the mature stage, and the AT with more gears will be mainly applied to high-end models. The overall market share of AT is flat or slightly decreased; the market share of the AMT will increase slightly due to fuel consumption and cost. The self-owned brand AT is increased mainly due to the decrease of DCT and MT, but the reduction rate will not be too fast due to the limitation of consumers' purchasing power. The substantial incentives taken by the government in taxation, subsidies and other aspects are the main reason for the rapid growth of the NEV market. In order to meet the fuel consumption standard of 5 L/100 km in 2020, enterprises need to vigorously develop new energy vehicles.

Bibliography

1. Yong Chen (2008) New development and trend of automatic transmission technologies. Automot Eng 30(10):938–945
2. Ge A (2001) Automatic transmission (I)—overview of automatic transmission. Automob Technol (5):1.3
3. Guangqiang Wu, Weibin Yang, Datong Qin (2007) Key technique of dual clutch transmission control system. Chin J Mech Eng 43(2):13–21
4. Li J, Zhang J, Feng J, et al (2000) Development, current situation and forecast of automated mechanical transmission. Automob Technol (3):1.3
5. Mingkui Niu, Xiusheng Cheng, Bingzhao Gao et al (2004) A study on shifting characteristics of dual clutch transmission. Automot Eng 26(4):453–457
6. Yang W, Wu G, Qin D (2007) Drive line system modeling and shift characteristic of dual clutch transmission powertrain. Chin J Mech Eng 43(7):188–194
7. Niu M, Gao B, Ge A, et al (2004) Dual-clutch type automatic transmission system. Automob Technol (6):1.3
8. Guo L, Ge A, Zhang T et al (2003) AMT shift process control. Trans Chin Soc Agri Mach 34(2):1.3
9. Yongjun Li, Shuxin Chen, Yong Cui et al (2003) Integrated control of the starting process of automated mechanical transmission. Automot Eng 25(2):178–181
10. Ge A (2001) Automatic transmission (II)—hydraulic torque converter. Automob Technol (6):1.5
11. Cao G, Ge A, Zheng L, et al (2005) Clutch engagement control during gear shifting process in automated manual transmission. Chin J Mech Eng 41(12):234–238

13. Lei Y, Yi Y, Ge A (2001) Integrated and intelligent shifting control of automated mechanical transmission. Automot Eng 23(5):311.314
14. Lun Jin, Xiusheng Cheng, Li Sun et al (2005) Simulation and studies on dual-clutch automatic transmission. Automob Technol 8:4–7
15. Liao C, Zhang J, Lu Q (2005) Coordinated powertrain control method for shifting process of automated mechanical transmission in the hybrid electric vehicle. Chin J Mech Eng 41(12):37–41
16. Qin D, Liu Y, Hu J, et al (2010) Control and simulation of launch with two clutches for dual clutch transmissions. Chin J Mech Eng 46(18):121.127
17. Yong C, Wenjiang Z, Wenzhong L (2009) Present situation and future trends of automatic transmission in China. SAE-China Congr Proc
18. Bing Zhou, Qinghua Jiang, Yi Yang (2011) Transmission ratio optimization with dual objectives of power performance and economy for a two-speed electric vehicle. Automot Eng 33(9):792–797
19. Xiaoming Weng (2009) A study on the shift quality of wet double clutch transmission. Automot Eng 31(10):927–931
20. Justin K, Anne-Catrin U, Lowy FD (2015) Staphylococcus aureus infections: transmission within households and the community. Trends Microbiol 23(7):437–44
21. Haydon PG, Giorgio C (2006) Astrocyte control of synaptic transmission and neurovascular coupling. Physiol Rev 86(3):1009–1031
22. Oh HR, Song H (2012) Energy efficient MAC protocol for delay-sensitive data transmission over wireless sensor network. Wireless Commun Mob Comput 12(9):755–766
23. Kumnuan U, Prasert A, Dowell SF, et al (2005) Probable person-to-person transmission of avian influenza A (H5N1). New England J Med 352(4):333–340
24. Clayton D (2010) A generalization of the transmission disequilibrium test for uncertain haplotype transmission. Am J Hum Genet 65(4):1170–1177
25. Jackson JB, Musoke P, Fleming T et al (2003) Intrapartum and neonatal single-dose nevirapine compared with zidovudine for prevention of mother-to-child transmission of HIV-1 in kampala, uganda: 18-month follow-up of the HIVNET 012 randomised trial. Lancet 362(9387):859–868
26. Seto WH, Tsang D, Yung RWH, et al (2003) Effectiveness of precautions against droplets and contact in prevention of nosocomial transmission of severe acute respiratory syndrome (SARS). Lancet 361(9368):1519–1520
27. Han SH, Lee JH (2005) An overview of peak-to-average power ratio reduction techniques for multicarrier transmission. IEEE Wirel Commun 12(2):56–65

Chapter 2
Manual Transmission

2.1 Overview

Despite the rapid development of automatic transmission technology, the traditional engine-matched manual transmission (MT) will continue to play an important role for some time to come, mainly because of its low cost, high efficiency and reliable operation. At present, 5 speed MT is mainly used in China, and some cars start to carry 6 speed MT, covering SUV, middle-sized and compact vehicles. The 6 speed MT will gradually become the mainstream. ZF and BMW are developing the 7 speed MT.

By the engine and drive shaft connection mode, the MT may be classified into FF layout, FR layout and RR layout; by the form and arrangement form of the shaft, the MT can be classified into two-shaft and three-shaft MT; by the gear type, the MT can be classified into spur gear and helical gear types; by the synchronizer type, the MT can be classified into constant pressure, inertial and inertial boost types and the most widely used type currently is the inertial synchronizer of the structures including slide, lock pin, lock ring and multi-cone types; by the number of gears, the MT can usually be classified into 5, 6 and 7 speed transmissions. The FF layout is mostly used for the cars. With the multi-speed development and restricted by the spatial and axial dimensions of the transmissions, the three-shaft 6 speed or 7 speed transmission will gradually become the mainstream transmission.

2.2 Transmission Drive Mechanism

An MT mainly consists of the powertrain, shift system, lubrication system and case. The powertrain consists of the input shaft, output shaft, gears, bearing, synchronizer and differential mechanism; the shift system consists of the shift fork, gear shifter shaft, self-lock device and interlock device. Figure 2.1 is the structural diagram of MT. The transmission has 6 forward gears and 1 reverse gear (R gear) and is of

Y. Chen, *Automotive Transmissions*, Springer Tracts in Mechanical Engineering,
https://doi.org/10.1007/978-981-15-6703-2_2

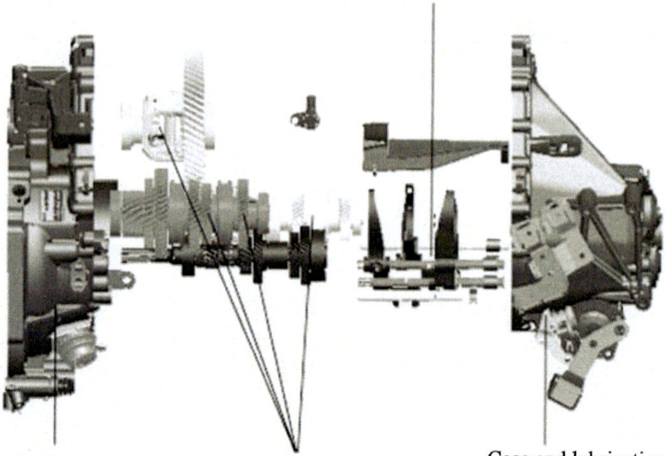

Fig. 2.1 Structural diagram of MT

3-parallel shaft structure (1 input shaft and 2 output shafts). It has been widely used in the FF passenger vehicles because of its small axial distance and easy spatial arrangement. Figure 2.2 shows the MT drive diagram, where, g1–g17 are gears and S1–S4 are synchronizers. The drive lines of all gears are as follows:

1 speed drive line: clutch C → input shaft → gear g2 → gear g9 → left shift of synchronizer S1 → output shaft 1 → gear g7 → gear g17 → differential mechanism → wheel.
2 speed drive line: clutch C → input shaft → gear g3 → gear g10 → right shift of synchronizer S1 → output shaft 1 → gear g7 → gear g17 → differential mechanism → wheel.
3 speed drive line: clutch C → input shaft → gear g5 → gear g12 → right shift of synchronizer S2 → output shaft 1 → gear g7 → gear g17 → differential mechanism → wheel.
4 speed drive line: clutch C → input shaft → gear g4 → gear g11 → left shift of synchronizer S2 → output shaft 1 → gear g7 → gear g17 → differential mechanism → wheel.
5 speed drive line: clutch C → input shaft → gear g4 g4 → gear g15 → left shift of synchronizer S3 → output shaft → gear g13 → gear g17 → differential mechanism → wheel.
6 speed drive line: clutch C → input shaft → gear g6 → gear g16 → right shift of synchronizer S3 → output shaft 2 → gear g13 → gear g17 → differential mechanism → wheel.

Fig. 2.2 MT drive diagram

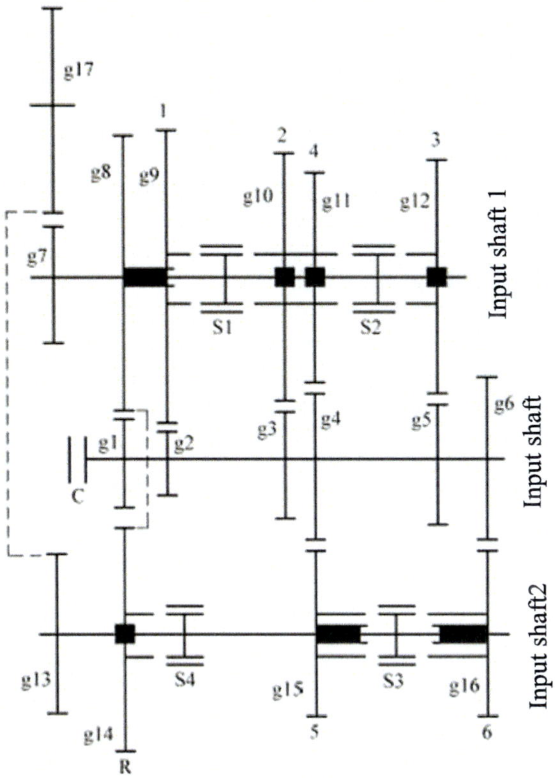

R speed drive line: clutch C → input shaft → gear g1 → gear g8 → gear g14 → left shift of synchronizer S4 → output shaft 2 → gear g13 → gear g17 → differential mechanism → wheel.

Based on the above analysis, the gear ratios of all gears are shown in Table 2.1. The number of teeth of each gear is represented by Z_i, and the subscript i corresponds to the gear label in Fig. 2.2. In reverse gear, the input shaft rotates in the opposite direction to the output gear of the main reducer; in other gears, the input shaft rotates in the same direction as the output gear of the main reducer.

Table 2.1 Gear ratio of each gear

Gear K	1	2	3	4	5	6	R
Gear ratio i_{gk}	$\frac{Z_9 \times Z_{17}}{Z_2 \times Z_7}$	$\frac{Z_{10} \times Z_{17}}{Z_3 \times Z_7}$	$\frac{Z_{12} \times Z_{17}}{Z_5 \times Z_7}$	$\frac{Z_{11} \times Z_{17}}{Z_4 \times Z_7}$	$\frac{Z_{15} \times Z_{17}}{Z_4 \times Z_{13}}$	$\frac{Z_{16} \times Z_{17}}{Z_6 \times Z_{13}}$	$\frac{Z_{14} \times Z_{17}}{Z_1 \times Z_{13}}$

2.3 Synchronizer

As an important part of MT, the synchronizer has a very important impact on the main technical indicators of the transmission such as shift portability and smoothness, can reduce the shift force on the shift knob, and reduce the shift impact and driver fatigue. An ideal synchronizer should have good synchronization performance and locking performance, which can not only achieve the shortest synchronization time with small shift force, but also ensure that the synchronizer clutch and lock ring will not enter the joint when the angular velocity is not consistent at the input and output ends.

At present, all synchronizers adopt the principle of friction, that is, the friction moment is generated on the working surface to overcome the inertia moment of the meshed part, so as to reach the synchronization status in a short time and change gears easily.

I. Structure and characteristics of synchronizer

There are constant pressure, inertial and inertial boost synchronizers, of which, the inertial synchronizer is most widely used. The lock ring type inertial synchronizer is mainly used for the passenger cars and light trucks, as shown in Fig. 2.3, and mainly consists of the mating spline, synchronous ring, synchronizer splined hub, synchronizer clutch, locking slider and spring.

The mating spline and the drive gear are combined into one. When the gear is engaged, the mating spline meshes with the spline inside the synchronizer clutch to drive the driving moment. The cone angle of its external conical surface is equal to that of the synchronous ring. When the gear is engaged, the external conical surface and the synchronous ring generate a synchronous friction moment.

The synchronous ring is a conical ring with grooves on the internal conical surface that can destroy the oil slick and rapidly discharge the lubricating oil. These grooves can be radial or circumferential. On the external conical surface, there are usually uniformly distributed notches equal to the number of sliders to hold and push the locking slider; in addition, there are short lock teeth and the included angle between the bevels of the lock teeth is the lock angle. When the mating spline synchronizes with the synchronizer hub, these bevels allow the synchronizer clutch to slip in with a slight rotation.

The synchronizer splined hub is connected to the shaft through the spline on the inner diameter, and the spline is also processed on the outer diameter. The spline is connected to the synchronizer clutch, and the loading lock sliders are evenly and uniformly distributed on the outer side of the synchronizer hub.

The splines connected to the synchronizer hub are processed on the inner ring of the synchronizer clutch; large circumferential grooves are processed on the outer ring to place the shift fork. The whole gearshift operates the shift fork axially and then acts on the synchronizer clutch.

The locking slider and spring constitute a central locking and positioning mechanism. Usually, there are three or four pairs of locking and positioning mechanisms

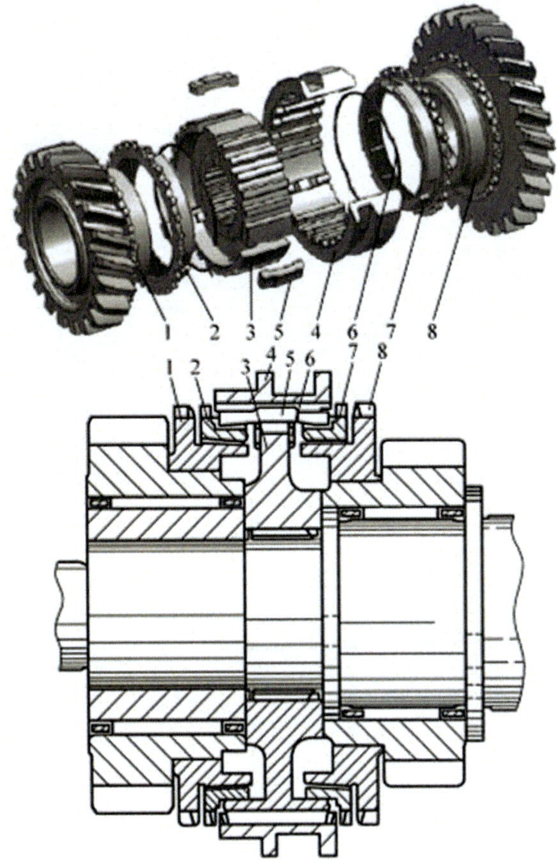

Fig. 2.3 Lock ring type inertial synchronizer. 1, 8—mating spline, 2, 7—synchronizer splined hub, 4—synchronizer clutch, 5—locking slider, 6—spring

evenly distributed, whose function is to keep the synchronizer clutch in the middle of the synchronizer hub between two gears in the neutral position and keep the synchronizer clutch in the neutral position under the action of certain axial force. When the axial force does not reach a given value, the synchronizer clutch is prevented from sliding axially in relation to the synchronizer hub.

II. Shift process of synchronizer

The shift process of the synchronizer is mainly divided into the following 6 stages, and the states of each stage are shown in Fig. 2.4.

Stage 1: the synchronizer clutch is not meshed with the mating spline and is in the neutral position, as shown in Fig. 2.4a. At this point, even if the clutch is engaged, the transmission cannot drive the torque, the wheel cannot rotate and the vehicle remains stationary. The locking slider is pressed on the inner circumference of the synchronizer clutch under the action of the spring force and the bump on the surface of the locking slider inhibits the axial movement of the synchronizer

(a) Stage 1: neutral position

(b) Stage 2: locking slider and synchronizer clutch action

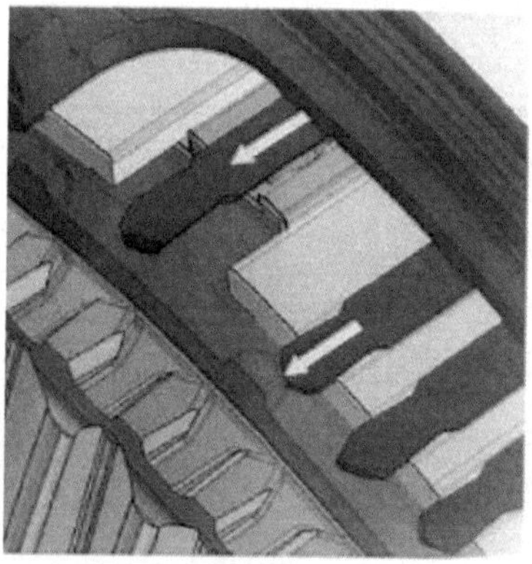

(c) Stage 3: synchronizer action and speed synchronization

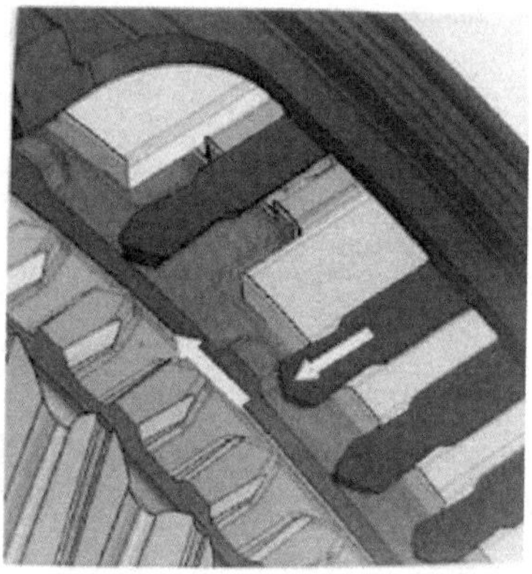

(d) Stage 4: synchronizer staggering after synchronization

(e) Stage 5: synchronizer clutch in contact with mating spline

(f) Stage 6: synchronizer clutch and mating spline are engaged

Fig. 2.4 State of each stage in the synchronizer shift process clutch. The synchronizer clutch remains in its natural state under the external force within a certain limit.

Stage 2: The driver manipulates the gear shift lever towards the selected gear, the action of the shift fork is transferred to the synchronizer clutch, and the synchronizer clutch and locking slider act, as shown in Fig. 2.4b. When moving towards the selected gear, the locking slider is pressed on the side of the synchronous ring. Under this force, the synchronous ring is in contact with the integrated cone of the mating spline. Because of the speed difference between the synchronous ring and the mating spline, the synchronous ring can only rotate within the range allowed by the locking slider mounting groove, and the lock chamfer on the end face of the synchronizer clutch and that of the synchronous ring turn to opposite positions. This state is called the flag state of synchronization.

Stage 3: The locking slider stops here, the synchronizer clutch further moves towards the mating spline and the lock chamfer on the end face of the synchronizer clutch is in contact with that of the synchronous ring, as shown in Fig. 2.4c. At this time, the synchronous ring presses on the conical surface of the mating spline, producing the friction moment. The driving moment of the input shaft → synchronizer splined hub → synchronizer clutch is transferred to the mating spline and the speed starts to synchronize. In order to ensure reliable contact and smooth shift, there are many tricks for the shape selection of lock chamfer on the end face of the synchronizer clutch and that of the synchronous ring, which are the key of MT design.

Stage 4: the synchronizer clutch is fully synchronous with the mating spline, as shown in Fig. 2.4d. The friction moment between the synchronous ring and the conical surface of the mating spline gradually disappears. In this state, the synchronizer clutch continues to move towards the mating spline. While the contact surface between the chamfer of the spline on the end face of the synchronizer clutch and the chamfer of the synchronous ring slides, the synchronizer clutch moves towards the meshing direction with the mating spline and enters the so-called turning stroke (stagger). The moment generated by the synchronizer clutch chamfer and synchronous ring chamfer in the turning process affects the shift feel. It is one of the MT design parameters and the key of MT design.

Stage 5: the synchronizer clutch continues to move towards the mating spline, and the spline chamfer of the synchronizer clutch comes in contact with the mating spline, as shown in Fig. 2.4e. During the shift, the control force of the gear shift lever is small and the synchronizer clutch has been fully synchronous with the mating spline. In order to prevent noise and ensure relatively smooth engagement, it is still critical to design the chamfer shape.

Stage 6: the synchronizer clutch continues to move towards the mating spline, the synchronizer clutch is fully engaged with the mating spline and the shift finishes, as shown in Fig. 2.4f. The drive shaft → synchronizer splined hub → synchronizer clutch → synchronous ring are mechanically connected with the mating spline and are in the integral rotation state. The engine torque and speed are converted to the corresponding transmission output torque and speed according to the gear ratio of the selected gear and transmitted to the wheel via the transmission output shaft and then via the differential mechanism. In the state of torque drive, the contact surface between the synchronizer clutch and mating spline needs to bear a large drive torque, which shall be easy to mesh and cannot be out of gear.

III. Working principle of synchronizer and determination of main parameters

1. Working principle of synchronizer

The gear shall be disengaged before shift, so that the transmission is in neutral position. The output end of the transmission is connected to the vehicle and has a considerable rotational inertia, so the speed at the output end of the synchronizer connected to the output end of the transmission is constant at the moment of shift, while the speed at the input end of the synchronizer is different from that at the output end. The speed at the input end of the synchronizer can achieve consistent with that at the output end through the friction moment of the synchronizer. Assuming that the churning loss at the input end of the synchronizer and the bearing friction moment are ignored, the following moment equation can be obtained by using the synchronizer diagram shown in Fig. 2.5:

$$T_m = J_i \frac{d\omega_i}{dt} \tag{2.1}$$

where: T_m—friction moment of synchronizer;

J_i—rotational inertia at the input end of the synchronizer;

ω_i—rotational angular velocity at the input end of the synchronizer;

t—time.

The axial force acting on the synchronizer through the gear shift lever has the following relationship with the friction moment of the synchronizer

$$T_m = \frac{F\mu R}{\sin\alpha} \tag{2.2}$$

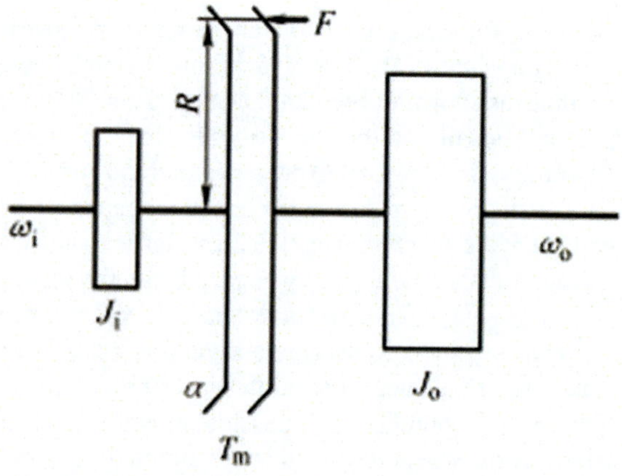

Fig. 2.5 Synchronizer diagram

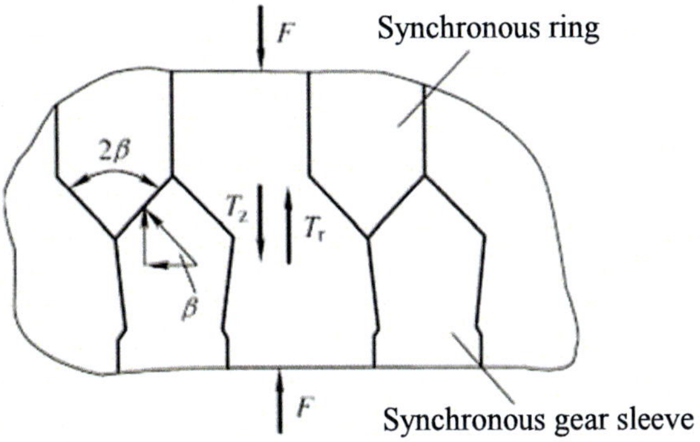

Fig. 2.6 Force analysis of locking surface

where: F—axial force acting on the synchronizer;
μ—friction factor between working faces;
R—mean radius of conical surface;
α—cone angle.

Assuming that the difference between the rotational angular velocities at the input and output ends of the synchronizer is $\Delta\omega$ and the velocity is synchronous at the time Δt, the friction moment equation at the time of synchronization is

$$\frac{F\mu R}{\sin\alpha} = J_i\frac{\Delta\omega}{\Delta t} \tag{2.3}$$

At the synchronizer action shown in Fig. 2.4c, i.e. the speed synchronization stage, the lock chamfer on the end face of the synchronizer clutch is in contact with that of the synchronous ring. The force analysis of the locking surface is shown in Fig. 2.6 and the axial force F and tangential force F_z are as follows:

$$F = N\sin\beta + \mu_s\cos\beta \tag{2.4}$$

where: N—positive pressure acting on the locking surface;
β—lock angle;
μ_S—static friction factor between the locking surfaces.

$$F_z = N(\cos\beta - \mu_s\sin\beta) \tag{2.5}$$

where: F_Z—tangential force acting on the synchronous ring.

From formulas (2.4) and (2.5)

$$F_z = \frac{F(\cos\beta - \mu_s\sin\beta)}{\sin\beta + \mu_s\cos\beta} \tag{2.6}$$

The tangential force F_Z acting on the synchronous ring forms a ring toggle moment M_T

$$M_T = F_z R_s \tag{2.7}$$

where: R_S—pitch radius of synchronous ring lock chamfer.

To ensure synchronization before shift, the locking condition of $T_m \geq M_T$ must be met, i.e.

$$\frac{\mu R}{\sin \beta} \geq \frac{R_s(\sin \beta - \mu_s \cos \beta)}{\cos \beta + \mu_s \sin \beta} \tag{2.8}$$

If $\mu_S = 0$, formula (2.8) is

$$\frac{\mu R}{R_s \sin \beta} \geq \tan \beta \tag{2.9}$$

The formula (2.9) is used to determine the lock angle required for locking. For easy shift of cars, $\beta = 105°–125°$, so that a larger ring toggle moment can be generated.

2. **Calculation of rotational inertia**

The rotational inertia at the input end of each gear of synchronizer shall be calculated during the shift. Taking the MT drive diagram in Fig. 2.2 for example, this part of rotational inertia includes the rotational inertia of the input shaft and its connecting gear, the rotational inertia of the clutch driven plate and the rotational inertia of the constant mesh gear in the input shaft gears.

The basic relation of the rotational inertia conversion is as follows

$$J_Z = J_B\left(\frac{Z_Z}{Z_B}\right)^2 \tag{2.10}$$

where: J_Z—rotational inertia after conversion;
J_B—converted rotational inertia;
Z_Z—number of teeth in the converting shaft;
Z_B—number of teeth in the converted shaft.

The rotational inertia converted to the input shaft in the neutral position is

$$J_{rk} = J_L + J_{sr} + J_1 + J_2 + J_3 + J_4 + J_5 + J_6 \tag{2.11}$$

where: J_L—rotational inertia of the clutch driven plate;
J_{sr}—rotational inertia of the input shaft and its connecting gear;
J_1—rotational inertia converted by the gears g8 and g14 to the input shaft;
J_2—rotational inertia converted by the gear g9 to the input shaft;
J_3—rotational inertia converted by the gear g10 to the input shaft;
J_4—rotational inertia converted by the gears g11 and g15 to the input shaft.
J_5—rotational inertia converted by the gear g12 to the input shaft;
J_6—rotational inertia converted by the gear g16 to the input shaft.

$$J_1 = J_{g8}\left(\frac{Z_1}{Z_8}\right)^2 + J_{g14}\left(\frac{Z_1}{Z_{14}}\right)^2 \tag{2.12}$$

$$J_2 = J_{g9}\left(\frac{Z_2}{Z_9}\right)^2 \tag{2.13}$$

$$J_3 = J_{g10}\left(\frac{Z_3}{Z_{10}}\right)^2 \tag{2.14}$$

$$J_4 = J_{g11}\left(\frac{Z_4}{Z_{11}}\right)^2 + J_{g15}\left(\frac{Z_4}{Z_{15}}\right)^2 \tag{2.15}$$

$$J_5 = J_{g12}\left(\frac{Z_5}{Z_{12}}\right)^2 \tag{2.16}$$

$$J_6 = J_{g16}\left(\frac{Z_6}{Z_{16}}\right)^2 \tag{2.17}$$

where: J_{g8}—rotational inertia of gear g8;
J_{g9}—rotational inertia of gear g9;
J_{g10}—rotational inertia of gear g10;
J_{g11}—rotational inertia of gear g11;
J_{g12}—rotational inertia of gear g12;
J_{g14}—rotational inertia of gear g14;
J_{g15}—rotational inertia of gear g15;
J_{g16}—rotational inertia of gear g16.

See Table 2.2 for the rotational inertia at the input end of the synchronizer at each gear.

3. **Calculation of angular velocity difference**

The angular velocity of the parts at the input end of the synchronizer changes before and after the shift, while the vehicle speed may be deemed unchanged before and after shift. Therefore, the angular velocity of the parts at the input end of the synchronizer before and after shift can be calculated by the speed. Combined with Fig. 2.2, when the vehicle speed is v (m/s) and the wheel radius is r (m), taking the conversion of gear 3 to another gear for example, the angular velocity difference of the parts at the input end of the synchronizer before and after shift is analyzed, as shown in Table 2.3.

Table 2.2 Rotational inertia at the input end of the synchronizer

Gear	1	2	3	4	5	6	R
Rotational inertia at the input end ($\times J_{rk}$)	$\left(\frac{Z_9}{Z_2}\right)^2$	$\left(\frac{Z_{10}}{Z_3}\right)^2$	$\left(\frac{Z_{12}}{Z_5}\right)^2$	$\left(\frac{Z_{11}}{Z_4}\right)^2$	$\left(\frac{Z_{15}}{Z_4}\right)^2$	$\left(\frac{Z_{16}}{Z_6}\right)^2$	$\left(\frac{Z_{14}}{Z_1}\right)^2$

Table 2.3 Angular velocity difference of the parts at the input end of the synchronizer before and after shift

Gear k	1	2	3	4	5	6
(a) Angular velocity of the parts at the input end of the synchronizer before shift $\left(\times \frac{Vi_{g3}}{r}\right)$	$\frac{Z_2}{Z_9}$	$\frac{Z_3}{Z_{10}}$	$\frac{Z_5}{Z_{12}}$	$\frac{Z_4}{Z_{11}}$	$\frac{Z_4}{Z_{15}}$	$\frac{Z_6}{Z_{16}}$
(b) Angular velocity of the parts at the input end of the synchronizer after shift $\left(\times \frac{Vi_{g3}}{r}\right)$	$\frac{Z_5}{Z_{12}}$	$\frac{Z_5}{Z_{12}}$	$\frac{Z_5}{Z_{12}}$	$\frac{Z_5}{Z_{12}}$	$\frac{Z_5}{Z_{12}}$	$\frac{Z_5}{Z_{12}}$
(c) Angular velocity difference of the parts at the input end of the synchronizer before and after shift $\left(\times \frac{Vi_{g3}}{r}\right)$	(a)–(b)	(a)–(b)	(a)–(b)	(a)–(b)	(a)–(b)	(a)–(b)
(d) Angular velocity rise and fall at the input end of the synchronizer	Rise	Rise	–	Fall	Fall	Fall

4. **Maximum line velocity of synchronous ring**

The friction line velocity u of the synchronous ring plays an important role in the thermal stress. The friction surface temperature increases exponentially with the increase of the line velocity and the maximum line velocity under the maximum angular velocity difference $\Delta\omega_{max}$ is

$$u_{max} = R_s \Delta\omega_{max} \tag{2.18}$$

5. **Resistance moment T_v of synchronous ring**

It is difficult to calculate the resistance moment T_V of the synchronous ring at each gear. Table 2.4 shows the resistance moment T_{Vi} measured at the input shaft when the fuel temperature is 80 °C and their relationship is

$$T_v = T_{vi} i_g z \tag{2.19}$$

where: i_g—gear ratio from the input shaft to the synchronizer;

z—fuel temperature correction factor, taken 1 at the room temperature. The factor will be greater than 1 with the decrease of the temperature.

Table 2.4 Resistance moment T_{Vi} measured at the input shaft at the fuel temperature 80 °C

Empirical value	Passenger vehicle	Commercial vehicle	Vehicle with auxiliary transmission
Resistance moment T_{Vi}/(N m)	2–5	4–8	10–141

6. **Synchronous ring friction moment T_r considering resistance moment T_V**

In the synchronization process, the synchronous ring friction moment T_r considering resistance moment T_V may be expressed as

$$T_R = -J_{rk}\frac{\Delta\omega}{t_R} - T_v \tag{2.20}$$

where: t_R—synchronization time.

In the upshift, $\Delta\omega < 0$, the resistance moment T_v and the synchronous ring friction moment T_R have the same direction and work together, favorable for shifting; in the downshift, $\Delta\omega > 0$, the resistance moment T_v and the synchronous ring friction moment T_R are in different directions and work together, not favorable for shifting.

7. **Sliding friction work W**

The relationship of the sliding friction work W with $\Delta\omega$, T_v and t_R in the synchronization

$$W = -\frac{1}{2}(J_{rk}\Delta\omega^2 + T_v\Delta\omega t_R) \tag{2.21}$$

In the upshift, $\Delta\omega < 0$, the resistance moment T_v reduces the sliding friction work W; in the downshift, $\Delta\omega > 0$, the resistance moment T_v increases the sliding friction work W.

8. **Sliding friction power P_m**

$$P_m = \frac{W}{t_R} \tag{2.22}$$

9. **Stress per unit area σ**

The synchronous ring contact pressure N is generated under the action of the shift force F during synchronization. The stress per unit area σ under the contact pressure N is

$$\sigma = \frac{N}{A_R} \tag{2.23}$$

where: A_R—sum of all friction areas of the synchronizer.

10. **Lock Angle**

There is a drag torque between the MT clutch and its drive mechanism. Even if the input and output of the synchronizer are fully synchronized, the synchronizer clutch

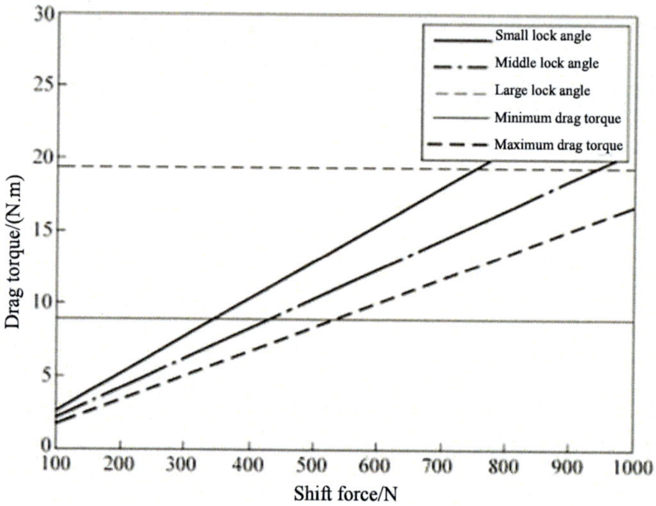

Fig. 2.7 Relationship between shift force and drag torque T_Z

shall pass through the synchronous ring and overcome the rotational input of the drag torque under the shift force to fully engage with the soldered teeth. Therefore, the drag torque T_z and the shift force F must satisfy the following relationship

$$T_z < \frac{FR_s(\cos\beta - \mu_s \sin\beta)}{\sin\beta + \mu_s \cos\beta} \tag{2.24}$$

Figure 2.7 shows the relationship between the shift force F and the drag torque T_z. With respect to a lock angle, only when the rotating torque generated by a certain shift force F is greater than the drag torque, can the synchronizer clutch be fully engaged with the soldered teeth. That is, the gear can be engaged only when the drag torque is below the curve. With the decrease of the lock angle, the drag torque T_z can be overcome increases.

11. **Shift force**

The actual shift force applied by the driver is largely dependent on the driving style, temperature and road conditions. When the temperature is low, the transmission drag torque is high, the shift force is increased and the shift time is longer. Table 2.5 lists

Table 2.5 Standard values of value shift force and synchronization time

Standard value	Gear	Passenger vehicle	Commercial vehicle
Shift force/N	Gear 1—top gear	80–120	180–250
Synchronization time/s	Gear 1—top gear	0.15–0.25	0.25–0.40

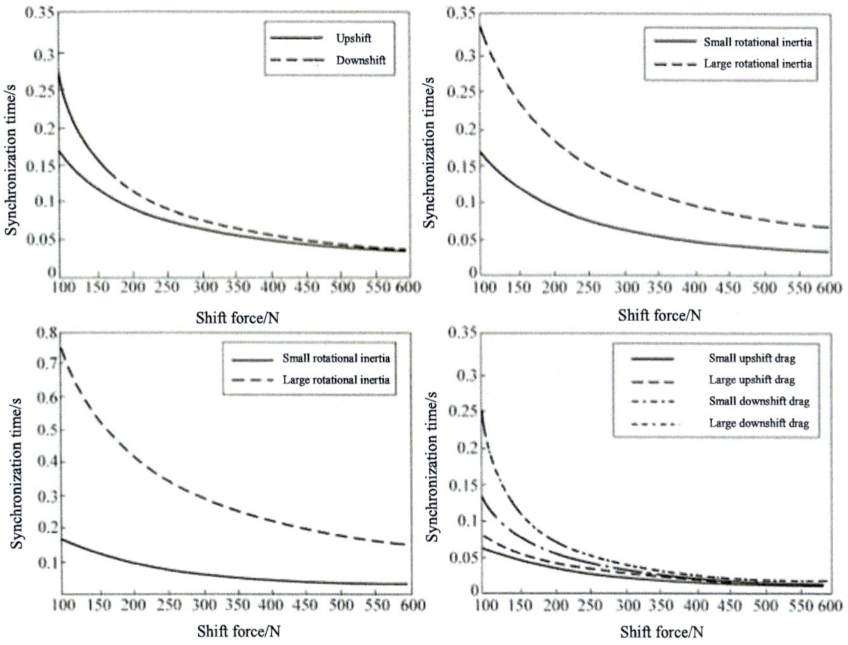

Fig. 2.8 Relationship between shift force and synchronization time

the acceptable standard values of shift force and synchronization time, and Fig. 2.8 shows the relationship between the shift force and synchronization time.

IV. Main evaluation indicators of synchronizer

(1) Synchronization time: conform to Table 2.5.
(2) Shift force: conform to Table 2.5.
(3) Maximum line velocity of synchronous ring: conform to Table 2.6.
(4) Compressive stress per unit area: conform to Table 2.6.
(5) Sliding friction work: conform to Table 2.6.

Table 2.6 Designed allowable value for the formation of friction pairs between steel and certain materials

Synchronous ring	Friction factor	Maximum line velocity/(m/s)	Sliding friction work per unit area/(J/mm^2)	Sliding friction power per unit area/(W/mm^2)	Compressive stress per unit area/(N/mm^2)
Special brass	0.08–0.12	5	0.09	0.45	3
Mo-sprayed	0.08–0.12	7	0.53	0.84	6
Sintered material	0.08–0.12	9	1.00	1.50	7

(6) Sliding friction power: conform to Table 2.6.
(7) Reliability: the synchronizer works reliably under various operating conditions. In particular, it is necessary to ensure reliable shift at low temperature, prevent forced engagement when the input and output speeds are out of sync, and prevent spontaneous out-of-gear after full engagement.
(8) Life: generally more than 150,000 km for passenger vehicles and more than 1,200,000 km for commercial vehicles.
(9) Cost: including development costs and manufacturing costs.
(10) Quality and installation space requirements.

2.4 Transmission Operating Mechanism

The transmission operating mechanism shall ensure that the driver can accurately and reliably engage the transmission into any required gear and can make it back to neutral at any time. The transmission operating mechanism shall meet the following requirements:

(1) Only one gear is allowed when shifting, usually by means of the interlock device.
(2) In the shift, the gears shall be engaged on the full length of the teeth and prevented from automatic spontaneous out-of-gear.
(3) A reverse lock is usually used to prevent accidental engagement of reverse gear. When the transmission begins to engage the reverse gear, it draws attention of the driver that obvious hand feel is generated from large resistance as a result of the role of the reverse lock.

The transmission operating mechanism can be divided into two types: direct and indirect control. In the direct control transmission operating mechanism, the transmission is arranged in the lower part of the cab and the gear shift lever is extended from the cab floor besides the driving seat. It is of simple structure and convenient operation, but it requires the equal stroke of each gear shift. It has been applied in the FR vehicles. In the indirect control transmission operating mechanism, the transmission is arranged far away from the driving seat. The operating mechanism shall have sufficient stiffness and the gap at the connection points shall be small, otherwise it will affect the feel when shifting gears. Figure 2.9 shows the indirect control transmission operating mechanism, which mainly consists of the gear shift lever, cable and select-shift actuator.

I. Gear shift lever

The gear shift lever, as a part of the transmission assembly and a component of the human-machine interaction device, is to change the gear ratio of the automotive transmission through the driver's control to realize the change of the vehicle driving gear. The gear number and position of MT of various models are not the same, but the gear arrangement principle is basically the same: the forward gear progressively

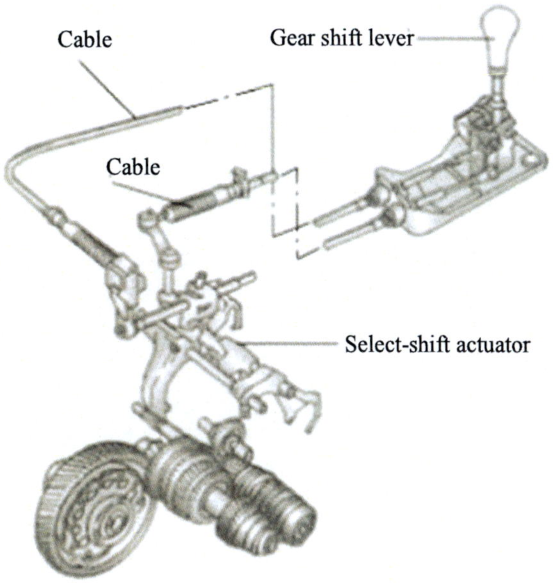

Fig. 2.9 Indirect control transmission operating mechanism

increases from top down, first left and then right; the reverse gear is in the upper left or lower right corner; the neutral is in the horizontal middle position. Figure 2.10 shows the gear arrangement of the commonly used hand shift lever.

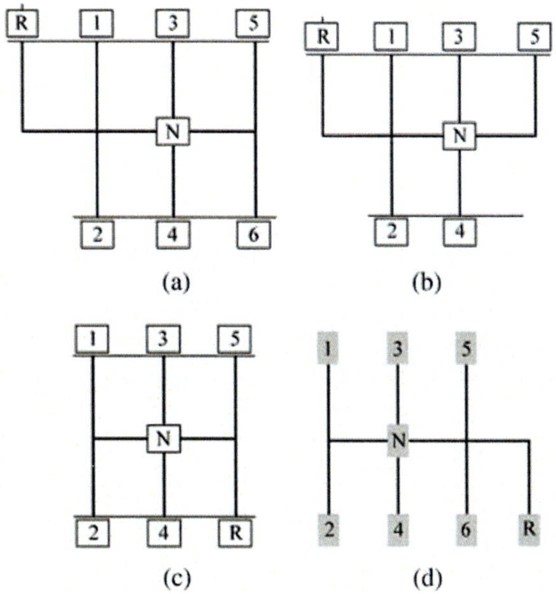

Fig. 2.10 Gear arrangement of the commonly used hand shift lever

Table 2.7 Product curvature radius

Outer diameter of protective tube/mm	Minimum working curvature radius/mm
<8	150
≥8	200

II. Cable

The transmission control cable assembly is to change the displacement of the gear shift lever action into the angular or linear displacement of the select-shift actuator. In order to accurately realize the correspondence between the gear of the gear shift lever and the actual gear, the cable assembly needs to meet the following requirements in terms of performance:

(1) In terms of the efficiency, the push and pull load efficiency of the cable shall not be less than 80% and the stroke efficiency shall not be less than 90%.
(2) The gap between cables shall not be greater than 1.6 mm before durability and not be greater than 2 mm after durability.
(3) The pulling-out force between the core wire and the core wire joint shall not be less than 1500 N, and between the protective tube and the protective tube joint shall not be less than 1500 N.
(4) The cable can still be pulled freely and its sliding resistance is not more than 5 N when it is bent 360° under the minimum working curvature radius specified in Table 2.7.
(5) The cable is tested at 130 °C (150 °C in hot region) and −40 °C (−60 °C in cold region) respectively after bending 360° under the minimum working curvature radius specified in Table 2.7. No melting, cracking or other abnormal phenomena can occur after the test.
(6) After the waterproof performance test, the dust cover is removed and there should be no water vapor, water drops and other phenomena in the cable.

III. Select-shift actuator

The select-shift actuator of the longitudinal MT of the engine and transmission is relatively simple and the gear shift lever is mounted directly on the transmission. However, the select-shift actuator of the horizontal MT of the engine and transmission is away from the gear shift lever and the driver's control of the gear shift lever must be transmitted to the select-shift actuator The usual approach is to divide the driver's control of the gear shift lever into front and rear, left and right halves, which are connected to the transmission select-shift actuator with the shift cable and the select cable, respectively. For example: as shown in Fig. 2.10, the gear shift lever has two actions, i.e. front and rear action for shifting (gear engagement and disengagement), left and right action for gear selecting; with respect to the select-shift actuator in Fig. 2.11 (with the gear arrangement shown in Fig. 2.10a), to engage the gear shift lever in the gear 1 from the neutral, the driver shall first transfer the action through the select cable and change the direction through the select linkage. The select-shift shaft

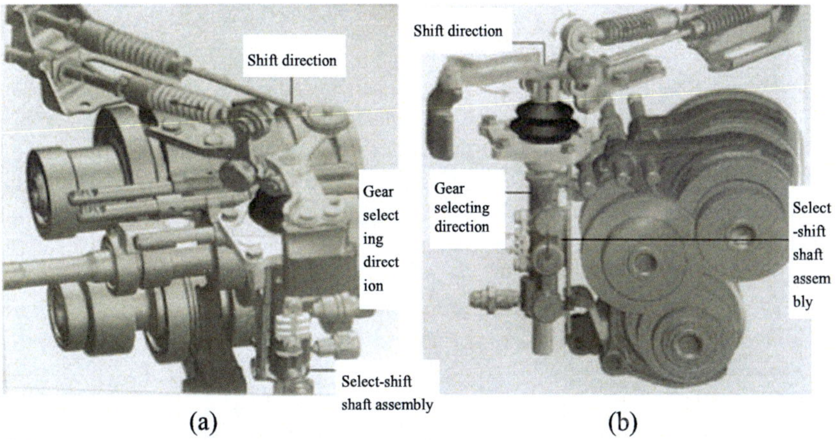

Fig. 2.11 Transmission select-shift actuator

of the transmission moves up and down and selects the gear 1. Then the shift cable transfers the action and the select-shift shaft in the shift linkage rotates to engage the gear 1.

Figure 2.12 shows the structure of the transmission select-shift actuator. The select-shift shaft assembly of the select-shift actuator may be moved up and down

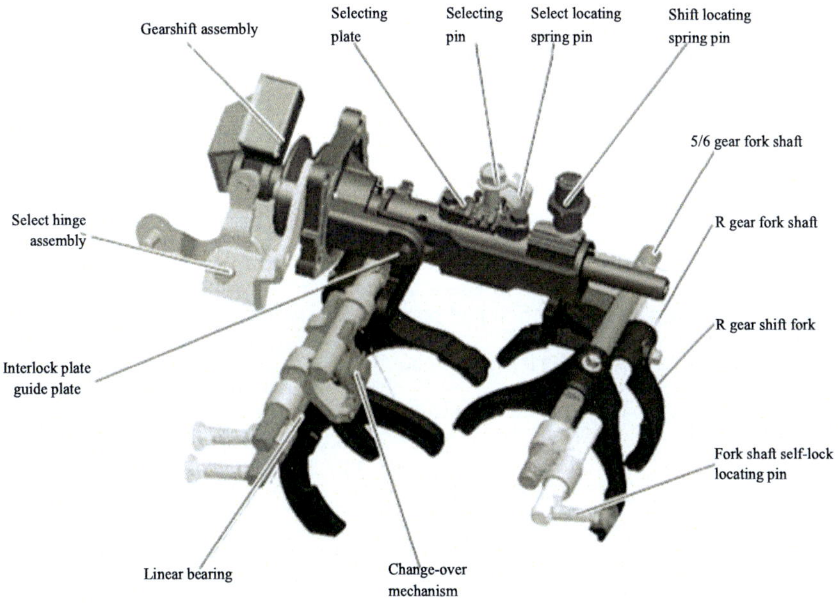

Fig. 2.12 Structure of select-shift actuator

and may also be rotated. Around the assembly is equipped with the select locating ball spring seat assembly, shift locating ball spring seat assembly and shift locating plate. The locating ball in the select locating ball spring seat assembly is pressed in the V-groove on the side of the select-shift shaft under a certain spring force. When the select-shift shaft moves up and down, the locating ball rolls up and down along the V-groove, forming a sense of resistance in the direction of the gear selection. The locating ball in the shift locating ball spring seat assembly is pressed in the axial V-groove below the select-shift shaft under a certain spring force. When the select-shift shaft moves left and right, the locating ball rolls up and down along axial V bevel, forming the shift feel in the shift direction. The select locating ball spring seat assembly and the shift locating ball spring seat assembly also play a role in limiting the up-and-down movement and rotation of the select-shift shaft. The shift locating plate is installed in the select-shift shaft assembly near the case. A locating pin on one side of the case is loaded into the groove of the shift locating plate to play an interlock role.

Move up and down to select the shift fork. The gear shift lever shown in Fig. 2.10a has 4 shift forks including reverse gear. The number of gear shifter shafts corresponds to the gear arrangement form of the gear shift lever, and 4 shift forks are installed in the respective gear shifter shafts.

During shift, the shift fork selected by the select-shift actuator moves horizontally to achieve gear disengagement or engagement.

2.5 Determination of Gear Ratio

The vehicle is subject to various drag forces in the process of driving. In order to overcome these drag forces, the wheels must be provided with a certain traction and power, which is accomplished by the driving system. The driving system consists of engine, clutch, transmission and main reducer. Different tractions are required under different driving conditions and different gear ratios are used to adapt to different requirements. The gear ratio has a great influence on the dynamic and economy performance of the vehicle, so it is necessary to choose the appropriate gear ratio to make the characteristics of traction and power closer to the ideal state.

I. Analysis of ideal characteristics of vehicle

In the case of low road grade, the vehicle driving equation is

$$F_t = \mathrm{mgf} + \mathrm{mgi} + \frac{C_D A V^2}{21.25} + \delta ma \tag{2.25}$$

where: F_t—traction (N);

m—vehicle mass (kg);

g—acceleration of gravity (9.8 m/s^2);

f—rolling resistance coefficient;

i—slope;
C_D—coefficient of air resistance;
A—frontal area (m^2);
V—vehicle speed (m/s);
δ—correction coefficient of rotating mass;
a—vehicle acceleration (m/s^2).
The power required for the vehicle driving is

$$P_t = F_t V \tag{2.26}$$

The traction and power required for the vehicle are shown in Fig. 2.13, where, the value in the curve is $i + \frac{\delta a}{g}$ The higher the value, the greater the gradeability or acceleration required by the vehicle, and the greater the traction and power required by the vehicle.

The traction and power that the driving system can provide are limited by the following conditions: maximum engine power, maximum vehicle speed, and adhesion coefficient of road surface. Based on the above limitations, the ideal traction and power that the driving system can provide are shown in Fig. 2.14. Obviously, Fig. 2.13 is very different from Fig. 2.14. In order for the driving system to meet the vehicle needs, a transmission must be added.

II. Selection of maximum gear ratio

The adhesion coefficient of road surface, maximum gradeability, maximum acceleration and minimum stable engine speed shall be considered for determination of the maximum gear ratio.

The maximum driving force that can be transferred between the tire and the road surface is limited by the adhesion coefficient of road surface, which must meet the following equation

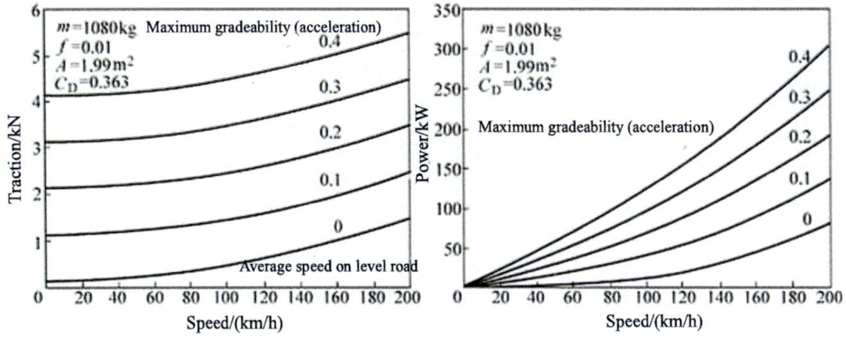

Fig. 2.13 Traction and power required by the vehicle

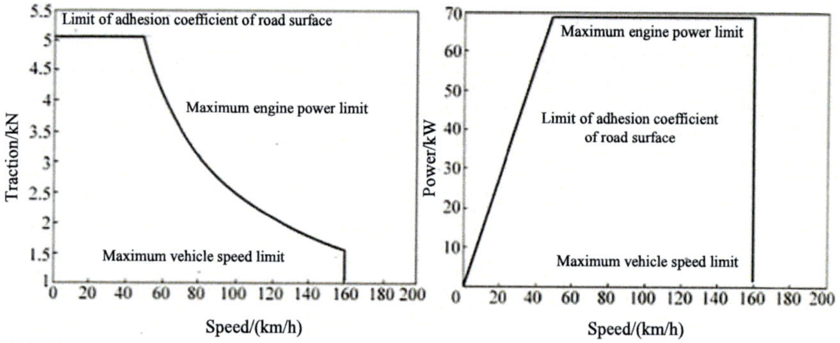

Fig. 2.14 Ideal traction and power that the driving system can provide

$$F_t \leq F_\varphi = F_z\varphi \tag{2.27}$$

式中 F_φ—maximum driving force that can be transferred by the road surface;
F_z—normal reaction of driving wheel;
φ—adhesion coefficient of road surface.

In the low-gear and low-speed driving, the air resistance is negligible, and the maximum driving force obtained is equal to that required, i.e.

$$\frac{T_e\max i_g\max \eta}{r} = mg(f\cos\alpha + \sin\alpha) + \delta ma \tag{2.28}$$

where: $T_{e\max}$—Maximum engine torque;
$i_{g\max}$—maximum gear ratio;
η—powertrain efficiency;
r—wheel radius;
α—slope angle.

The following two limiting cases shall be considered for the maximum gear ratio $i_{g\max}$: maximum gradient that can be passed at an acceleration of 0; maximum acceleration of a level road. According to formula (2.28), the maximum gear ratio designed from the maximum gradeability is

$$i_g\max = \frac{mgr(f\cos\alpha + \sin\alpha)}{T_e\max\eta} \tag{2.29}$$

When the driving force on the driving wheel is large and the vehicle must drive steadily at a low speed, the maximum gear ratio of the powertrain is

$$i_g\max = 0.377\frac{n_{\min} r(1-s)}{V_{\min}} \tag{2.30}$$

where: $n_{\min}$—minimum stable engine speed;

$V_{\min}$—minimum stable vehicle speed;

s—slip ratio of driving wheel.

In formula (2.30), s = 0 in the actual design and the actual maximum gear ratio takes the larger value above. Then the maximum acceleration on the level road under the maximum gear ratio is

$$a_{\max} = \frac{T_{e\max} i_{g\max} \eta}{\delta m r} - \frac{g f}{\delta} \quad (2.31)$$

III. Selection of minimum gear ratio

If the driving wheel does not slip at high speed,

$$i_{g\min} = 0.377 \frac{n r}{V_{\max}} \quad (2.32)$$

where: n—engine speed.

Many factors shall be considered when choosing the minimum gear ratio. With respect to a passenger vehicle, the running time at this gear may be up to 80%. According to the differences in design types, there are three design schemes as follows:

(1) Optimum gear ratio design: in order to convert the maximum engine power into the highest performance, the resistance power must pass through the maximum power point of the engine, namely the intersection point A of curves 2 and 4 as shown in Fig. 2.15. In this special case, it is very simple to combine the maximum speed, engine power and gear ratio. It is only required to change the engine speed n in formula (2.32) to the speed at the maximum engine power.

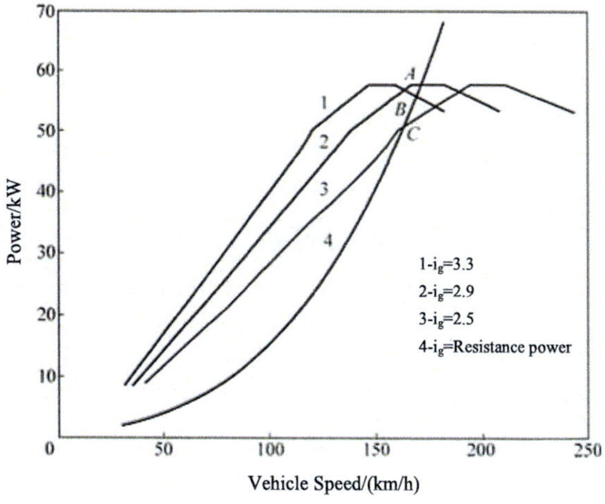

Fig. 2.15 Selection of minimum gear ratio

This design scheme can well take into account the acceleration power reserve and fuel economy at the top gear.

(2) Sports design: the sports design requires a large amount of power reserve at the top gear. The available power intersects the resistance power at the descending section of the engine, as shown at point B in Fig. 2.15. At this point, the minimum gear ratio is larger than the optimum gear ratio.

(3) Economic design: the economic design requires that the resistance power be close to the engine power at the top gear and pays attention to the fuel economy of the vehicle. The available power intersects the resistance power at the ascending section of the engine, as shown at point C in Fig. 2.15. At this point, the minimum gear ratio is smaller than the optimum gear ratio. The minimum gear ratio determined by this method is called energy saving gear, which shall make the resistance curve close to the lowest fuel consumption curve. The highest speed gear is the second top gear.

To sum up, among the three schemes to select the minimum gear ratio, the first one gives consideration to both dynamic and economy performance, and can achieve the highest speed; the second pays more attention to dynamic performance with poor economy and can achieve a higher speed; the third pays more attention to economy performance with poor dynamic performance and can achieve a lower speed than the first two. The minimum gear ratio is selected as required in practical application. The minimum gear ratio obtained from the economic design scheme may be deemed as the top gear, while the minimum gear ratio obtained from the other two design schemes can be selected as the second top gear.

IV. Determination of gear ratio spread

The gear ratio spread is the ratio of the maximum gear ratio to the minimum gear ratio, mainly depending on the vehicle purpose, engine speed spread and specific power of vehicle. Generally, the maximum gear ratio is determined by the starting conditions, and the minimum gear ratio is determined by the fuel economy range or maximum speed on the engine characteristic diagram. The formula of the gear ratio spread is

$$i_{G,tot} = \frac{i_{g\max}}{i_{g\min}} \tag{2.33}$$

V. Selection of middle gear ratio

Different types of vehicles have different number of gears. Increasing the number of gears will improve the dynamic and economy performance of the vehicle, but will complicate the transmission structure and increase the cost and weight of the transmission. With the increase of the fuel economy requirement, the transmission tends to be multi-speed.

The relationship between the gear ratios of two adjacent gears (speed ratio range Ψ) can be expressed as

$$\Psi = \frac{i_{g,k-1}}{i_{g,k}} \leq \frac{n\max}{nT\max} \tag{2.34}$$

The following shall be considered for the selection of middle gear ratio:

(1) If there are more gears, the working range of the low fuel consumption area of the engine can be used more fully, so as to better give consideration to both dynamic and economy performance of the vehicle. However, if the number of gears is increased, the shift frequency, the weight and size of the transmission will be increased accordingly.
(2) The driving distance at the low gear accounts for a small proportion and the driving distance at the high gear accounts for a large proportion.
(3) The percentage of driving distance at each gear depends on the specific power, road distribution, traffic conditions and driving style.
(4) The smaller the speed ratio range Ψ, the easier the shift and the more fun it is to drive.
(5) The thermal load of the synchronous ring is in direct proportion to the square of the speed ratio range Ψ.

Some of the above aspects are contradictory, so it is required to make a compromise in the transmission design. Practice has proved that the geometric gear ratio range and gradient gear ratio range are two effective calculation methods.

(1) Geometric gear ratio range: in the gear design, the theoretical value of the gear ratio range among the gears is always equal, i.e.

$$q = \frac{i_{g1}}{i_{g2}} = \frac{i_{g2}}{i_{g3}} = \ldots \tag{2.35}$$

Giving the maximum gear ratio i_{gmax}, the minimum gear ratio i_{gmin} and m gears,

$$q = \sqrt[m-1]{(i_{gmax}/i_{gmin})} \tag{2.36}$$

$$i_{g1}=i_{gmax} \quad i_{g2}=i_{g1}/q \quad i_{g3}=i_{g1}/q^2 \quad i_{gm}=i_{g1}/q^{(m-1)}=i_{gmin} \tag{2.37}$$

The geometric speed ratio range is mostly used in the commercial vehicle transmissions, which avoids the overlap of the gears.

(2) Gradient speed ratio range: the gradient speed ratio range is used for the passenger vehicle transmissions. The higher the gear, the smaller the range.
Table 2.8 shows the calculation results of the 6 speed transmission with the maximum gear ratio of 13.752 and the minimum gear ratio of 2.637 under two speed ratio ranges. Figure 2.16 shows the calculation results of the change in the engine speed and driving force of different gears with the vehicle speed

Table 2.8 Calculation results of 6 speed transmission in two algorithms

Gear	1	2	3	4	5	6
Geometric gear ratio (theoretical)	13.752	9.883	7.103	5.105	3.669	2.637
Adjacent gear ratio range		1.3915	1.3914	1.3914	1.3914	1.3914
Gradient gear ratio (theoretical)	13.752	7.463	5.120	3.897	3.146	2.637
Adjacent gear ratio range		1.8428	1.4575	1.3139	1.2389	1.1929
Gradient gear ratio (actual)	13.752	7.954	5.303	4.036	3.173	2.637
Adjacent gear ratio range		1.7289	1.4999	1.3139	1.2720	1.2033

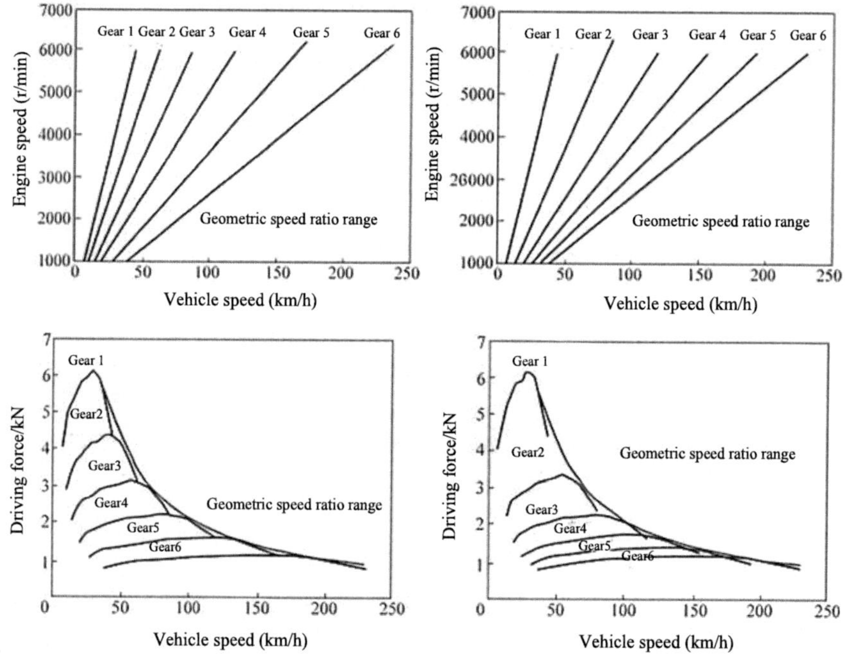

Fig. 2.16 Calculation results of the change in the engine speed and driving force with the vehicle speed under two speed ratio ranges. Calculation results of the change in the engine speed and driving force with the vehicle speed under two gear ratio ranges

under two speed ratio ranges. Obviously, in the geometric speed ratio range, the deviation between each gear and the engine power curve is small and this range is suitable for the case where the use time of each gear is uniform; while in the gradient speed ratio range, the deviation between each gear and the engine power curve is different, large at low gear and small at high gear. Therefore, this speed ratio range is suitable for the case where the use time is short at low gear and long at high gear.

2.6 Joint Working of Engine and Transmission

The engine and transmission shall be subject to joint working performance calculation after matching: calculation of vehicle dynamic performance, mainly including the maximum speed, gradeability and acceleration capacity; calculation of vehicle fuel economy, i.e. the ability of the vehicle to drive economically with minimal fuel consumption under the condition of dynamic performance guaranteed. Generally, the fuel economy is measured by the fuel consumption of a vehicle driving 100 km under certain operating conditions or the mileage of the vehicle driving under a certain amount of fuel.

I. Vehicle dynamic performance

1. **Maximum speed**

The maximum speed is the maximum speed a vehicle can travel on a level and good road surface. Theoretically, the maximum speed refers to the intersection point of the driving force curve and the driving resistance curve in the driving force diagram (Fig. 2.17) or the intersection point of the driving power curve and the driving resistance power curve. The actual measurement method is to maintain the average of the maximum speed in both directions at the measurement distance of 1 km. The main test conditions are: load the vehicle to half the load; level and dry road surface, and favorable adhesion coefficient of road surface; maximum wind speed not exceeding 3 m/s; the vehicle must travel along the test track in both directions.

2. **Maximum gradeability**

The maximum gradeability refers to the road grade that a vehicle can climb when all the remaining driving force is used for climbing at constant speed, which is used to measure the climbing performance of the vehicle. As shown in Fig. 2.18, with the

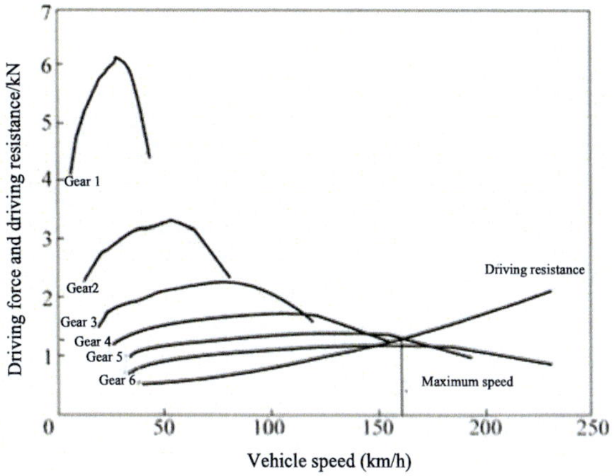

Fig. 2.17 Maximum driving speed determined by the driving force diagram

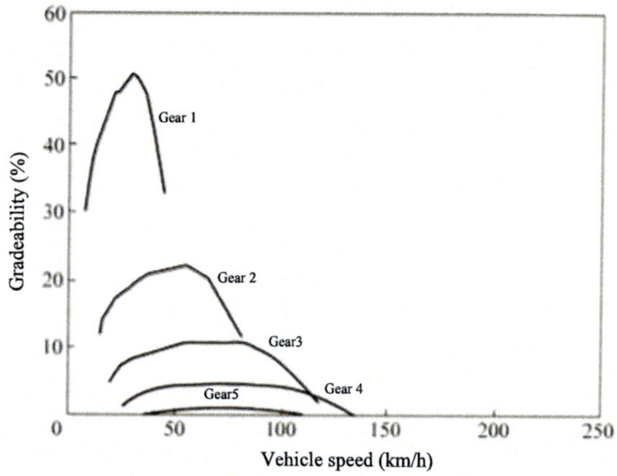

Fig. 2.18 Vehicle gradeability at all gears

increase of the gear, the gradeability gradually decreases, and there is no gradeability at gear 6.

3. **Acceleration capacity**

Acceleration capacity refers to the capacity of the vehicle to accelerate with all of its remaining driving force on a level road, as shown in Fig. 2.19.

II. Vehicle fuel economy

The specific fuel consumption b_e at the working point at any time can be read from the engine fuel consumption characteristic curve in Fig. 2.20. It is necessary to know the engine speed and torque at this time, where the engine speed can be obtained directly from the engine data or through formula (2.38); the engine torque can be obtained either directly from engine data or through formula (2.39). Point A in Fig. 2.20 means that the fuel consumption rate is 250 g/(kW h) when the engine speed is 3100 r/min

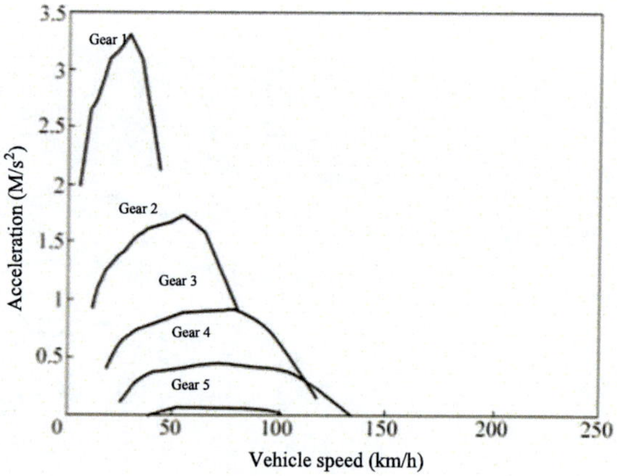

Fig. 2.19 Vehicle acceleration performance at each gear

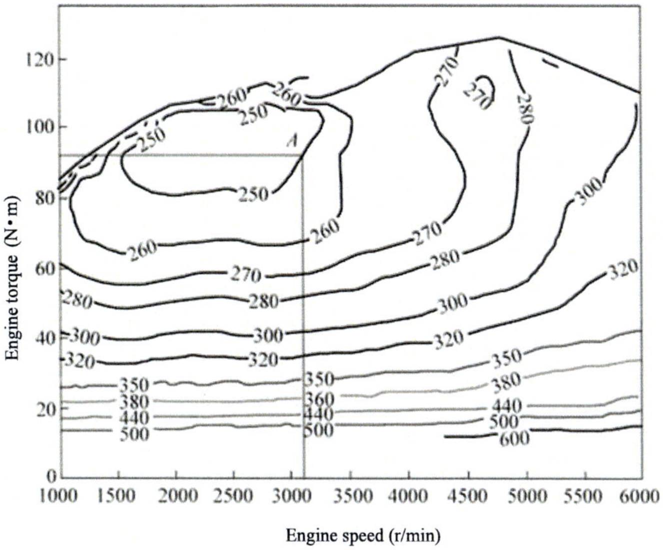

Fig. 2.20 Engine fuel consumption characteristic curve

and the torque is 92 N m.

$$n_e = \frac{i_g V}{0.377r} \tag{2.38}$$

$$T_e = \frac{F_t r}{i_g \eta} \tag{2.39}$$

The engine power P_e may be obtained through formula (2.40)

$$P_e = \frac{F_t V}{\eta} \tag{2.40}$$

Fuel consumption per hundred kilometers Q_s

$$Q_s = \frac{100 b_e P_e}{\rho v} = \frac{100 b_e F_t}{\rho \eta} \tag{2.41}$$

where: b_e—specific fuel consumption;

ρ—fuel density.

Calculation example of fuel consumption per hundred kilometers: with respect to point A in Fig. 2.20, $b_e = 250$g/kw h, $P_e = \frac{T_e n_e}{9549} = 29.9\,\text{kw}$, $\rho = 755\,\text{g/L}$, and vehicle speed $v = 0.377\frac{rn}{i_g} = 0.377\frac{0.269\times3100}{2.674}$ km/h $= 117\,\text{km/h}$ at gear 6, then the fuel consumption per hundred kilometers is $Q_s = 8.46\,\text{L}/100\,\text{km}$.

The fuel economy calculation is to calculate the fuel economy under the test operating cycle as shown in Fig. 2.21, which mainly includes the calculation of fuel consumption under the conditions of driving at constant speed, driving at constant acceleration, driving at constant deceleration and idle shutdown.

1. **Fuel consumption at constant speed**

Under the driving condition at constant speed, the power provided by the engine needs to overcome the rolling resistance and air resistance. If the speed is v, the power to be provided by the engine is

$$P_e = mgf + \frac{C_D A v^2}{21.25} v/\eta \tag{2.42}$$

The abscissa of the engine fuel consumption characteristic curve in Fig. 2.20 is changed from the engine speed to the vehicle speed and the ordinate is changed from the engine torque to the engine power to form the engine characteristic performance curve as shown in Fig. 2.22. Then the fuel consumption per hundred kilometers under the driving condition at constant speed can be calculated according to formulas (2.41) and (2.42). The fuel consumption $Q_{\text{constant speed}}$ of this section can be calculated according to the given mileage or time.

2. **Fuel consumption at constant acceleration**

Under the driving condition at constant acceleration, the power provided by the engine needs to overcome the rolling resistance and air resistance, as well as the acceleration resistance. The power to be provided by the engine at a certain speed is

$$P_e = mgf + \frac{C_D A v^2}{21.25} + \delta m \frac{dv}{dt} V/\eta \tag{2.43}$$

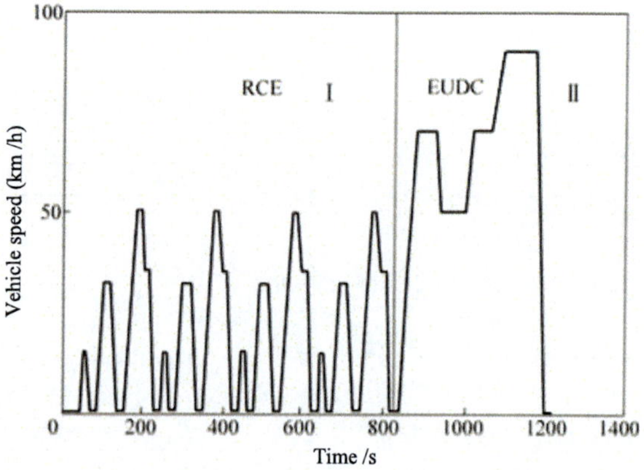

Fig. 2.21 Test operating cycle

Fig. 2.22 Engine characteristic performance curve

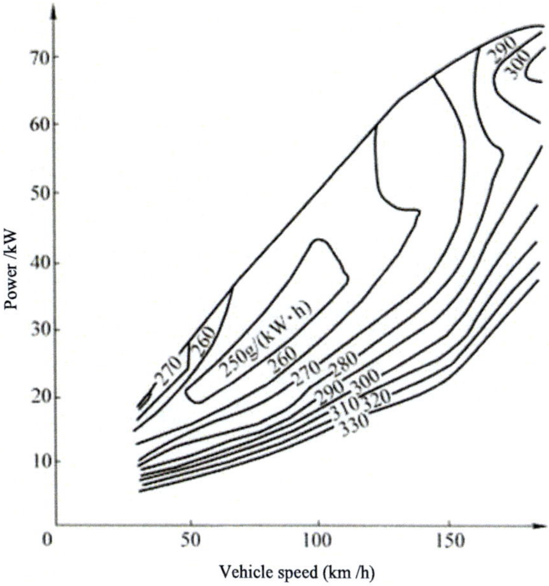

Taking the example of vehicle speed V_a at the start, v_b at the end and the acceleration of dv/dt, the fuel consumption of this section is calculated.

(1) The driving section is divided into $n = \frac{V_b - V_a}{1} = V_b - V_a$ sections by 1 km vehicle speed.
(2) Average speed $V_i = V_a + (i - 0.5)\frac{dV}{dt}$ at the section i $(i = 1, 2 \ldots\ldots n)$
(3) Time $t_i = \frac{1}{v_i}$ of the section i (i = 1, 2, ..., n).
(4) Power $P_{e,i}$ of the section i $(i = 1, 2 \ldots\ldots n)$, calculated from the formula (2.43).
(5) Calculate the fuel consumption per hundred kilometers in the section i $(i = 1, 2 \ldots\ldots n)$ according to the formula (2.41).
(6) Calculate the fuel consumption Q_i in the section i.
(7) Total fuel consumption $Q_{\text{acceleration}} = \sum_{i=1}^{n} Q_i$ in the driving section with constant acceleration.

3. **Fuel consumption at constant deceleration**

When decelerating, the throttle is released and slightly braked. The engine is in the forced idle state, and its fuel consumption is that in the normal idle state. Taking the example of vehicle speed v_a at the start, v_b at the end and the deceleration of dv/dt, the fuel consumption of this section is calculated.

(1) Total deceleration time $t = \frac{v_a - v_b}{dv/dt}$;
(2) If the idle fuel consumption rate is q_d, the fuel consumption during deceleration is $Q_{\text{deceleration}} = q_d t$.

4. **Fuel consumption in idle shutdown**

If the idle shutdown time is t_s(s), the fuel consumption during idling is $Q_{\text{idle}} = q_d t_s$.

With respect to a vehicle with start-stop system, the fuel consumption during idle shutdown is mainly related to the temperature and not related to the idle shutdown time. The fuel consumption in this period should be a function of the coolant temperature and may be expressed as $Q_{\text{idle}} = f(t_{\text{emperature}})$.

5. **Fuel consumption per hundred kilometers in the whole drive cycle**

The fuel consumption per hundred kilometers in the test operating cycle as shown in Fig. 2.21 is

$$Q_s = \frac{Q_{\text{constant speed}+} Q_{\text{acceleration}} + Q_{\text{deceleration}} + Q_{\text{idle}}}{S} \times 100 \tag{2.44}$$

where: S—driving distance in the whole cycle.

2.7 Shift Performance Evaluation

The shift performance evaluation mainly includes the evaluation of the shift performance under static, dynamic and low temperature conditions and the evaluation criteria range from 1 to 1, with 1 being the worst and 10 being the best. The subjective shift performance evaluation items listed in Table 2.9 are weighted and scored according to the subjective shift performance evaluation criteria listed in Table 2.10. The evaluation results of all items are represented by the spider diagram shown in Fig. 2.23. Figure 2.24 shows the measurement results of the dynamic shift force. There is a certain unsmooth resistance from gear 2 to the front segment of the neutral, and a certain inhalation feel in the rear segment; from the neutral to gear 1, there is a slight step sense, with large gear engagement force and obvious second impact.

2.8 New Technologies of MT

From the perspective of energy saving, MT mainly has three new technologies to be applied: start-stop function, gear shifting indicator, and active speed control of the engine during the shift.

1. **Start-stop function**

MT needs to realize the start-stop function, that is, the engine can be stopped automatically when the vehicle stops and the engine idles, and the engine can be restarted automatically when needed, so as to save fuel and reduce emissions. To achieve this

Table 2.9 Subjective shift performance evaluation items

Item	Static	Dynamic	Low temperature
Shift noise	O	O	
Free clearance of gear	O	O	O
Shift smoothness		O	O
Overshoot		O	O
Select force	O	O	O
Accuracy		O	O
Shift safety		O	O
Static shift force	O		
Dynamic shift force		O	O
Shift force mode	O	O	O
Second impact		O	O
Inhalation	O	O	O
Irregular shift resistance	O	O	O
Shift viscous resistance	O	O	O
Abuse of the shift		O	O
Low temperature shift			O
Upshift scratch		O	O
Low temperature scratch			O
Clutch release	O	O	O
Reverse locking	O	O	O
Control force to restore the neutral		O	O
Gear shift lever vibration		O	O
Retention	O	O	O
Ergonomics	O	O	O
Shift stroke	O	O	O

Note ○ is the item requiring subjective evaluation

function, it is required to install a neutral position switch on the transmission that can detect whether it is in neutral. Figure 2.25 is the schematic diagram of the MT start-stop function. The engine stops automatically in case of the vehicle speed below a certain value, full of battery, brake vacuum large enough, allowed engine shutdown (satisfactory water temperature and catalyst temperature), satisfactory safety conditions (driver on the seat and engine hood closed), air conditioning defogging state not requested, accelerator pedal release or neutral transmission; the engine may be restarted automatically in case of low battery, air conditioning defogging state requested and accelerator pedal pressed under neutral transmission or clutch release.

Table 2.10 Subjective shift performance evaluation criteria

Evaluation index	1	2	3	4	5	6	7	8	9	10
Evaluation	Unacceptable		Poor		Unclear	Acceptable	Relatively good	Good	Very good	Excellent
Customer satisfaction	Dissatisfied				Partially dissatisfied	Satisfied		Very satisfied	Fully satisfied	
Improvement requirement	All customers			General customer		Picky customer				
NVH	Very loud			Loud	Disturbing sound	Clear audible	Audible	Partially audible	Just audible	Inaudible

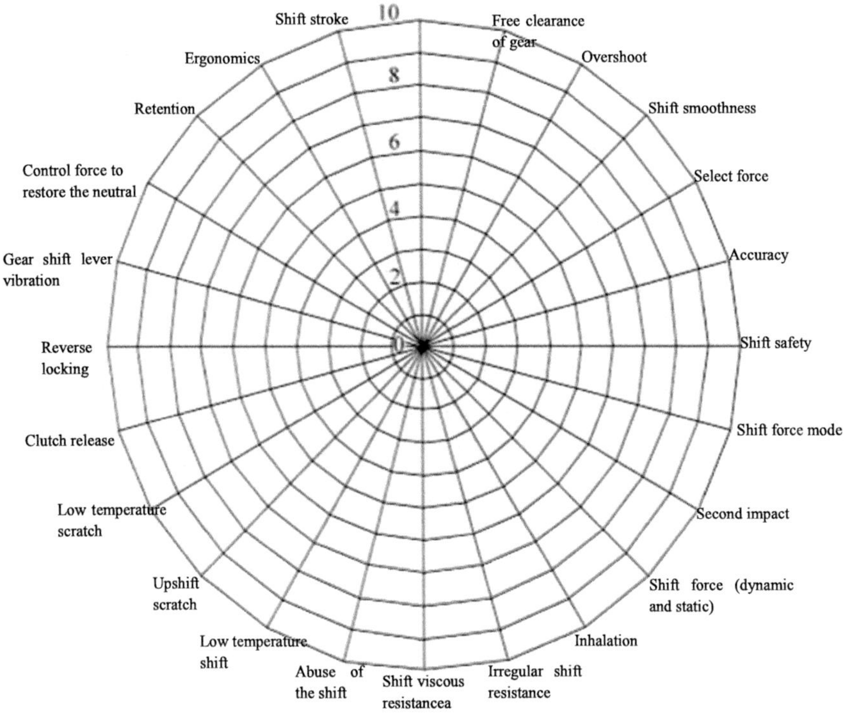

Fig. 2.23 Spider diagram of shift performance evaluation results

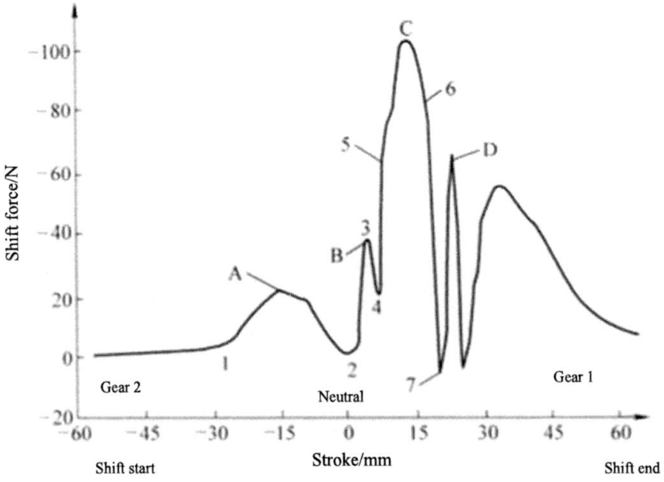

Fig. 2.24 Dynamic shift force. 1—Gear disengagement, 2—neutral, 3—synchronizer locating pin off point, 4—pre-synchronization point, 5—start of synchronization, 6—end of synchronization, 7—meshing point between synchronizer clutch and soldered teeth. A—Maximum out-of-gear force, B—maximum gear engagement force, C—maximum synchronizing force D—second impact force

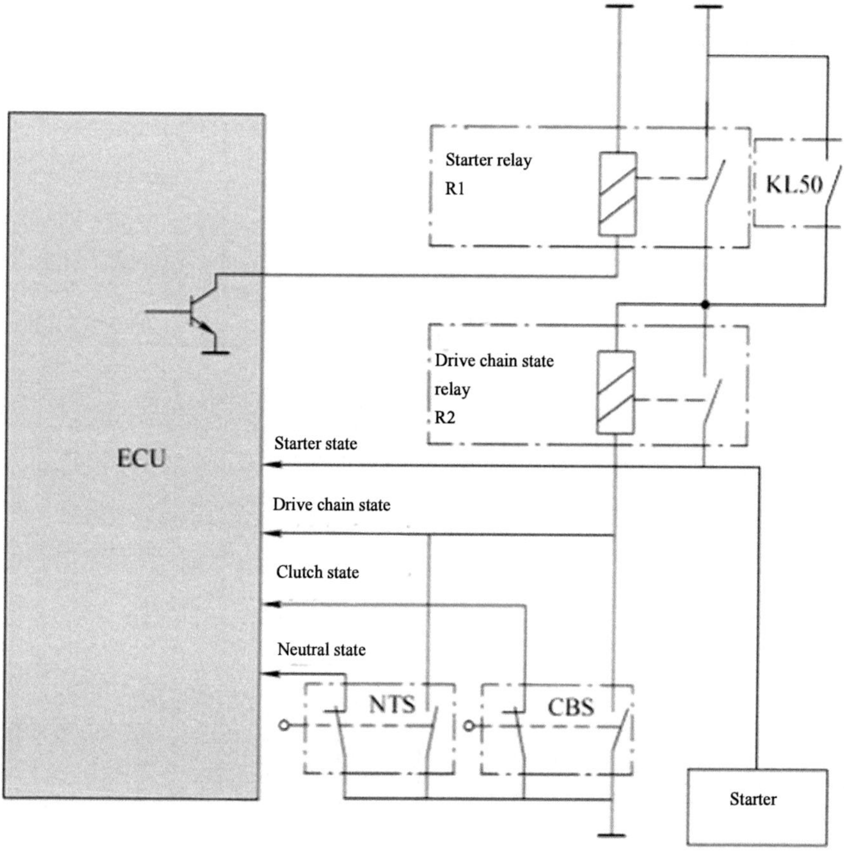

Fig. 2.25 Schematic diagram of the MT start-stop function

2. Gear shifting indicator

The EU Parliament and Council amended and added the provisions in the "Requirements for Gear Shifting Indicator (GSI) for M1 vehicles", which came into force on November 1, 2012 for newly finalized vehicles and on November 1, 2014 for all new vehicles in production. Although there are no relevant provisions in China currently and the gear shifting indicator is only visible in the joint venture and imported high-end cars, it is still more worthy of promotion as the energy saving function of the vehicle. Over the past few decades, the cutting-edge technologies have been constantly used to improve fuel economy in vehicles, but few studies have been conducted to improve fuel economy based on driving habits. The gear shifting indicator is an indicator that suggests or prompts the driver when to shift gears while driving. Shifting gears at the right time can mean lower fuel consumption, ensure more efficient use of fuel, and reduce emissions, which can help novice drivers grasp the right time to shift gears as soon as possible and improve driving pleasure.

Fig. 2.26 Information on MT gear shifting indicator

Figure 2.26 shows the information on the MT gear shifting indicator. As shown in the figure, the current gear is 3, and the upshift prompt message prompts the driver to upshift. To achieve this function, information about the engine, transmission, etc., is acquired, decisions are made according to certain control rules, and signals are sent to the gear shifting indicator for the driver to operate.

3. **Active speed control of engine**

An automated function is provided in 7 speed MT of Chevrolet Corvette that the engine is subject to active speed control through the electronic throttle valve instead of running freely after the clutch release during shift, so that the engine speed can track the target speed of the new gear, so as to reduce the sliding friction time and sliding friction work of the clutch engagement, and reduce the shift impact brought by the clutch engagement. In the power downshift condition, the usual MT vehicle will give the passenger the feeling of power replenishment delay, which can be significantly solved by the active speed control of the engine during the shift. With this technology, the engine speed can be increased to be equal to or slightly larger than the target speed at the new gear in the clutch release time, so that the friction moment is in the direction of the driving force at the clutch engagement without generating a drag force, which greatly improves the shift acceleration and comfort. In the upshift condition, this function can also shorten the shift time and reduce the shift impact. The difference is that, in the upshift condition, the engine speed control is reduce the speed of the engine to the synchronous speed at the new high gear after the clutch release, but the engine speed shall be guaranteed to rise to slightly higher than the target speed at a new gear at the moment of the clutch engagement after the new gear is engaged, so that the clutch friction moment plays a driving role and shortens the power interruption time.

Bibliography

1. Chen Yuxiang, Zang Mengyan, Chen Yong, et al (2012) Shift force analysis of manual transmission based on virtual prototyping technology. China Mech Eng 23(8)
2. Wang F, Fuqiang Z, Guanghui L (2007) MT modeling example tutorial: PRO/ENGINEER Wildfire (with CD). China Machine Press, Beijing
3. Zhou F, Hu Z (2012) The testing methods of manual gear shift. Straits Sci (12):9–11
4. Ma X, Peng G, Yu B (2011) Study and optimization of automotive manual transmission idle knock noise. Chin J Automot Eng 1(z1):167–167
5. Yue X (2010) Implementation of automobile clutch and manual transmission repair. China Machine Press, Beijing
6. Longyang X, Shuguang Z, Qing S et al (2013) The experimental study of NVH characteristics of automotive manual transmissions. Manuf Autom 35(6):50–53
7. Fan W, Fang W (2015) Development and optimization of shift performance of manual transmission in passenger vehicles. In: 2015 SAE-China Congress Proceedings
8. Junlin W (2015) Methods for detection and diagnosis of manual transmission. Mach China 20:131–132
9. Hu Z, Chen Y, Chen Y, et al (2011) The method of multi-body dynamics evaluation for the shiftability of manual transmission. In: 2011 SAE-China congress proceedings
10. Zheng L, Liu J, Xu L (2014) Application of MT neutral gear sensor in start-stop system. In: TM symposium China IEC, (P) HEV and EV transmissions and drives
11. Li Y, Fang W, Chen A (2010) Development of high-efficiency manual transmission. In: 2010 SAE-China congress proceedings
12. Li Y, Fang W, Liu Q (2008). How to evaluate the manual transmission shift feeling. In: 2008 SAE-China congress proceedings
13. Jiang F, Han Y, Huang L (2012) A control method for starting vehicles with manual transmission: CN102518519A. 2012.06.27
14. Huifang Z, Lirong Z (2011) manual transmission efficiency of factor analysis and improvement measures. Manufact Autom 33(4):31–33
15. Longyang X, Shuguang Z, Qing S et al (2013) The experimental study of NVH characteristics of automotive mManual transmissions. Manufact Autom 35(6):50–53
16. Zhang H, Bai Y, Wang W, et al (2015) Simulated analysis of shift force of manual transmission. Technol Bus (19):251–251
17. De l CM, Theodossiades S, Rahnejat H. An investigation of manual transmission drive rattle. Proc Inst Mech Eng, Part K: J Multi-body Dyn 224(2):167–181
18. Lucente G, Montanari M, Rossi C (2007) Modelling of an automated manual transmission system. Mechatronics 17(2–3):73–91

Chapter 3
Automatic Transmission

3.1 Overview

Automatic transmission (AT), as the earliest AT, mainly consists of the hydraulic torque converter, planetary gear train and shift control system. Its variable speed and variable torque function is mainly accomplished through different gear combinations and hydraulic transfer. The AT is mainly classified into FR layout and FF layout. The gear change of the AT is mainly accomplished by two-parameter control based on speed and load, namely the stroke of the accelerator pedal. As one of the most important parts of AT, the hydraulic torque converter mainly consists of impeller, turbine and guide wheel. It can change the torque through the change of liquid moment of momentum, and it has good adaptability and automatic adjustment function for external load as well as the roles of clutch, stepless change and torque adjustment. Another important part of AT is planetary gear train. The gear ratio of AT is determined by the number of planetary gear trains and their different combinations controlled by the clutch and brake.

The AT technology has been applied to the automobile industry for more than 80 years and achieves high passability, good driving safety, good adaptability, simple and easy control of vehicles. With the emergence of more new automotive technologies, the structure of AT is constantly improved, and its technology is gradually mature. Compared with the traditional manual transmission, the AT not only can better adapt to the driving requirements of the vehicles, but also has the following significant advantages.

(1) Simple and easy control and excellent driving performance. One the one hand, the hydraulically or electronically controlled vehicle with AT realizes the automatic shift. When AT is used, the shift can be easily realized by simply manipulating the slide valve in the hydraulic control device, while the MT can complete shift only by using the shift fork to toggle the shifting slide gear tediously. It is obvious that the AT is easier to manipulate in terms of shift. On the other hand, in addition to the influence of the vehicle structure, the vehicle driving performance is also related to the correct choice of the control and manipulation

Y. Chen, *Automotive Transmissions*, Springer Tracts in Mechanical Engineering,
https://doi.org/10.1007/978-981-15-6703-2_3

mode. The AT may achieve the best dynamic and economy performance of the vehicle because of its automatic shift under the established optimum shift rule. The driver's skills and experience are unrelated to the shift.

(2) Make the driving safer and reduce the fatigue strength of the driver. It has been surveyed that a car shifts the gears an average of five to nine times per kilometer while driving downtown, with the shift frequency of 3–5 times/min, the maximum shift frequency up to 4 times/20 s and generally 4–6 concerted actions in each shift, which will inevitably generate two results: first, speed up the driver's metal and physical fatigue speed and distract the driver's driving attention, which will inevitably cause hidden trouble of traffic over time; second, the driver reduces the shift times due to fatigue and changes the speed by controlling the accelerator pedal, thus reducing the vehicle fuel economy. After the vehicle realizes the automatic speed changing, the driver only needs to control the accelerator pedal to realize the automatic speed changing easily, thus canceling the control of the clutch pedal and gear shift lever, greatly reducing the driver's control fatigue strength and making the driving safer.

(3) Improve the dynamic performance of the vehicle. The automatic shift characteristic of AT and the torque conversion characteristic of the hydraulic torque converter can effectively improve the vehicle starting acceleration. The power transferred by the whole powertrain is not interrupted during automatic shift. The fuel supply is not controlled as in the manual shift and the gear is always shifted when the engine power is fully utilized, thus improving the dynamic performance of the vehicle.

(4) Reduce exhaust emissions. The engine of a vehicle equipped with AT may always work in its economic speed range, i.e. in the speed range with less emissions and pollution. This is because AT can automatically adjust the gear ratio according to the speed and load within a certain range, thus greatly reducing the exhaust emission and pollution.

(5) High fuel economy. It is generally believed that AT has low transmission efficiency and is not conducive to the realization of fuel economy. However, considering the whole driving condition of the vehicle, the AT always automatically takes the best match with the engine, and can automatically adapt to the change of resistance at the best shift moment, so that the vehicle can have better fuel economy.

(6) Prolong the service life of the powertrain parts. The AT adopts hydrodynamic drive, which can reduce or even eliminate the vibration in the powertrain, especially in the case of bad driving conditions. Relevant tests show that when the vehicle equipped with AT is driving on bad roads, the maximum dynamic load torque borne by its transmission shaft is 20–40% of that of the vehicle equipped with MT; while in the standing start, the maximum dynamic load torque is 50–70% of that of the MT, which prolongs the service life of the auto parts.

Of course, AT also has many disadvantages: low efficiency in transmission without stepless change; high machining accuracy and manufacturing process of the parts, resulting in the increase of the manufacturing cost, complex mechanism and repair difficulties.

3.2 Composition and Control Principle of AT

Figure 3.1 shows an 8-speed AT composed of a hydraulic torque converter and a planetary transmission, with the working principle as follows: the throttle position sensor and vehicle speed sensor convert the engine throttle percentage and vehicle speed into an electrical signal that is sent to the electronic control unit (ECU) of the electronic control system along with other sensor signals that reflect the performance of various parts of the vehicle and the system. Then the AT compares the input signal with the control law stored in the ECU memory. The ECU issues instructions to a number of corresponding electro-hydraulic proportional valves to regulate the hydraulic oil flowing to the control valve, so that the actuator, clutches and brakes are under control, thus accurately controlling the shift gear moment and locking the clutches. The control principle of the AT is shown in Fig. 3.2. It mainly consists of four parts: hydrodynamic transmission (hydraulic torque converter), mechanical transmission, hydraulic control unit and electronic control unit.

I. Hydrodynamic transmission

The hydrodynamic transmission is classified into the hydraulic coupler and hydraulic torque converter and the hydraulic torque converter is used in most hydrodynamic drives of modern automatic transmissions.

Fig. 3.1 8-speed AT (picture source: ZF)

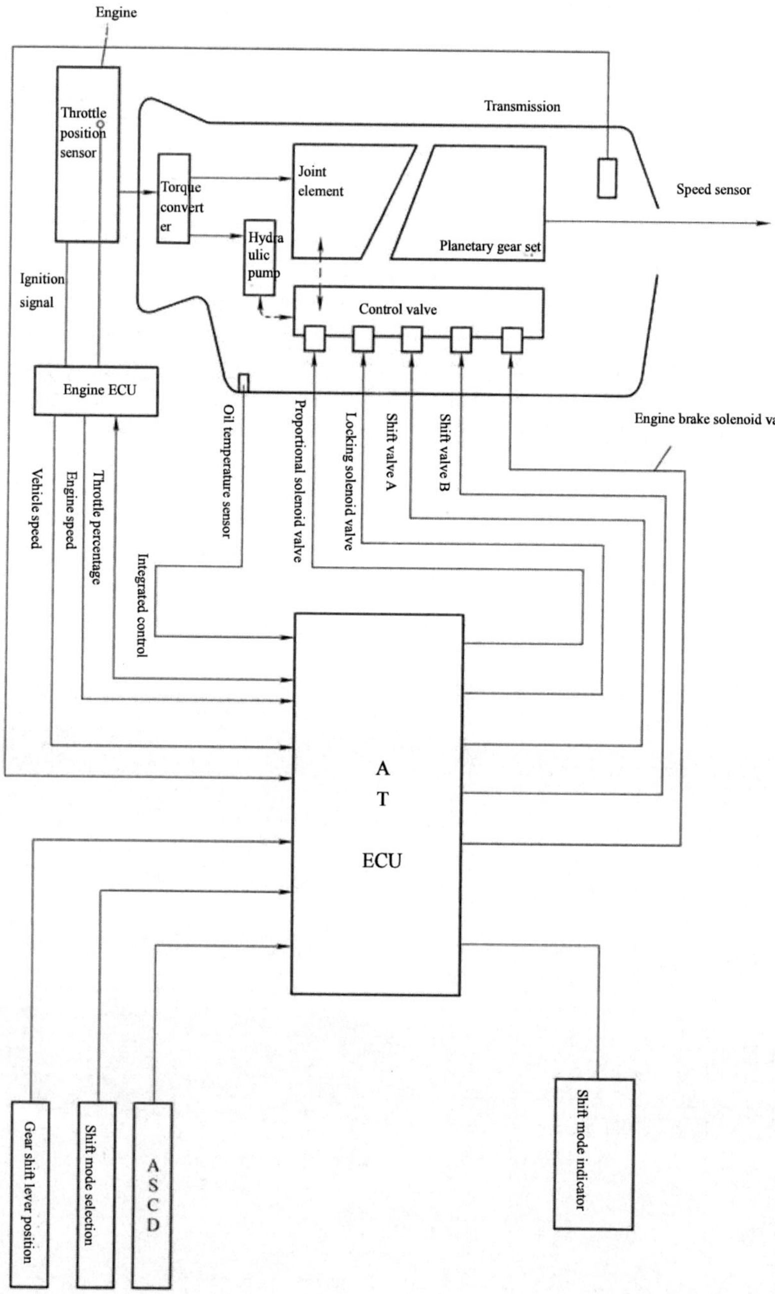

Fig. 3.2 Control schematic diagram of AT

The hydraulic torque converter is located at the most significant end of the AT and is mounted on the engine flywheel, acting like a clutch with MT. It is a vane transmission with the liquid as the working medium. When working, the interaction between the active wheel blade and working liquid is used to realize the interconversion of mechanical energy and liquid energy (that is, the impeller continuously absorbs the power of the internal combustion engine and transfers it to the turbine), and the transmitted torque is changed through the change of the liquid moment of momentum. With the ability to continuously change the speed and torque, it is a continuously variable transmission (CVT) most successfully used in the vehicles, fundamentally simplifies the control procedure; makes the vehicle start smoothly, accelerate quickly and gently with excellent vibration damping performance, thus prolonging the life of the powertrain and improving the ride comfort, average vehicle speed and driving safety; during stall, it has the maximum torque ratio, which prevents engine flame-out and more importantly, greatly improves the vehicle passability.

The drawback of the hydraulic torque converter is its low efficiency and narrow efficiency range. The loss is emitted as heat, further increasing the burden on the radiator, so the fuel economy is not high.

II. Mechanical transmission

The mechanical transmission includes planetary gear train and shift actuator.

The planetary gear train, as one of important parts of AT, mainly consists of sun gear, gear ring, planetary carrier, planetary gear and other elements. The planetary gear train is a mechanism for realizing speed change. The change of gear ratio is realized by taking different elements as the driving elements and restricting the movement of different elements. During the process of gear ratio changing, the whole planetary gear set is in motion, and the transmission of power is not interrupted, thus realizing the power shift.

The shift actuator is mainly used to change the driving elements in the planetary gear train or restrict the movement of an element and change the power transmission direction and gear ratio. It mainly consists of multi-disk clutch, brake and one-way overrunning clutch. The clutch is to transmit the power to an element of the planetary gear train to make it a driving element. The brake is to hold an element in the planetary gear train to keep it motionless. The one-way overrunning clutch is also one engaging element of the planetary transmission, with basically the same role as the multi-disk clutch and brake. It is also used to fix or connect some of the sun gear, planetary carrier, gear ring and other basic elements in several planetary gear sets, so that the planetary transmission consists of the gears with different gear ratios.

III. Hydraulic control unit

The hydraulic control unit is mainly composed of hydraulic pump, main pressure regulator valve, throttle valve, manual valve and electro-hydraulic proportional valve, playing the role of transmission, control, manipulation, cooling and lubrication. Controlled by the electronic control system, the hydraulic control system feeds the hydraulic fluid into the clutch or brake cylinder that needs to work, so as to change the speed ratio and meet the driving requirements.

IV. Electronic control unit

The electronic control unit includes three parts: signal input device, electronic control unit and actuator.

The signal input device, composed of the sensor and signal switching device, is responsible for converting the vehicle driving state information into an electrical signal for the control circuit to receive. There are three types of sensor signals: analog, pulse and switch.

The electronic control unit (ECU), as the core of the electronic control system, receives the vehicle driving state information detected by the sensor and the intervention information given by the driver, makes a comparison operation and then issues an instruction according to a rule to automatically control the powertrain. The ECU mainly consists of the input channel, controller and output channel. The input channel receives various input signals, and the controller compares these signals with the data in memory. Based on the comparison results, the controller makes the decision whether to shift gears or not. The output device processes the control signal or directly sends it to the actuator such as the solenoid valve.

The actuator mainly refers to electro-hydraulic proportional valve, which adjusts the pressure in the hydraulic circuit according to the command of the ECU.

3.3 Mechanical Structure of AT

I. Mechanical structure of hydraulic torque converter

The hydraulic torque converter mainly consists of the rotatable impeller B, turbine W and fixed guide wheel D, as shown in Fig. 3.3, all of which are precisely cast with aluminum alloy, or made by stamping and welding with steel plates. The impeller B is integrated with torque converter housing 4 and bolted to the flange at the rear end of the engine crankshaft 1. The torque converter housing 4 is made into two halves, which are connected together after assembly. The turbine W is installed on the driven shaft 2, and the guide wheel D is fixed on the guide wheel fixing casing 3. After all the active wheels are assembled, a circular body with circular section is formed.

When the integrated torque converter is in the working area of the coupler, even at the maximum efficiency point, there is still about 5% slippage between the impeller and the turbine, i.e. corresponding energy loss. The multi-speed AT reduces the dependence on the function of the hydraulic torque converter. In order to reduce fuel consumption, a lockup clutch can be used to integrate the impeller and turbine under certain conditions. It is mounted on the spline of the turbine shaft. At low vehicle speed or the low speed ratio of the torque converter, the operating oil pressure flows into the front of the lockup clutch through the cavity between the turbine shaft and the impeller shaft, $F_2 > F_1$, making it in the state of disengagement, as shown in Fig. 3.4. At the medium and high vehicle speed or the torque converter condition of $i > i_M$, the operating oil pressure flows into the rear through the oil circuit between the turbine and the guide wheel shaft bush. The oil in the front is discharged through the

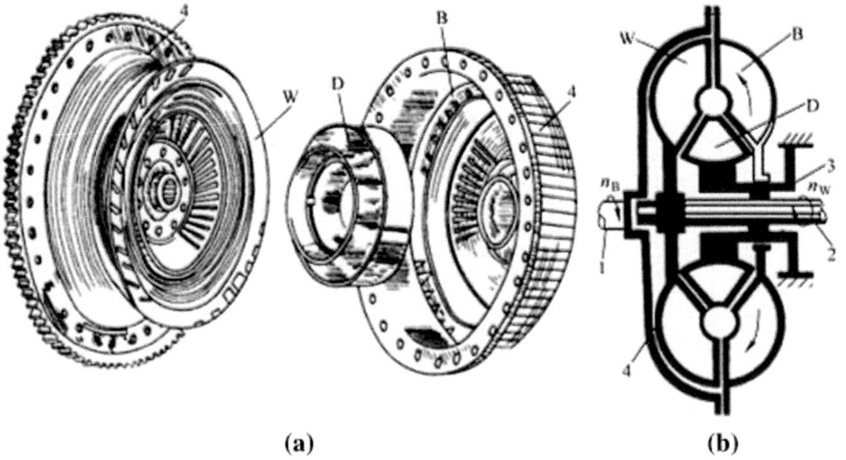

Fig. 3.3 Mechanical structure of hydraulic torque converter. 1—engine crankshaft, 2—driven shaft, 3—guide wheel fixing casing, 4—torque converter housing. B—impeller, W—turbine, D—guide wheel

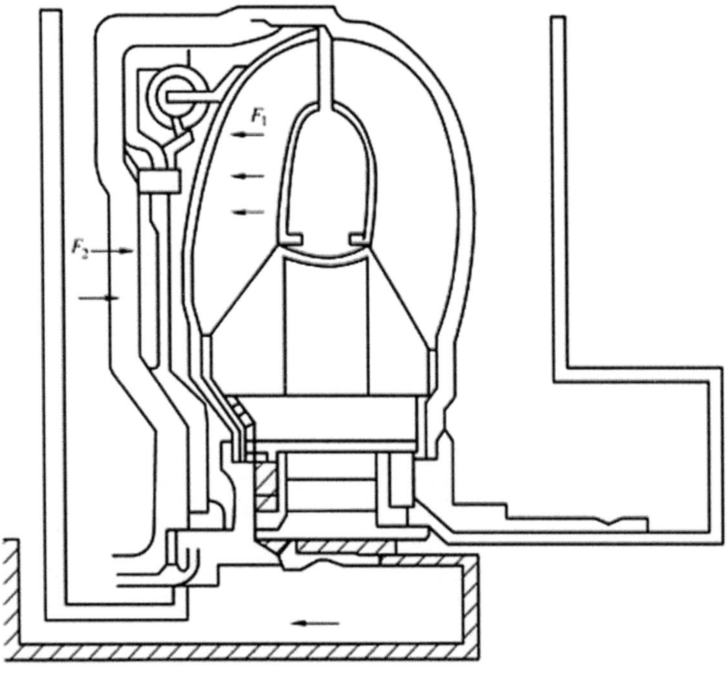

Fig. 3.4 Locking of hydraulic torque converter

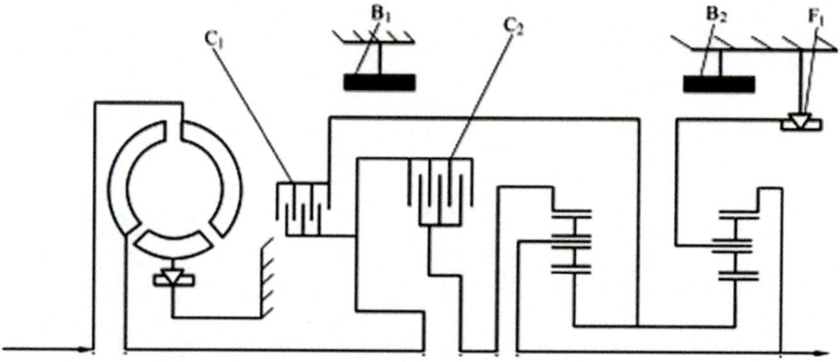

Fig. 3.5 3 speed Simpson planetary gear train. B_1—2 speed brake, B_2—low and reverse gear brake, C_1—reverse and high gear clutch, C_2—forward clutch, F_1—low-gear OWC

turbine shaft cavity, $F_1 > F_2$, and the hydraulic torque converter is locked. Therefore, there is no power loss and no oil temperature rise and there is no need to send the hydraulic torque converter oil to the radiator for cooling.

II. Mechanical structure of planetary gear train

1. Simpson planetary gear train

The 3 speed Simpson planetary gear train, as shown in Fig. 3.5, consists of two planetary gear sets with completely same gear parameters. It is characterized by few gear types, small amount of finish, good manufacturability and low cost; input and output with gear ring, high strength and high transmitted power; high efficiency without power cycling; low component speed and smooth shift; for a three-degree-of-freedom transmission, although two elements need to be operated for each shift, they are the controls with the adjacent gear if arranged reasonably, that is, only one control is actually replaced. Since its invention, the Simpson planetary gear train has been widely adopted by various countries around the world.

When the structure shown in Fig. 3.5 changes from gear 2 to gear 3, the alternation of the release brake B_1 and engaging clutch C_1 shall be timely and accurate. If C_1 is not engaged prematurely, there will be motion interference between the independent components; if B_1 is released too fast, the engine will run idle and the shift impact will increase (Table 3.1). In order to improve the shift quality, an OWC F_2 is connected in series between B_1 and sun gear assembly to make the gear shift smooth (Fig. 3.6), and its working law is shown in Table 3.2.

To further improve fuel economy and reduce noise, vehicles have developed to multi-speed, with the 4 speed becoming standard facility. The front and rear planetary gear sets are connected by an auxiliary member and all the others are independent, forming a Simpson mechanism with five independent components (the above are four independent components), so a control (clutch or brake) may be added to achieve

Table 3.1 Working law of shift actuator of 3 speed Simpson planetary gear train

Control lever position	Gear	Shift actuator					Control lever position	Gear	Shift actuator			
		C_1	C_2	B_i	B_2	F_1			C_1	C_2	B_i	B_2
D	Gear 1		O			O	R	Reverse gear	O			O
	Gear 2		O	O			S, L	Gear 1		O		O
	Gear 3	O	O					Gear 2		O	O	

Note O means engagement, brake or locking

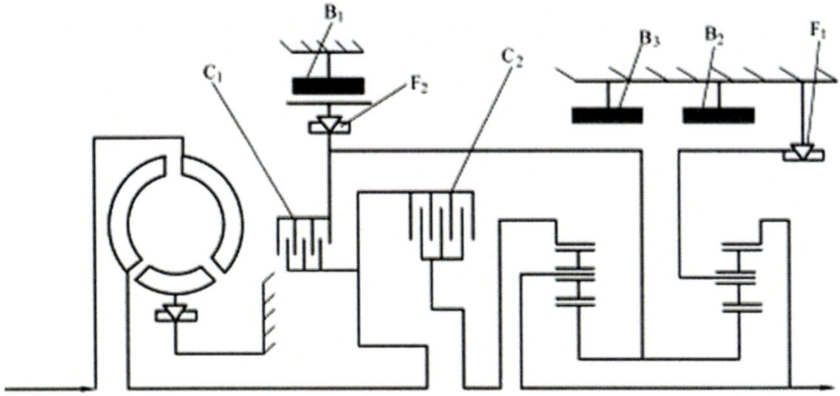

Fig. 3.6 Improved 3 speed Simpson planetary gear train. B_1—2 speed brake, B_2—low and reverse gear brake, B_3—2 speed force brake, C_1—reverse and high gear clutch, C_2—forward clutch, F_1—low-gear OWC, F_2—2 speed OWC

4 speed transmission (Fig. 3.7 and Table 3.3). It is characterized by small size and small mass.

On the basis of the original 3 speed Simpson planetary gear train, the building block composition method, together with a parameter and the first two rows of planetary gear sets, is used to obtain a 4 speed AT. The planetary gear sets added can be prepositioned or post-positioned to achieve overspeed or speed reduction and to obtain 4 schemes. The parts with high universality are conductive to reduce the cost.

2. **Ravigneaux planetary gear train**

The Ravigneaux planetary gear train is a compound planetary gear train composed of a single planetary gear set and a dual-planetary gear set, which share a planetary carrier, long planetary gear and gear ring, so the gear train has only 4 independent components (Fig. 3.8 and Table 3.4). It is characterized by few components,

Table 3.2 Working law of shift actuator of improved 3 speed Simpson planetary gear train

Control lever position	Gear	Shift actuator					Control lever position	Gear	Shift actuator						
		C_1	C_2	B_i	B_2	F_1			C_1	C_2	B_i	B_2	B_3	F_1	F_2
D	Gear 1		O			O	R	Reverse gear	O		O	O			
	Gear 2		O	O			S, L	Gear 1		O		O			
	Gear 3	O	O					Gear 2		O	O				

Note ○ means engagement, brake or locking

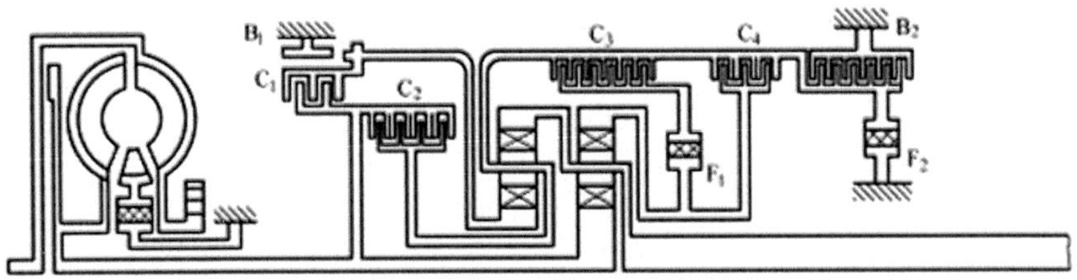

Fig. 3.7 4 speed Simpson planetary gear train. B_1—2 speed and 4 speed brake, B_2—low and reverse gear brake, C_1—reverse clutch, C_2—high gear clutch, C_3—forward clutch, C_4—forward force clutch, F_1—forward OWC, F_2—low-gear OWC

Table 3.3 Working law of shift actuator of 4 speed Simpson planetary gear train

Control lever position	Gear	Shift actuator							
		C_1	C_2	C_3	C_4	B_1	B_2	F_1	F_2
D	Gear 1			○				○	
	Gear 2			○		○		○	
	Gear 3		○	○				○	
	Overdrive		○	●		○			
R	Reverse gear	○					○		
S, L	Gear 1			●	○		○		
	Gear 2			●	○	○	○		
	Gear 3		○	●	○				

Note ○ means engagement, brake or locking; ● means engagement, brake or locking without transmitting power

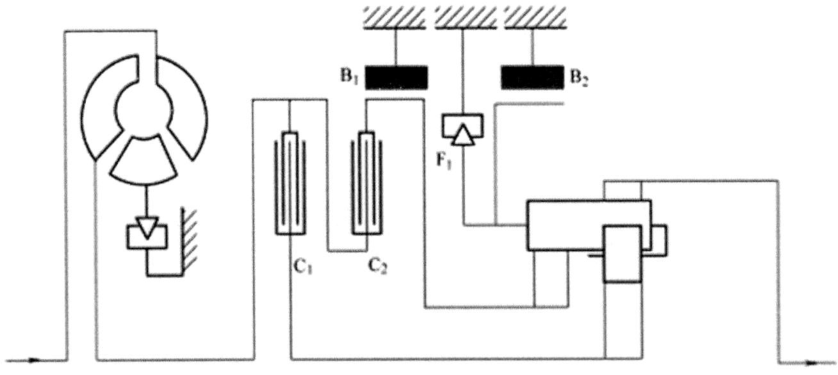

Fig. 3.8 Ravigneaux planetary gear train. B_1—2 speed brake, B_2—low and reverse gear brake, C_1—forward clutch, C_2—reverse and high gear clutch, F_1—1 speed OWC

Table 3.4 Working law of shift actuator of 3 speed Ravigneaux planetary gear train

Control lever position	Gear	Shift actuator					Control lever position	Gear	Shift actuator					
		C_1	C_2	B_1	B_2	F_1			C_1	C_2	B_1	B_2	F_1	F_2
D	Gear 1	O				O	R	Reverse gear		O		O		
	Gear 2	O		O			S, L	Gear 1	O			O		
	Gear 3	O	O					Gear 2	O		O			

Note ○ means engagement, brake or locking

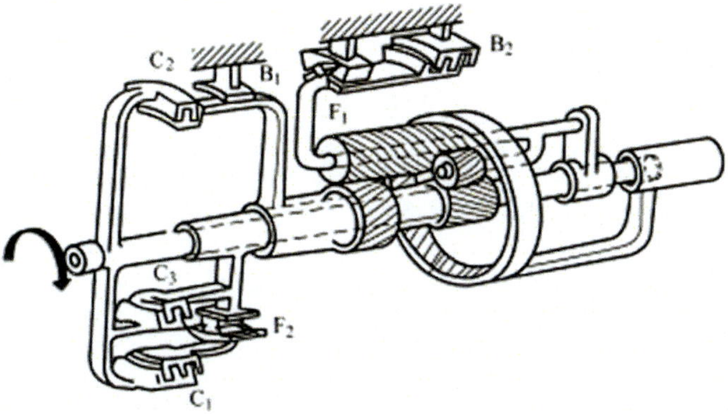

Fig. 3.9 Improved Ravigneaux planetary gear train. B_1—2 speed brake, B_2—low and reverse gear brake, C_1—forward clutch, C_2—reverse and high gear clutch, C_3—forward force clutch, F_1—low-gear OWC, F_2—forward OWC

low speed; compact structure, small axial dimensions; large and flexible gear ratio variation range; suitable for FF layout.

The improved 3 speed Ravigneaux planetary gear train, as shown in Fig. 3.9, is added a force clutch C_3 and OWC F_2 between the input shaft and the rear sun gear, improving the shift quality and achieving engine braking in gears 2 and 3. The working law of the actuator is shown in Table 3.5.

The improved 3 speed Ravigneaux planetary gear train is added a clutch C_4 between the input shaft and the planetary carrier to change to a 4 speed transmission. Or a brake B_3 may be added to achieve 4 speed transmission (Fig. 3.10 and Table 3.6). Meanwhile, the power dividing scheme is adopted at the gear 3, with 60% engine power transmitted mechanically and 40% transmitted through the hydraulic torque converter; the torque converter is locked in gear 4 to improve the efficiency.

III. Mechanical structure of shift actuator in transmission

All the gears of the planetary gear train are normally engaged and the gear shift is achieved by constraining the basic elements of the planetary mechanism (fixing or connecting some elements) instead of using the shift fork to change the speed in the MT. The shift actuator consists of multi-disk clutch, brake and OWC, and, with the connection, fixing or locking function, may make the transmission achieve different gear ratios, so as to shift the gear.

1. **Structure of multi-disk clutch**

As shown in Fig. 3.11, the drive hub 8 is connected to the turbine shaft 2 by splines, and the axial displacement is controlled by the snap ring 12. The clutch driven hub 14 is welded to the sun gear of the planetary gear train. The clutch drive friction plate (steel plate) 10 is connected with the keyway of the clutch drive hub 8 through the internal splines, and the driven friction plate 13 is connected with the clutch driven hub 14 through the internal splines. The drive friction plates 10 and driven friction

Table 3.5 Working law of shift actuator of improved Ravigneaux planetary gear train

Control lever position	Gear	Shift actuator							Control lever position	Gear	Shift actuator						
		C_1	C_2	C_3	B_1	B_2	F_1	F_2			C_1	C_2	C_3	B_1	B_2	F_1	F_2
D	Gear 1	O					O		R	Reverse gear		O			O		
	Gear 2	O			O			O	S, L	Gear 1			O		O		
	Gear 3	O	O					O		Gear 2			O	O			

Note ○ means engagement, brake or locking

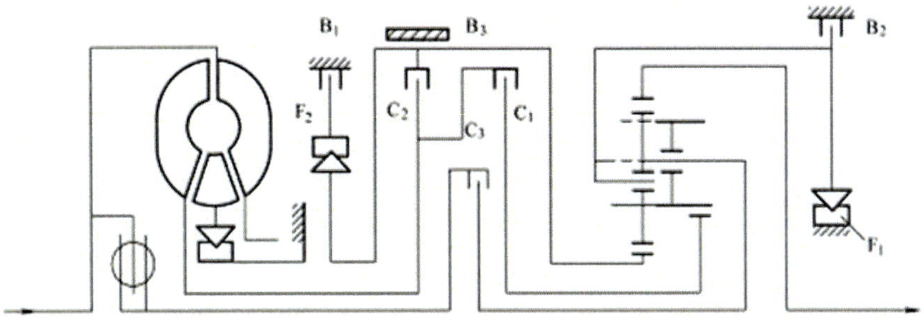

Fig. 3.10 power-split 4 speed Ravigneaux planetary gear train. B_1—2 speed brake, B_2—low and reverse gear brake, B_3—2 speed pre-brake, C_1—forward clutch, C_2—reverse clutch, C_3—lockup clutch, F_1—1 speed OWC, F_2—2 speed OWC

plates 13 are placed alternately in the same amount. When the electronic control unit controls the corresponding shift solenoid valves on and off, the hydraulic oil enters through the hydraulic oil supply hole 3, so that the piston 5 overcomes the force of the piston reset spring 15 to compress the drive friction plate 10 and the driven friction plate 13 in a jointed state. The power of the turbine shaft 2 is transferred to the sun gear of the planetary gear train. Conversely, the hydraulic oil on the right side of piston 5 is decompressed, and the reset spring 15 makes the piston 5 reset. The drive and driven friction plates release from each other, forming a gap and cutting off the power transmitted to the sun gear by the turbine shaft 2.

2. **Structure of brake**

As shown in Fig. 3.12, the band brake in the AT uses a brake band with internal friction material wrapped on the cylindrical surface of the brake drum. One end of the band is fixed to the transmission case and the other end is connected to the piston in the brake hydraulic cylinder. When the brake oil enters the brake hydraulic cylinder, the compression piston reset spring pushes the piston, thus moving the free end of the brake band and tightening the brake drum. Since the brake drum is integrated with a part of the planetary gear train, the brake drum clamping means that the part is clamped so that it cannot rotate. After the brake oil pressure is removed, the reset spring resets the piston in the brake hydraulic cylinder and pulls back to the free end of the brake band, thus releasing the brake drum and removing the brake.

3. **Structure of one-way clutch (OWC)**

The OWC, as shown in Fig. 3.13, realizes the fixation or connection based on the principle of one-way locking. With high transmission torque capacity and small friction when idling and without control mechanism, the OWC is completely controlled by the direction of force applied to the connected element and can be engaged or disengaged instantly, automatically cut off or switch on the torque when changing speed, so as to ensure smooth shift without impact and greatly simplify the structure of the hydraulic control system. The most common OWC types are roller and wedge types.

Table 3.6 Working law of shift actuator of power-split Ravigneaux planetary gear train

Gear lever handle position	Gear	Shift solenoid valve A	Shift solenoid valve B	Shift actuator							
				C_o	C_2	C_3	B_o	B_2	B_3	F_o	F_2
P	Park	OFF	ON								
R	Reverse gear	OFF	OFF		O			O			
N	Neutral	OFF	ON								
D	Gear 1	OFF	ON	O						O	
	Gear 2	ON	ON	O			O				O
	Gear 3	ON	OFF	O		O					
	Overdrive	OFF	OFF			O			O		
3	Gear 1	OFF	ON	O						O	
	Gear 2	ON	ON	O			O				O
	Gear 3	ON	OFF	O		O					

Gear lever handle position	Gear	Shift solenoid valve A	Shift solenoid valve B	Shift actuator						
				C_1	C_2	C_3	B_1	B_2	B_3	F_1
2	Gear 1	OFF	ON	○						○
	Gear 2	ON	ON	○			●		○	
1	Gear 2	OFF	ON	○				○		

Note ○ means engagement, brake or locking; ● means engagement, brake or locking without transmitting power

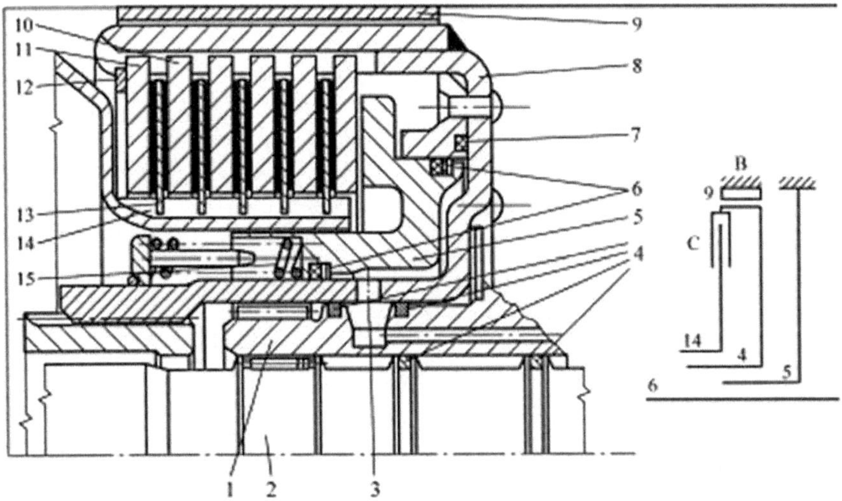

Fig. 3.11 Multi-disk clutch and band brake of Benz W5 A5 gear AT. 1—Case, 2—turbine shaft, 3— hydraulic oil supply hole, 4—oil ring, 5—piston, 6—slotted ring, 7—O ring, 8—steel plate carrier (drive hub) and brake drum, 9—brake band, 10—drive friction plate (steel plate), 11—end plate, 12—snap ring, 13—driven friction plate, 14—friction plate carrier (driven hub), 15—piston reset spring

Fig. 3.12 Belt brake

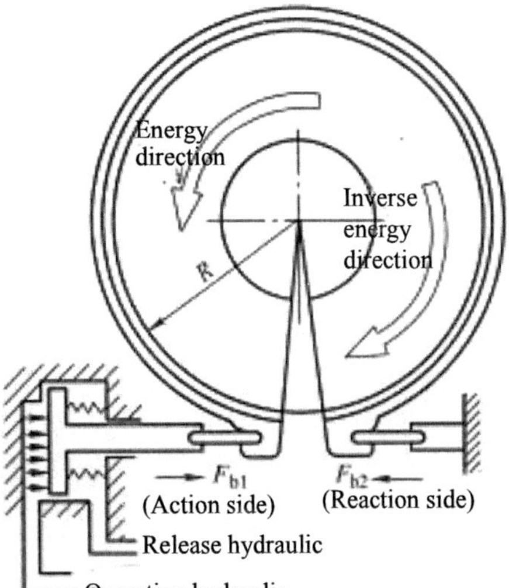

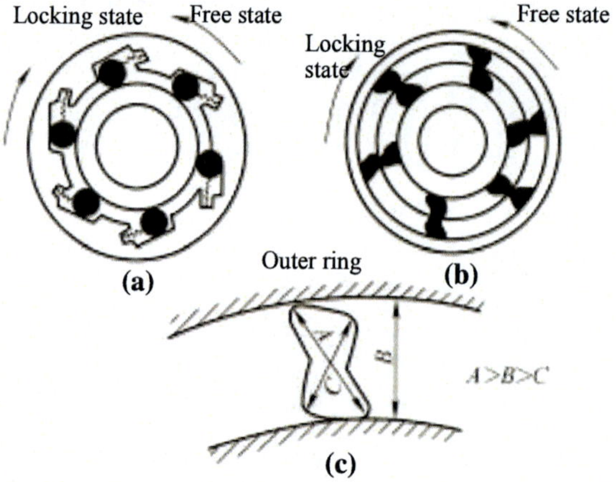

Fig. 3.13 OWC

The OWC can be used to prevent the transmission of power from the output shaft to the engine (prevent the engine braking), but when the engine braking is needed in the downhill and other conditions, additional auxiliary brakes are required.

3.4 AT Speed Change Process Analysis

U340E AT is an electronically controlled 4 speed AT driven by the front wheel. The planetary gear train and the shift actuators of the U340E AT are arranged in Fig. 3.14a and the drive route is shown in Fig. 3.14b. There are a total of 8 shift actuators, including 3 clutches, 3 brakes and 2 OWCs in the U340E AT. The roles of the shift actuators are shown in Table 3.7 and the worksheet is shown in Table 3.8.

I. Power transmission route at reverse gear

The power transmission route at reverse gear is shown in Fig. 3.15. In the reverse gear, the reverse clutch C_3 is engaged to drive the rear sun gear clockwise, and then the rear planetary gear rotates anticlockwise; the 1 speed/reverse gear brake B_3 works to fix the rear planetary carrier and front annular gear, the rear planetary gear drives the rear annular gear, and the front planetary carrier and rear annular gear slow down reversely.

II. Power transmission route in gear 1

1. **Power transmission route in gear D_1**

The power transmission route in gear D_1 is shown in Fig. 3.16. In the gear D_1, the forward clutch C_1 is engaged to drive the front sun gear clockwise and the front planetary gear anticlockwise. The front planetary carrier is connected with the vehicle

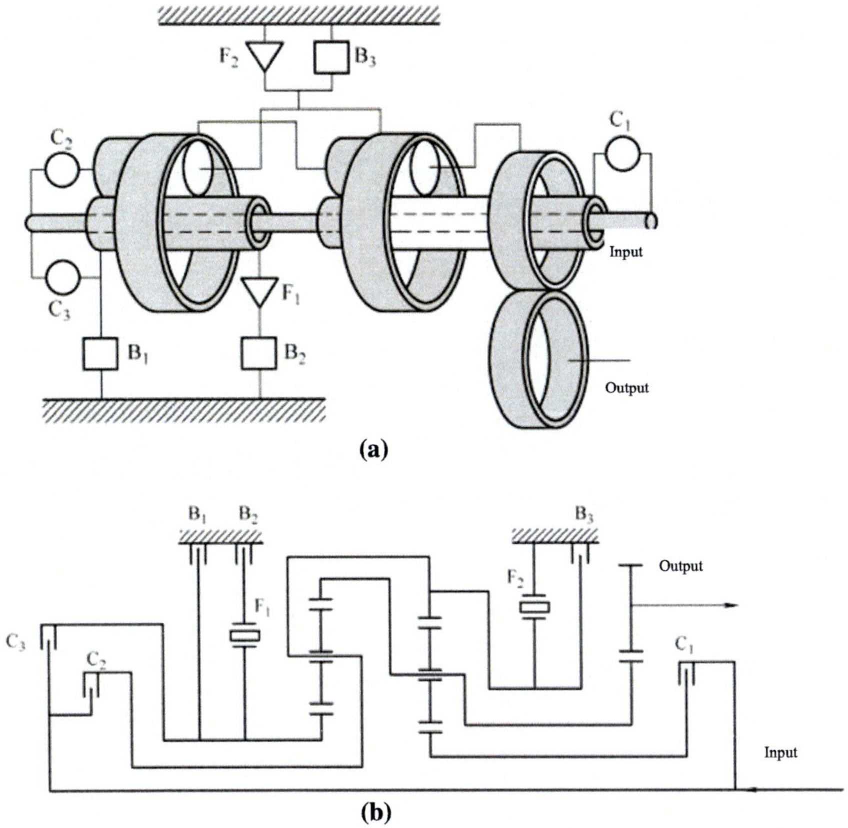

Fig. 3.14 Structure and drive route of U340E AT. B_1—Overdrive/2 speed brake, B_2—2 speed brake, B_3—1 speed/reverse gear brake, C_1—forward clutch, C_2—direct gear clutch, C_3—reverse clutch, F_1, F_2—OWC

body, and the resistance to motion is relatively large, so it can be temporarily regarded as fixed. Then the front annular gear has the tendency of anticlockwise rotation. At this time, the low-gear OWC F_2 is locked to prevent anticlockwise rotation of the front annular gear. While the planetary gear rotates anticlockwise, it drives the front planetary carrier clockwise around the front annular gear. That is, the front planetary carrier and rear annular gear slow down in the same direction. In gear D_1, the low-gear OWC F_2 is locked to unidirectionally fix the front annular gear and the rear planetary carrier, which is an essential condition for the power transmission. It cannot transmit the power reversely, so there is no engine braking in D_1.

2. **Power transmission route in gear L_1**

The power transmission route in gear L_1 is shown in Fig. 3.17. In gear L_1, in addition to the working parts in D_1, the 1 speed/reverse gear brake B_3 works to bidirectionally

Table 3.7 Roles of shift actuators

Shift actuator	Role
C_1 (forward clutch)	Drive the front row of sun gears
C_2 (direct gear clutch)	Drive the rear planetary carrier and front annular gear
C_3 (reverse clutch)	Drive the rear sun gear
B_4 (overdrive/2 speed brake)	Fix the rear sun gear in gears 2 and 4
B_2 (2 speed brake)	Work in gear 2 and unidirectionally fix the rear sun gear through the OWC F_i
B_3 (4 speed/reverse gear brake)	Fix the front annular gear and the rear planetary carrier in the reverse gear
F_4 (OWC)	The brake B_2 is locked in gear 2. Unidirectionally fix the rear sun gear
F_2 (OWC)	Unidirectionally fix the front annular gear and the rear planetary carrier

Table 3.8 Worksheet of actuators

Gear		C_1	C_2	C_3	B_1	B_2	B_3	F_1	F_2
R			O			O	O		
D	1	O							O
	2	O				O		O	
	3	O	O			O			
	4		O		O	O			
2	1	O							O
	2	O			O	O		O	
L	1	O					O		O

Note O means engagement, brake or locking

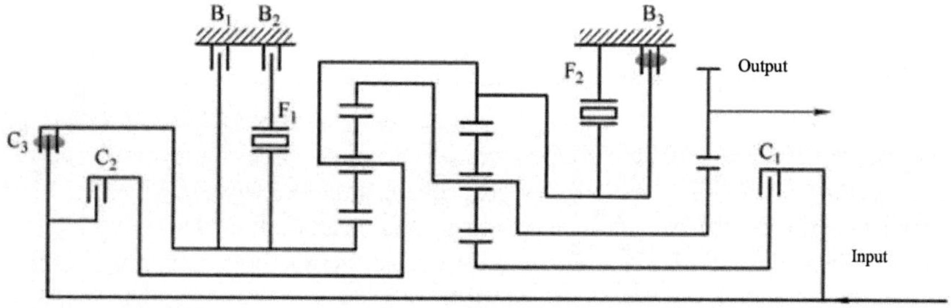

Fig. 3.15 Power transmission route at reverse gear

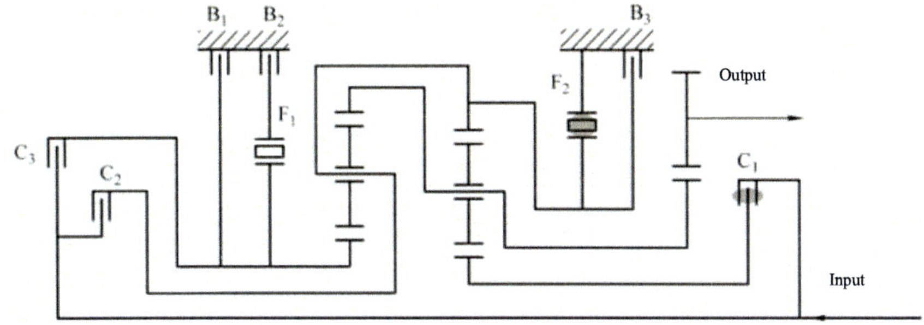

Fig. 3.16 Power transmission route in gear D_1

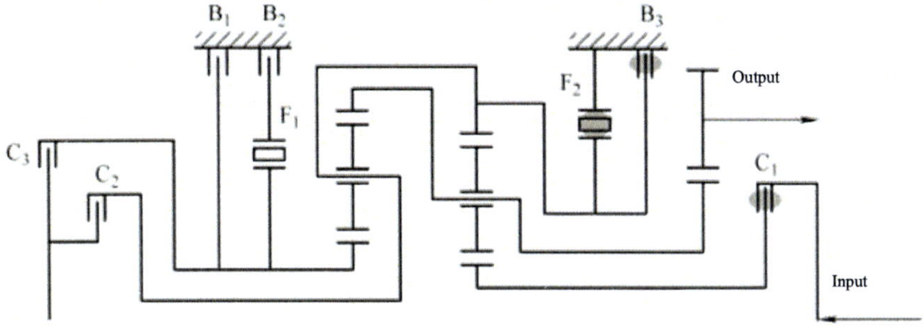

Fig. 3.17 Power transmission route in gear L_1

fix the front annular gear and the rear planetary carrier. The low-gear OWC locking is no longer an essential condition for the power transmission. The power can be transmitted reversely, so there is engine braking in L_1.

III. Power transmission route in gear 2

1. Power transmission route in gear D_2

The power transmission route in gear D_2 is shown in Fig. 3.18a. In gear D_2, the forward clutch C_1 is engaged to drive the front sun gear; the 2 speed brake B_2 works, the OWC F_1 is locked to unidirectionally fix the rear sun gear, and the front planetary carrier and rear annular gear slow down in the same direction. In gear D_2, the OWC F_1 locking is an essential condition for the power transmission. It cannot transmit the power reversely, so there is no engine braking in D_2.

The gear ratio in gears 1 and 2 is subject to qualitative analysis below. With respect to the planetary gear train, the rear planetary carrier and front annular gear are fixed, and the rear annular gear and front planetary carrier (power output end) slow down and rotate clockwise. After rising to gear 2, the rear planetary carrier is relaxed, the rear sun gear is fixed and the rear annular gear is connected with the vehicle body

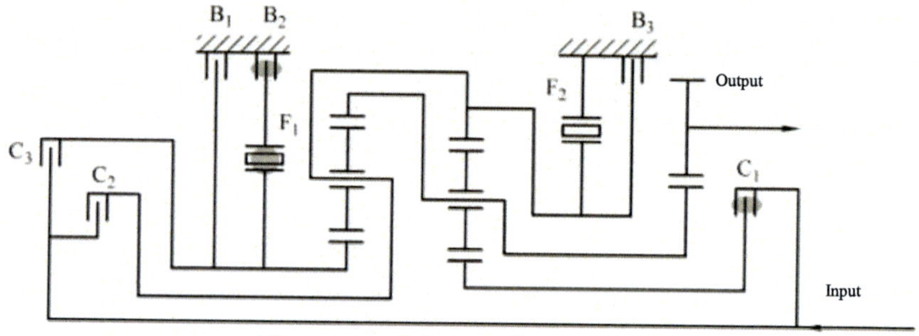

a) Power transmission route in gear D2

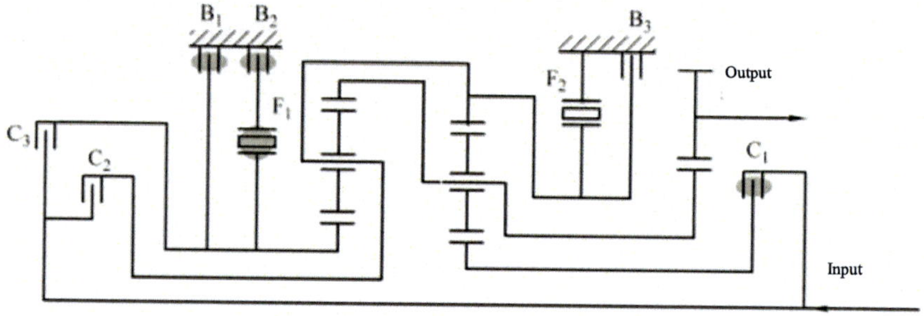

b) 2-position power transmission route in gear 2

Fig. 3.18 Power transmission route in gear 2

and still slows down and rotates in the same direction. Then the rear planetary carrier and front annular gear rotate at reduced speed in the same direction from static state. With respect to the front planetary gear train, if the front sun gear and the input shaft rotate in the same direction and at the same speed and the front annular gear is fixed, the front planetary carrier rotates at reduced speed in the same direction, which is gear 1; if the front annular gear can also rotate at the input shaft speed in the same direction, the front planetary carrier rotates in the same direction and at the same speed, which is gear 3; if the front annular gear rotates at reduced speed in the same direction, the speed of the front planetary carrier is between the above two cases, which is gear 2. In other words, in the gears 1 and 2, the annular gear of the front planetary gear train rotates at reduced speed in the same direction from the fixed state when the input element (sun gear) is in unchanged state, so the speed of the front planetary carrier increases.

2. **2-position power transmission route in gear 2**

The 2-position power transmission route in gear 2 is shown in Fig. 3.18b. In the 2-position gear 2, in addition to the working parts in D_2, the overdrive/2 speed brake B_1 works and is connected in parallel with $(B_2 + F_1)$. The rear sun gear is fixed bidirectionally. At this time, the OWC F_1 locking is no longer the only condition for the power transmission, so there is engine braking in 2-position gear 2.

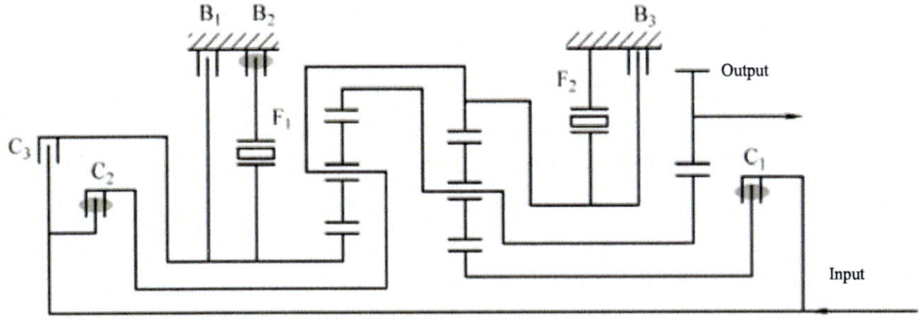

Fig. 3.19 Power transmission route in gear D_3

IV. Power transmission route in gear D_3

The power transmission route in gear D_3 is shown in Fig. 3.19. In gear D_3, the forward clutch C_1 is engaged to drive the front sun gear; the direct gear clutch C_2 is engaged to drive the rear planetary carrier and front annular gear. Since two parts in the planetary gear train are driven simultaneously, the whole planetary gear train rotates as a whole, which is the direct drive gear. In gear D_3, the brake B_2 is still engaged, but it does not work since the OWC F_1 is overrunning (slipped). No OWV is involved in the power transmission in the gear D_3, so there is engine braking.

V. Power transmission route in gear D_4

The power transmission route in gear D_4 is shown in Fig. 3.20. In gear D_4, the direct gear clutch C_2 is engaged to drive the rear planetary carrier and front annular gear; the overdrive/2 speed brake B_1 works to fix the rear sun gear, and then the front planetary carrier and rear annular gear speed up in the same direction. Similarly, there is engine braking in gear D_4.

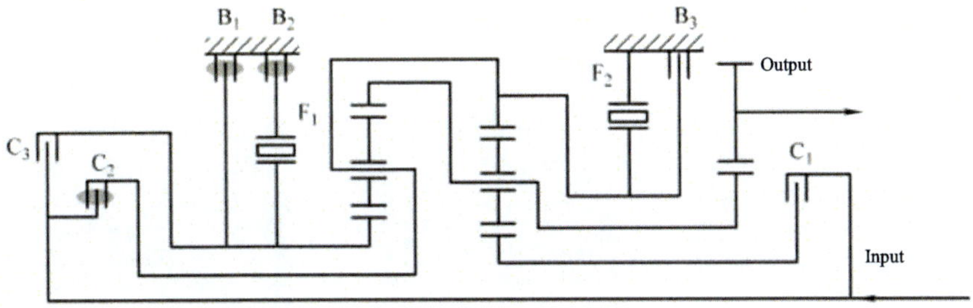

Fig. 3.20 Power transmission route in gear D_4

3.5 Hydraulic Control System of AT

The hydraulic control system is divided into four parts: main hydraulic control system, shift control system, shift quality control system and hydraulic torque converter control system. The hydraulic control system provides the required hydraulic oil for all parts of AT. Its specific functions are:

(1) Supply oil to the control system and maintain the operating oil pressure of the main oil circuit to ensure the smooth operation of the control mechanisms.
(2) Ensure the oil supply of the shift actuators to meet the need of shift and other operations.
(3) Provide the lubricating oil for all moving parts of the transmission, such as gear, bearing, thrust washer and clutch friction plate and ensure normal lubricating oil temperature.
(4) Dissipate the heat of the whole AT through the circular cooling of the oil and keep the transmission working in the reasonable temperature range.
(5) Control the work of the hydraulic torque converter and timely take away the heat from the hydraulic torque converter in time to maintain normal operating temperature.

I. Main hydraulic control system

The main hydraulic control system is the power source of the whole hydraulic system, including hydraulic pump and hydraulic regulating system.

The hydraulic pressure supplied by the hydraulic pump is distributed to the control actuator through the control valve. The hydraulic circuit consists of the valve body shown in Fig. 3.21 and the spool valve for hydraulic switching and adjustment is placed in the valve body, as shown in Fig. 3.22.

The current is delivered to the solenoid valve for hydraulic control (Fig. 3.23) through the A/T control unit for hydraulic control. The solenoid valves can be divided

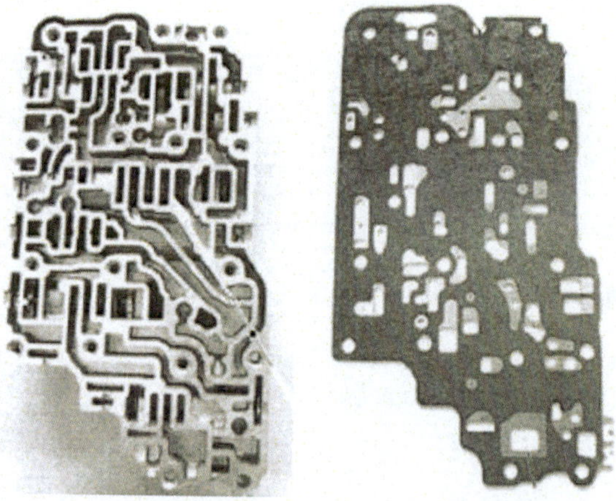

Fig. 3.21 Valve body appearance

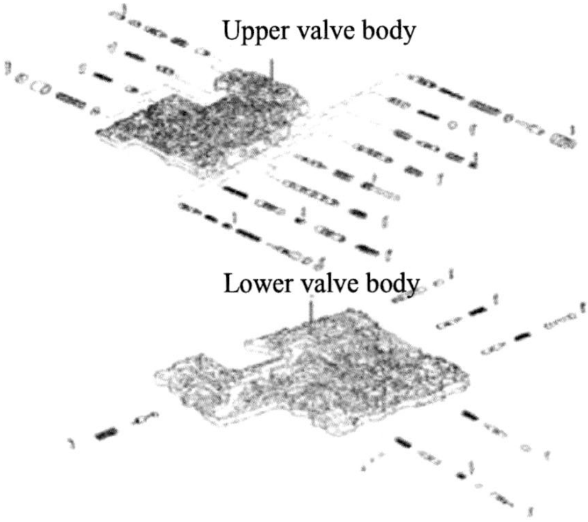

Fig. 3.22 Hydraulic valve assembly

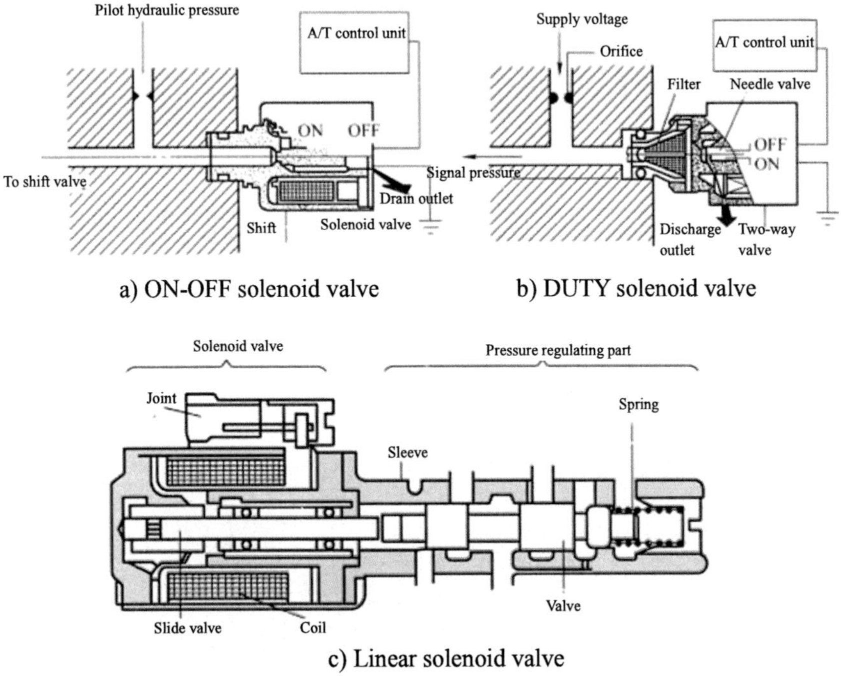

Fig. 3.23 Solenoid valve for hydraulic control

into three categories: ON-OFF solenoid valve, which changes the hydraulic pressure through ON-OFF switching according to the electrical wave signals; DUTY solenoid valve, which regularly supplies the pulse current and simulates and adjusts the hydraulic pressure by changing the current additional width (DUTY) of each pulse; linear solenoid valve, which is equipped with a pressure regulator valve to convert the electromagnetic force into hydraulic pressure.

The above solenoid valves can be combined to form a hydraulic control circuit with different functions shown in Fig. 3.24. All control functions are integrated to form the valve assembly unit shown in Fig. 3.25.

Fig. 3.24 Main pressure control circuit

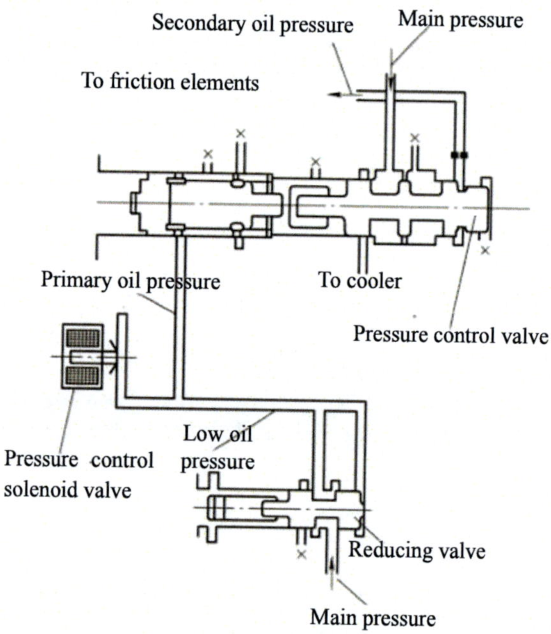

Fig. 3.25 Appearance of valve assembly unit

(I) Hydraulic pump

1. Function of hydraulic pump

As the power source of the AT, the hydraulic pump is mounted between the hydraulic torque converter and the planetary gear train and driven by the shaft sleeve at the rear end of the hydraulic torque converter housing. The technical performance of hydraulic pump has a great influence on the working performance of AT.

2. Structure and principle of hydraulic pump

There are three types of hydraulic pumps in AT: internal gear pump, rotor pump and vane pump.

(1) Internal gear pump: the internal gear pump mainly consists of the pinion, internal gear, external gear, crescent-shaped separator and case, as shown in Fig. 3.26. The pinion is a drive gear driven by the hydraulic torque converter shaft sleeve and the internal gear is a driven gear. The crescent-shaped separator separates the working chamber between the drive and driven gears into oil suction chamber and oil pressure chamber. When the hydraulic pump works, the drive gear drives the driven gear to rotate together. As the drive and driven gears are constantly out of mesh in the oil suction chamber, the volume increases, forming partial vacuum, which generates suction to draw the hydraulic oil from the oil inlet. With the rotation of the gear, the oil in the oil suction chamber is brought into the oil pressure chamber through the backlash. In the oil pressure chamber, the gear

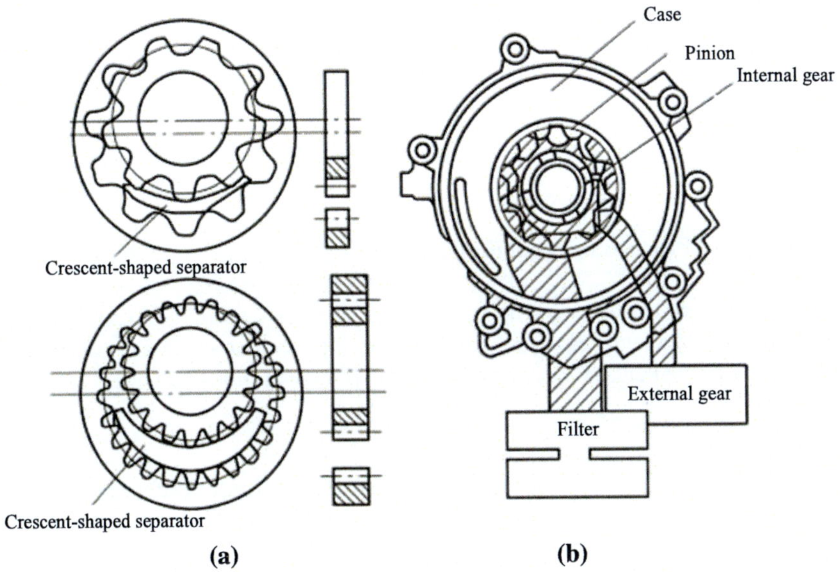

Fig. 3.26 Internal gear pump

Fig. 3.27 Vane pump. 1—Rotor, 2—stator, 3—case, 4—vane

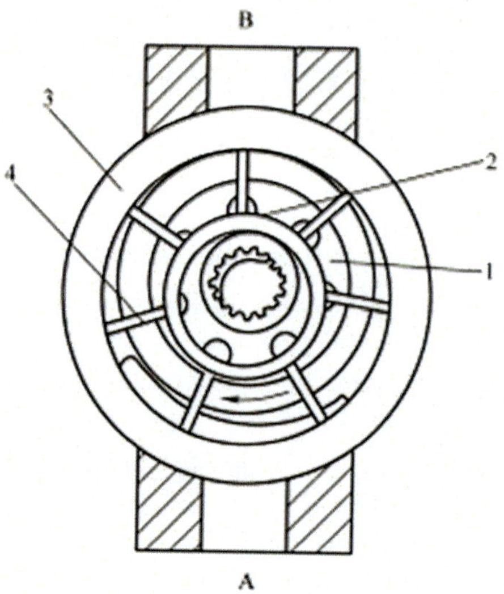

is continuously meshed and its volume decreases, so that the oil is discharged from the oil outlet with a certain pressure.

(2) Vane pump: the structure of vane pump is shown in Fig. 3.27, which is composed of stator, rotor, vane, case and pump cover. When the rotor rotates, the vane opens outward under the action of the centrifugal force and oil pressure at the bottom, close to the internal surface of the stator, and reciprocates in the rotor blade slot with the rotation of the rotor, thus forming a sealed working chamber between every two adjacent vanes. The vane pump has the advantages of smooth operation, low noise, uniform pump oil volume and high volume efficiency, but it is complex in structure and sensitive to the pollution of the hydraulic oil.

The hydraulic system of the AT is a low pressure system and the operating oil pressure does not exceed 2 MPa generally, so the gear pump is most widely used.

(II) Hydraulic regulating system

1. **Function of hydraulic regulating system**

The hydraulic regulating system of the AT liquid feeding system is composed of the main pressure regulator valve and auxiliary pressure regulator valve to control the oil pressure in the main oil circuit and the flow of the hydraulic pump.

2. **Structure and principle of hydraulic regulating system**

(1) Main pressure regulator valve: the oil pumped by the hydraulic pump first enters the main pressure regulator valve for oil pressure regulation. The structure of the main pressure regulator valve is shown in Fig. 3.28, which is mainly composed of the valve element, spring, plunger and plunger sleeve.

When the hydraulic pump does not rotate, under the action of spring force, the valve plugs 1–3 in the main pressure regulator valve move up to the top, and

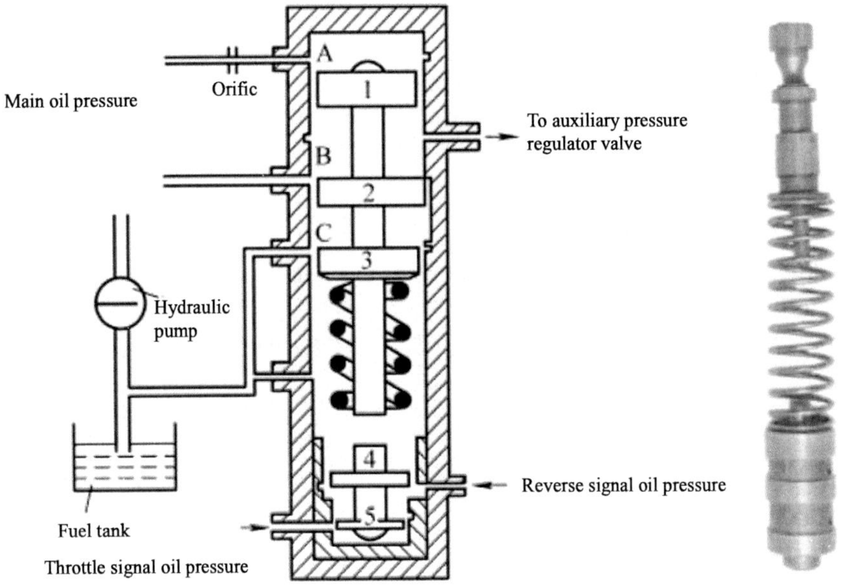

Fig. 3.28 Structure of main pressure regulator valve. A—Oil inlet of signal oil pressure, B—oil inlet to auxiliary pressure regulator valve, C—drainage port

the valve plugs 4 and 5 move down to the bottom. When the hydraulic pump is working, the oil pumped out enters the top of the main pressure regulator valve through valve port A, and the valve plugs 1–3 are pressed down 1–3 against the spring force. When the oil pressure acting on the top of the valve element is equal to the spring force, the valve plug will stop in the corresponding position and the valve port B is maintained at a certain opening: a certain degree of opening: the oil enters through the valve port B and one way enters the auxiliary pressure regulator valve through the outlet valve port to supply oil for the hydraulic torque converter and oil circuit; the other way returns to the fuel tank through the valve port C to keep the oil pressure in the main oil circuit at a certain value.

The oil pumping volume of the hydraulic pump varies with the engine speed. When the engine speed rises, the oil pumping volume of the hydraulic pump increases, the oil pressure acting on the upper end of the main pressure regulator valve element rises instantaneously, the valve plugs 1–3 are down, the opening of the valve port C increases and the oil discharge increases, so that the oil pressure drops rapidly, the valve plugs 1–3 rise again, and the opening of the valve port C decreases. The main oil pressure remains stable in the process of continuous automatic adjustment. The valve body works similarly when the engine speed decreases. Even if the hydraulic pump speed changes in a large range, the oil pressure in the main oil circuit of the hydraulic control system can also be kept stable after regulation by the main pressure regulator valve.

The torque transmitted by the AT is different at different engine load, so there are different requirements for oil pressure of the main oil circuit. The load of the engine and the torque transmitted by the AT increase with the throttle percentage. The oil pressure of the main oil circuit shall rise to prevent slippage of the shift actuators such as the clutch and brake. The oil pressure of the main oil circuit shall be reduced when the throttle percentage is small. Moreover, the throttle signal oil pressure acts at the bottom of the main pressure regulator valve. The engine speed and the oil pressure generated by the hydraulic pump increase with the throttle percentage. At this time, the downward oil pressure on the valve plugs 1–3 in the main pressure regulator valve increases, but the oil pressure signal in the lower end of the main pressure regulator valve from the throttle percentage also increases, so that the upward force exerted on the valve element increases accordingly, which reduces the opening of the valve port C, reduces the oil discharge, and increases the oil pressure in the main oil circuit. Similarly, when the throttle percentage decreases, the oil pressure in the main oil circuit decreases.

In the reverse gear, the main pressure regulator valve introduces the reverse signal oil pressure (controlled by manual valve), the oil pressure acts on the valve plug 4, and the upward force increases, making the opening of the valve port C reduced, the oil drainage reduced and the main oil circuit pressure increased compared with the forward gear, to prevent the slippage of the reverse actuators when engaged.

(2) Auxiliary pressure regulator valve: the structure of the auxiliary pressure regulator valve is shown in Fig. 3.29. It adjusts some oil flow from the main pressure regulator valve to a certain oil pressure, lubricates the transmission parts of the

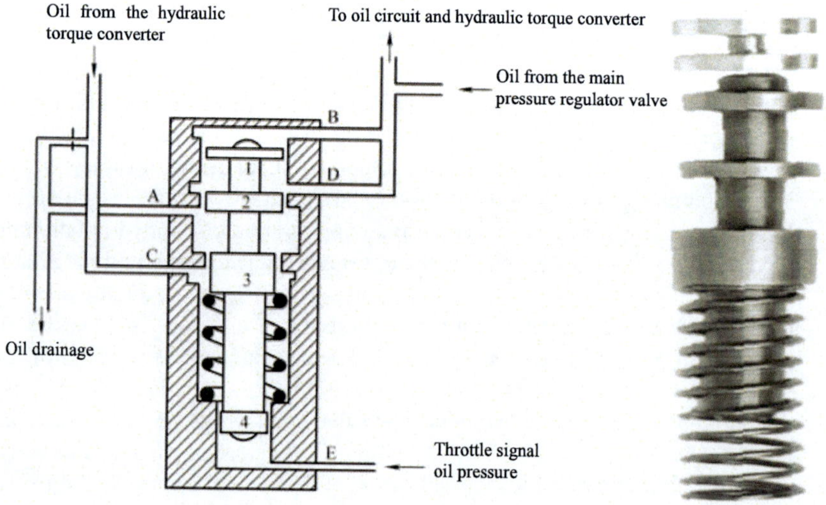

Fig. 3.29 Structure of auxiliary pressure regulator valve

transmission and supplies oil for the hydraulic torque converter; in case of the engine stall, it cuts off the drainage circuit of the hydraulic torque converter to keep sufficient oil in the hydraulic torque converter.

The valve port A is connected to the cooler and forms a drainage port with varying opening with the valve plug 2; the valve port B, as a pressure valve port, applies the pressure of the oil circuit on the valve plug 1; the valve port C is an oil inlet; the valve port D and the valve plug 2 form the switching valve of the drainage circuit of the hydraulic torque converter and in case of the engine stall, the valve plug 2 blocks the valve port D; the valve port E is connected to the throttle signal oil pressure and the upward force increases, making the opening of the valve port D reduced, the oil drainage reduced and the main oil circuit pressure increased compared with the forward gear, to prevent the slippage of the reverse actuators when engaged.

II. Shift control system

1. **Composition of shift control system**

The shift control system is composed of manual valve, shift valve and other hydraulic valves and solenoid valves as well as the corresponding oil circuit. The oil circuit makes the AT in different gears according to the gear lever handle position In addition, in each forward gear, according to the engine load, vehicle speed and other signals, the shift control system can automatically control the upshift or downshift, so that the AT gear adapts to the driving state.

2. **Structure and principle of shift actuators**

(1) Manual valve: the manual valve stem is connected to the gear lever handle. When the driver controls the gear lever handle in different positions, the manual valve element is also moved to the corresponding position, to achieve the switching of gear and oil circuit.

As shown in Fig. 3.30, the left end of the manual valve is connected to the gear lever handle by a connecting rod. The gear lever handle is controlled to make the manual valve in P, R, N, D, 2 and L respectively. As shown in Fig. 3.30a, when the gear lever handle is in N, the oil passage from the manual valve to

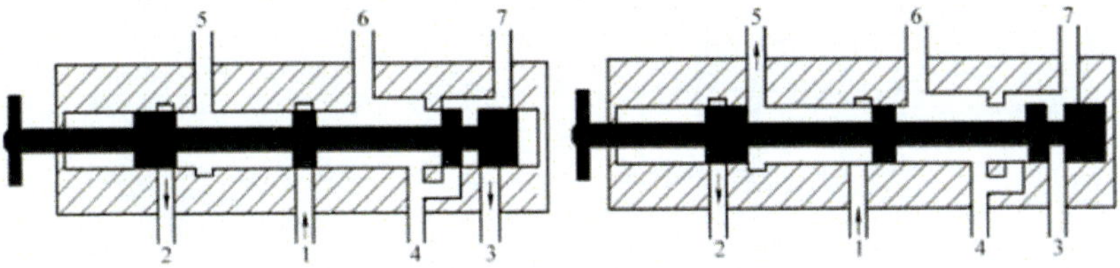

a) N b)R

Fig. 3.30 Structure of manual valve. 1—Oil inlet of main control oil pressure, 2, 3—drainage port, 4—oil outlet in gear, 2, 5—oil outlet in gear R, 6—oil outlet in forward gear, 7—oil outlet in gear L the control circuit is closed and there is no oil pressure in the control oil circuit. When the gear lever handle is in P, D, 2, L and R respectively, the manual valve connects the main control oil circuit from the hydraulic pump to the control oil circuit of each gear to complete the corresponding work. For example, when the gear lever handle is in R (Fig. 3.30b), the manual valve connects the main oil circuit to the reverse oil circuit, and controls the oil through the shift valve to control the high-gear/reverse clutch and low-gear/reverse gear brake.

(2) Shift valve: as a reversing valve, the shift valve is used to change the flow direction of the control oil, so that the oil pressure of the main oil circuit controls different shift actuators to achieve each gear ratio of the AT. A 4-speed AT is usually provided with 3 shift valves, i.e. 1–2 gear, 2–3 gear and 3.4 gear shift valves respectively.

Figure 3.31 shows the 1–2 gear shift valve of the AT in the electro-hydraulic Simpson planetary gear train, which is controlled by the solenoid valve 2. When the solenoid valve is powered on, the oil of the main oil circuit is drained through the drainage port of the solenoid valve, there is no control oil pressure in the upper end of the shift valve element, and the valve element is at the top end under the spring force; when the solenoid valve is powered off, the oil of the main oil circuit acts on the top end of the shift valve element and overcomes the lower spring force, making the valve element at the bottom.

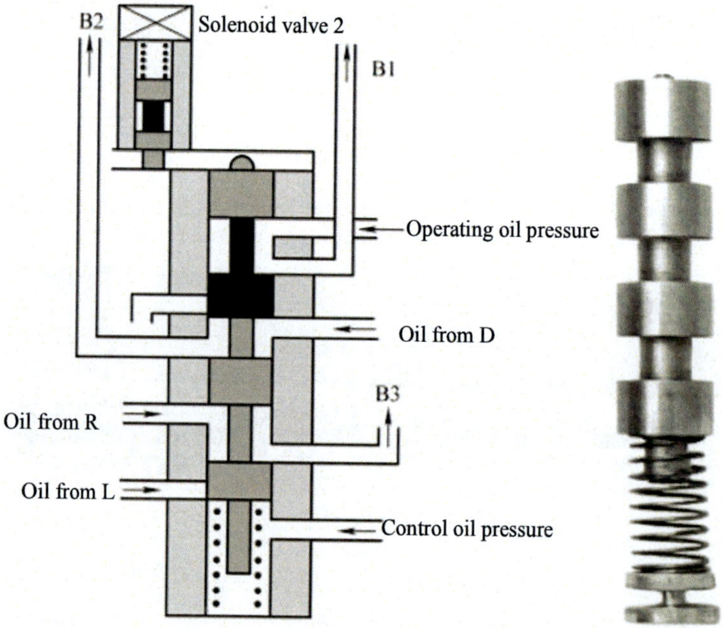

Fig. 3.31 1–2 gear shift valve

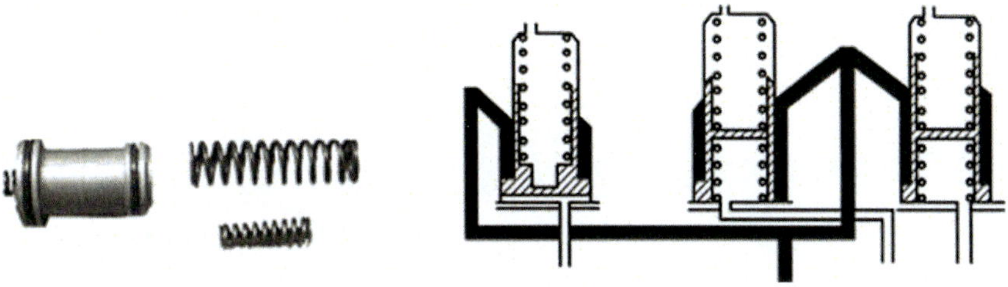

Fig. 3.32 Structure diagram of accumulator

III. Shift quality control system

1. **Basic principle of shift quality control**

AT realizes gear shift by controlling the work of shift actuators, and the work of shift actuators when engaged or separated will directly affect the shift quality. If the oil pressure is established too fast, the clutch will be engaged with the brake too fast, easy to produce shift impact; if the oil discharge is too slow, the clutch will be disengaged from the brake too slow, resulting in slippage. Therefore, the hydraulic control system of the AT shall be equipped with the buffer and safety devices, mainly including accumulator, regulating valve and one-way valve.

2. **Shift quality control elements**

(1) Accumulator: generally, a corresponding accumulator is configured in each forward oil circuit of the AT and located in the oil circuit between the shift valve in each gear of control oil circuit and the shift actuators. It is generally composed of the cylinder tube, damping piston and spring, as shown in Fig. 3.32. During the shift, the control oil from the shift valve acts on the clutch, brake and other shift actuators and flows into the corresponding accumulator. At the beginning of the work, the oil pressure cannot promote the action of the accumulator piston, so the oil pressure in the working cylinder of the controlled actuator increases quickly, so that the actuator can quickly eliminate the free clearance. With the increase of oil entering the accumulator, the oil pressure overcomes the spring force to press the piston down, the volume increases and controls the slow increase rate of the oil pressure in the pipeline, so that the actuator is engaged gently, reducing the shift impact and absorbing the impact from the oil to ensure the stability of working pressure in the hydraulic control system. When the actuator is released from work, the oil will release quickly under the spring action of the accumulator, which can ensure the complete separation of the actuator and prevent the actuator from slipping.

(2) Regulating valve: a low-gear slide regulating valve and 2 speed intermediate regulating valve are configured in the oil circuit and mainly consist of the slide valve and spring, as shown in Fig. 3.33. When the gear lever handle is in L and 2, the regulating valve reduces the pipeline pressure from the main oil circuit to reduce the oil pressure acting on the piston of the actuator. Figure 3.33 shows

Fig. 3.33 Structure of regulating valve

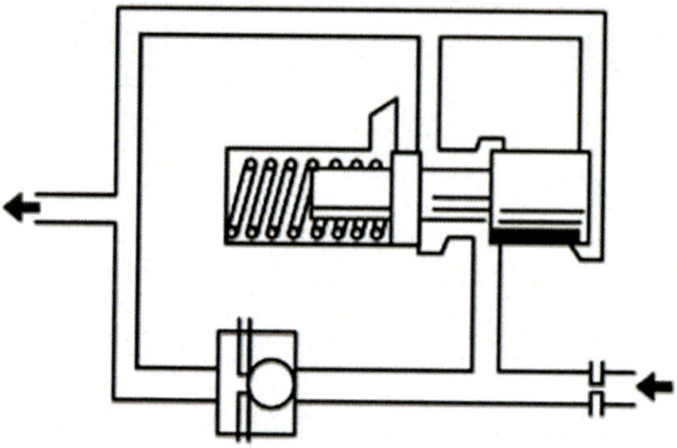

2 speed slide regulating valve, with the oil inlet receiving the oil pressure from the 1–2 gear shift valve and the oil outlet connected to the control oil circuit of the brake B_1. The oil pressure also acts on the right of the valve body and overcomes the spring force to push the valve element to the left, reducing the cross-sectional area of the oil inlet, thereby reducing the control oil pressure at the oil outlet and reducing the shift impact.

(3) One-way valve: many ball valves are installed in the AT valve plate to use as one-way valves and have other functions:

(1) The safety valve, as shown in Fig. 3.34, is generally connected in the main oil circuit in parallel and blocks the drainage port when the oil pressure in the main oil circuit is below the spring force. When the oil pressure in the main oil circuit is larger than the spring force, the drainage port is opened and part of the oil in the main oil circuit is discharged from the drainage port, ensuring the safety of the hydraulic control system.

(2) The reverse quick release valve is shown in Fig. 3.35. During oil filling, the ball valve is closed to reduce the shift impact; during oil return, the ball

Fig. 3.34 Structure of safety valve

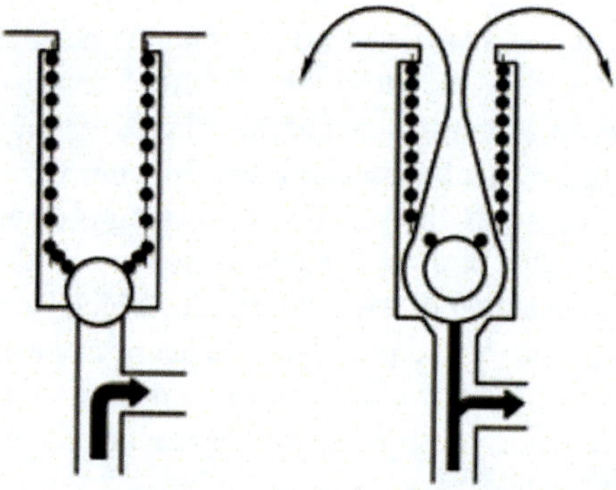

a) Pressure below the set value b) Pressure above the set value

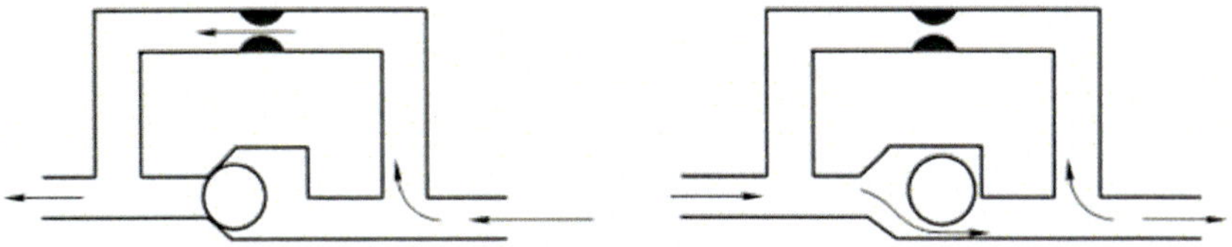

a) Slow in b) Fast out

Fig. 3.35 Reverse quick release valve

valve is open without throttling, which accelerates the oil drainage process and quickly separates the shift actuator.

IV. Hydraulic torque converter control system

1. Hydraulic torque converter control system functions

The hydraulic torque converter control system mainly has two functions: provide the hydraulic oil with a certain pressure for the hydraulic torque converter; control the work of the lockup clutch in the hydraulic torque converter. The hydraulic torque converter control elements include hydraulic torque converter valve (including pressure regulator valve, pressure limiting valve and oil return valve), lockup signal valve, lockup relay valve, lockup clutch control valve and corresponding oil circuit.

2. Structure and principle of hydraulic torque converter control elements

(1) Hydraulic torque converter valve: the hydraulic torque converter valve is to feed the hydraulic oil in the main oil circuit into the hydraulic torque converter after reducing pressure. The heated hydraulic oil in the hydraulic torque converter is sent to the hydraulic oil radiator outside the AT through the torque converter outlet, and the cooled hydraulic oil is sent to the gear transmission to lubricate the planetary gears and their bearings. A pressure limiting valve is set in the oil inlet access of the hydraulic torque converter for some hydraulic torque converter control elements to prevent oil seal from leaking due to excessive oil pressure in the hydraulic torque converter; an oil return valve is often set in the oil outlet access of the hydraulic torque converter to prevent the oil pressure too low or too high, as shown in Fig. 3.36.

(2) Lockup signal valve and lockup relay valve: the work of the lockup clutch in the hydraulic torque converter is controlled by the lockup signal valve and lockup relay valve, as shown in Fig. 3.37.

When the vehicle is in overdrive and the speed and corresponding governor valve oil pressure rise to a certain value, the lockup signal valve is pushed to the position shown in Figure under the role of the governor valve oil pressure, so that the hydraulic oil of the main oil circuit from the overdrive oil circuit enters the lower end of the lockup relay valve.

(3) Lockup clutch control valve: the pulse solenoid valve, i.e. lockup clutch control valve, is used in the new electronic AT to control the work of the lockup clutch,

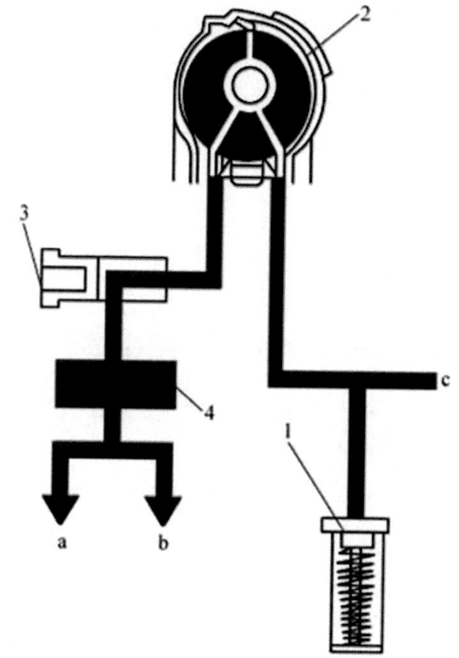

Fig. 3.36 Pressure limiting valve and oil return valve. 1—Pressure limiting valve, 2—hydraulic torque converter, 3—oil return valve, 4—hydraulic oil radiator, a—to front lubrication channel

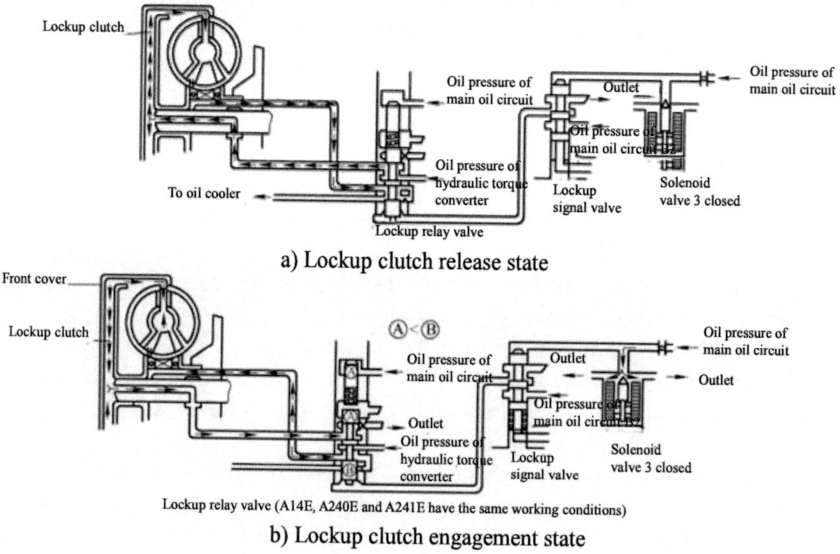

Fig. 3.37 Lockup signal valve and lockup relay valve

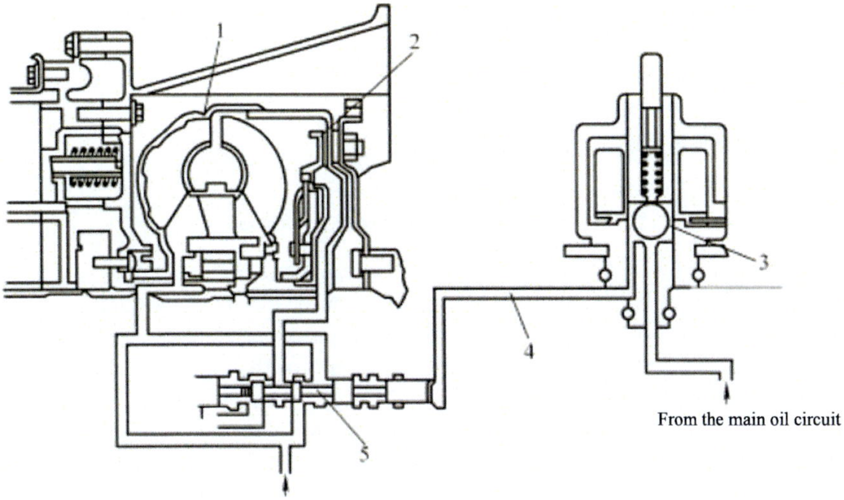

Fig. 3.38 Working principle of the lockup clutch control valve of the hydraulic torque converter. 1—Torque converter, 2—lockup clutch, 3—pulse linear lockup solenoid valve, 4—adjustable control pressure, 5—lockup clutch control valve

as shown in Fig. 3.38. The ECU adjusts the lockup clutch control valve opening through the duty cycle of the pulsed electrical signal to control the oil pressure acting on the right end of the lockup clutch control valve and the opening of the oil drain port when the lockup clutch valve moves to the left, thus controlling the oil pressure in the right of the lockup clutch piston.

Working condition analysis.

(1) Separation control. When the duty cycle of the pulsed electrical signal acting on the lockup clutch control valve is 0, the solenoid valve is closed, no oil pressure acts on the right end of the lockup clutch control valve, the oil pressure is the same on the left and right sides of the lockup clutch pressure plate and the lockup clutch is separated.
(2) Trackslip control. When the duty cycle of the pulse signal acting on the lockup clutch control valve is small, the opening of the solenoid valve, the oil pressure acting on the right end of the lockup clutch control valve, the opening of the oil drain port when the lockup clutch valve moves to the left, the hydraulic difference on left and right sides of the lockup clutch piston and the resulting lockup clutch engagement force are small, so that the lockup clutch is half-engaged.
(3) Lockup control. The greater the duty cycle of the lockup electrical signal, the greater differential pressure on the left and right sides of the lockup clutch pressure plate and the engagement force of the lockup clutch. When the pulsed electrical signal reaches a certain value, the lockup clutch is fully engaged.

When the lockup clutch is engaged, the electronic control unit can make the engagement process more softened to reduce impact.

3.6 Electronic Control System of aT

I. Composition and function of electronic control system

The electronic control system consists of sensors, electronic control units and actuators, as shown in Fig. 3.39. The basic function of the electronic control system is to input the detection signal of the vehicle speed sensor and throttle position sensor and the selector switch signal of the shift program to the electronic control unit for computing, determine the gear and shift point and output the shift instruction according to the preprogrammed shift program, control the on and off of the current in the solenoid valve coil and automatically switch the oil circuit of the shift actuator to achieve automatic shift. In addition, the electronic control system also has the functions of hydraulic torque converter lockup control, oil pressure adjustment, fault self-diagnosis and fail safe protection.

II. Basic tasks of electronic control system

1. **Control shift law**

In order to meet the dynamic or economy performance requirements of the vehicle, the vehicle is usually set to the sports (power) shift or economic shift pattern, which

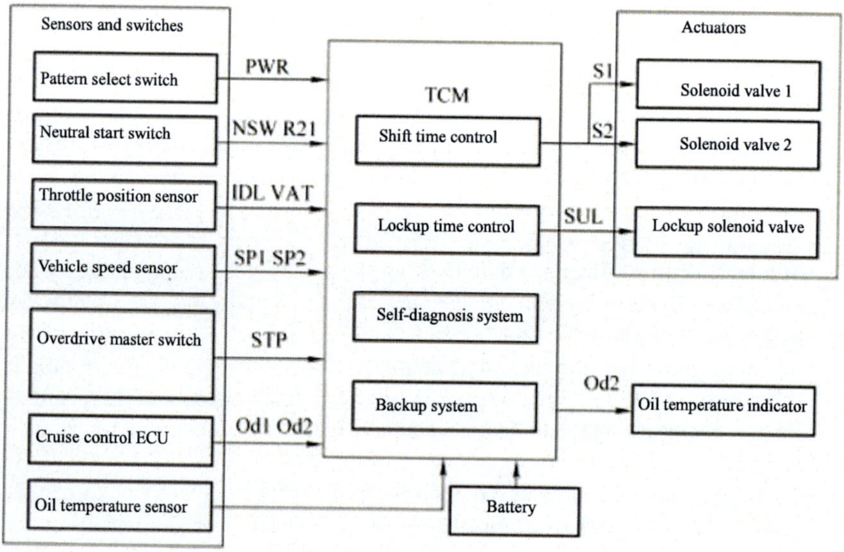

Fig. 3.39 Block diagram of AT electronic control system

can be selected through the shift pattern select switch or through fuzzy logic control. In the economic shift pattern, the transmission shift point and the hydraulic torque converter lockup time are determined on the basis of improving the fuel economy during the vehicle driving; in the sports shift pattern, the upshift point and hydraulic torque converter lockup time shall be postponed to meet the vehicle acceleration performance requirements.

The electronic control system of some AT adopts the fuzzy logic control for shifting, which is selected by the electronic control unit according to the opening change rate of the accelerator pedal. The electronic control unit has two shift control programs: shift time program related to the driver and driving conditions, and shift time program related to the driving resistance. The former, also called DSP, uses the fuzzy logic control to meet the driving requirements of different drivers; in the latter, the electronic control unit calculates the driving resistance according to the vehicle speed, throttle position, engine speed and acceleration and then determines the shift time.

2. **Hydraulic torque converter lockup control**

The control program for the hydraulic torque converter lockup is stored in the electronic control unit. According to the detection signal of the throttle position sensor and vehicle speed sensor and the gear shift lever position, the electronic control unit controls on and off of the hydraulic torque converter lockup control valve, changes the pressure of hydraulic torque converter clutch oil circuit and controls the engagement and release of the hydraulic torque converter clutch, so as to achieve the hydraulic torque converter lockup control.

3. **Shift process control**

The shift process control is mainly to control the coordination between the engine and the transmission, thus completing the shift process quickly and smoothly. The shift quality is mainly reflected by the shift impact, clutch sliding friction work and other parameters. In order to achieve high shift quality, a lot of calibration work is needed, including prefilling control, torque phase control, inertial phase control, etc.

This chapter gives a brief introduction to the electronic control system of AT. For details, please refer to "Electronic control system of AT".

3.7 AT Performance Tests

1. **Bench performance test**

The bench performance test is to evaluate the power transmission performance of each transmission. The test items include:

(1) General performance test for extensive research on the performance of the gears in the driving state.

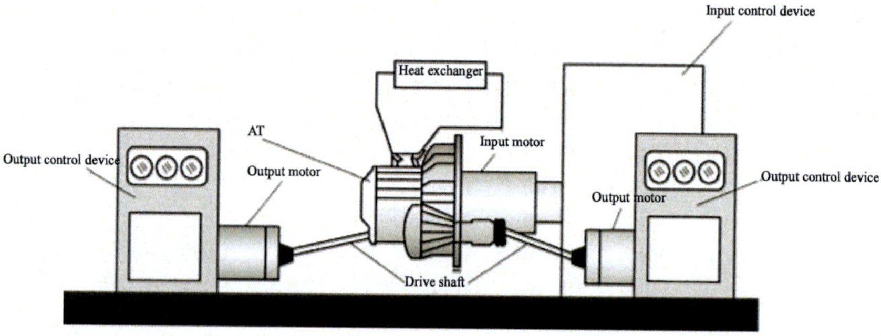

Fig. 3.40 Bench performance test device

(2) Full open throttle torque performance test (mainly check the engine performance in the full open operation of the throttle).
(3) Road load performance test to study the driving performance at constant speed.
(4) Inertial performance test to study the performance in the reverse drive.
(5) No-load loss test to study the torque loss of each gear under no-load of the output shaft.

Moreover, the lockup clutch engagement state is also measured sometimes for the AT with lockup clutch although it is not mentioned in the standard.

The bench performance test device, as shown in Fig. 3.40, determines the state of the input or output shaft and evaluates its performance by detecting the torque and speed changes of the input or output shaft. In addition, the oil temperature and pipeline oil pressure of the inlet and outlet and oil sump (inside the oil pan) of the hydraulic torque converter as well as the operating oil pressure of all clutches are determined generally to master the running state of the AT. In recent years, the electronically controlled AT control system is widely used, the signals from the vehicle speed, engine speed, throttle position and other sensors and various command signals of the computer shall be stored and the evaluation test starts for the oil pressure characteristics and response characteristics from the computer command signals and for the control of the electric mode used in the uphill, downhill and other running states.

The above test is carried out under normal operating temperature. For practical evaluation, the equipment with the environmental conditions set freely starts to be subject to the oil absorbency and other oil surface evaluation at low temperature, torque loss evaluation at low temperature in neutral position, oil pressure performance evaluation at low temperature and other performance tests in the practical fields at low temperature in recent years, but such tests have not been standardized.

In addition to the above test, there are also tests to evaluate the drive torque, friction engagement device, sliding torque and oil agitation resistance of hydraulic pump.

2. Speed change performance test

The speed change performance test is to evaluate the transient characteristics (impact and time lag) at the speed change, the lockup clutch engagement and release, including:

(1) Test carried out on a bench using a dynamometer that can be set (inertial mass of the engine and the vehicle mass as well as the driving resistance load can be set).
(2) Actual test using a vehicle in the environmental laboratory.
(3) Actual driving test on the test track.

In order to improve the speed change performance of the AT in various complex environments, it is becoming increasingly important to carry out relevant tests and evaluations in environmental laboratories that simulate low temperatures in winter and low atmospheric pressure at high altitudes. The transition performance is generally evaluated based on the vehicle acceleration or the output shaft torque (drive shaft torque) and the engine speed. AT the same time, oil pressure signals acting on the friction connection device and various control signals of electronic AT are collected to improve the variable speed characteristics. Signals from the engine are also collected when coordinated with the engine. In addition, acceleration, torque changes and other sense of speed change that cannot be captured as well as the abnormal noise caused by speed change shall be compensated by sensory evaluation.

The AT speed change generally includes the automatic speed change caused by the change of engine load (throttle percentage, etc.) and the output shaft speed, as well as the manual speed change caused by the gear shift lever shift selection control and the overspeed switch operation. In recent years, emphasis has been placed on the feel of driving a car with a manual transmission, with an increasing number of manual transmissions shift gears through paddles on the steering wheel and front and rear control of the gear shift lever. The test method is to master these speed change conditions in advance and to change the speed consciously under various operating conditions.

3. Performance test of hydraulic control system

The performance test of hydraulic control system, a separate test for the valve body assembly, evaluates the static and dynamic characteristics of the valve and the stability of the hydraulic control circuit by using hydraulic devices capable of controlling the pressure and flow. The actuator of the electronic automatic transmission of the mainstream product is tested by a variety of solenoid valves, sometimes also by the drive device.

4. Hydraulic pump performance test

The hydraulic pump performance test, as a separate test for the hydraulic pump assembly, evaluates the output performance, pulse, drive torque and noise of the hydraulic pump by the device driver which can control the speed and adjust the output performance of the hydraulic pump.

5. **Shell strength test**

The shell strength test is to evaluate the influence of the driving force, driving reaction, external input force of the suspended part, vibration and temperature change generated by the rotating body on the strength and durability of the case and shell, as well as the sealing of joint surface.

6. **Parking brake test**

The parking brake test is to evaluate the performance, strength and durability of the fixing and releasing functions of the output shaft of the parking brake device, including the misuse conditions such as vehicle stop and misoperation on the ramp.

7. **Gear shift lever control feel test**

The gear shift lever control feel test is to confirm and evaluate the controllability, rhythm sensation of the gear shift lever and the control force of the external joy stick of the AT according to the control force of the gear shift lever or the control force change mode (Fig. 3.41) of the external joy stick of the AT.

8. **Oil level and ventilator performance test**

The oil level and ventilator performance test is conducted for the automatic transmission fluid (ATF) to evaluate the required oil volume setting, oil absorbency in actual operation and the function of ventilator.

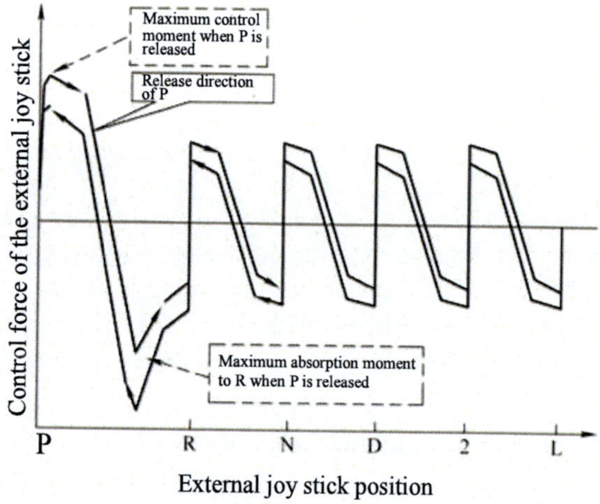

Fig. 3.41 Gear shift lever control force and external joy stick position

9. **Friction test**

The friction test is used to detect the torque loss of each gear, which can be roughly divided into two types: the no-load loss test and the transmission efficiency test under the no-load state of the output shaft. The no-load loss test is used to evaluate the torque loss caused by the hydraulic pump drive torque, friction element slip torque and hydraulic oil agitation resistance. The transmission efficiency test is aimed at the torque loss mentioned above to further evaluate the torque loss caused by the gear efficiency.

10. **Other performance tests**

The AT performance tests also include cooling system test, lubrication performance test, adaptability evaluation test to environmental conditions, vibration noise test and quietness evaluation test.

3.8 Development Direction of aT

With the rapid development of microelectronics technology in recent years, the electronic automatic transmission comes into being, bringing a more ideal powertrain for the automobiles. The mechatronics technology has entered the field of automobile, which has promoted the great reform of the automotive transmissions. The AT trends to be electronic. The development direction of modern AT is as follows.

1. **Use of universal lockup clutch with electronic control**

In order to improve transmission efficiency and economic performance, the hydraulic torque converter lockup clutch is widely used in the AT of passenger cars, and electronic control is carried out to keep the gear shift smooth. The functional characteristics of the lockup hydraulic torque converter determine the AT process of hydraulic coupler, hydraulic torque converter and lockup hydraulic torque converter. The hydraulic torque converter can increase the engine torque and absorb torsional vibration in addition to transferring torque, but the hydraulic coupler cannot. The hydraulic torque converter with lockup clutch overcomes the low transmission efficiency of the ordinary hydraulic torque converter caused by slip between the output and input shafts. The hydraulic torque converter will not work at the lockup of the lockup clutch. At this time, the transmission input and output are rigidly connected and the electronic control of the ECU makes the shift smoother. The lockup clutch is widely used in the hydraulic torque converter of the modern AT, which improves the fuel economy and lowers the transmission temperature.

2. **Multi-speed**

The trend of multi-speed AT development has been a consensus in the industry. From the early 4AT to the common 5AT, 6AT, the number of automatic transmission gears is increasing. The increase in the number of gears has the following advantages:

(1) The increase in the number of gears enables the engine to work in the most fuel-efficient conditions for more time and achieves better economy.
(2) The high-gear AT has a larger gear ratio range and a more detailed gear ratio difference between gears, which can improve the performance of the vehicle to some extent.

For example, the core structure of the ZF9HP transmission (Fig. 3.42) is composed of four planetary gear sets and six shift actuators, one shift actuator more than the ZF8HP transmission. It controls the different combinations of the planetary gears by means of the shift actuators, realizing the setting of 9 forward gears and 1 reverse gear. It shortens the transmission length by nested planetary gears and controls the overall mass of the transmission by optimizing the transmission structure and using lightweight parts on the basis of meeting the intensity requirement.

The comparison of ZF9HP transmission with horizontal 8 speed and 6 speed AT gear ratio is shown in Table 3.9. As can be seen from the table, the number of overdrive gears (gear ratio less than 1) of the ZF9HP transmission has reached 4. The overdrive gear is from the gear 6 and the gear ratio of the gear 9 is only 0.48, far below that of the general overdrive gear. The total gear ratio range even reaches 9.81 (ratio of maximum gear ratio to minimum gear ratio). It can be seen that the 9 speed transmission is more economical in its gear ratio setting, enabling the vehicle to travel at a more economical speed over a wider speed range. Similarly, a wide

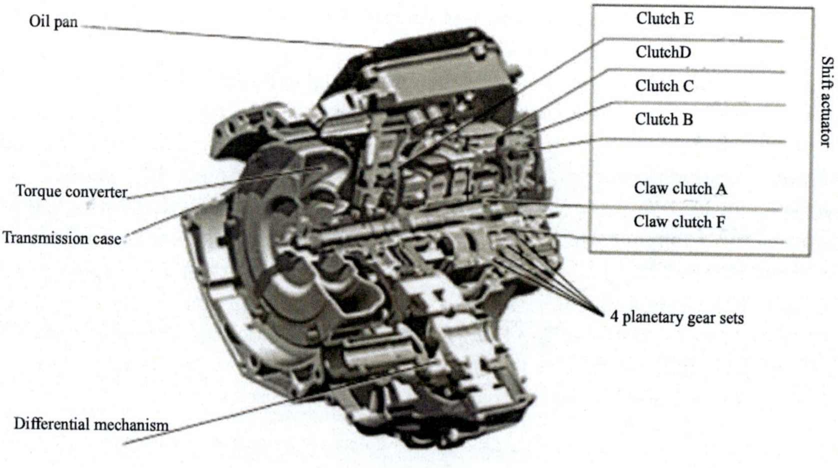

Fig. 3.42 Structure of ZF9HP transmission

Table 3.9 Comparison of ZF9HP transmission with horizontal 8 speed and 6 speed AT gear ratio (horizontal)

Transmission	ZF9HP 9 speed AT	Aisin AW TG80—LS 8 speed AT	Aisin AW TF80—SC 6 speed AT
Gear 1	4.70	5.20	4.15
Gear 2	2.84	2.97	2.37
Gear 3	1.90	1.95	1.56
Gear 2	1.38	1.47	1.16
Gear 5	1.00	1.22	0.86
Gear 6	0.80	1.00	0.69
Gear 7	0.70	0.82	–
Gear 8	0.58	0.69	–
Gear 9	0.48	–	–
Reverse gear	0.80	4.25	3.39
Gear ratio range	9.81	7.59	6.05

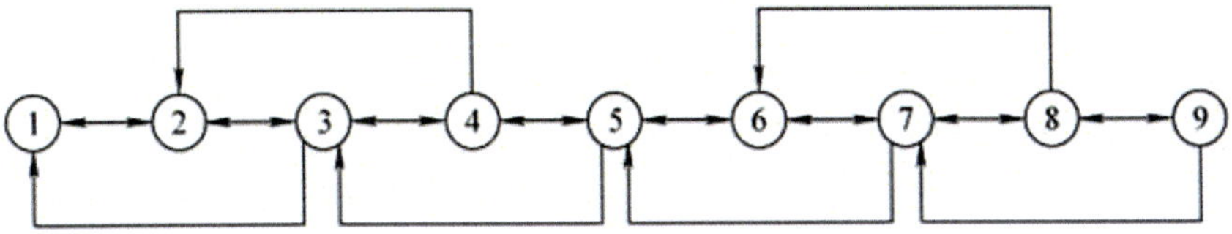

Fig. 3.43 Shift logic of ZF9HP transmission

gear ratio is beneficial for dynamic performance, allowing the vehicle to maintain maximum performance output speed over a wider speed range.

As shown in Fig. 3.43, the combinational logic of the ZF9HP transmission control elements is also very clever and only two shift actuators are required either for shifting on alternate gears or for sequential shift (Table 3.10), ensuring that the transmission has fast response times in both modes of execution, and its acceleration time from 0 to 100 km/h can be reduced by at most about 2 s compared with the 6 speed AT.

3. **Shift control by multiple solenoid valves**

The AT shift is controlled by multiple solenoid valves, which greatly improves the shift quality.

The AT actuators have been developed from one or two solenoid valves in the past to multiple solenoid valves. The increasing number of solenoid valves has led to the gradually development of the upshift and downshift control of the gear D from throttle oil pressure and speed oil pressure to the shift solenoid valve control. Many new solenoid valves are used in transmissions, such as forced downshift solenoid valve, timing solenoid valve, torsion buffer solenoid valve, reverse solenoid valve and

Table 3.10 Combination of ZF9HP transmission control element

Gear	Brake		Clutch		Claw clutch	
	C	D	B	E	F	A
1		O			O	O
2	O				O	O
3			O		O	O
4				O	O	O
5			O	O		O
6	O			O		O
7		O		O		O
8	O	O		O		
9		O	O	O		
R		O	O			

Note O means engagement, brake or locking

torsion transfer solenoid valve, further expanding the control range of the electronic control system. Currently, the control of some electronically controlled ATs over the manual mode, gear D and reverse gear is in the charge of the shift solenoid valve. The fuzzy control technology humanizes the vehicles, allows the transmission ECU to simulate and learn the driver's driving habits, and automatically corrects ECU control commands.

Previously, the pressure control valves of the hydraulic control system were the pilot valve and the main valve, both of which jointly controlled the oil pressure. Now a new type of proportional valve (Fig. 3.44) has emerged. A single proportional valve can control the oil pressure, thus reducing the mass of the hydraulic system.

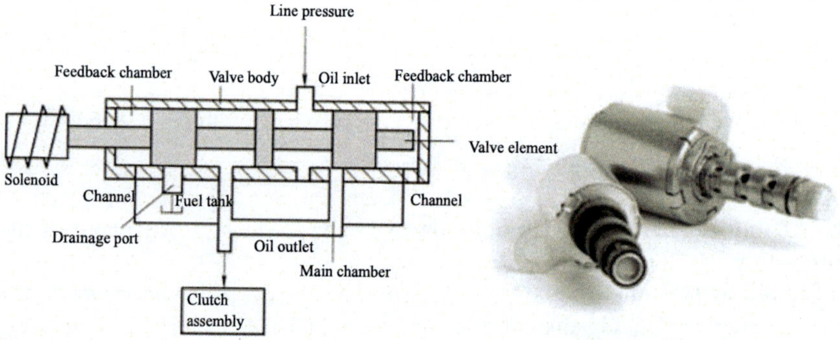

Fig. 3.44 New proportional valve

4. **Transmission miniaturization**

The miniaturization of AT is helping to reduce weight and shorten the power transmission routes to make vehicles more fuel-efficient. Since the 1970s, the number of minicars has increased dramatically, thus providing a prerequisite for the miniaturization of the AT. In addition, the trend of using automatic drive axle (that is, combining the transmission and the drive axle as a whole) is very obvious, and the miniaturization of the transmission promotes the development of FF and automatic drive axle. Compared with hydraulic control, the electronic control has the following obvious advantages: achieve more complex and diverse control functions that the hydraulic control is difficult to achieve, so that the performance of the transmission is improved; greatly simplify the structure of control system and reduce productive investment; realize the control function by means of the combination of software and hardware. Because the software is easy to modify, the product can adapt to the change of structural parameters. With the development of automobile electronization, the electronic control of the automotive drivetrain can share resources with engine, brake system, SRS and other systems through the bus networking to realize the overall control, so as to further simplify the control structure.

5. **Reduction of work noise**

At present, the research on noise reduction at home and abroad mainly focuses on the following aspects: improving the geometric precision and surface quality of parts and improving the assembly quality; obtaining the dentation that can achieve the low noise effect through the finite element software simulation technology; optimizing and improving the helical angle and tooth width of the gears to increase the contact area of the meshing gears, so as to reduce the noise caused by the gear friction.

6. **Optimization of shift control**

The shift control is optimized and improved from two aspects: the intelligence of the shift point control and the high quality of the shift transition process. In terms of intelligent control of shift point, the vehicle speed and throttle percentage are used as the basic control parameters, the factors such as the driving conditions, driver's personality and willingness to maneuver are used as auxiliary conditions to determine the shift law, and intelligent control techniques such as robust control, model control, adaptive control and neural network control are adopted; in terms of the high quality of the shift process, improving the stability of the transition and reducing the thermal load of the joint elements will be two key issues in the future research.

Bibliography

1. Automotive Handbook Editorial Board (2001) Automotive handbook: design. Beijing: China Communications Press
2. Chen J (2009) Automobile structure: volume two. China Machine Press, Beijing
3. Naunheimer H et al (2011) Automotive transmissions. Spring Heidelberg, New York
4. Cao L (2009) Analysis of power transmission route of automatic transmission (28)—power transmission route of GAC-Toyota YARiS U340EAT. Motor China (5):33.34
5. JSAE (2010) Automobile engineering manual 4: powertrain design. Translated by SAE-China. Beijing Institute of Technology Press, Beijing
6. Zhaoying Y (2013) Research of the hydraulic control system of automatic transmission. Hebei University of Science and Technology, Shijiazhuang
7. JSAE (2010) Automobile engineering manual 6: powertrain test evaluation. Translated by SAE-China. Beijing Institute of Technology Press, Beijing
8. Zhao Q, Xie F, Yu T (2010) Development status and technical trend of automatic transmission. Machinery 37(12):1–5
9. Kubalczyk R, Gall R, Liang Z (2013) instant analysis on automatic transmission planetary gear system. Drive Syst Tech 3:001

Chapter 4
Continuously Variable Transmission

4.1 Overview

The continuously variable transmission (CVT) is a transmission which is driven by a non-geared drive mechanism and can continuously change the gear ratio.

A lot of ingenious and well-structured drive mechanism forms (Fig. 4.1a) have emerged in the development process of the variable speed technology. The widely used metal belt CVT was developed by Dr. H. Van Doorne of the Netherlands in the late 1960s and only used in the minicar at that time.

In the modern era, due to the improvement of materials science and technology and manufacturing level, CVT can be applied to large displacement 3.5L cars.

CVT has mechanical and hydraulic types. The mechanical CVT is applied in the vehicles actually, with the main transmission types of belt type and annular contact type. In terms of popularity, the belt CVT, which changes the gear ratio by changing the width of pulley groove, has been widely used. The belt CVT also includes VDT belt type, chain type, composite belt type and rubber belt type structure, among which VDT belt type occupies the mainstream position.

This chapter mainly introduces the VDT belt CVT for automobile. In the mid to late 1970s, the Dutch VDT company developed the VDT belt; in 1987, Subaru first fitted the VDT belt CVT on its Justy model; in 1995, the Dutch VDT company was acquired by BOSCH. By 2015, the annual output of the CVT belt had exceeded 40 million.

The gear ratio of the automatic transmissions, such as AT and DCT, cannot change continuously. To change the speed continuously, the engine must work at different speeds, thus resulting in shift impact. In addition, given a certain driving speed, the engine may not be able to work at the appropriate speed, which will reduce the vehicle's fuel economy and power performance.

Compared with AT, CVT can obtain continuous gear ratio. Figure 4.2 shows the relation curve between driving speed and engine speed Fig. 4.3 shows the corresponding power and fuel consumption at the engine operating point. It can be seen that in the process of changing the driving speed of a car equipped with CVT, the

Y. Chen, *Automotive Transmissions*, Springer Tracts in Mechanical Engineering,
https://doi.org/10.1007/978-981-15-6703-2_4

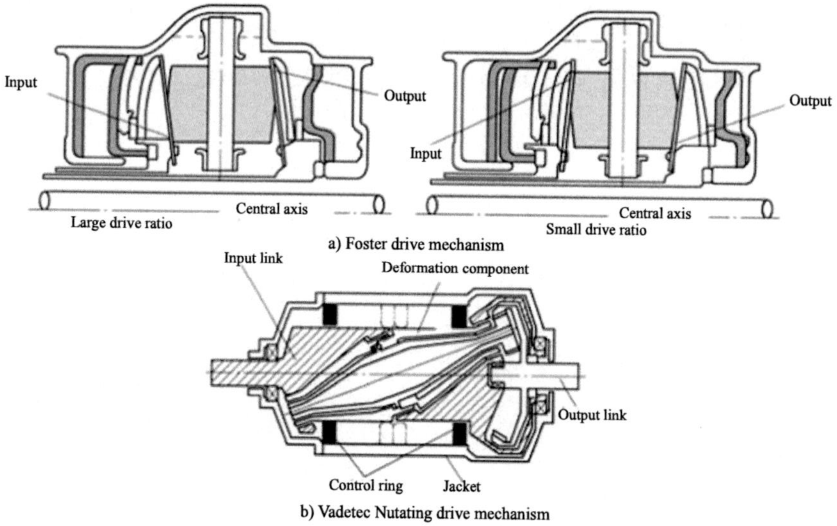

Fig. 4.1 Two drive mechanisms

Fig. 4.2 Relation curve between driving speed and engine speed

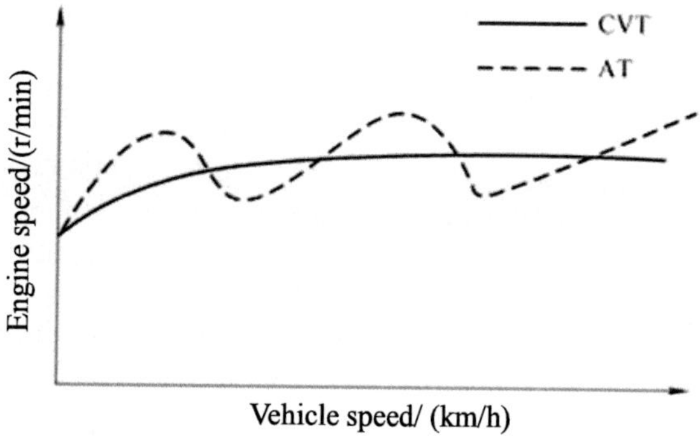

change of engine speed is relatively small without obvious fluctuation, and can make the engine operating point close to the curve of the minimum specific fuel consumption, which ensures that the car has good power performance, comfort and fuel economy.

However, compared with AT, CVT has a higher cost and a greater friction loss. In particular, in the process of changing speed, the belt and the pulley shall move relative to each other in the radial direction, and the efficiency will be reduced to about 60%. Limited by the structure, the range of CVT gear ratio is smaller than AT, and the torque capacity is not high. Because of the size of the pulley, the arrangement of the belt CVT is also limited.

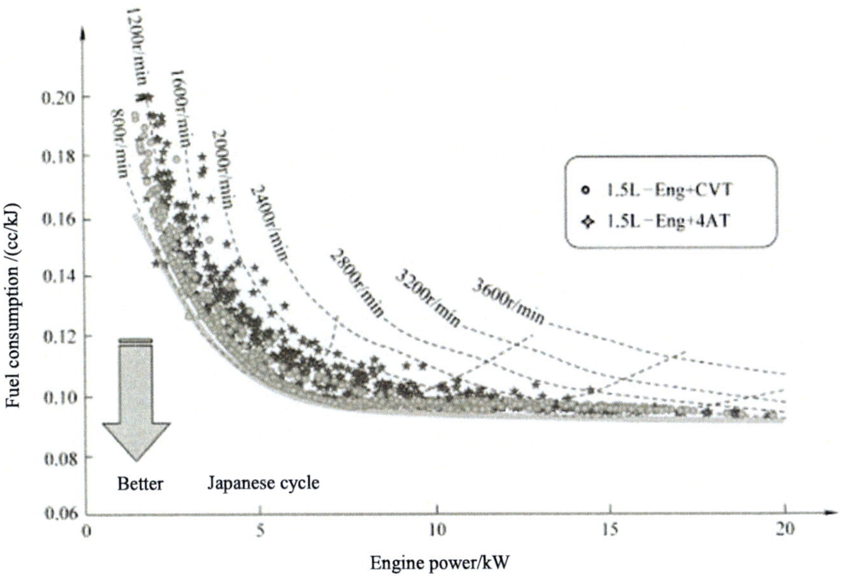

Fig. 4.3 Power and fuel consumption at the engine operating point

4.2 Composition of CVT

According to the power transmission sequence, the main components the metal belt CVT (Fig. 4.4) and their functions are as follows:

(1) Hydraulic torque converter: start; reduce impact.
(2) Planetary gear train: forward/reverse/neutral switching.
(3) Belt drive mechanism: speed change; parking (lock the driven active wheel).
(4) Main reducer: deceleration.
(5) Differential mechanism: differential motion.

The torque output by the engine is transmitted to the drive pulley through the hydraulic torque converter and the forward/reverse/neutral switching mechanism; the power is transmitted to the driven pulley in the form of belt drive and finally drives the wheel through the main reducer and transmission shaft.

1. Hydraulic torque converter

The hydraulic torque converter of CVT only runs within 5–10 m after starting at the speed below 20–25 km/h, beyond which it is immediately locked, with low utilization rate. Limited by the layout space, the axial dimension of the hydraulic torque converter is smaller than that of AT. Although this sacrifices the advantage of hydrodynamic drive the actual impact on CVT is not significant.

In addition to the hydraulic torque converter, other forms of starters include wet clutch and electromagnetic clutch.

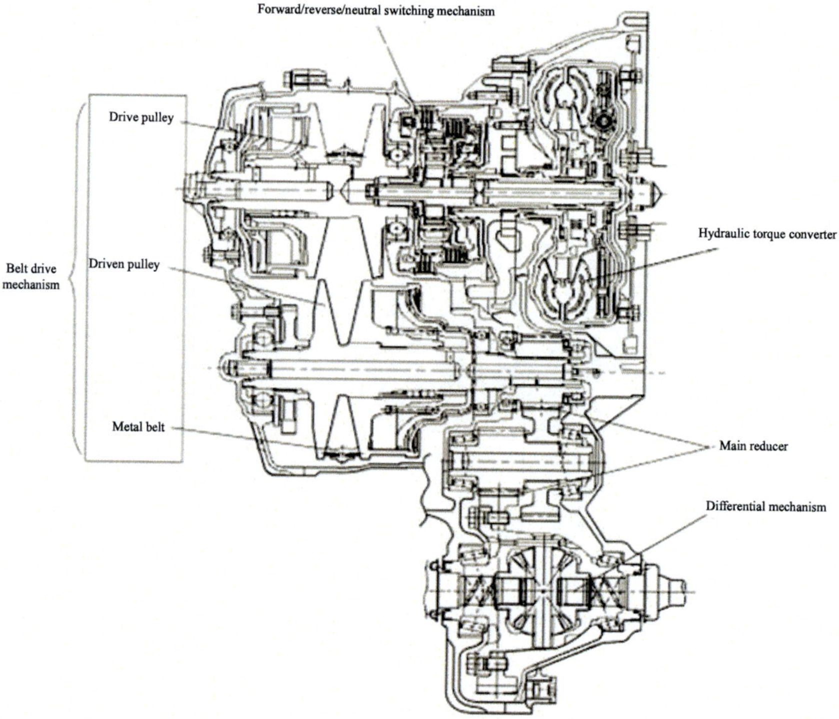

Fig. 4.4 Composition and power transmission of metal belt CVT

2. **Forward/reverse/neutral switching mechanism**

In the CVT transmission shown in Fig. 4.5, there is a single row of planetary gear train between the hydraulic torque converter and the drive pulley, the annular gear is connected to the drive pulley shaft, and the center gear is connected to the output shaft of the hydraulic torque converter; a clutch is set between the center gear and the planetary carrier, and a brake between the planetary carrier and the transmission case. The forward and reverse gears are switched by controlling the engagement and disengagement of the clutch and brake in the planetary gear set.

When moving forward, the clutch is engaged and the brake is disengaged. The annular gear and the center gear have the same angular velocity direction. In reversing, the clutch is disengaged and the brake is engaged. The annular gear and the center gear have the opposite angular velocity direction.

3. **Belt drive mechanism**

The belt CVT which has been widely used mainly includes VDT belt type, chain type, composite belt type and rubber belt type. The structure and working principle of different belt CVTs are basically similar: the belt drive mechanism consists of the belt, drive and driven pulleys; each set of pulley consists of two sets of cone disks,

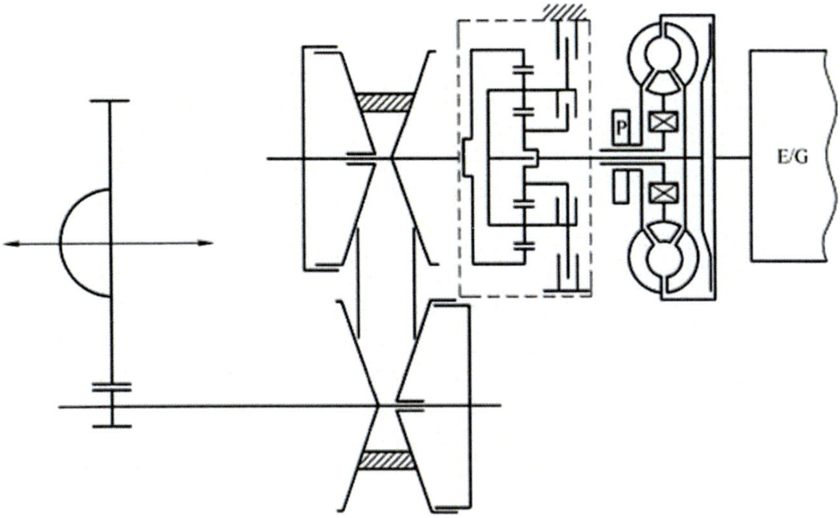

Fig. 4.5 Forward/reverse/neutral switching mechanism (from Japanese patent JP2012-241745)

each consisting of a sliding cone disk that can move along the shaft and a fixed cone disk that is fixedly connected to the shaft; the belt is located between two cone disks.

As shown in Fig. 4.6, when the sliding cone disk moves along the shaft, the groove amplitude formed by the cone surfaces of the two cone disks changes, and the pitch

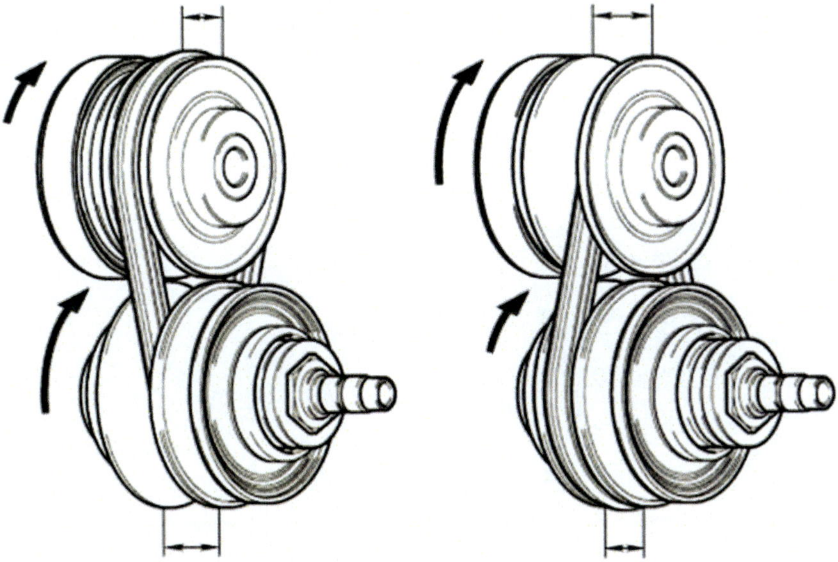

Fig. 4.6 Change in the pulley groove amplitude in the speed change

radius of the belt also changes accordingly. Since the length of the belt and the center distance of the pulley are unchanged, the pitch radius on the other side shall also change correspondingly.

The gear ratio of the belt drive is

$$i = \text{pitch radius on the driven side/pitch radius on the drive side} \tag{4.1}$$

The main form of the belt in the belt CVT is shown in Fig. 4.7.

The VDT belt structure shown in Fig. 4.7a will be detailed in Sect. 4.3.

Figure 4.7b shows the chain structure. The chain plates between adjacent chain links are connected by a pair of roll pin shafts. When the adjacent links rotate each other, there is relative rolling motion between the roll pin shaft pairs connected on the cambered surface, which reduces the friction loss. The chain belt has a large torque capacity, but large vibration noise.

Figure 4.7c shows the composite belt structure, which is similar to the VDT belt structure, but with different materials, functions and operating environment. The unit block is molded with heat-resistant resin with aluminum alloy reinforced sheet to withstand the pulley packing force; embedded with an aramid core wire. Two tension belts are slotted and inserted from both sides of the unit block to form a composite belt with the unit block. The composite belt works in an open environment with

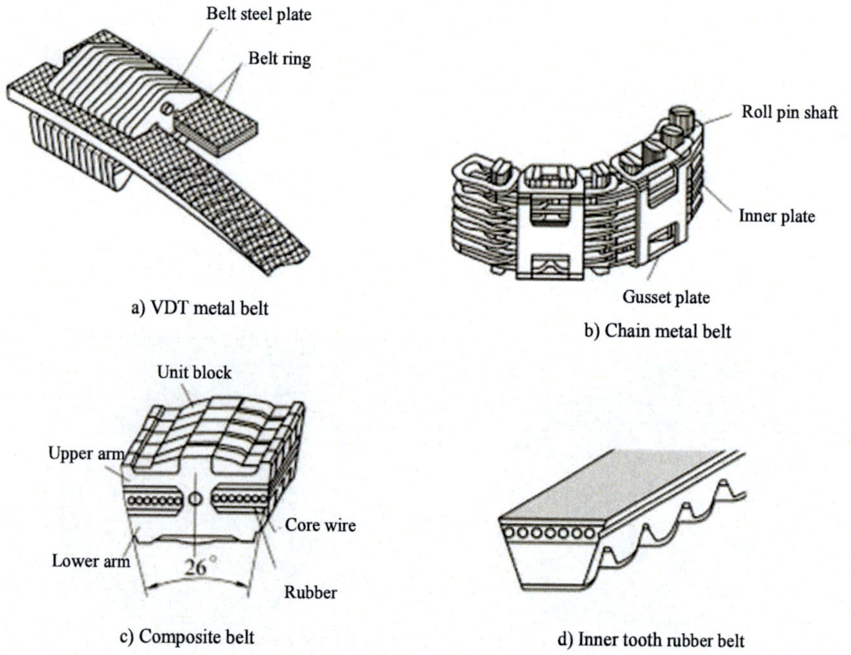

Fig. 4.7 Main forms of belt

no lubrication, large friction factor and light transmission weight, but poor heat resistance and large noise.

Figure 4.7d shows the inner tooth rubber belt, which is divided into three layers: the top layer is a stretching layer made of high-strength fiber cloth to protect the second and third layers; the second layer is a tensile layer, containing nylon fiber bundles, mainly used to withstand the tension of the belt; the third layer is a compression layer made of rubber material to bear the packing force of the pulley and generate friction to transmit the power. The axial stiffness of rubber is increased by the use of fibers with uniform axial direction. The design of the inner teeth can reduce the bending stiffness and provide a large contact area, so as to reduce the loss of belt bending, reduce heat and improve the transmission efficiency.

4. **Parking brake**

The parking gear in the parking brake is directly arranged on the output shaft of the transmission to achieve parking by occluding with the parking brake ratchet.

5. **Reduction gear and differential mechanism**

In a CVT-equipped vehicle, the engine flywheel is generally designed to rotate in the same direction as the forward wheel, thus reducing the impact of variable speed. As shown in Fig. 4.8, in order for CVT to achieve the above rotation direction relationship, a two stage reduction gear is set between the driven pulley shaft and the output shaft, which increases the distance between the driven pulley shaft and the output shaft, ensuring the space requirement of the transmission shaft.

The differential mechanism and differential limited slip mechanism used in the CVT-equipped vehicle are not significantly different from those used in the vehicles equipped with other transmissions and not described in detail here.

6. **Hydraulic, cooling and lubrication systems**

The automatic transmission fluid (ATF) used in the CVT mainly has the following functions:

(1) Control the axial motion of the sliding cone disks of the drive and driven pulleys.
(2) Control the engagement and disengagement of the lockup clutch.
(3) Control the engagement and disengagement of the clutch and brake in the forward/reverse mechanism.
(4) Act as the medium of the hydraulic torque converter.
(5) Lubrication.
(6) Cool the metal belt, clutch, brake and hydraulic torque converter.

Requirements for hydraulic pump: when the hydraulic pump pumps the ATF, the ATF pressure and ATF output fluctuation, and ATF pressure changes on the ATF pumping amount shall be small impact; the hydraulic pump shall have low operating noise and small size and mass; compared with AT, the hydraulic pump used in CVT can pump a large amount of ATF, so it is necessary to ensure small ATF leakage of

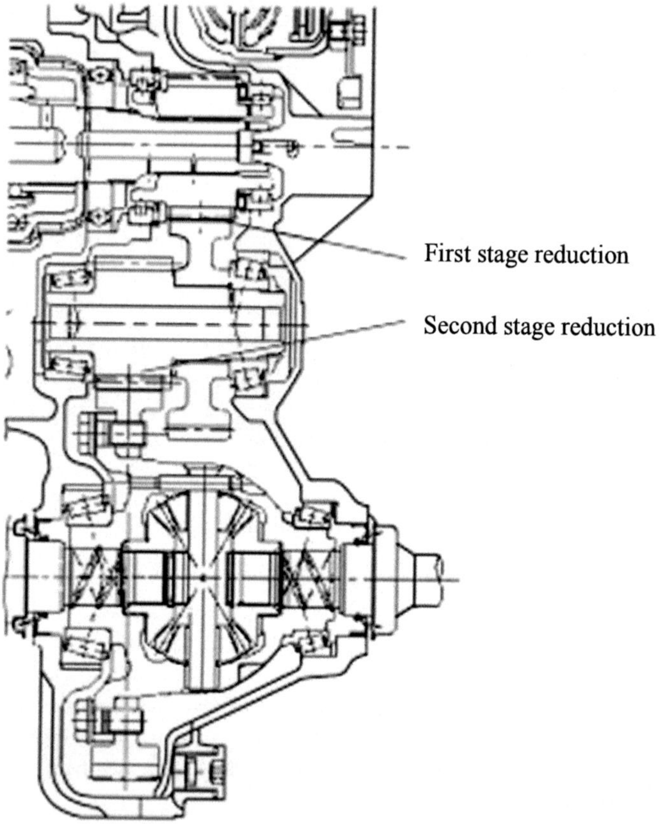

Fig. 4.8 CVT two stage main reducer

the pump when working under high pressure. Hydraulic pump types that meet the application requirements of CVT include gear pump and vane pump.

There are multiple slip contacts and rolling contacts in CVT, as well as oil agitation and high pressure leakage. In order to keep CVT operating below the necessary temperature, a cooling system shall be set to reduce the transmission temperature. There are usually two cooling methods: one is to draw cooling water from the engine cooling system and flow through the transmission for cooling; the other is to set a separate cooling circulatory system in the transmission.

4.3 Composition and Drive Theory of VDT Belt

I. Pulley

1. Structure of pulley

The pulley consists of fixed cone disk and sliding cone disk. The pitch radius of the drive and driven pulleys is changed by adjusting the axial position of the sliding cone disk, so as to change the gear ratio of the belt drive, as shown in Fig. 4.9. The sliding cone disk can realize the continuous change of the pitch radius during the axial movement.

2. Bending moment tolerated by pulley

The sliding cone disk and the pulley shaft are connected by ball spline (Fig. 4.10), which can reduce the limiting static friction and sliding friction of the sliding cone disk in the axial direction, so as to reduce the sudden acceleration change after the push over the limiting static friction, reduce the friction resistance, and make the speed change smoother and more convenient for manipulation. On the part in contact with the belt, the pulley will tend to expand outward and will bear a certain bending moment, which is balanced by the contact surface pressure between the sliding cone disk hub and the pulley shaft. In fact, this bending moment is so large that the shaft and hub are designed into a stepped structure to avoid its direct effect on ball spline steel balls (Fig. 4.11). However, the shaft hole surface spacing still changes periodically during pulley rotation, so the ball spline steel balls will still be squeezed, resulting in alternating radial pressure. The three change rules are shown in Fig. 4.12.

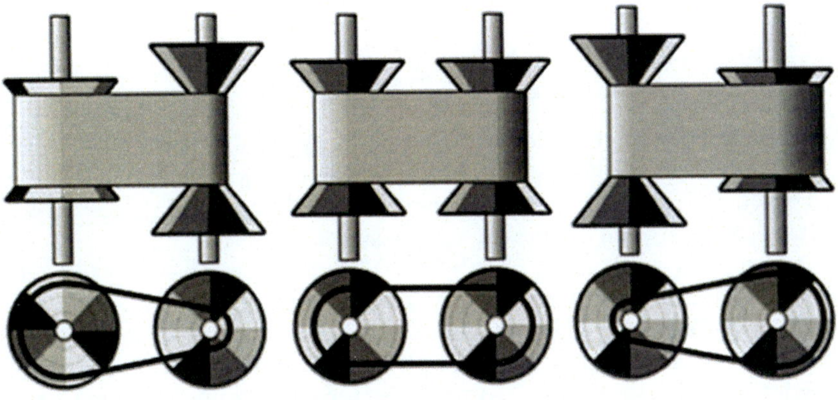

Fig. 4.9 Change in axial position of sliding cone disk and pitch radius

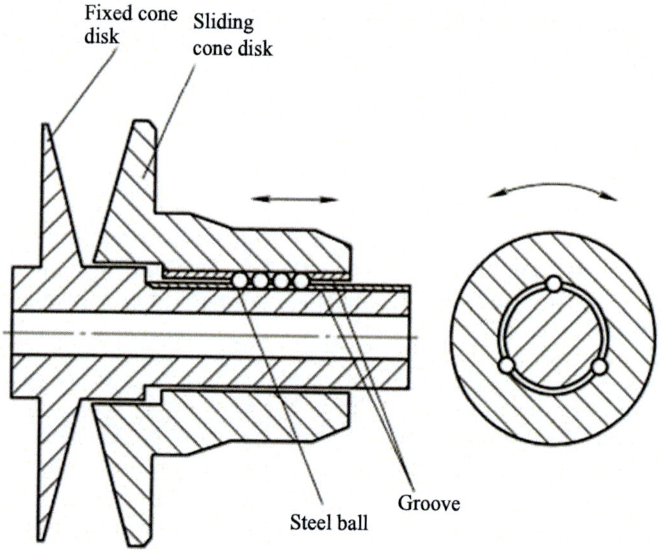

Fig. 4.10 Stress on ball spline and steel balls

Fig. 4.11 Stepped structure of shaft hub and radial pressure of steel balls

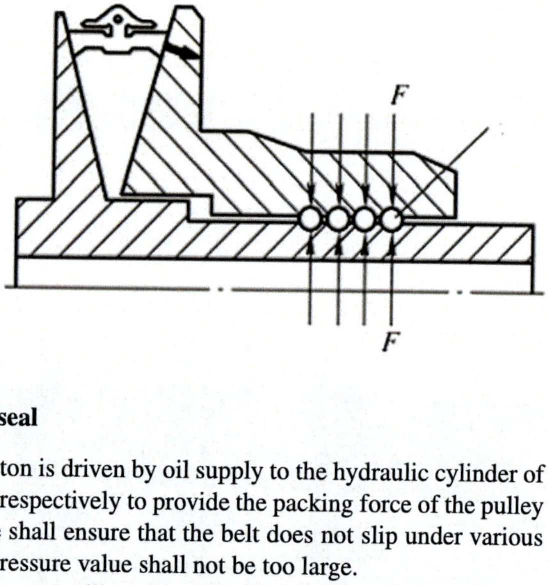

3. **Hydraulic cylinder and seal**

As shown in Fig. 4.13, the piston is driven by oil supply to the hydraulic cylinder of two sets of sliding cone disks respectively to provide the packing force of the pulley to the belt. The packing force shall ensure that the belt does not slip under various working conditions, and the pressure value shall not be too large.

To change the gear ratio by controlling the ratio of pitch radius is to adjust the pressure balance of the hydraulic cylinder pressure of the drive and driven pulleys.

The cone disk moves under the action of the pressure in the hydraulic cylinder. A hydraulic cylinder consists of a cylinder block and a piston. The cylinder block is integrated with the active wheel, and the oil seal is mounted on the piston rod. There is a certain clearance between the cylinder block and the piston, but the deformation

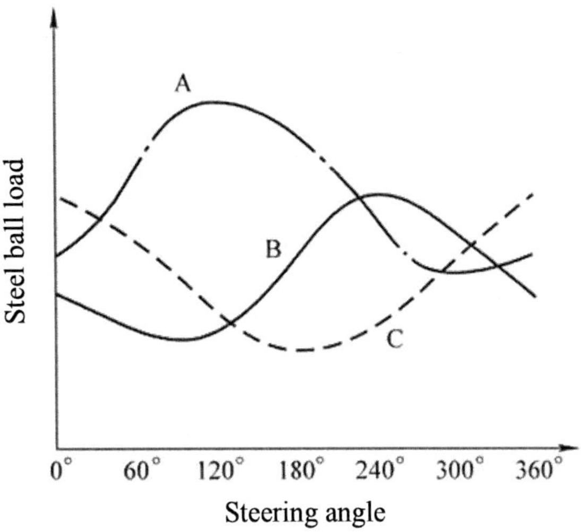

Fig. 4.12 Change law of radial pressure of steel balls

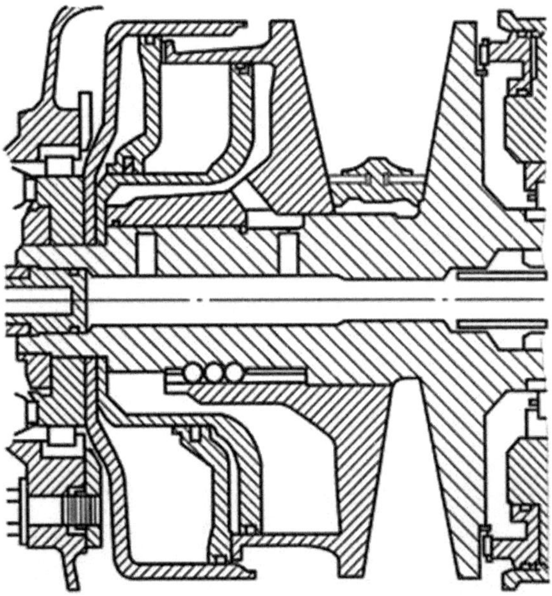

Fig. 4.13 Structure of hydraulic cylinder

of the pulley caused by a large bending moment on the pulley will cause the clearance between the cylinder block and the piston to change. If the static clearance is designed to be too small, the clearance can disappear after the pulley deformation, resulting in interference between the cylinder and piston and damaging the oil seal. If the clearance is too large, the oil seal will be pressed out in case of excessive pressure in the cylinder. Therefore, it is necessary to ensure that CVT pulley has sufficient stiffness to resist deformation, reasonably design the static clearance value, and select the appropriate oil seal material to meet the durability requirements.

4. **Deformation and stress of pulley**

Pulley produces a large bending moment under the action of the belt. The change of load and the deformation of the oil seal or belt caused by pulley deformation will affect the relationship between the oil pressure and gear ratio, so the stiffness of pulley shall be increased. The pulley does not require enough strength under a large alternating load, but the strength will be increased with the contour size and mass when the stiffness is increased. During the design process, the finite element method was used to analyze pulley of different structural forms. On the premise that the stress and deformation requirements were met, the design scheme with the minimum contour size and mass was selected.

5. **Pulley central plane dislocation and its elimination**

As shown in Fig. 4.14, when changing speed, the belt moves radially along the cone as well as along the axis. By arranging the two sliding cone disks of the drive and driven pulleys relative to each other, it can prevent the dislocation of the central plane of the drive and driven pulleys during the speed change.

However, the relative arrangement of the two fixed cone disks in the left and right only offsets part of the central plane dislocation in the speed change. When the belt length is constant and the gear ratio varies, the increment of the pitch radius of the

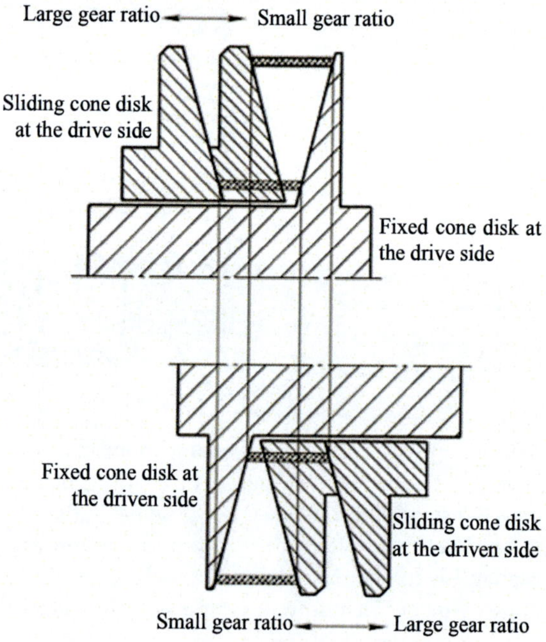

Fig. 4.14 Production mechanism of pulley central plane dislocation

drive pulley is not equal to the decrement of the pitch radius of the driven pulley, which causes the dislocation of the central plane of the pulley and the offset of the metal belt.

The relationship of the gear ratio i, fixed cone disk spacing U and offset δ is shown in Fig. 4.14. At high and low speeds, the offset δ is on the same side (−); when the gear ratio is 1:1, the offset δ is on the other side (+). In general, the shim is used to adjust the axial position of the fixed cone disk and change the spacing U of the two fixed cone surfaces so that the value of 0 is near the center of the variation range of the offset to optimize the offset.

The effect of offset on the service life of the belt is very significant. There are generally two methods to improve the dislocation: designing the tessellation lines of the cone and adjusting the axial position of the fixed cone disk. The former method has too high requirement for the cone disk processing technology and affects the contact between the cone and the steel plate, so the latter is often used to improve the dislocation. As shown in Fig. 4.15, the dislocation will change accordingly when the spacing U between two fixed cone disks is adjusted. This principle is usually used to optimize the position of dislocation 0.

In addition to the above gear ratio and the fixed cone disk spacing, the central plane dislocation is also affected by the bearing manufacturing errors, thermal expansion of the box or shaft, etc.

6. **Pulley wedge angle**

If a large pulley wedge angle is taken, the axial movement range of the cone disk will be increased, resulting in the width of the belt and the axial size of the pulley being too large. Moreover, the radial pressure of the pulley on the belt will also increase, resulting in increased tension and shortened life.

If a small pulley wedge angle is taken, the radial component of the packing force of the pulley on the belt will be too small, which may make it difficult to remove the

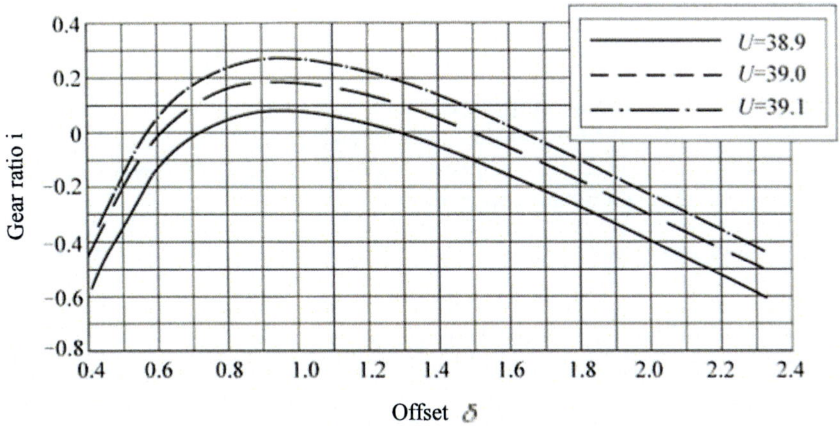

Fig. 4.15 Relationship of the gear ratio i, fixed cone disk spacing U and offset δ

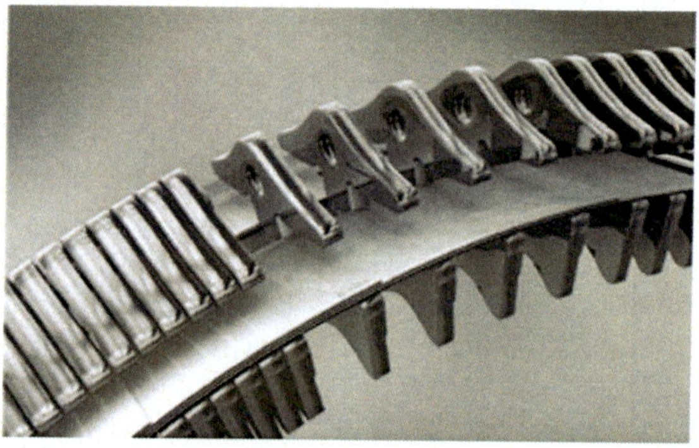

Fig. 4.16 Structure of VDT belt

belt from the pulley. Moreover, too small wedge angle will lead to the belt loose side tension smaller.

II. **VDT belt**

1. **Structure of VDT belt**

VDT belt is composed of steel plate and belt ring, as shown in Fig. 4.16. The thickness of the steel plate is about 2 mm. In order to prevent the dislocation of two adjacent steel plates, there are protrusions on one side and pits on the other side. The bevel angles on both sides match the pulley cone. Hundreds of steel plates are stacked when working, depending on the thrust between them.

The belt ring is nested by several seamless steel ring pieces with a thickness of only 0.2 mm, which are extremely tough and can withstand considerable bending stress cycles.

The belt ring composed of two sets of nested ring pieces is embedded in the slot on the left and right sides of the steel plate, and the steel plate is sandwiched in the middle, which is combined into a flexible drive belt that can be bent. The packing force of pulley on the belt is borne by the steel plate, while the radial component of the packing force is balanced by the tension of the belt ring.

In the circumferential direction, there is friction between the belt ring and the steel plate. The belt ring deforms at the intersection of the circular part and the straight part of the belt. If the belt ring is composed of a ring piece, a large bending stress will occur due to the excessive thickness of a single belt ring. Therefore, the steel belt is composed of several thin ring pieces.

2. **Torque transmission**

The ability of the VDT belt to transmit torque is determined by the friction factor, pitch radius, packing force, angle between the tessellation lines of the cone and bottom surface, etc. Generally, the friction factor is 0.08–0.10, and the included angle between the tessellation lines of the cone and bottom surface is about 11°.

In order to ensure that the belt does not slip, a safety factor shall be multiplied when designing the cone disk packing force, usually 1.2–1.5. If the value of safety factor is too small, the belt will still slip due to working conditions, machining error of parts, temperature change, friction surface deformation after continuous working and other factors; if the value of safety factor is too large, the friction loss would be so large that it would affect the transmission efficiency and the service life of the belt and pulley components.

The packing force F_s required for the torque transmission shall satisfy the following relationship

$$F_s > \frac{KT_e \cos\alpha}{\alpha\mu R} \tag{4.2}$$

where,

K—safety factor, usually 1.2–1.5;
T_e—transmitted torque;
μ—friction factor, generally 0.08–0.10;
α—included angle between the tessellation lines of the cone and bottom surface;
R—pitch radius.

In general, μ is 0.08–0.10 and K is 1.2–1.5.

When the belt drives the load to work, the belt will have loose side (upper of Fig. 4.17) and tight side (lower of Fig. 4.17). The VDT belt mainly transmits the torque relying on the thrust between the steel plates of the loose side, so it is called the compression belt drive. The force analysis of VDT belt is shown in Fig. 4.17. The torque calculation formula is as follows.

$$\left.\begin{array}{l}\text{Torque at the power input side } T_i = (T_i - T_s)\cdot\eta + Q\cdot\eta \\ \text{Torque at the power output side } T_o = (T_i - T_s)\cdot r_o + Q\cdot r_o\end{array}\right\} \tag{4.3}$$

Although this metal belt drive is known as compression belt drive, there is only a squeezing force between the steel plates. The loose side tension of the whole belt is

Fig. 4.17 Force analysis of VDT belt

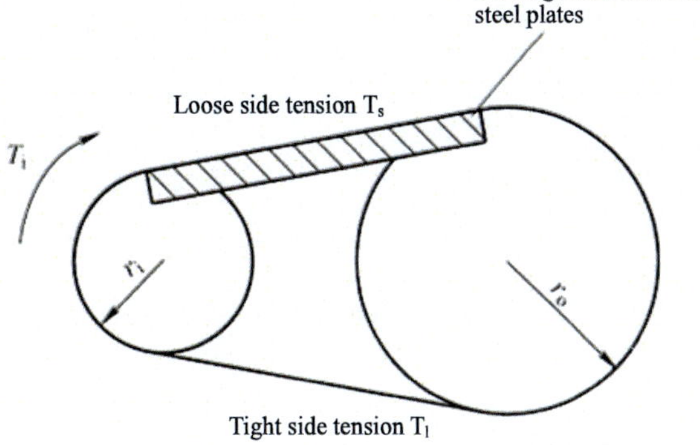

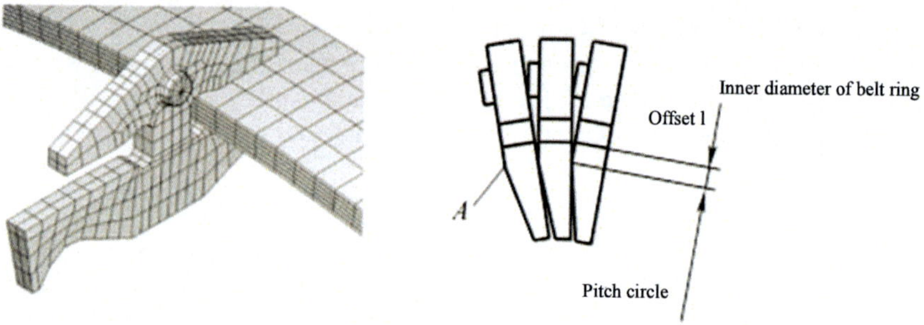

Fig. 4.18 Determination of pitch circle of VDT belt

the difference between the thrust of the steel plates and the loose side tension of the belt ring and must be positive.

The motion of the VDT belt is analyzed. When the belt enters the circular motion state, the steel plates are arranged into an arc. The steel plates could swing with point A in Fig. 4.18 as the fulcrum, and the distance from the fulcrum to the pulley center is the pitch radius. The gear ratio is determined by the pitch circle of the drive and driven sides.

The contact radius (inner diameter of the belt ring) between the steel plate and the inner layer of the steel belt offsets the pitch radius outward by 1 mm. The thrust between adjacent steel plates is very great, which requires high parallelism between the front and back of steel plates.

In order to ensure the smooth operation of the system, in addition to ensuring that the parallelism error between the front and back of the steel plates meets the requirements, it is also necessary to ensure that the contact area of each unit is as large as possible, which requires that the contact surface between the steel plate and the belt ring, as well as the 1 mm wide steel plate surface in contact between adjacent steel plates have high machining accuracy.

3. **Centrifugal force of belt**

Since each component of the VDT belt is made of metal, it has a large mass and a large centrifugal force is generated at high-speed revolution, i.e.

$$F_e = m_e \cdot R \cdot \omega_e^2 \tag{4.4}$$

where,

F_e—centrifugal force on a single steel plate (N);
m_e—mass of a single steel plate (kg);
R—turning radius of center of mass of steel plate (m);
ω_e—angular velocity of working pulley (rad/s).

The centrifugal force affects the friction between the pulley and the belt. Figure 4.19 shows the equilibrium relationship between the centrifugal force acting

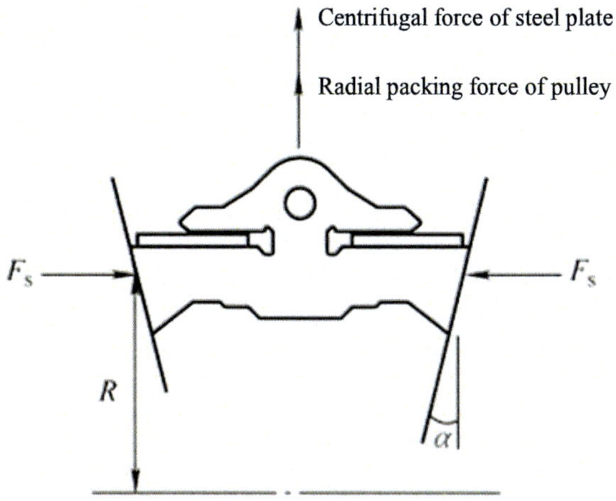

Fig. 4.19 Centrifugal force produced by steel plate

on the steel plate and the radial component of the packing force. It can be seen that under the action of centrifugal force, the radial component of the packing force will decrease, that is, the packing force will decrease, causing the decrease in the friction between pulley and the belt.

4. **Slip and friction moment in the belt**

When the gear ratio is 1:1, the steel plate and the belt ring rotate at the same angular velocity, and there is no slip between the steel plate and the innermost ring. When the gear ratio is not 1:1, the minor diameter side always slips.

As shown in Fig. 4.20, the tension of two pulleys on the belt is equal at the varying speed. Given the wrap angle θ on the minor diameter side, the wrap angle on the major diameter side is $\theta + 2\Delta\theta$. The central angle of the shaded sector in the figure is $2\Delta\theta$ and the force acting on the outer edge of this part is a pair of balance forces of the belt. The wrap angle θ on the minor diameter side is the same as that on the major diameter side minus the central angle of the shaded sector, so the packing force is the same on both sides of the angle θ. While on the major diameter side, the friction of the shaded part of $2\Delta\theta$ shall be provided and the packing force on this side is greater than that on the minor diameter side, so there is no slip on the major diameter side.

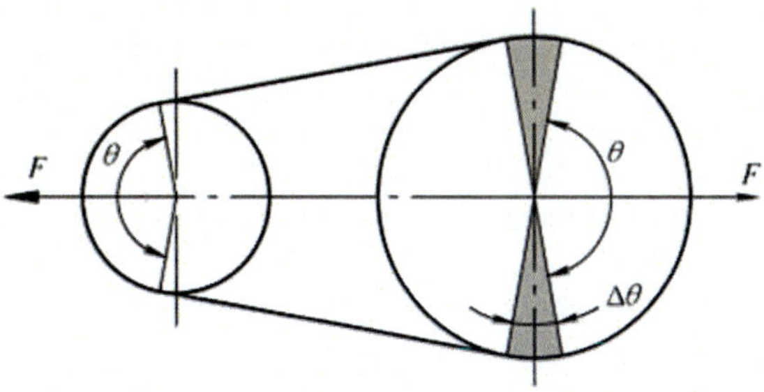

Fig. 4.20 Wrap angle and force at varying speed

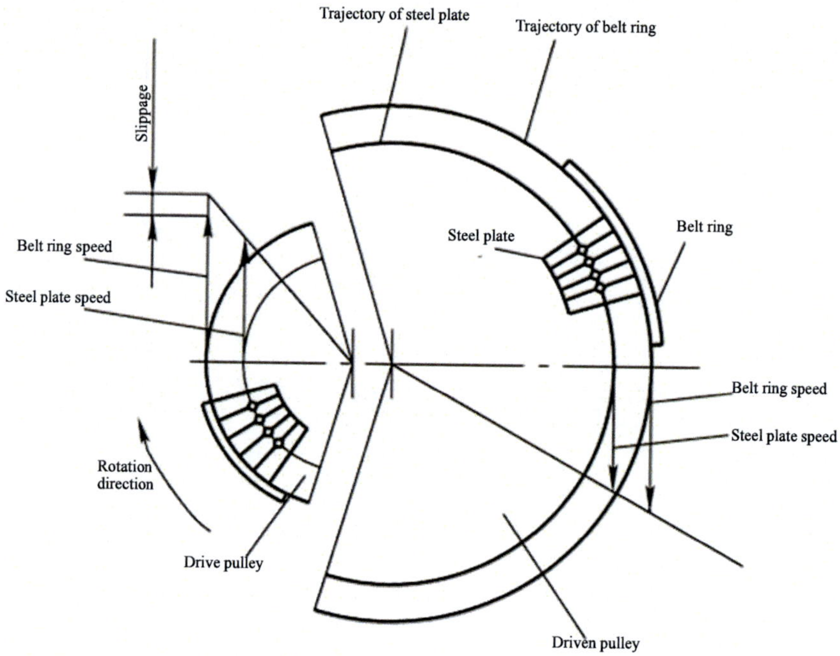

Fig. 4.21 Production mechanism of slip in the belt

The mechanism of the slip in the belt is shown in Fig. 4.21. It is assumed that the steel plate and belt ring speed on the major diameter side have been determined, and the slippage (speed difference) is

$$\Delta v = \Delta v_1 - \Delta v_2 = \left(\frac{2\pi n_p \cdot t}{t} - 2\pi n_p \cdot t \right) \cdot \frac{1}{60} \tag{4.5}$$

where:

n_p—speed of active pulley (r/min);
i—gear ratio;
t—offset between the steel plate and inner side of steel belt (m);
$\Delta v_1, \Delta v_2$—speed difference between the steel plate and the inner ring piece (m/s).

In case of slip, friction moment will be generated, which is equal to the sum of the product of slippage, slip surface pressure and friction factor between slip surfaces. That is

$$\text{Friction moment} \approx \Sigma(\text{slippage} \times \text{slip surface pressure} \times \text{friction factor}) \tag{4.6}$$

The slippage quantity is the least at constant speed and increases at high speed. The positive pressure on the slip surface is positively correlated with the packing force and negatively correlated with the centrifugal force, that is, the friction is affected by the packing force and the speed. Although the slippage between the ring pieces is small, the contact area is large, so the friction is greatly affected.

5. **Stress analysis of belt ring and steel plate**

In this metal belt drive, there are two kinds of proportional relations: the ratio of pitch radius and the ratio of belt ring radius, which determine two kinds of gear ratios respectively. In order to determine the true gear ratio of the metal belt drive, it is necessary to analyze the force relationship within the belt.

Generally speaking, the characteristics of belt drive are relative to the flat belt (that is, the tension belt composed of belt rings), while the V-belt (that is, the thrust belt composed of steel plates) has a larger torque capacity. Therefore, if the torque transmitted by the metal belt exceeds the torque capacity of the belt ring, the gear ratio should be determined by the ratio of pitch radius; otherwise, it should be determined by the ratio of belt ring radius.

As indicated above, slip occurs on the minor diameter side and not on the major diameter side during the operation of the metal belt; the inner ring radius is 1 mm larger than the pitch radius; the inner belt ring will slip on the minor diameter side. Therefore, it can be inferred that the steel strip on the minor diameter side has a higher angular velocity, and the steel strip is subject to a greater centrifugal force, so the tension in the belt ring here is greater. In the case of large gear ratio, if the transmitted torque is small, the steel plates squeeze each other at the lower side of the belt; if the transmitted torque is large, the steel plates squeeze each other at the upper side of the belt, as shown in Table 4.1. The test results show that when the torque increases, the slippage becomes larger and the gear ratio decreases by about 1%. Further increasing the torque will even cause the separation of adjacent steel plates, further reducing the gear ratio. In the case of small gear ratio, the tension of the upper side belt ring is relatively large, while the compression of the upper side steel plate is not affected by the transmitted torque, so the slip rate is relatively small when the gear ratio is large.

The reasons for belt ring stress are as follows:

(1) Belt ring stress caused by the belt tension from the packing force of the pulley.
(2) Belt ring stress caused by the loose and tight sides from the torque transmission.
(3) Belt ring stress caused by the slip friction between the steel plate and the belt ring.
(4) Belt ring stress caused by the centrifugal force of the steel plate.
(5) Belt ring stress caused by the belt bending.

The first four terms above are tensile stresses, and the ring piece is thick; the fifth item is bending stress, and the ring piece is thin.

If the central plane dislocation of two pulleys exceeds the allowed amount, the steel plate plane will have great dislocation from the tessellation lines of the cone and a great stress will generate inside the steel plate, as shown in Fig. 4.22. Excessive dislocation may even damage the steel plate. At this point, a bending moment also

Table 4.1 Relationship among load, gear ratio and belt tension

	The transmitted torque is greater than the torque capacity of the belt ring	The transmitted torque is smaller than the torque capacity of the belt ring	Note
Large drive ratio	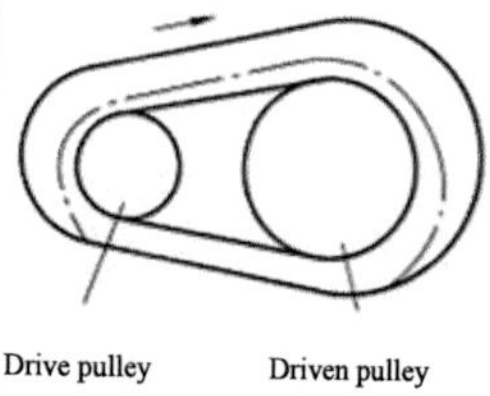 Both the belt ring and steel plate transmit torque	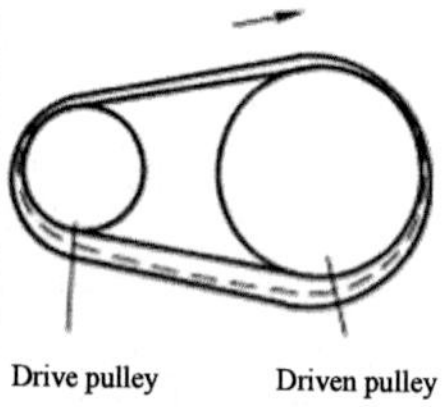 The belt ring transmits the torque and the steel plate acts as a hindrance	The lower side tension of belt ring is large The compression between the steel plates is transferred from the lower side to the upper side due to increased torque
Small drive ratio	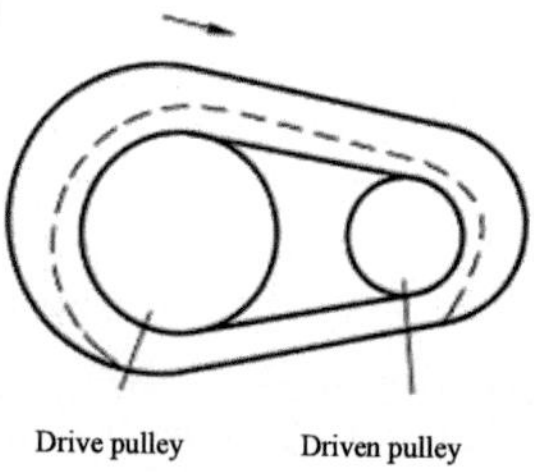 The steel plate transmits the torque and the belt ring acts as a hindrance	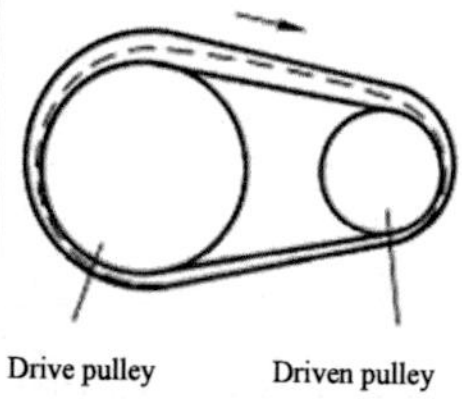 The steel plate transmits the torque and the belt ring acts as a hindrance	The lower side tension of belt ring is large The compression between the steel plates is always on the upper side

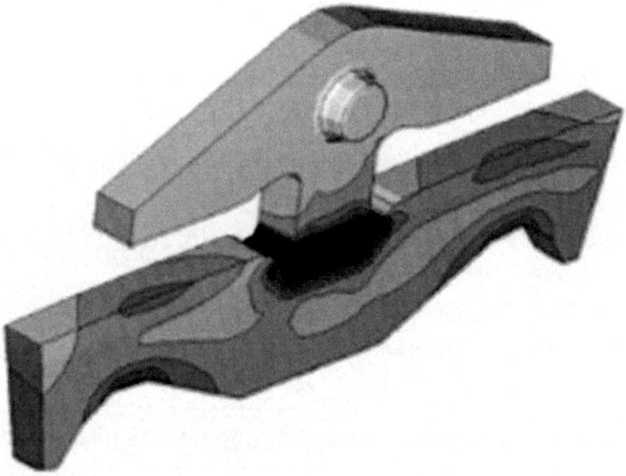

Fig. 4.22 Steel plate stress under uneven force

acts on the steel plate in addition to the cone disk pressure. Uneven force may cause one or a few steel plates to assume the full packing force of the pulley, resulting in considerable stress on the steel plate.

In order to reduce the influence of this phenomenon, it is necessary to ensure that the plane of the steel plate and the tessellation lines of the cone are coplanar as far as possible, and conduct the test at the maximum speed of each gear ratio. In the design, the variable speed dislocation, assembly error, axial jitter, thermal expansion and other factors should be considered to control the dislocation within the allowable range.

6. **Belt noise**

The noise frequency is proportional to the speed of the belt and inversely proportional to the pitch. The pitch of the VDT belt is the distance between two adjacent steel plates and the pulley junction point.

Although the pitch of VDT belt is small, the displacement transfer error reaches 90 m at the pitch radius of 35 mm, and the noise is obvious. However, the force on the metal belt is more uniform than that on the gear; the belt is less rigid than the gear, which is beneficial to the vibration absorption; moreover, the large rotational inertia of the pulley can effective reduce the vibration noise.

The method to reduce the noise is to change the thickness of the steel plate and to control the noise frequency in the non-sensitive zone by analyzing the mode.

4.4 Composition and Principle of Hydraulic Control System

I. **Composition of hydraulic control system**

CVT is an electronically controlled hydraulic automatic transmission and its hydraulic system consists of the power element, the actuator and the control mechanism. The power element is a hydraulic pump, generally gear pump or vane pump; the actuator includes the clutch, brake and pulley hydraulic cylinder in the CVT; the control mechanism includes pressure regulator valve, solenoid valve, shift valve and lockup clutch control valve, as shown in Fig. 4.23.

Some oil circuits and hydraulic control elements in Fig. 4.21 are integrated into a control valve body (main valve body), as shown in Fig. 4.24. The main valve body contains oil channel, valve element, solenoid coil or stepping motor, accumulator, separator, flowmeter and other components. The main valve body is a two-piece combination to meet the requirements of CVT complex oil circuit layout. The valve body has high requirement for the design and processing level: necessary oil amount and pressure are guaranteed in the oil circuit, so that the hydraulic oil can be reliably transmitted to the actuators, which also puts forward higher requirements for sealing; the oil channel shall have reasonable width; the profile size and mass of the valve body are limited by space and lightweight requirements. Moreover, the valve body

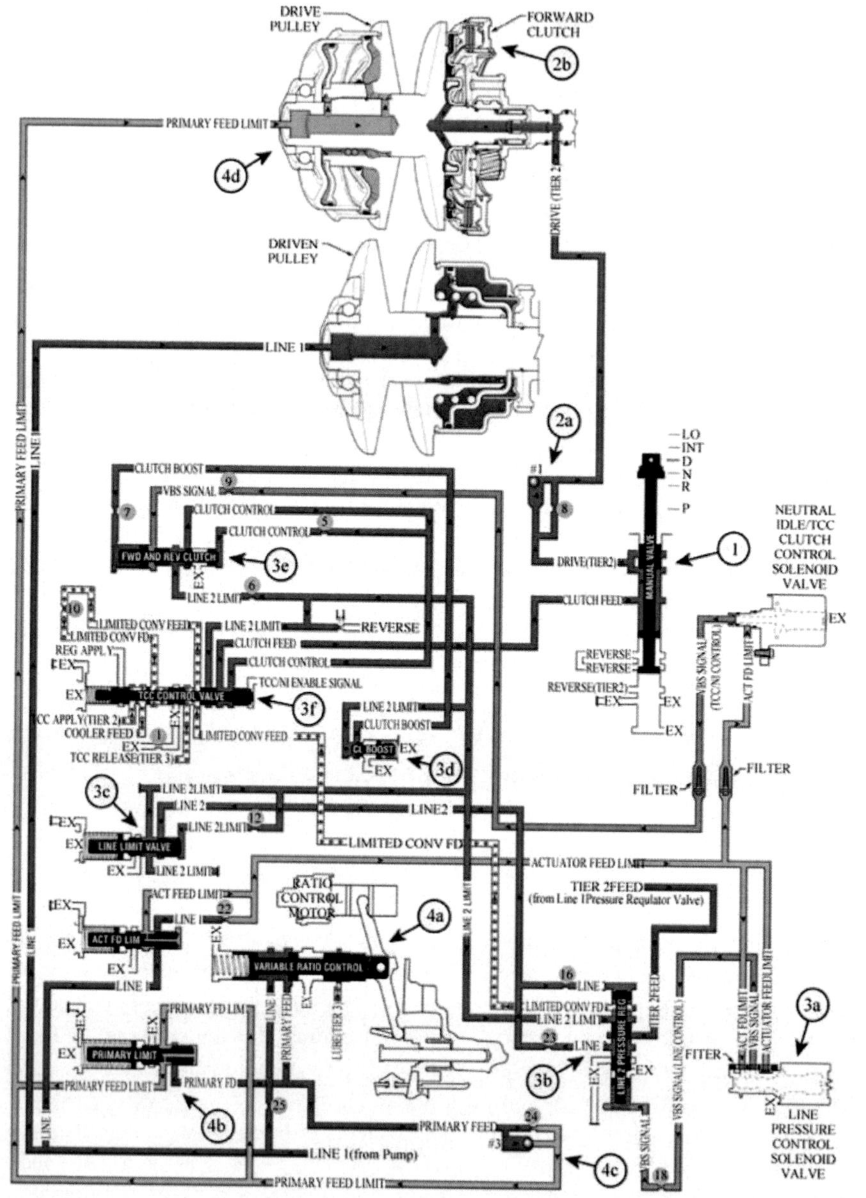

Fig. 4.23 Composition of CVT hydraulic control system

Fig. 4.24 Main valve body

processing technology shall be satisfactory and easy for disassembly, repair and adjustment.

II. **Working principle of CVT hydraulic control system**

The CVT hydraulic control system mainly realizes the following functions:

(1) Provide a proper packing force for the drive and driven pulleys to prevent the belt from slipping.
(2) Adjust the balance of the packing force of the drive and driven pulleys and get the required gear ratio.
(3) Control the operating of the clutch and brake in the CVT to achieve forward and reverse.
(4) Control the start and the hydraulic torque converter lockup.
(5) Lubrication and cooling.

Its working principle is described below according to the functional requirements.

1. **Packing force control to prevent belt slip**

Insufficient packing force of the pulley will lead to slip, thus intensifying the friction surface wear, even damaging the contact surface and reducing the belt life. To achieve sufficient torque capacity of the CVT, the hydraulic system shall provide sufficient packing force for the pulley. The packing force of the drive and driven pulleys must satisfy the in Eq. (4.2) to prevent the belt from slipping.

The packing force of the drive and driven pulleys is provided by the respective hydraulic cylinder. The hydraulic cylinder of the driven pulley is provided with a return spring and an oil chamber to eliminate the centrifugal effect, as shown in Fig. 4.25.

The packing force of the working pulley is

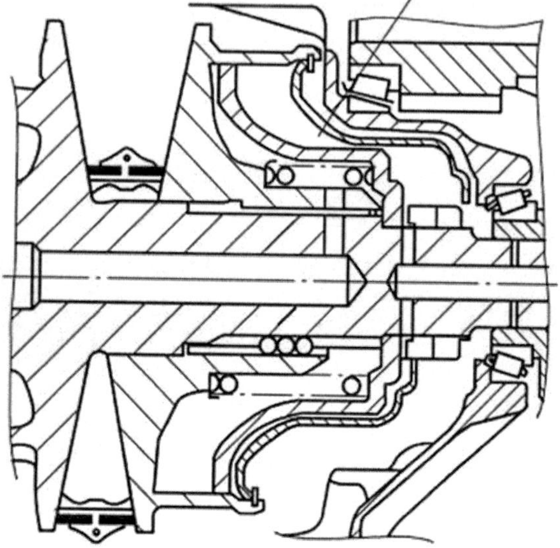

Fig. 4.25 Oil chamber to eliminate the centrifugal effect

$$F_s = F_p + F_b + F_c \tag{4.7}$$

where:

F_p—packing force provided by hydraulic force;
F_b—packing force provided by the spring;
F_c—packing force caused by centrifugal effect.

The friction at the oil seal, ball spline, sliding cone disk hub and other locations can also cause changes in the packing force of the working pulley, but such friction is small and can usually be ignored.

(1) Packing force provided by hydraulic force: the force driving the cone disk is mainly provided by the hydraulic force. By controlling the pressure in the hydraulic cylinder, the belt is guaranteed not to slip and the gear ratio is adjusted. The packing force provided by hydraulic force is

$$F_p = S_p \times p \tag{4.8}$$

where,

p—in-cylinder pressure (PA);
S_p—piston area, with the formula as

$$S_p = \frac{\pi(d_o^2 - d_i^2)}{4} \tag{4.9}$$

where,

d_o—piston outside diameter (m);
d_i—piston inside diameter (m).

For the double-piston hydraulic cylinder, the packing force can be divided into two parts, i.e.

$$F_p = F_{p1} + F_{p2} \tag{4.10}$$

where, F_{p1}, F_{p2}—packing force (N) provided by two pistons respectively of the double-piston hydraulic cylinder.

Under the same packing force, the larger the piston area, the lower the oil pressure required. Although the hydraulic pump is required to increase the pumping amount under large piston area, the pumping pressure is reduced, the leakage reduced and the driving force and size of the hydraulic pump are also reduced. For the overall reducer, the friction loss will also be reduced.

Due to the limitation of layout space, a smaller double-piston hydraulic cylinder is usually used on the side of the drive cylinder. However, the single-piston hydraulic cylinder is used in fact to make the structure simple. Therefore, the hydraulic cylinder in the CVT of the low-emission vehicles is generally a single-piston hydraulic cylinder.

(2) Packing force provided by the spring: there is no oil pressure in the loop at downhill in case of engine stall, but the belt is not allowed to slip and the interference caused by the relaxation of the belt shall be avoided. To keep the belt tensioned, a spring is installed in the hydraulic cylinder to provide a portion of packing force, while also reducing the burden on the hydraulic system, reducing the driving force of the hydraulic pump and reducing the friction loss of the transmission.

(3) Packing force caused by centrifugal effect: the hydraulic cylinder is filled with the ATF. When working, the ATF rotates with the hydraulic cylinder and the additional centrifugal pressure generated by the centrifugal effect will become part of the pulley packing force. The centrifugal pressure of the hydraulic torque converter can even cause the casing to expand, affecting the internal pressure in the forward/reverse clutch piston. Therefore, in terms of the CVT design, the influence of centrifugal force is not negligible. The packing force caused by centrifugal effect is

$$F_c = \frac{\pi p \cdot (d_o^2 - d_i^2)^2}{4} \tag{4.11}$$

where,

ρ—oil density (N/m^3);
ω—speed (rad/s);
d_o—outside diameter of hydraulic chamber (m);
d_i—inside diameter of hydraulic chamber (m).

It can be seen from the above equation that if the speed is too high, the centrifugal force may cause the packing force to be in an excessive state, and the friction between the belt and the pulley increases, leading to the decrease of the service life. Moreover, the pressure change caused by the centrifugal force will destroy the balance of the packing force of the drive and driven pulleys, resulting in the failure to obtain the expected gear ratio. To eliminate the effect of centrifugal action, an oil chamber to eliminate the centrifugal effect, as shown in Fig. 4.23, is provided in the hydraulic cylinder of the driven pulley, which is usually at a high speed.

Even with the above oil chamber, it is structurally impossible to reduce d_o to the same level as d_i, that is, centrifugal force cannot be completely eliminated. Therefore, the accurate adjustment of pulley packing force to the target value depends on electronic control, and the speed and gear ratio are taken into account to appropriately reduce the oil pressure.

(4) Regulation of line atmospheric pressure: the atmospheric pressure regulator valve of the pipeline is shown in Fig. 4.26. The high-speed on-off solenoid valve is used as the pilot valve, and the pulse width modulation (PWM) technology is applied to get the low pressure signal to change the valve position of the regulating valve to control the line atmospheric pressure. The spring force and signal pressure act on the right side of the regulating valve stem; the hydraulic pressure related to the line atmospheric pressure acts on the left side. If the force on the right side is large, the stem is pushed to the left, the unloading loop is closed, and the line atmospheric pressure is increased. When the line atmospheric pressure is too high, the valve stem is pushed to the right, the unloading loop is opened, and the line atmospheric pressure drops.

The line atmospheric pressure is adjusted according to

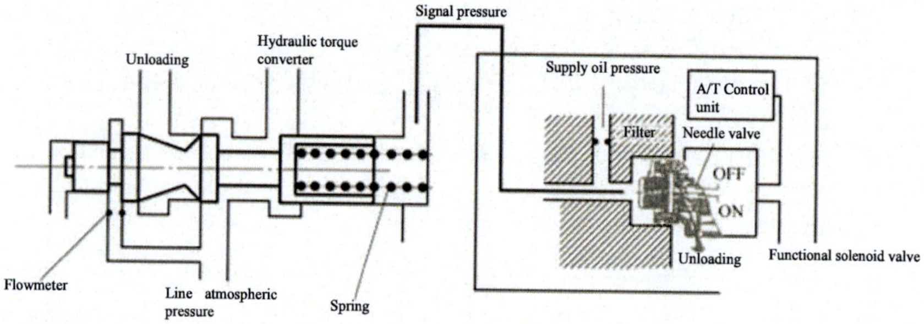

Fig. 4.26 Line atmospheric pressure regulator valve

$$P_L = \frac{F_s + P_S S_S}{S_L} \tag{4.12}$$

where,

P_L—line atmospheric pressure (Pa);
F_S—spring force (N);
P_S—signal pressure (Pa);
S_S—action area of signal pressure (m^2);
S_L—action area of line atmospheric pressure (m^2).

2. **Hydraulic control of gear ratio regulation**

To change the gear ratio of belt drive, it is necessary to adjust the balance of the packing force between the drive and driven pulleys.

Packing force formula

$$F_x = \frac{T_e \cos \alpha}{2\mu k r_x} \tag{4.13}$$

where,

F_x—packing force of pulley (N);
α—included angle between the tessellation lines of the cone and bottom surface (°);
μ—friction factor between pulley and steel plate;
k—distribution coefficient of friction in circumferential tangential direction and normal direction, $0 \leq k \leq 1$;
r_x—pitch radius (m).

Meanwhile, the pitch radius on both sides satisfies the following relation

$$\begin{cases} 2A\cos\dfrac{\pi-\theta}{2} + r_1\theta + r_2(2\pi-\theta) = L \\ A\sin\dfrac{\pi-\theta}{2} = r_1 - r_2 \\ r = r_2/r_1 \end{cases} \tag{4.14}$$

Based on the above relation, the packing force of the two pulleys can be determined respectively when the target gear ratio and load are known.

(1) Regulation characteristic of pulley packing force: in the actual working process, the tension at the belt connection to the drive pulley is large and the operating radius tends to decrease; on the side of driven pulley, the tension at the belt connection is small and the operating radius tends to increase. If no adjustment is made, the pitch radius on the drive pulley side will decrease with the rotation angle, while the pitch radius on the driven pulley side will increase with the rotation angle, resulting in a smaller gear ratio. In order to prevent the gear ratio

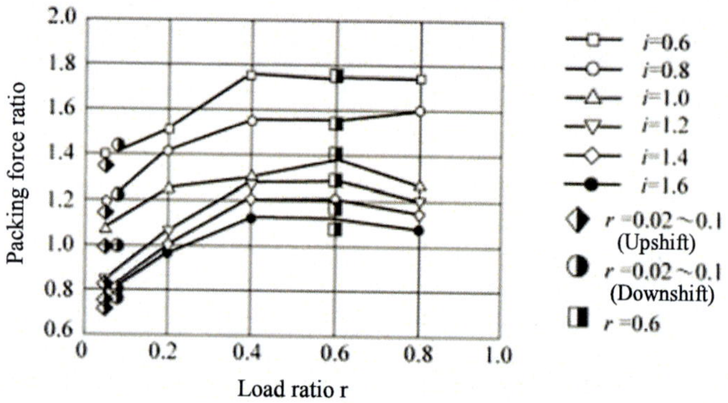

Fig. 4.27 Regulation characteristic of packing force

from changing, it is necessary to increase the packing force on the drive side and reduce the packing force on the driven side, so as to increase the gear ratio and balance the decreasing trend of gear ratio. This trend will become more significant as the load increases.

Figure 4.27 shows the regulation characteristic of the pulley packing force on both sides to obtain stable gear ratio under different loads. It is affected by the load, the belt stiffness, geometric profile and many other complex factors., Meaning of load ratio in the abscissa axis: 0 is no load and 1.0 is the state in which the belt happens to slip; the packing force ratio in the ordinate axis indicates the ratio of the packing force of the drive pulley to that or the driven pulley. To obtain the stable gear ratio, the packing force of the drive pulley shall be increased according to the load.

(2) Pressure provided for each pulley: to make the packing force of the drive and driven pulleys meet the gear ratio regulation requirements, the hydraulic system shall supply the corresponding pressure to the hydraulic cylinders of two pulleys respectively. In terms of hydraulic function, the pressure of the hydraulic cylinder of the pulley can be considered as the sum of the line atmospheric pressure and shift pressure. Adjusting the gear ratio is to adjust the hydraulic cylinder pressure of two sliding cone disks based on the premise of non-slip.

There are two common speed regulation modes:

(1) Single pressure regulation. The driven pulley supplies the line atmospheric pressure and the drive pulley supplies the shift pressure. In this mode, the speed is regulated only by changing the pressure of the pulley on one side.
(2) Dual pressure regulation. The case is the same as the single pressure regulation in the case of small gear ratio. The pressure supply mode is reversed in the case of large gear ratio.

The comparison of two pressure regulation modes is shown in Table 4.2.
In general, the dual pressure regulation is better and more widely used.

Table 4.2 Comparison of two pressure regulation modes

	Oil pressure of drive pulley	Oil pressure of driven pulley	Advantages	Speed range
Single pressure regulation	Shift pressure	Line atmospheric pressure	(1) The pressure regulator valve structure is relatively simple (2) The piston area on the drive side is large when the line atmospheric pressure is not restricted (3) Reduced oil pressure and high economical efficiency in the case of large gear ratio	Full gear ratio range
Dual pressure regulation	Shift pressure	Line atmospheric pressure	(1) Little difference in the piston area of two pulleys (2) Large piston area on the driven side and reduced oil pressure (3) Achieve high economical efficiency	Small drive ratio
	Line atmospheric pressure	Shift pressure		Large drive ratio

(3) Hydraulic circuit that controls the variable speed: the piston area ratio is closely connected with the pressure supply mode of the two hydraulic cylinders mentioned above.

In the single pressure regulation, the driven side is set as the line atmospheric pressure. At this time, the piston area on the drive side shall be at least twice that on the driven side, so that the enough packing force can be generated on the drive side and the gear ratio can be adjusted to the maximum. The piston area ratio is usually set at about 2.1.

In the dual pressure regulation mode, it is free to select the piston area ratio of two pulleys and it is often taken 1.6 or 1 in the actual application. When 1.6 is taken, the oil pressure at large gear ratio (small load) can be reduced to improve economical efficiency. When 1 is taken, the piston of two pulleys can be generalized.

Generally speaking, it is better to adopt dual regulation mode and choose area ratio 1 than 1.3–1.6.

To change the gear ratio, the balance of the packing force of the pulleys on both sides should be adjusted. The variable speed control can be achieved by the solenoid valve or stepping motor to control the gearshift control valve. As mentioned above, there are two modes of pressure regulation: single pressure regulation and dual pressure regulation. In this way, the following four pressure regulation modes can be combined.

(1) Single pressure regulation stepping motor control. The oil pressure on the driven side is line atmospheric pressure to prevent the belt from slipping. The structure of the drive side is shown in Fig. 4.28.
The stepping motor and the sliding cone disk are connected by a connecting rod, with the middle of the connecting rod connected to the gearshift control valve stem. The gearshift control valve is a slide valve.
The process of adjusting to low gear ratio (high speed) is explained below. The TCU electronic signal controls the push rod of the stepping motor upward and drives the connecting rod to move the gearshift control valve stem upward; at this time, the hydraulic cylinder is connected to the line atmospheric pressure circuit, the internal pressure of the hydraulic cylinder of the drive pulley increases, and the sliding cone disk moves downward to make the pitch radius larger; after the sliding cone disk moves down, the connecting rod will be driven to make

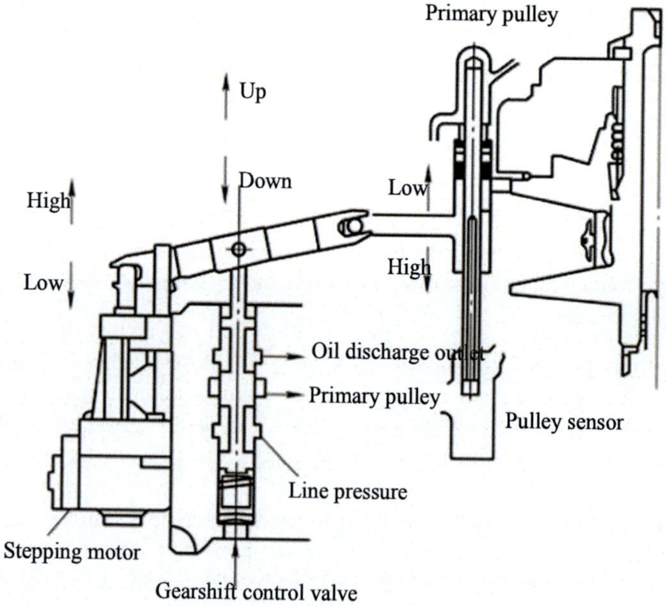

Fig. 4.28 Single pressure regulation stepping motor control

the gearshift control valve stem move down. When the sliding cone disk moves to make the valve position in the holding position, the hydraulic cylinder is disconnected from the line atmospheric pressure circuit, and the shift process will end.

If the stepping motor position remains unchanged and the gear ratio becomes smaller due to some external reasons, that is, the sliding cone disk on the drive side moves down, the connecting rod will be driven to pull the gearshift control valve stem move down; at this time, the hydraulic cylinder is connected to the unloading loop, the pressure in the cylinder drops, the sliding cone disk returns to the left side to restore the set gear ratio and restore the holding state.

The stepping motor control can be used to make the hydraulic system work smoothly and respond quickly without any speed deviation.

(2) Dual pressure regulation stepping motor control. The dual pressure regulation stepping motor control has basically the same working principle with the single pressure regulation stepping motor control. The difference is that the dual pressure regulation stepping mode control has two sets of hydraulic circuits, which are connected with the drive and driven pulleys respectively; in the pressure regulation, the pressure of the hydraulic cylinder on one side is changed and the other side is set to the line atmospheric pressure; the pressure of the pulleys on both sides is adjustable.

This circuit can achieve speed regulation by reducing the driven pulley pressure.

(3) Single pressure regulation solenoid valve control. The hydraulic cylinder on the driven side is kept at the line atmospheric pressure to prevent the belt from slipping. The gear ratio is adjusted by changing the pressure in the hydraulic chamber on the drive side.

The composition and working principle of the single pressure regulation solenoid valve control circuit are shown in Fig. 4.29. The gearshift control valve can be controlled by the electromagnetic signal, so that the circuit pressure is arbitrary value less than line atmospheric pressure.

Compared with the stepping motor control form, the solenoid valve control can cancel the linkage and make the structure simple and compact.

(4) Dual pressure regulation solenoid valve control. The dual pressure regulation solenoid valve control circuit is shown in Fig. 4.30. The pressure regulator valve adjusts the hydraulic circuit pressure to prevent the belt from slipping. The combination of two groups of solenoid valve and slide valve controls the sliding cone disk on the drive and driven sides respectively, and the pressure of the pulleys on both sides is adjusted respectively according to the TCU instructions. In the dual pressure regulation mode, it is free to select the piston area ratio of two pulleys.

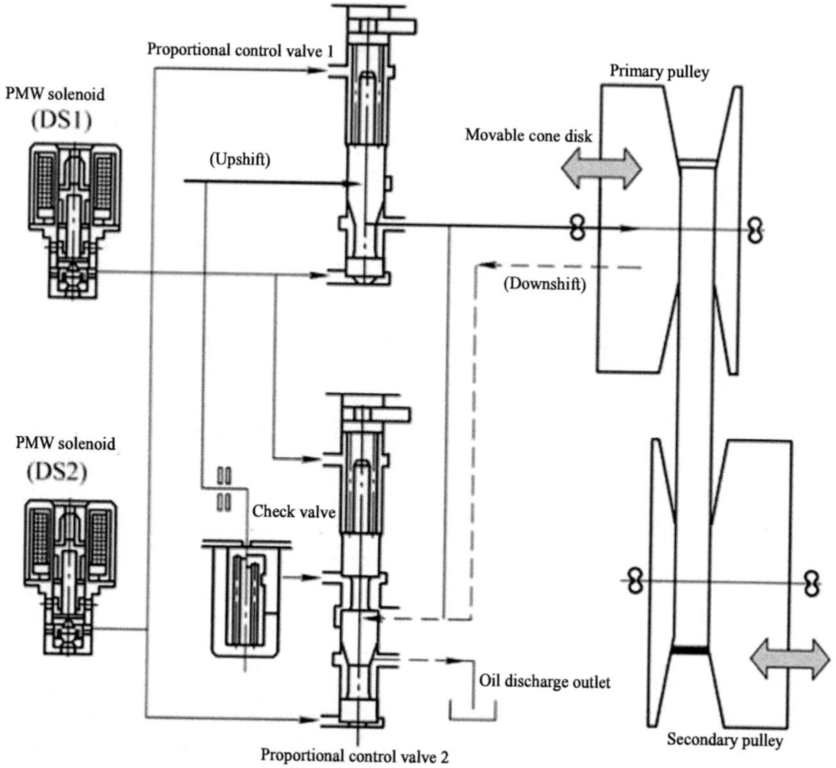

Fig. 4.29 Single pressure regulation solenoid valve control circuit

4.5 CVT Electronic Control System

I. Development and application of CVT

Although the number of gears of the modern electronic automatic transmission is increasing and has changed from the traditional 4 speed to 5, 6 and 7 speed, it is still a stepped transmission (with only a few fixed gear ratios). To achieve flexible driving, low fuel consumption and low noise, the more gears the better. A further extension of this idea is the CVT, which can realize the stepless change of gear ratio.

Audi, Nissan, Honda and many other models have adopted ECVT and achieved manual/stepless integration. The emergence of ECVT has made a qualitative leap in the dynamic and economy performance and handling stability of the vehicle, but there are also some problems that need to be improved, such as drive impact and noise, start control, variable speed control, drive belt (pulley) life problems.

Currently, the common domestic models that adopting the ECVT include Audi A6, Honda Fit and Cowin.

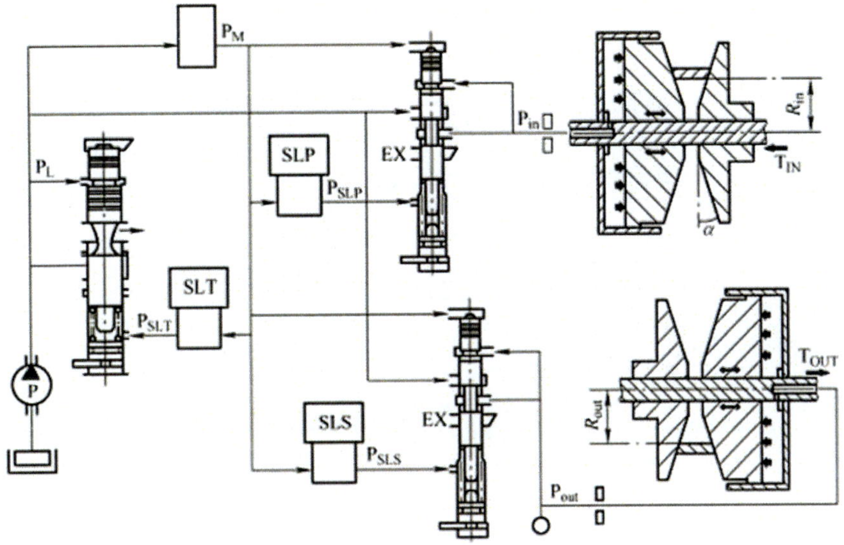

Fig. 4.30 Dual pressure regulation solenoid valve control circuit

1. **Multitronic stepless/manual transmission of Audi A6**

Audi Multitronic transmission is installed with a drive group called multi-disk chain belt on the basis of the original CVT. This group greatly expands the use of the CVT, transmits and controls the power output with a peak value of up to 310 N m and its gear ratio exceeds the limits of previous automatic transmissions. Multitronic also uses a new electronic control system to overcome the shortcomings of the original CVT. For example, when going uphill or downhill, the system can automatically detect the slope and assist the vehicle by adjusting the speed ratio to increase the power output or increase the braking torque of the engine.

2. **Honda Fit CVT**

Honda Fit CVT is a new-generation steel belt CVT designed for small cars and allows for high torque transmission between two pulleys. With smooth operation and high transmission efficiency, it is a better type of transmission in small cars.

3. **Cowin CVT**

VT1F CVT produced by ZF is used as Cowin CVT. Together with its outstanding engine, this powertrain is from the BMW MINI Cooper. The CVT has four driving modes: continuously variable transmission, automatic cruise, sports mode and 6-speed manual mode.

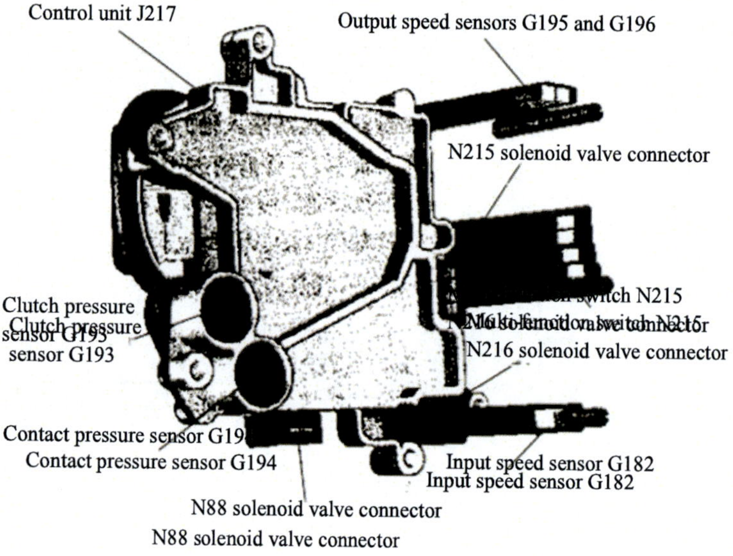

Fig. 4.31 Audi 01JCVT electronic control system

II. Composition and control principle of electronic control system

Figure 4.31 shows the Audi 01JCVT electronic control system, which is mainly composed of an electronic control unit, an input device (sensor, switch) and an output device (solenoid valve). The electronic control unit is integrated in the speed ratio converter, and the control unit is directly bolted to the hydraulic control unit. Three pressure regulator valves are directly connected to the control unit through the plug (S joint) without connecting wires. The control unit is connected to the automotive wiring harness by a small 25-pin plug. The electronic control system is characterized by its integrated sensor technology in the control unit and its housing accommodates all the sensors, so the wire harness and plug are no longer required. This structure has greatly improved the work efficiency and reliability. In addition, the engine speed sensor and multi-function switch are designed as Hall sensor, which is free from mechanical wear and electromagnetic interference, further improving its reliability. Its disadvantage is that the sensor is an integrated part of the control unit. If a sensor is damaged, the whole electronic control unit must be replaced.

Audi 01JCVT electronic control unit has a dynamic control program (DRP) that calculates the rated transmission input speed. In order to obtain the best gear ratio for each driving state, the input information of the driver and the actual working state of the vehicle need to be taken into account. The rated input speed of the transmission is calculated according to the boundary condition dynamic control program (DRP). The transmission input speed sensor G182 monitors the actual speed at the drive sprocket. The electronic control unit will compare the actual value with the set value and calculate the control current of the pressure regulating solenoid valve N216, so that N216 will generate the control pressure of the hydraulic gearshift control valve, which is almost proportional to the control current. The control unit monitors the shift by checking the transmission input speed sensor G182, transmission output speed sensor G195 and the engine speed signal.

III. Overhaul of ECVT

(I) Maintenance requirements and precautions

(1) When the engine is running, be sure to engage the gear shift lever in the P position and pull the parking brake before repairing the vehicle to prevent accident.
(2) Du not open the throttle by mistake after engaging D gear when the vehicle is stationary (e.g. accidentally touch the throttle for the operation in the engine compartment). In this case, the car will start running immediately and cannot be stopped to move even if the parking brake is pulled.
(3) Ultrasonic cleaning devices are not allowed to clean the hydraulic and electronic control units.
(4) Never start the engine or drag the vehicle when the cover is removed or ATF is not added.
(5) If repairs are not made immediately, open parts must be carefully covered or sealed.

(II) Maintenance and overhaul methods

1. Maintenance method

Maintenance work mainly includes checking CVT for ATF leakage, lubricating oil level of the CVT and main reducer, adding and replacing CVT ATF.

(1) Check the ATF level

1) Preconditions for checking ATF level:
   ① The transmission is not allowed to be in emergency operation.
   ② The vehicle must be in a horizontal position.
   ③ Connect the vehicle diagnosis, measurement and information system VAS5051, and then select the vehicle self-diagnosis and vehicle system "02-transmission electrical equipment".
   ④ The engine must be idle.
   ⑤ The air conditioning and warm air must be turned off.
   ⑥ The ATF temperature is not allowed to exceed 30 °C before the inspection, and the transmission is cooled first if necessary.

2) ATF level inspection steps:
   ① Read the ATF temperature in the vehicle diagnosis, measurement and information system VAS5051 and operate at the transmission temperature 30–35 °C.
   ② When the engine is idle, press the brake pedal and stop at all gears (P, R, N, D), idle the engine for about 2 s at each position, and finally engage the gear shift lever in the P position.

③ Lift the vehicle and unscrew the check bolt to check whether ATF has spilled from the check hole. If so, add ATF.

(2) Replace ATF

1) Open the drain bolt at the bottom of the transmission to drain out the old ATF.
2) Open the ATF filling bolt at the bottom of the transmission and add the new ATF to the transmission using a special ATF filler.
3) Check the ATF level.

2. **Overhaul method**

(1) Inquiry: inquire the owner to diagnose the source of fault information, confirm the time of fault occurrence and fault symptoms, etc., which is the first step of fault maintenance.
(2) Basic check: mainly some peripheral checks, including engine idle, ATF level, oil quality and gear shift control mechanism.
(3) Self-diagnostic examination: The CVT electronic control system, with the function of fault self-diagnosis, can indicate the fault by flashing the fault indicator light, and store the fault in the control unit. The fault indicator light of 01JCVT has three fault diagnosis modes: minor fault, the fault indicator light is normal and the driver can sense the fault according to the vehicle running condition; general fault, the fault indicator light is fully on, the alternative program makes the vehicle driving and the fault is stored; serious fault, the fault indicator light is fully on and flashing, the fault is stored and the repair is required immediately. The fault indicator light can be used for preliminary diagnosis. If any fault is stored, a special detector can be applied to retrieve the fault code and repair according to the maintenance instructions.
(4) Overhaul of electronic hydraulic control system: some CVT hydraulic control systems can check the fault cause directly through the oil pressure test. Most CVT hydraulic systems use the oil pressure sensor to reflect the operating oil pressure inside the transmission, so a special detector must be used to further confirm the fault information by reading the dynamic data from the running state of the vehicle. The hydraulic control elements (valve body) and hydraulic actuators (clutch or brake) can be hydraulically tested and disassembled.
The troubleshooting of the CVT electronic control system is almost the same with that of other electronic automatic transmissions. The fault guidance function of the special detector may be used for the fault code analysis, dynamic data flow analysis, waveform analysis, computer circuit and network data communication analysis, electronic element (sensor, switch and solenoid valve) testing and replacement, so as to achieve troubleshooting.
(5) Overhaul of mechanical elements: as for CVT mechanical elements, only the disassembling inspection or faulty part repair and replacement can be carried out.

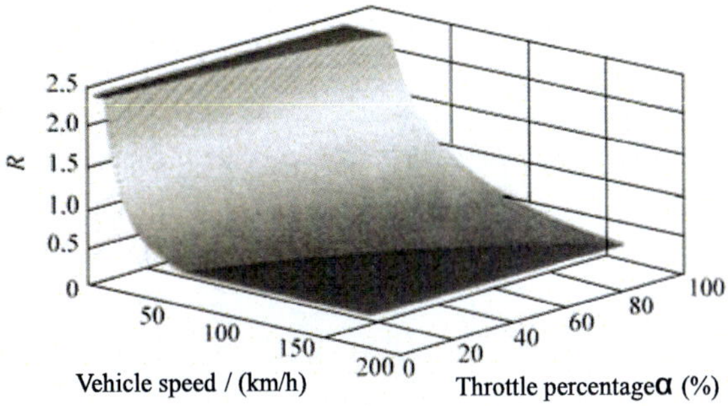

Fig. 4.32 Speed ratio tracking control strategy

4.6 Control of CVT

The control items of CVT can be divided into the following two categories according to the control structure: items (1)–(5) are the gear ratio control of the belt drive part; items (6)–(8) are the control of components other than belt drive.

(1) Provide the working pulley chamber with oil pressure to control the gear ratio.
(2) Control the clutch and brake in the forward/reverse mechanism to realize forward/reverse switching.
(3) Control the engagement and disengagement of the lockup clutch.
(4) Maintain pressure in the hydraulic torque converter.
(5) Supply oil to the cooling system.
(6) Supply oil to the lubrication system.
(7) Make the wet clutch start smoothly.
(8) Ensure the smoothness of forward, reverse and slide switching process.

I. **Speed ratio control strategy**

(1) Single parameter (speed ratio) tracking control strategy (Fig. 4.32).
(2) Integrated control strategy (Fig. 4.33).
(3) Mode tracking control strategy (Fig. 4.34).

II. **Oil pressure control strategy of actuators**

(1) Basic control strategy.
(2) Slip control strategy (Fig. 4.35).
(3) Control strategy based on maximum efficiency (Fig. 4.36).

4.7 Main Performance Tests of Metal Belt CVT

The performance tests of the metal belt CVT include the component performance test and the whole performance test, including the following.

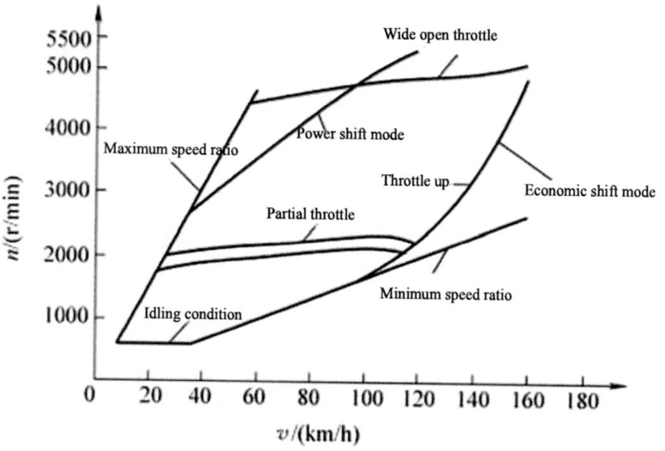

Fig. 4.33 Integrated control strategy

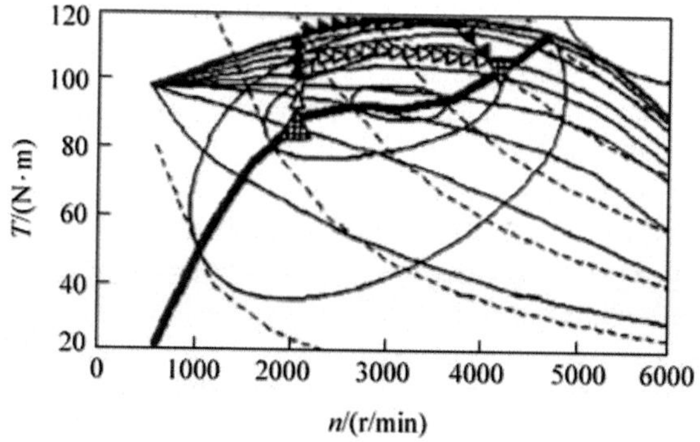

Fig. 4.34 Mode tracking control strategy

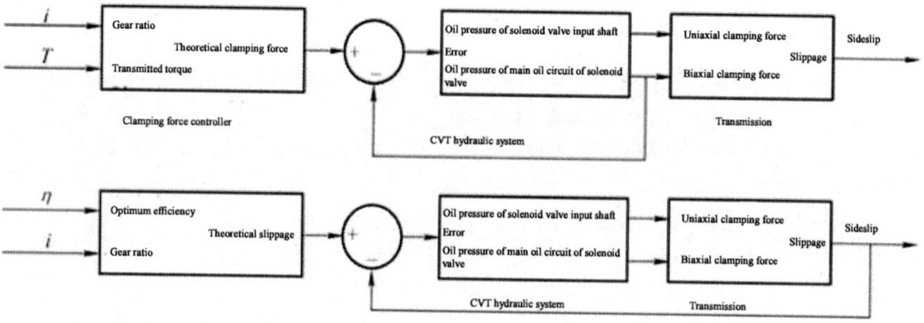

Fig. 4.35 Slip control strategy

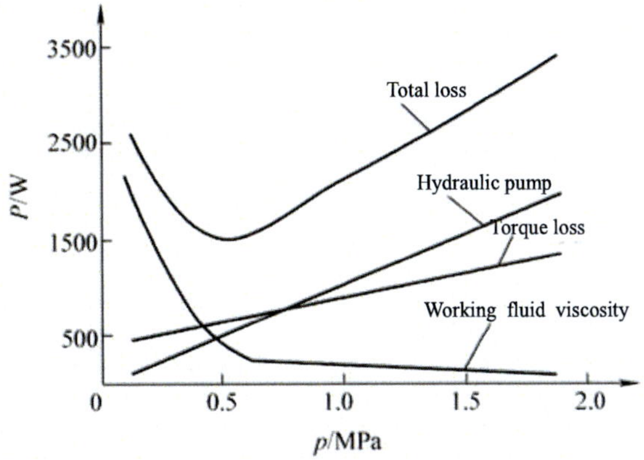

Fig. 4.36 Control strategy based on maximum efficiency

1. **Torque transmission characteristic test**

The torque transmission characteristic test is to test the torque transmission characteristics of the belt drive, i.e. torque capacity and transmission efficiency. The biaxial test bed is shown in Fig. 4.37. The clamping hydraulic pressure of the drive and driven pulleys is controlled respectively to set the clamping force and pulley ratio (or fix a certain pulley ratio for testing). The torque loss of the belt drive may be calculated by measuring the torque of the input and output shafts.

2. **Transmission control test**

The transmission control test is to evaluate the performance of the metal belt CVT, including the steady-state characteristics and dynamic characteristics. The tested CVT is loaded on the motor and operated without load. After the variable speed instruction is given, the steady-state characteristic is mainly evaluated as the deviation between the actual pulley ratio and the given pulley ratio; the evaluation indexes of dynamic characteristics mainly include time lag, variable speed, overshoot and

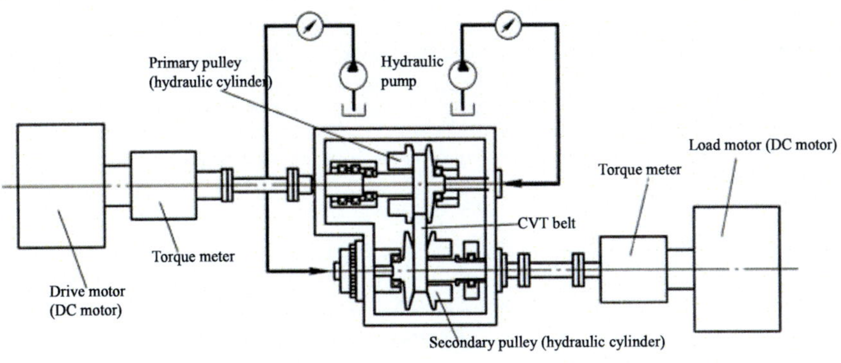

Fig. 4.37 Torque transmission characteristic test device

impact. In addition, the stability of the system (i.e., the pulley ratio change) is tested with disturbances such as fluid pressure fluctuations.

The toroidal traction CVT is mainly tested to correctly control the speed change of the toroidal traction CVT and obtain the kinetic characteristics of the transmission (roller). The toroidal traction CVT achieves speed change by making a pair of rollers rotate or move up and down. The steady-state and transient characteristics of the rotor motion corresponding to the variable speed instruction are detected by their Y-THETA sensors, and the variable speed characteristics of the toroidal traction CVT are obtained by the relationship between roller motion and gear ratio.

3. **Performance test of hydraulic control system**

The performance test of the hydraulic control system of the metal belt CVT is the same as that of the AT. All kinds of solenoid valves or stepping motors used as actuators are tested jointly with the drive.

4. **Noise test**

The high frequency noise of CVT can easily cause the noise problem in the vehicle. To set the target performance of the interior dark noise level, the method of taking the measured emission sound near the powertrain related to the interior dark noise as the target performance is usually applied. With respect to the metal belt CVT, the noise test can not only improve the overall noise sensitivity of the powertrain and reduce the deviation between components, but also correctly grasp the proportion of belt noise in the overall noise.

5. **Reliability test**

The reliability test of CVT includes the reliability test of hydraulic actuator, CVT, gear and clutch, and control system. CVT components will be checked before the installation of the whole CVT. Before the prototype loading, the whole CVT shall run continuously for 60 h to verify the reliability of the CVT.

In addition, the CVT shall also be tested for driving conditions, including control system simulation test, hardware-in-loop test and real vehicle test, which is a test method to test the comprehensive performance of the powertrain.

The tests of special parts in CVT mainly include friction endurance test of steel plate and belt ring and fatigue test of metal belt, as shown in Fig. 4.38.

4.8 CVT Upgrade

AT present, although CVT is better than AT in performance, it still has disadvantages in boundary dimensions, weight and cost. Due to high transmission development cost, high early investment of production equipment and high risks, it is still difficult to the CVT to become mainstream since it only has 20 years of industrialization history although its application is expanding.

Fig. 4.38 Friction endurance test of steel plate and belt ring

Nowadays, the development of CVT mainly focuses on increasing the gear ratio range, increasing torque capacity, reducing friction loss and reducing weight and cost. In some of the new CVT technologies, the main means of application is to upgrade and change the CVT mode based on belt CVT. The following describes a metal belt CVT with an auxiliary transmission and a non-belt drive CVT—toroidal traction CVT.

1. **Metal belt CVT with auxiliary transmission**

To improve the start response and fuel consumption, JATCO CVT8 (Fig. 4.39) connects the metal belt drive in series with the auxiliary transmission, giving the CVT a wider range of speed ratios. The auxiliary transmission is of the planetary gear train structure to improve the shift quality of the auxiliary transmission. The forward/reverse gear is also shifted by the auxiliary transmission. The step-up gear is also arranged between the hydraulic torque converter and the drive wheel of the belt drive to reduce the load of the belt drive, reduce the loss of the hydraulic system, suppress the vibration when changing speed, and improve the dynamic and economy performance and shift smoothness.

2. **Toroidal traction CVT**

The toroidal traction CVT can achieve the fast speed change function that the traditional belt CVT cannot, making the speed change response more suitable to the

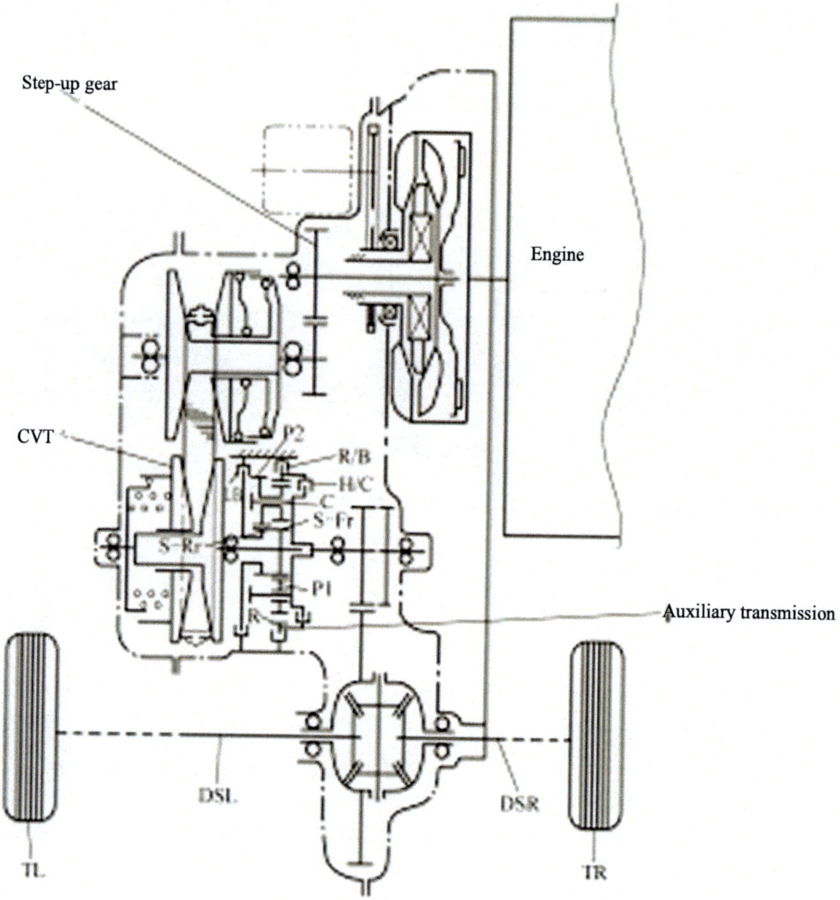

Fig. 4.39 Metal belt CVT with auxiliary transmission

driver's intention; compared with the traditional 4AT, the toroidal traction CVT saves 10% of the fuel consumption. The toroidal traction CVT is a development direction of CVT in the future.

The variable speed structural unit of the toroidal traction CVT, as shown in Fig. 4.38, is composed of a pair of toroidal metal disks and rollers opposite each other. The power is input from one side of the metal disk and output from the other side. Two rollers are arranged in the middle of the two metal disks that can rotate around their respective x axis. The rollers always keep contact with the two tapered metal disks.

The transmission of power in the variable speed structural unit is actually accomplished by the shear force in the oil film between the roller and the metal disk, and the relationship between the traction force T and the packing force N is

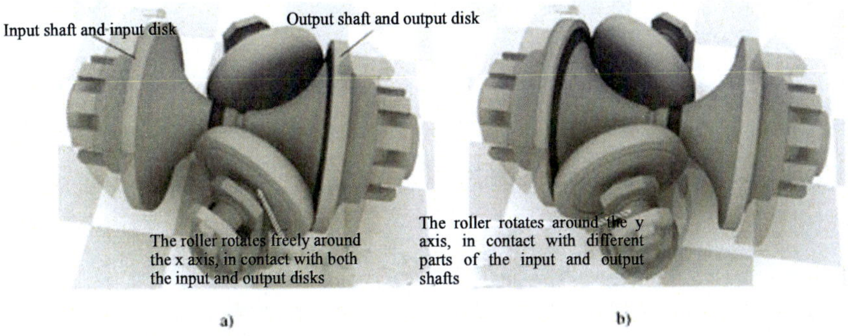

Fig. 4.40 Toroidal traction CVT

$$T = \mu N \tag{4.15}$$

where,

μ—traction coefficient, determined by the transmission lubricating oil.

The lubricating oil used in the toroidal traction CVT is special synthetic oil which can keep the oil film intact under the action of high pressure on the contact surface.

The gear ratio is adjusted by the control of the roller position through the hydraulic system. The two rollers rotate symmetrically around the z axis to change the position of the contact points of the two rollers on the tapered metal disk, thus changing the drive and driven pitch radius and adjusting the gear ratio. In the figure, the dark ring on the two metal disks represents the contact mark between the cone disk and the roller, and the gear ratio of the toroidal CVT is determined by the ratio of the radius of the two rings. The toroidal traction CVT shown in Fig. 4.38a is in the state of large gear ratio and the toroidal traction CVT in Fig. 4.40b is in the overdrive state.

For the single variable speed structural unit, the excessive load of transmission force will cause the corresponding increase of friction loss and reduce the economy. For this purpose, two variable speed structural units can be connected in parallel (power is input from the metal disks at both ends and output from the metal disk in the middle) to obtain greater torque capacity (Fig. 4.41). Such a structure would increase the longitudinal dimensions of the transmission, suitable for FR vehicles.

To avoid slipping during the transmission of power, the variable speed structural unit shall provide a sufficient traction corresponding to the input torque. The traction is generated by the compression load N required for the contact surface between the disk and the roller, which is provided by the front and rear disk springs and the self-force amplifier. The rear disk spring provides the minimum axial load value; the function of the front disk spring is to transfer the axial load from the state provided by the rear disk spring to the state provided by the self-force amplifier. The self-force amplifier, as shown in Fig. 4.42, generates axial compression load of the metal disk on the rollers through a pair of rollers between the cylindrical cam and cam while transmitting the torque to the metal disk.

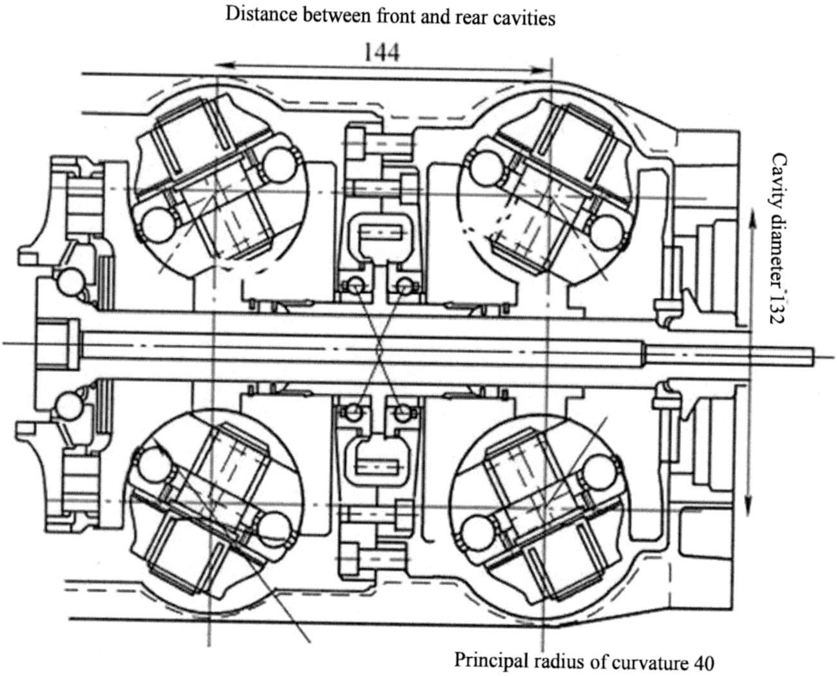

Fig. 4.41 Structure diagram of toroidal traction CVT

Fig. 4.42 Self-force amplifier of toroidal traction CVT

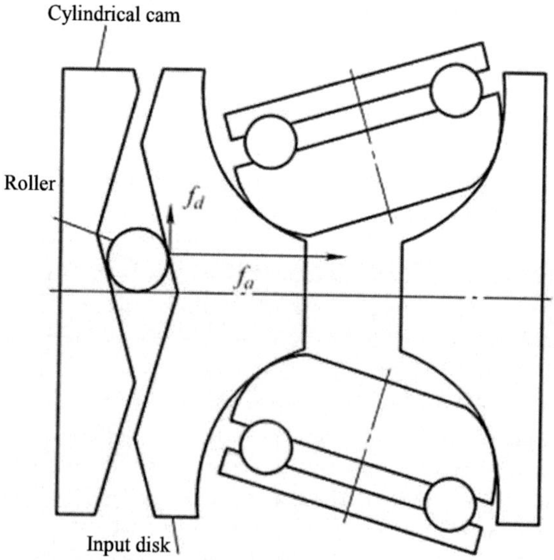

Bibliography

1. Anlin Ge (2001) Automatic transmission (VII)—Continuously variable transmission CVT (I). Autom Technol 11:1–4
2. Naishi Cheng (2007) Principle and design of metal V-belt CVT. China Machine Press, Beijing
3. Yong Chen (2008) New development and trend of automatic transmission technologies. Autom Eng 30(10):938–944
4. Ying An, Chuanxue Song, Shuai Gao (2009) Computation and optimization of CVT misalignment. Autom Technol 02:26–28
5. Crolla DA (2011) Automotive engineering: powertrain chassis system and vehicle body. Translated by Tian Chunmei, et al. China Machine Press, Beijing
6. Yinquan Yang (2010) The entire machine evaluation and experiment normative study of mechanical electronic CVT. Northeastern University, Shenyang
7. Yiyang Zhang, Shaobo Li, Yue Cheng, Naishi Cheng (2014) Design and experiments of driving system for chain CVT. China Mech Eng 25(18):2456–2460
8. Yunzhi Hao (2011) Study on CVT control system and hardware in loop simulation. Chongqing University, Chongqing
9. Jiajia Tian (2010) Study on drive characteristics of rubber belt type CVT. Chongqing University, Chongqing
10. Xiaohu Lu, Yueping Chen, Weilian He (2007) Metal belt structure and strength analysis for the metal belt CVT. Drive Syst Tech 21(2):20–27
11. Yang Yang, Datong Qin, Yalian Yang et al (2006) Power matching control and simulation on hydraulic system of CVT. China Mech Eng 17(4):426–431
12. Yuanchun Cai (2011) Research on the fuel economy and system reliability of metal V-belt CVT [D]. Hunan University, Changsha

Chapter 5
Dual Clutch Transmission

5.1 Overview

The dual clutch transmission (DCT) is different from other automatic transmissions in that it uses two clutches to control the clutch actuator and the gear selecting and shift actuator to achieve gear shift without power interruption. The DCT mainly includes dual clutch system, shaft-tooth system, control system, gear shift actuating system, clutch actuating system, cooling and lubrication system, parking system and other modules.

By the clutch cooling mode, DCT can be divided into DDCT and WDCT. The DDCT, with the air as the cooling medium, transmits the torque through the friction plates on the driven plate without additional forced cooling and lubrication; the WDCT, with the oil as the medium, makes forced cooling and lubrication and transmits the torque through the clutch friction plates immersed in the oil.

According to different control modes of clutch actuator, DCT can be divided into hydraulic DCT, electro-hydraulic DCT and motor DCT. In the hydraulic DCT, the dual clutch pushes the clutch friction plates to combine through the oil pressure generated by the hydraulic pump inside the transmission, thus achieving torque transmission; in the electro-hydraulic DCT, the dual clutch drives the hydraulic cylinder through the motor and forms a certain oil pressure in the fixed oil circuit to push the clutch friction plates to combine, thus achieving torque transmission; in the motor DCT, the dual clutch directly drives the mechanical structure through the motor to push the clutch friction plates to combine, thus achieving torque transmission.

By the shift actuator drive mode, DCT can be divided into hydraulic drive shift DCT and motor drive shift DCT. The hydraulic drive shift DCT controls the gear switching of the transmission shift actuator through hydraulic drive; the motor drive shift DCT controls the gear switching of the transmission shift actuator through motor drive.

By the arrangement of input shaft, the DCT can be divided into coaxial DCT and parallel DCT. The two input shafts corresponding to the coaxial DCT are arranged in

Y. Chen, *Automotive Transmissions*, Springer Tracts in Mechanical Engineering,
https://doi.org/10.1007/978-981-15-6703-2_5

coaxial position; the two input shafts corresponding to the parallel DCT are arranged in parallel position.

By the number of forward gears, the DCT can be divided into 6 speed DCT and 7 speed DCT. At present, some relevant research institutions are still developing 9 speed DCT or 10 speed DCT.

By the drive mode, the DCT can be divided into horizontal front drive DCT, longitudinal front drive DCT and longitudinal rear drive DCT. The corresponding car body of the horizontal front drive DCT is placed horizontally and transmits the power through the front-wheel drive; the corresponding car body of the longitudinal front drive DCT is placed longitudinally and transmits the power through the front-wheel drive; the corresponding car body of the longitudinal rear drive DCT is placed longitudinally and transmits the power through the rear-wheel drive.

By the number of driving wheels, the DCT can be divided into two-wheel drive DCT and four-wheel drive DCT. The two-wheel drive DCT transmits the power by driving the front or rear wheel of the vehicle; the four-wheel drive DCT transmits the power by driving the front and rear wheels of the vehicle simultaneously.

By the power source, the DCT can be divided into internal combustion engine power DCT and hybrid power DCT. In the case of internal combustion engine power DCT, the internal combustion engine provides power input; in the case of hybrid power DCT, two power sources provide hybrid power input.

By the drive torque, the DCT can be divided into low and medium torque and high torque DCT. The DDCT is mainly used in the vehicles with low and medium torque, and belongs to low and medium torque DCT. However, WDCT can be applied to vehicles with higher torque, which belongs to high torque DCT. This is mainly because the wet clutch uses active lube cooling and lubrication, while the dry dual clutch uses passive cooling like the ordinary clutch, which has limited heat capacity and is not suitable for high torque engines. However, with the development of the material technology, the bearing capacity of the dry dual clutch will gradually increase, and its drive torque will also gradually increase.

By the parking mechanism form, the DCT can be divided into mechanical parking DCT, electronic parking DCT and automatic parking DCT. The mechanical parking DCT is connected with the parking control mechanism through a certain mechanical structure to realize the parking and unlocking functions; the electronic parking DCT makes the parking control mechanism realize parking and unlocking functions through control signal; the parking mechanism of automatic parking DCT is normally closed, and the parking function is controlled enabled and disabled according to the hydraulic pressure, so as to enable parking during normal driving and disable parking during parking.

With the continuous maturity of DCT technology and the continuous improvement of driving performance, DCT has been recognized by more and more users. In recent years and in the next few years, most of the automatic transmissions in mass production will be DCT, and vehicles equipped with DCT will become the main vehicles in Chinese and European models.

5.2 System Composition and Working Principle of DCT

I. System composition

The DCT mainly consists of the dual clutch system, shaft-tooth mechanism, select-shift actuator and electronic control system. The composition of its mechanical system is shown in Fig. 5.1. The wet 6 speed DCT shown in Fig. 5.2 is introduced as an example below. With 6 forward gears and 1 reverse gear, it mainly consists of multi-disk wet dual clutch, three-shaft gear transmission, automatic shift actuator and electronically controlled hydraulic control system. The two hydraulically-driven wet clutches are concentric, giving the transmission assembly a smaller axial length. The clutch marked red on the outside in the figure is an odd gear clutch, or clutch C1 for short; the clutch marked green on the inside is an even gear clutch, or clutch C2 for short. The DCT has two coaxial input shafts, and the input shaft 1 is nested in the input shaft 2. The input shaft 1 is connected to the clutch C1, and the constant mesh gear on the input shaft 1 is meshed with the gear 1, gear 3 and gear 5, respectively; the input shaft 2 is a hollow shaft connected with clutch C2 and the constant mesh gear on the input shaft 2 is meshed with the gears 2, 4 and 6 respectively; the reverse gear is meshed with the constant mesh gear of input shaft 1 through the reverse shaft gear (not shown in the figure). In other words, clutch C1 is in charge of gears 1, 3, 5 and R, while clutch C2 is in charge of gears 2, 4 and 6. Corresponding to the two input shafts are the two output shafts, which are connected to the differential input gear through the front gear, so that both output shafts are output through the differential

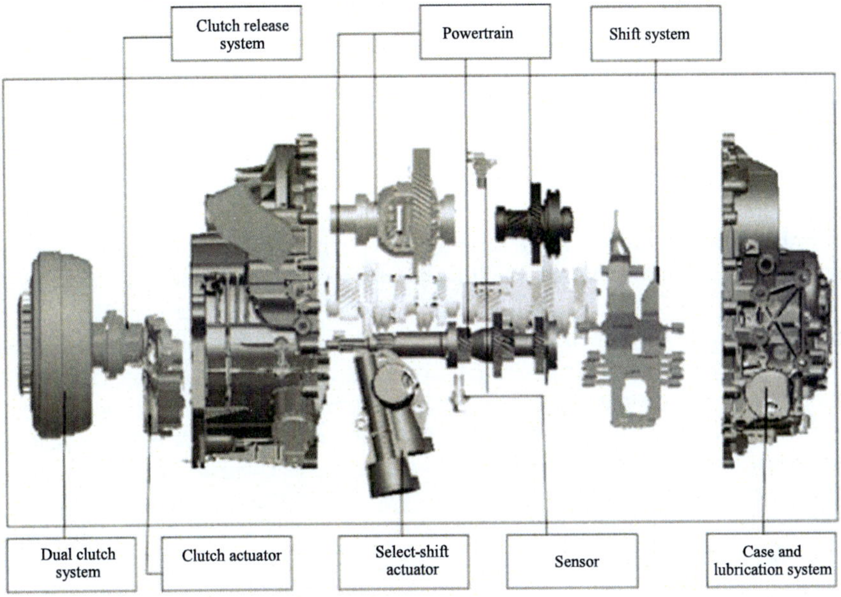

Fig. 5.1 Composition of DCT mechanical system

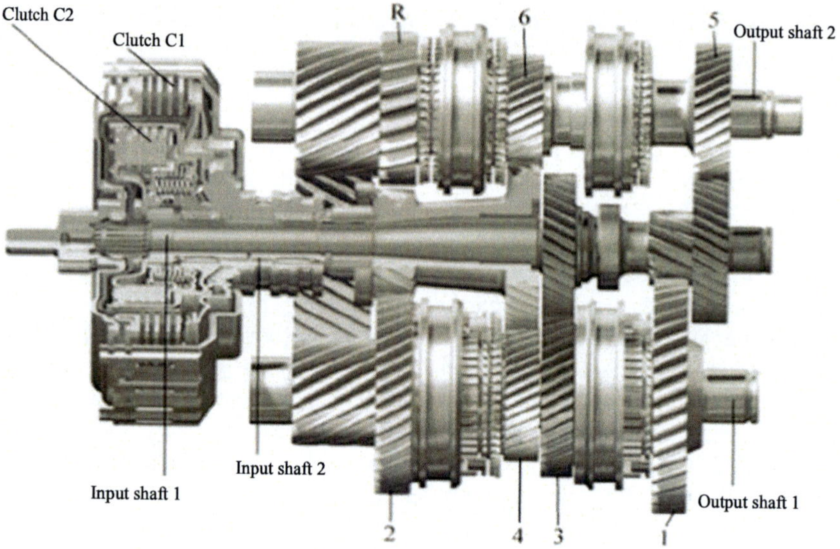

Fig. 5.2 Typical DCT transmission structure

mechanism. Each gear shift is achieved by a synchronizer, which is driven by a fork supported on the transmission case. The engagement and release of two clutches and the movement of the fork are completed by hydraulic actuators.

II. **Working principle**

Figure 5.3 shows the working principle of DCT. Assuming the DCT runs in gear 1, the power is transmitted through the active part of the clutch through the clutch C1 to the input shaft 1, through the gear 1 on the input shaft 1 to the gear 1 synchronizer, and then to the output shaft. The electronic control unit TCU of the transmission can judge whether the transmission can put in the gear 2 in advance. If so, the synchronizer of gear 2 can be engaged in advance. Since the corresponding clutch C2 of gear 2 is in the release state without transmitting power, the synchronizer of gear 2 enters synchronization under no-load condition. When the TCU decides to change from the current gear 1 to 2, the shift control essentially transfers the power transferred from clutch C1 to clutch C2 smoothly, which is the most critical clutch alternation process in the DCT shift process. When clutch C1 is completely released, the clutch C2 enters the working state completely and the vehicle is running in gear 2. When working in other gears, the control process is similar, except for the difference between upshift and downshift. During the shift, the power of the engine is transferred to the wheels continuously, so the completed shift is a power shift. When the vehicle achieves the power shift, it will greatly improve the riding comfort, as well as the economy and emission characteristics of the vehicle.

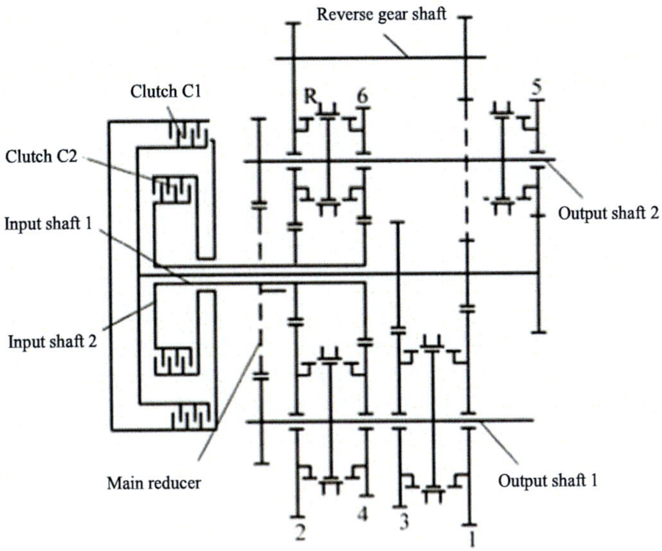

Fig. 5.3 Working principle of DCT

5.3 Typical Design Scheme of DCT

The DCT is divided into DDCT and WDCT and Table 5.1 compares the DDCT with WDCT in performance. The DDCT is mainly used in the vehicles with low and medium torque, while WDCT can be applied to vehicles with higher torque. The dry dual clutch uses passive cooling like the ordinary clutch, and its heat capacity is limited, so in the case of high power input, the system will soon reach its heat capacity

Table 5.1 Performance comparison of DDCT and WDCT

Comparison item	DDCT	WDCT
Heat load performance	Low	High
Load performance	Relatively high	High
Transmission efficiency	High	Relatively high
Fuel economy	Good	Relatively good
Cost	Relatively low	Relatively high
Maintenance performance	Easy	Not easy
Structure	Simple and compact	Complex
Mass	Small	Relatively large
Applicable model	Small and medium passenger vehicles	Large cars and off-road vehicles
Shift controllability	Good	Relatively good

limit, far below the limit that can be reached by the corresponding hydraulic torque converter or wet clutch. In addition, the wear of dry clutch friction plate is also the focus of DCT service life. The wet clutch uses the active cooling of lubricating oil with large heat capacity, but it must be combined with the control action actuator and cooling hydraulic system. It is complex and expensive. Meanwhile, the loss of the hydraulic pump will also lead to high fuel consumption.

The DCTs developed by various companies are analyzed below. Figure 5.4 shows Geely wet 7DCT, Fig. 5.5 shows Getrag wet 7DCT, Fig. 5.6 shows AC wet 6DCT and Fig. 5.7 shows Volkswagen wet 6DCT. Through comparative analysis, the WDCT is mainly arranged with 2 output shafts and Getrag DCT adopts 3 output shafts, beneficial to control the matching of axial length and reverse gear ratio, but it increases the number of parts and radial space. At present, most newly developed projects adopt the design principle of common drive gear (gear 4/6, gear 3/5, etc.), which is beneficial to shorten axial size, reduce weight, and improve transmission efficiency of the powertrain. In addition, the gears 1 and R shall be arranged on different input shafts as far as possible, so as to help balance the life of the clutch and shorten the shift time from the reverse gear to the gear 1.

Figure 5.8 shows Geely dry 7DCT, Fig. 5.9 shows Getrag dry 6DCT, Fig. 5.10 shows Hyundai dry 6DCT and Fig. 5.11 shows Volkswagen dry 7DCT. Due to the influence of the torque capacity and heat capacity of the dry dual clutch, the transmission capacity of the DDCT is limited. At present, the DDCT is generally mounted on vehicles with low and medium torque, and the space of matching models is small,

Fig. 5.4 Geely wet 7DCT

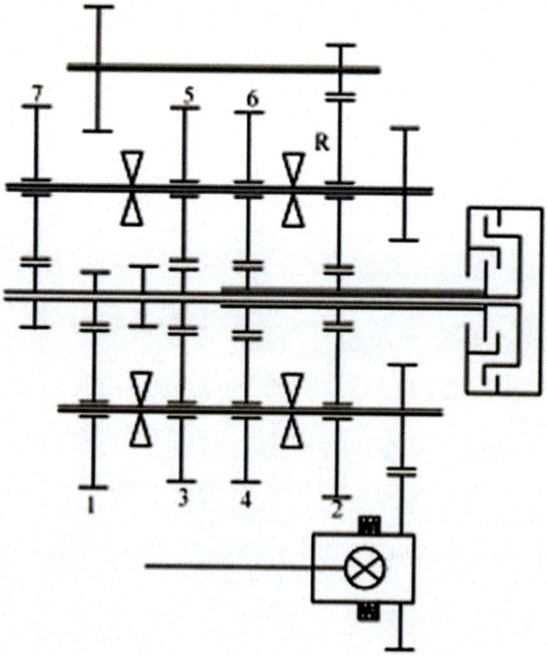

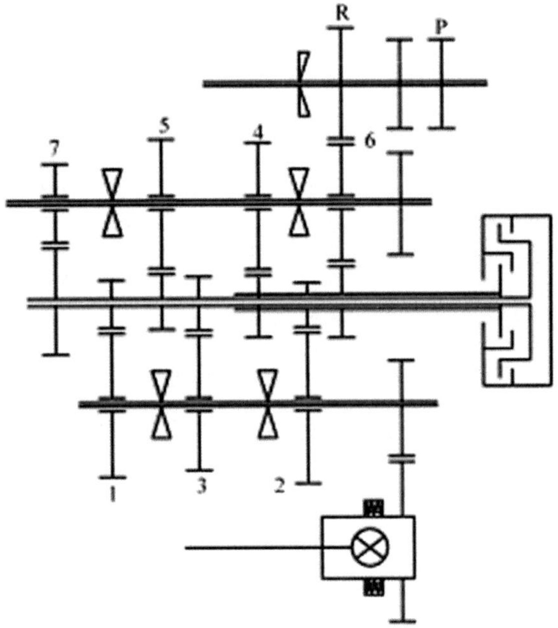

Fig. 5.5 Getrag wet 7DCT

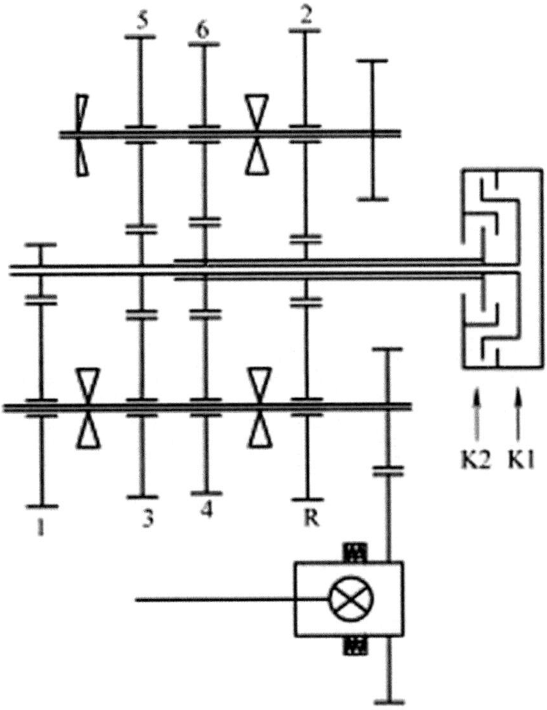

Fig. 5.6 JAC wet 6DCT

Fig. 5.7 Volkswagen wet 6DCT

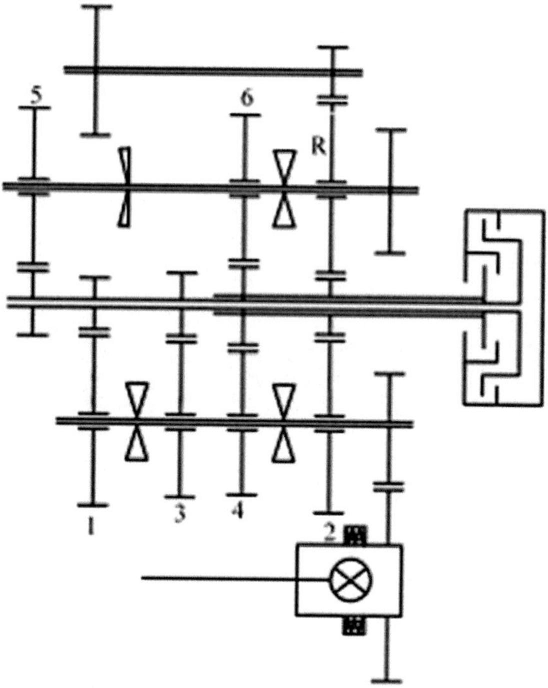

and most of them are in 6 gears, which is conducive to vehicle layout. The gears 1 and R are arranged on different input shafts, which is beneficial to improve the service life of the clutch and shorten the shift time. Since DDCT and MT are lubricated in the same way (splash lubrication), the low-speed gears shall be arranged together as far as possible, and their output shafts are arranged below. The parking ratchet is mostly arranged on the output shaft, which is conducive to reducing the size of parts and saving layout space.

5.4 Dual Clutch

The dual clutch is an important part of DCT. The performance characteristics of the DCT system mainly depend on the form of dual clutch. Which clutch system (dry or wet) can provide a better solution for the next generation of vehicles is a hot topic in the current technology field. At present, both dry and wet clutches have their applications in the market, but there are some differences in the structure and operating characteristics.

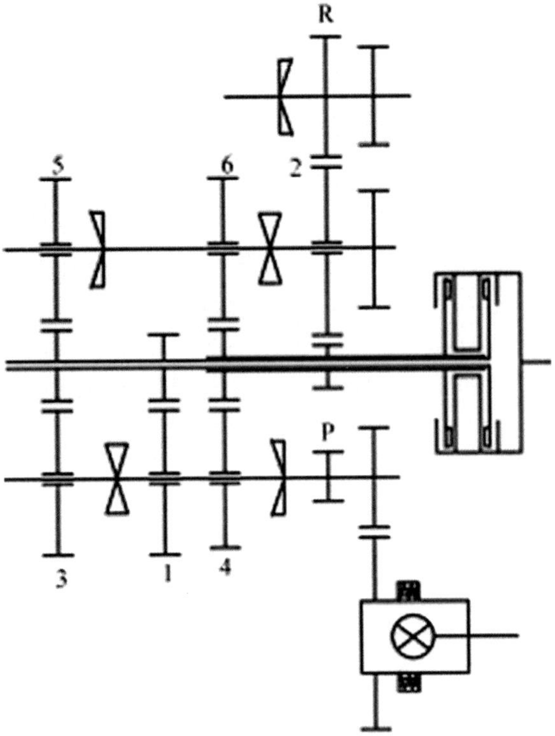

Fig. 5.8 Geely dry 7DCT

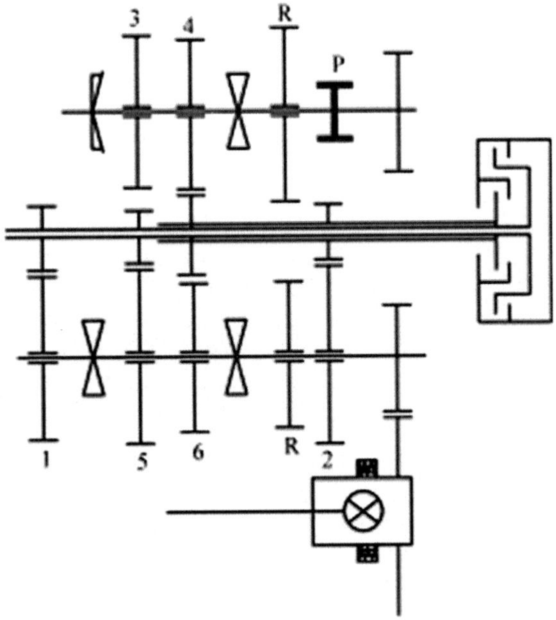

Fig. 5.9 Getrag dry 6DCT

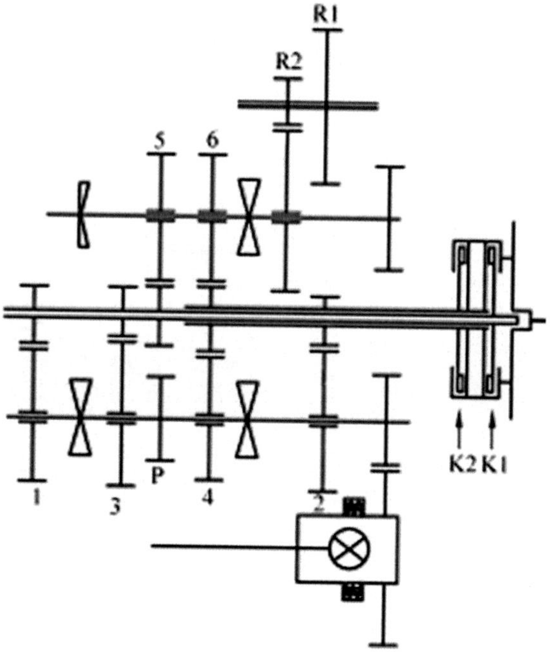

Fig. 5.10 Hyundai dry 6DCT

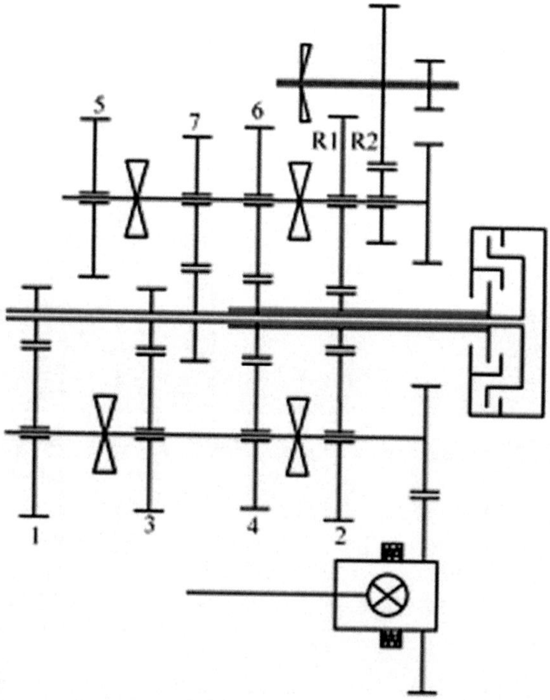

Fig. 5.11 Volkswagen dry 7DCT

I. Dry dual clutch

1. Structure of dry dual clutch

The biggest feature of the DDCT is the use of two dry clutches, thus avoiding the WDCT disadvantages of low system efficiency caused by wet clutch drag torque, clutch cooling and control oil, and further improving fuel economy. DDCT is typically represented by Volkswagen DQ200, which has seven forward gears and can transmit 250 N m torque. Its internal working principle is the same as other DCTs. Figure 5.12 shows the structure of the dry dual clutch and Fig. 5.13 shows the working principle of the dry dual clutch.

2. Actuator of dry dual clutch

The basic requirement for the dual clutch actuator is that it cannot self-lock and the motor must be continuously energized when the clutch is closed. To prevent overheating of the motor, the continuous load power limit of the motor is about 20 W; the clutch shall be able to release automatically when the power is cut off under the closed state of the clutch.

In terms of classification, the dry dual clutch actuators are mainly electrically operated actuator and electro-hydraulic actuator. The electrically operated actuator, with the advantages of simple structure, easy control, stable performance, low cost, low energy consumption and little pollution to the environment, is especially suitable for passenger vehicles. Figure 5.14 is the sketch of the electrically operated clutch actuator and Fig. 5.15 is the structure diagram of the electrically operated clutch actuator. According to the structure of DCT, for safety reasons, the clutch must be

Fig. 5.12 Structure of dry dual clutch

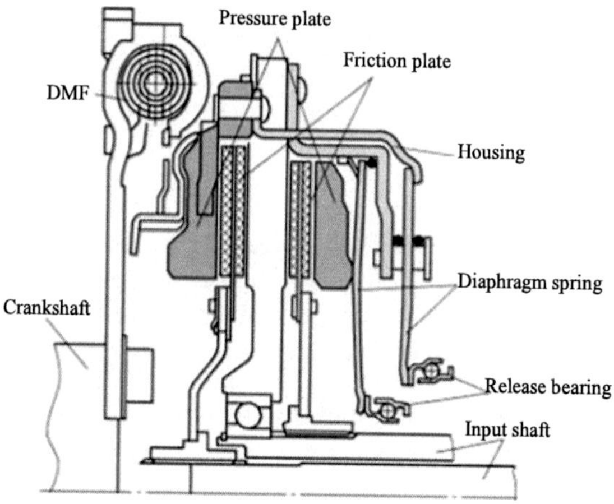

Fig. 5.13 Working principle of dry dual clutch

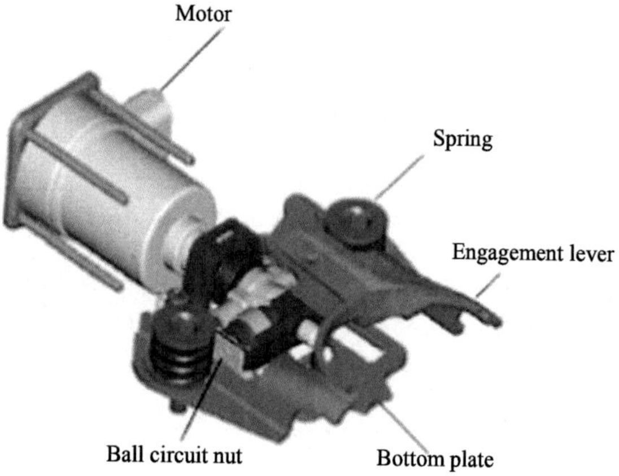

Fig. 5.14 Sketch of electrically operated clutch actuator

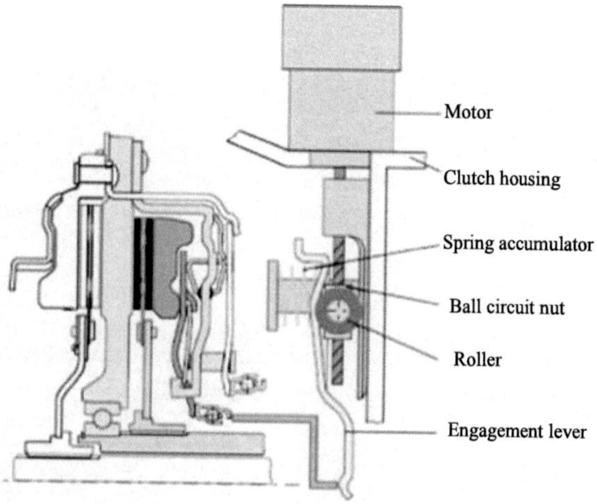

Fig. 5.15 Structure diagram of electrically operated clutch actuator

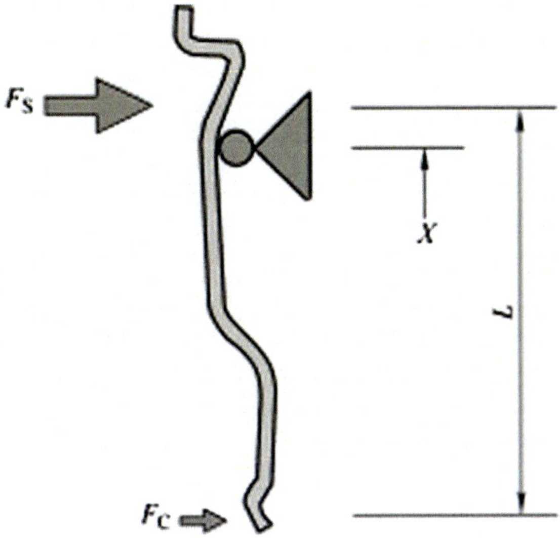

Fig. 5.16 Working principle of engagement lever

able to automatically release when the clutch actuator fails, which is realized through the normally open clutch. The normal close of the DCT leads to the interlock of two clutches, which is accompanied by a large amount of energy loss and the generation of the uncontrollable auxiliary torque at the gear output end.

The ball circuit nut converts the rotary motion into rectilinear motion; the roller acts as the engagement lever fulcrum; the spring accumulator has a pre-compression force acting on one end of the engagement lever; the release bearing acts on the other end of the engagement lever, and the engagement lever and the clutch housing bottom plate form a wide upper and narrow lower structure, with the roller mounted between the two. Figure 5.16 shows the working principle of engagement lever. The force on the clutch release bearing end is

$$F_C = F_S \frac{x}{L - x} \tag{5.1}$$

where,

F_C—force on the clutch release bearing end (N);
F_S—spring pre-compression force (N);
L—distance between F_C and F_S (m);
X—distance between the roller fulcrum and F_S (m).

Figure 5.17 shows the engagement by moving the fulcrum. When the motor rotates in a positive direction and drives the roller to move down, the lower end of the engagement lever moves to the left, the packing force increases, and the clutch transfers the torque; when the motor rotates in reverse and drives the roller to move up, the lower end of the engagement lever moves to the right, and the packing force

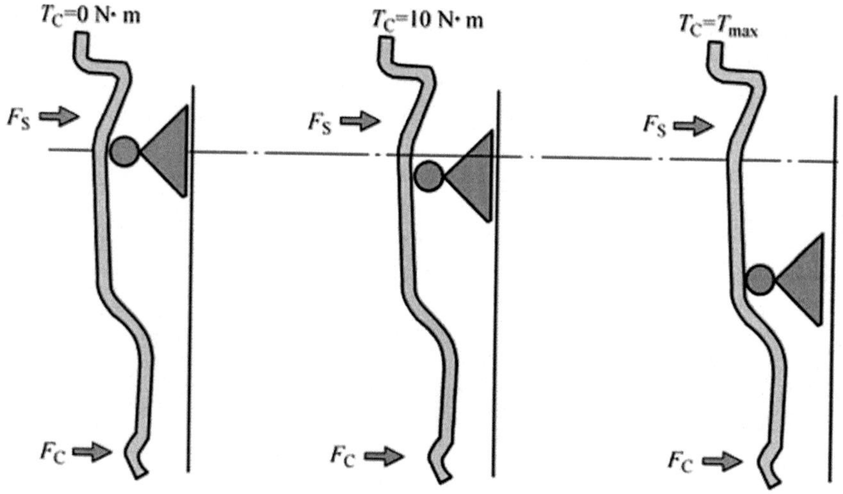

Fig. 5.17 Engagement by moving the fulcrum

gradually decreases, so does the torque transferred by the clutch; when the motor is powered off at the clutch engagement position, the axial component of the force exerted by the engagement lever on the roller is the return force. According to the relationship of the lead of the ball screw, the reverse torque acting on the motor will be generated to drive the roller to move up and return to the initial position near the motor.

3. **Torque transmitted by clutch**

As one of the most critical working parts in DCT, dual clutch is mainly used to realize the function of vehicle starting and transmission gear switching. The torque transmission characteristic of a single clutch is the same with that of the ordinary dry clutch and the torque transmitted by the clutch is

$$T_C = \frac{2\mu F_n Z}{3}\left[\frac{R_o^3 - R_i^3}{R_o^2 - R_i^2}\right] \tag{5.2}$$

where,

μ—clutch friction factor;
F_n—clutch packing force;
Z—number of clutch working surfaces, $Z = 2$ for DDCT;
R_o—outside diameter of clutch friction surface;
R_i—inside diameter of clutch friction surface.

There are three main working states of dual clutch in the working process, respectively, release, engagement and sliding friction. In the three states, the actual torque transmitted by the clutch is

$$T_C = \begin{cases} T_C = \dfrac{2\mu F_n Z}{3}\left[\dfrac{R_o^3 - R_i^3}{R_o^2 - R_i^2}\right] \text{ Sliding friction state} \\ T_e \text{ Total engagement} \\ 0 \text{ Total release} \end{cases} \tag{5.3}$$

where,

T_e—engine torque.

The sliding friction state of the clutch is most concerned. As shown in the formula (5.3), the torque transmitted by the clutch is mainly related to the clutch packing force and affected by the friction factor and friction area change.

4. **Clutch sliding friction work**

In the process of vehicle starting and shifting, the power output by the engine should overcome various resistances to drive the vehicle. Different working conditions between the engine and the vehicle shall be synchronized by the sliding friction of the clutch, and the sliding friction work generated by the sliding friction will be converted into heat to make the clutch temperature rise. The clutch sliding friction condition in the starting process is worse than that in gear shift, especially in the hill starting, the clutch sliding friction heat is usually the most serious, so the main concern is the clutch sliding friction work at the starting. The clutch sliding friction process at the vehicle starting, as shown in Fig. 5.18, includes the following two phases.

In the first phase, the drive and driven parts of the clutch contact from the start, the transmitted torque T_C gradually increases from 0 to T_n and cannot overcome the resistance torque to move the vehicle. In this phase, the vehicle is stationary, so all

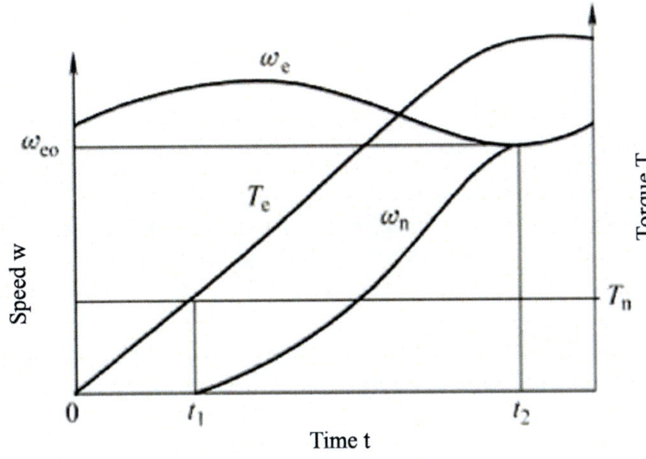

Fig. 5.18 Clutch sliding friction process at the vehicle starting

input energy is converted into sliding and friction work

$$L_{C1} = \int_0^{t_1} T_C \omega_e dt = \int_0^{t_1} T_e \omega_e dt \quad (5.4)$$

In the second phase, $T_C > T_n$, the torque transmitted by the clutch overcomes the resistance torque to move the vehicle and start the vehicle. At this time, the driven plate speed ω_n of the clutch gradually increases from 0 and is finally equal to the engine speed ω_e. The sliding friction work generated in this phase is

$$L_{C2} = \int_{t_1}^{t_2} T_C(\omega_e - \omega_n) dt \quad (5.5)$$

The total sliding friction work generated in the clutch engagement process is

$$L_C = L_{C1} + L_{C2} \quad (5.6)$$

Since the heat converted by the sliding friction work is the root cause for the clutch temperature rise, which affects system safety, control performance and life, measures must be taken to improve and control. It can be known from formula (5.6) that the engine speed and torque shall be reduced to reduce the clutch sliding friction work. To reduce the torque transmitted by the clutch, the speed ratio of the transmission gear 1 (or reverse) may be properly increased for the vehicle starting that may produce a lot of sliding friction work under certain objective conditions, such as the gross vehicle mass and the driving resistance of the vehicle to reduce the engine torque required for starting, so as to significantly reduce the sliding friction work and effectively reduce the clutch temperature rise, in line with the development trend of the AT gear number increase and speed ratio range increase currently.

Figure 5.19 shows the clutch temperature rise at different gear 1 ratio when the vehicle starts repeatedly on a 12% ramp with a full load. Curve 1 is the temperature rise of the vehicle installed with the DCT with the same gear ratio as manual transmission; Curve 2 is the temperature rise after the gear ratio of gear 1 increased by 20% under the same working condition. According to the friction materials currently in use, the critical temperature of the DDCT is between 300 and 400 °C, beyond which the friction system will be permanently damaged. As can be seen from Fig. 5.20, curve 1 (Fig. 5.19) can only realize 6 repeated starts, while curve 2 can realize more than 25 repeated starts. This means that the increase of the gear ratio of the gear 1 significantly reduces the speed of clutch temperature rise, and the clutch can withstand more frequent and longer climbing, uphill parking, starting sliding and other conditions, which is of vital significance to improve the safety and life of the dry dual clutch.

Figure 5.20 shows the relationship between the vehicle speed in gear 1 and the number of possible starts at the minimum stable engine speed (usually 1000 r/min).

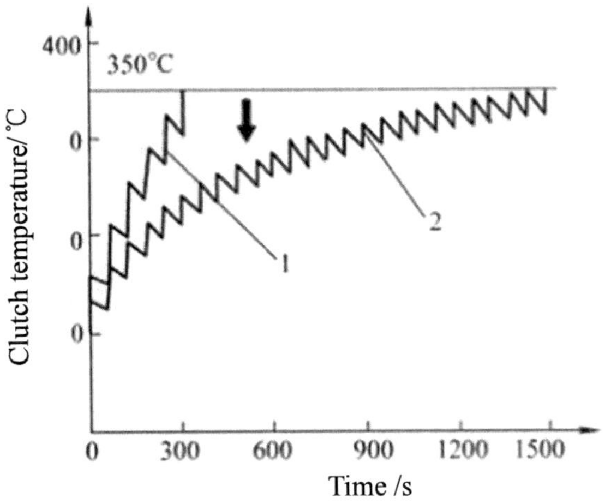

Fig. 5.19 Temperature rise when the vehicle starts repeatedly on a 12% ramp with a full load

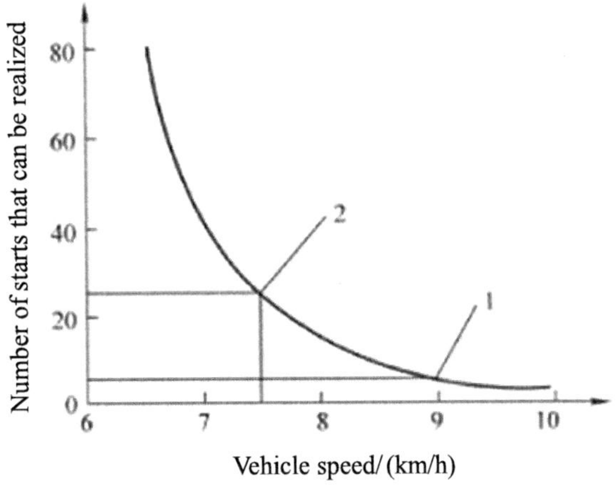

Fig. 5.20 Relationship between the vehicle speed in gear 1 and the number of possible starts at the minimum stable engine speed

Obviously, the lower the vehicle speed in gear 1 at the minimum stable engine speed, the more possible starting times, the more reliable the clutch works.

5. **NVH of dual clutch**

The NVH of the powertrain depends on a number of factors, such as damping, drive characteristics of the power system, vibration coupling, engine excitation, clutch excitation, transmission excitation, etc. For a dual clutch system, to improve the driving comfort, the time when the clutch is in the sliding friction state is significantly increased (creeping function). In the sliding friction process, especially in gears 1, 2 and R, the powertrain system operates at its inherent frequency and the fluctuation of the minimum torque at the clutch output end may cause large vehicle vibration or noise. Such vibration or noise is amplified because only a clutch is loaded and the other clutch is free to vibrate when the DCT starts, resulting in more noise. In order to ensure that the DCT can also achieve the noise level of AT, the excitation of the clutch must be greatly reduced. The friction characteristics of the dual clutch are optimized, so that the friction factor of the clutch friction pair increases slightly when the relative sliding speed increases, so that the friction plate can show the characteristics that are helpful to reduce the vibration in most of the time of the clutch sliding.

I. Wet dual clutch

1. Structure of wet dual clutch

At present, the most representative product of WDCT is the DQ250, which was jointly developed by BorgWarner and German Volkswagen and officially launched in 2003. It adopts two wet clutches and six forward gears, with a transmitted torque of 350 N·m. DQ250 was initially used in Golf R32 and Audi TT models, but was later promoted to other Volkswagen models such as Passat, Jetta and Beetle due to its superior performance, and was widely recognized by users. Figure 5.21 shows the structure of the wet dual clutch and Fig. 5.22 shows the structural schematic diagram of the wet dual clutch.

2. Actuator of wet dual clutch

Figure 5.23 shows the working principle of the hydraulic operated wet dual clutch. Under normal circumstances, the working pressure of the clutch C1 and clutch C2 is controlled by C1 proportional solenoid valve and C2 proportional solenoid valve respectively. The working pressure directly acts on the clutch working cylinder to produce the packing force to meet the working pressure requirements under different

Fig. 5.21 Structure of wet dual clutch

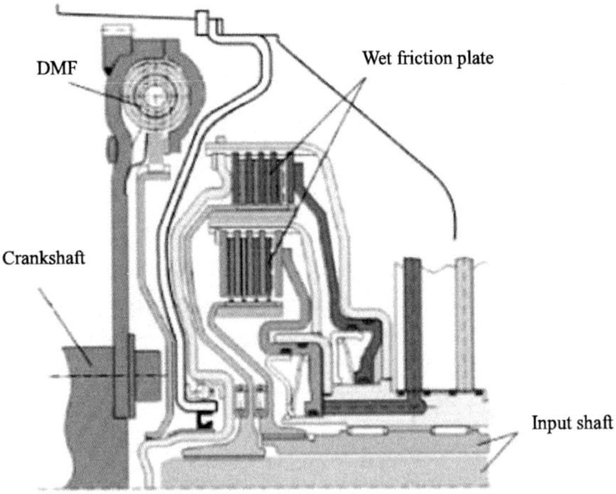

Fig. 5.22 Structural schematic diagram of wet dual clutch

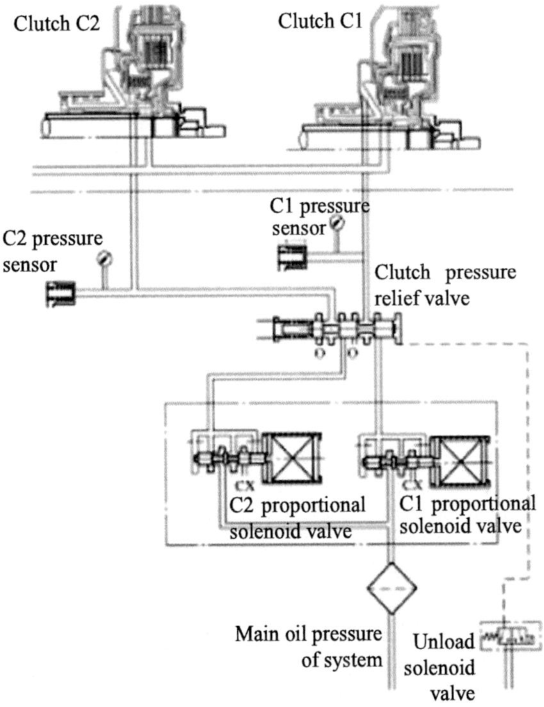

Fig. 5.23 Working principle of hydraulic operated wet dual clutch

working conditions. A safe hydraulic control circuit is added to ensure the safety of the DCT. When proportional solenoid valve fails at high oil pressure, even if the proportional solenoid valve element is stuck, whether in the oil circuit open or closed position, the oil pressure in the clutch cannot be released and shall be unloaded by unloading the clutch pressure relief valve. At this time, exit all control gears of the damaged proportional solenoid valve, and finally disconnect the on-off solenoid valve and drive home with the other clutch. Although the corresponding clutch of the damaged proportional solenoid valve is engaged, it cannot transmit torque because it is not put into the corresponding gear.

3. **Friction plate**

Friction plate is an important part of wet dual clutch and its friction factor is related to friction material and oil. Different friction materials have different surface properties and friction factors. The surface pressure also has an effect on the surface properties of friction materials. With the increase of surface pressure, friction factor decreases. The change of temperature will change the properties of the oil, thus affecting the friction factor. Generally, the friction factor decreases with the increase of temperature, but the specific situation depends on the oil. The friction factor is related to the pore size of the fiber texture. The larger the pore size, the greater the friction factor. After a period of use, impurities in the oil will be filled into the pores of the friction material, thus reducing the friction factor. Sometimes, after a period of use, the torque capacity of the friction plate is slightly increased, and the friction factor is increased, which is actually caused by sufficient contact after the friction surface is run in.

The friction plate used in the wet dual clutch is double-sided friction plate. Increasing the number of friction surfaces can increase the torque capacity. The friction plate and the dual disk are installed alternately. The friction factor is the friction factor between the friction plate and steel plate and is divided into static friction factor and dynamic friction factor.

(1) Static friction factor: the greater the static friction factor, the greater the torque capacity. Different friction materials have different friction factor. The friction factor of the paper base friction materials is higher than that of semi-metallic friction materials and sintered alloy friction materials. For paper base friction materials, factors influencing the friction factor include pore ratio, pore diameter, surface pressure, groove shape and oil.

1) Generally, the higher the pore ratio, the higher the friction factor. Because the more the pores, the more rough the friction action surface, and the greater the relative motion resistance, the greater the friction factor. Of course, the more the pores, the lower the structural strength of the friction material, the worse the wear resistance of the friction plate.
2) The larger the pore diameter, the higher the friction factor. Because the larger the pore diameter, the more rough the friction action surface, and the greater the relative motion resistance, the greater the friction factor. Of course, the larger the pore diameter, the lower the structural strength of the friction materials, the more easy to wear and peel.
3) The higher the surface pressure, the lower the friction factor. This is because when the friction material is compressed, the pore ratio in the material decreases and the pore diameter decreases.

4) Different groove shape has different friction factor. The groove scrapes off the thick oil film and the thin oil film becomes an elastomer that can bear the shear force under high pressure. The elastomer formed by the thin oil film can bear more shear force than that formed by thick oil film, so the torque transmitted by thin oil film is larger, showing larger friction factor.
5) Oil plays a very important role in the middle of friction plate and dual disk. The characteristics of oil affect the process of power transmission.
6) Usually, the friction factor decreases with the increase of temperature, because the increase of temperature will lead to the decrease of viscosity and shear force.

(2) Dynamic friction factor: the dynamic friction characteristic is required to be beneficial to control calibration and reduce vibration and noise in the process of gear switching and slip control. Therefore, for the lockup clutch used for the shift clutch and slip control, the dynamic friction characteristic is required to be $d\mu/dv > 0$. Whether $d\mu/dv$ is greater than or small than 0 mainly depends on the oil, because the oil is related to the friction material and the additives in the friction material also affect the oil. The most appropriate friction characteristic and oil are selected according to the actual operating conditions of the clutch used.
Friction stability refers to the property that dynamic friction characteristics change after a period of time of combined durability. After the clutch works for a period of time, due to impurities into the pores and surface pressure for long time, some of the physical properties of the friction material have changed, manifested as friction factor decrease and torque capacity decrease. Therefore, the performance requirements shall be put forward for the friction plate subject to durability test in the design. That is, the friction factor after the durability test is still higher than a certain minimum value.

4. **Friction plate ablation**

During the process of clutch engagement, a large amount of heat will be generated on the mating contact surface of the dual disk and friction plate. Because of the poor thermal conductivity of paper base materials, almost all the heat is absorbed by the dual disk, which makes the temperature of the dual disk rise. When the temperature exceeds the carbonization temperature of the friction plate, the friction materials will be carbonized, and the fiber texture in the friction materials (natural fibers and synthetic fibers) becomes carbon fibers. When the carbonization rate of the carbon fibers (the ratio of the mass of the carbonized carbon fibers to the mass of the fibers that can be carbonized initially) exceeds a certain value, the friction material cannot be used because the properties of the friction material have changed dramatically. Almost all the heat of the paper base friction material is absorbed by the dual disk, which is then cooled by the cooling oil in the groove of the friction plate and by the external cooling oil, taking the heat away. As for the multi-disk wet clutch, since the dual disks on both sides are more in contact with the external cooling oil, the two dual disks have the best cooling conditions. Assuming that the fiction work of each friction surface is the same, the paper base friction material has poor thermal conductivity and almost all the heat is absorbed by the dual disk. The middle dual disk absorbs heat, the temperature rises quickly, and the heat dissipation condition is poor, so the middle friction plate is most likely to burn. The heat dissipation area of the dual disks on both sides is the same, but the thick dual disk has large mass and large heat energy reserve. Therefore, after emitting the same heat, the temperature of the thick dual disk decreases relatively little and the friction plate close to the thick dual disk burns badly. Those with poor heat dissipation are more likely to burn than those with good heat dissipation, while those with high friction plate contact pressing plate temperature are more likely to burn than those with low temperature. The degree of burn loss presents the same trend according to the heat dissipation and temperature distribution. The degree of burn loss is the most serious in the middle and is reduced on both sides.

On both ends of the clutch, the thick dual disk is restrained in the approximately middle position of the dual disk on the other end. The pressure on the outer side of the friction plate is large, while the pressure on the inner side is small. The greater the surface pressure, the greater the friction work and the more heat generated during the slip. Therefore, the friction plate burns from the outer edge to the inner edge.

5. **Clutch drag torque**

The factors influencing the clutch drag torque include clutch clearance, oil groove design parameters (including groove width, groove depth, radial width of the friction plate, number and shape of oil grooves), lubricating oil flow, viscosity, temperature and speed (relative speed between the friction plate and the dual disk).

The clutch friction plate clearance shall be determined according to the specific situation, the larger the clearance, the smaller the drag torque, but the longer the dynamic response time of the clutch. When the separation clearance of the friction pair is small to a certain extent, the drag torque of the wet clutch will increase sharply, thus greatly reducing the transmission efficiency of the powertrain, increasing the heat load of the wet clutch and reducing the life of the friction plate. Therefore, the separation clearance should be prevented from too small in the design.

The larger the proportion of the friction surface area occupied by the friction plate groove, the greater the surface pressure of the friction plate, the more heat generated during the slip, and the smaller the drag torque. Some friction plates are also designed into a wavy texture (e.g. friction plate of the CVT19 clutch). The friction plate of wavy texture only forms an oil film with the dual disk at the bulge, while the friction plate of flat structure forms an oil film with the dual disk on the whole surface, so the wavy texture may reduce the drag torque; but the wavy texture requires large axial space and long dynamic clutch response time. In general, the drag torque of the friction plate with grooves is greater than that of the friction plate without grooves. The drag torque of the friction plate with radial grooves is the largest, while that of the friction plate with spiral grooves is the smallest and of other grooves is between them. Meanwhile, the grooves on the surface of the friction plate are beneficial to reduce the drag torque at high speed and low viscosity, but not conducive to reduce the drag torque at low speed and high viscosity. The more grooves there are on the friction plate, the smaller the peak value of drag torque is, and the lower the relative speed corresponding to the peak value is. However, when the relative speed is lower than a certain value, the change in the number of grooves will no longer affect the drag torque. The material change of the friction pair has little effect on the drag torque.

In the low speed range, the lubricating oil flow has little effect on the drag torque. In the high speed range, the drag torque increases with the lubricating oil flow. The drag torque also increases with the lubricating oil viscosity.

In the low speed range, the drag torque increases with the speed; the peak drag torque is reached at a speed; with the increase in the relative speed subsequently, the drag torque decreases.

5.5 Select-Shift Actuator

The DCT select-shift actuator can be classified into high-speed on-off valve controlled electrically operated actuator and hydraulic operated actuator. The electric shift actuator has the advantages of quick response, high precision and low price and the disadvantage of complex control system. The electrically operated shift actuator may be divided into single-motor, two-motor and multiple-motor drive modes according to the number of the select and shift control motors. In the single-motor actuator, only one motor is used to select and shift gears through the mechanical mechanism. This scheme is characterized by simple and compact structure and easy layout, but can only shift gears sequentially without large problem in the upshift. It cannot quickly put into the required gear in the downshift and is used for Getrag DCT. In the two-motor actuator, a motor is used respectively to control the selection and shift, which is simple and compact in structure and easy to arrange. The disadvantage is that the motor is difficult to control, especially the coordination between the select and shift motors. Geely DCT adopts this scheme. In the multiple-motor actuator, the select motor is canceled, and the number of control motors is determined according to the gear number of the transmission and its distribution form. The obvious disadvantages of the actuator are many motors required, excessive host, difficult layout and complex control, so it is seldom used.

The hydraulic operated select-shift actuator uses the hydraulic system to select and shift the gear. The select-shift actuator is arranged flexibly and the hydraulic oil in the hydraulic system is usually the same oil as the transmission oil. There are also cases where the hydraulic oil in the hydraulic system is not the same oil as the transmission oil. At this time, the separate hydraulic pump, accumulator and hydraulic control valve block are required. This scheme has high requirement for sealing and is used in Volkswagen DQ200.

1. **Electrically operated select-shift actuator**

Figure 5.24 shows an active interlock DCT select-shift actuator, which includes the select-shift assembly/reverse shaft and the drive unit. Figure 5.25 shows a shift finger

Fig. 5.24 Active interlock DCT select-shift actuator

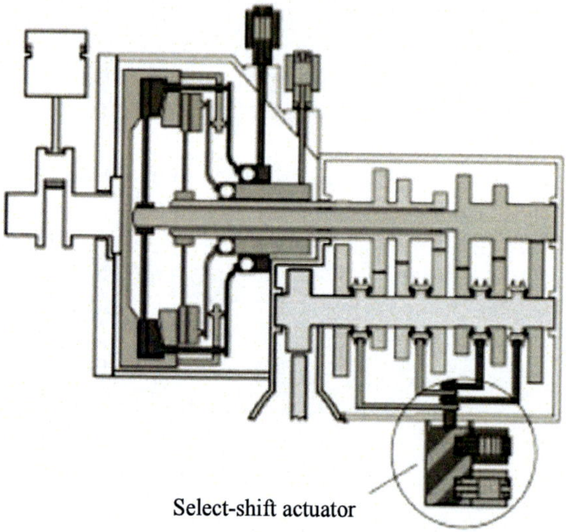

Fig. 5.25 Shift finger assembly

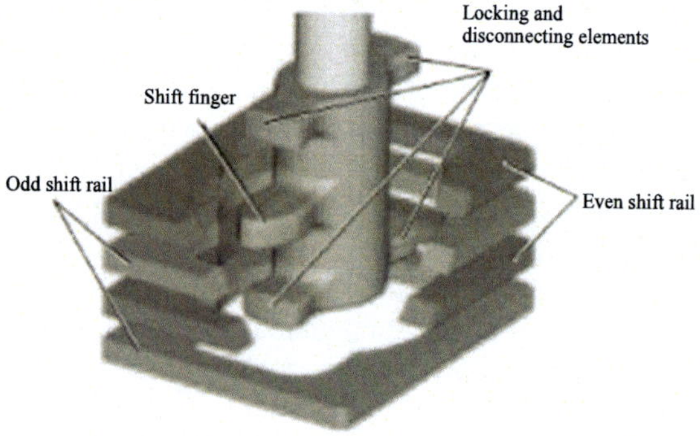

assembly consisting of shift finger, locking and disconnecting elements, which acts on the shift guide slot to drive the shift synchronizer.

The particularity of the active interlock DCT select-shift actuator lies in that the opening width of the shift guide slot is larger than the width of the shift finger. Therefore, even if one gear is already engaged, the shift shaft can be rotated in reverse and another shift guide slot can be selected by means of the shift finger (Fig. 5.26).

If a new gear is to be preselected, the locking and disconnecting elements will be disengaged from the gear that is driven by the same clutch and the rotation direction of the shift shaft is independent of the movement direction of the shift finger. Figure 5.27 shows the active interlock disengagement.

Figure 5.28 shows a double-motor select-shift actuator. The select motor is responsible for placing the shift finger in the required gear shift lever guide slot and is equipped with a sensor to identify the axial position of the shift finger assembly; the shift motor is responsible for gear engagement and disengagement and returns the

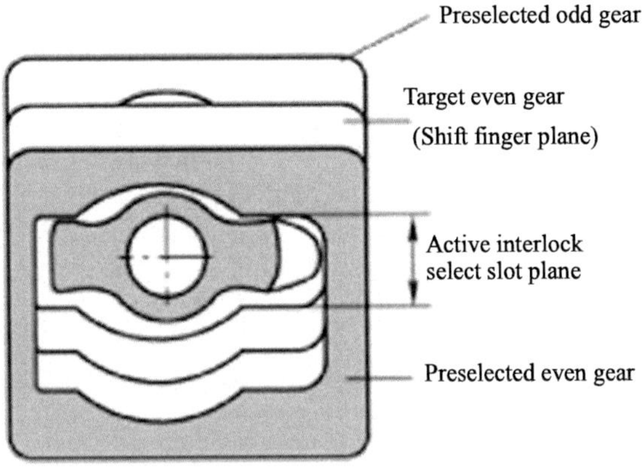

Fig. 5.26 Active interlock shift guide slot

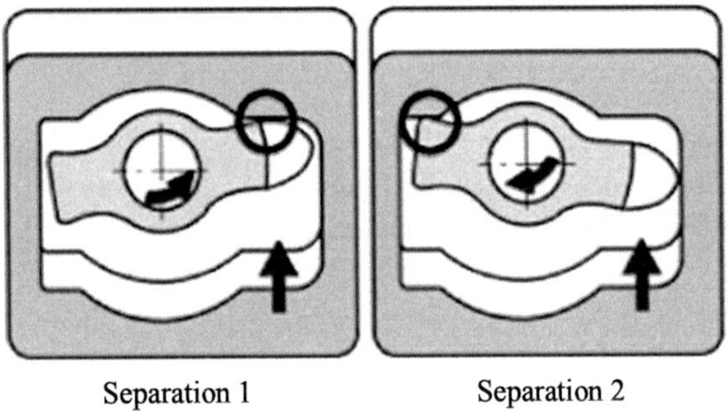

Fig. 5.27 Active interlock disengagement

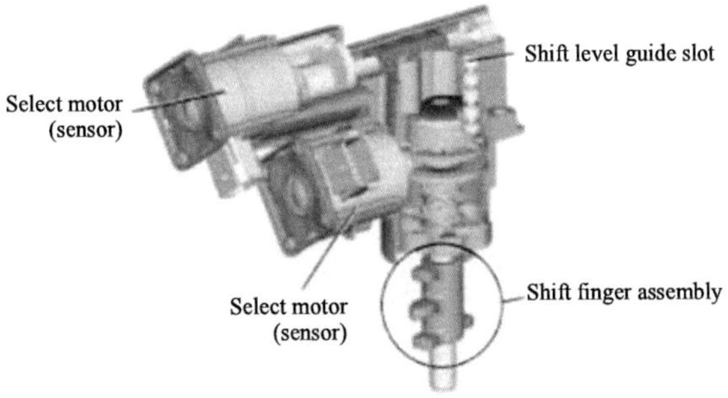

Fig. 5.28 Double-motor select-shift actuator

shift finger to the middle select position. The motor is also equipped with a sensor to identify the rotation angle of the shift finger assembly and judge the gear.

2. **Hydraulic operated select-shift actuator**

The hydraulic operated select-shift actuator mainly completes the gear selection, engagement and disengagement according to the instructions issued by the TCU. Position code of synchronizer: 0 neutral; 1 left; 2 right; 3 uncertain. The control state of the synchronizer in each gear is shown in Table 5.2. Figure 5.29 shows the hydraulic schematic diagram of the shift control module.

The solenoid valves S11 and S12 are two-position three-way on-off valves. The input end of the valves maintains the system pressure and the output end is connected with the return oil end in case of no power; the system pressure at the valve input end is connected with the output end and the system pressure is output when the solenoid valve is powered on. The solenoid valves S23 and S24 are PWM solenoid regulating valves. Different control pressure of the solenoid regulating valves is obtained depending on different PWM. The solenoid regulating valve 1 produces the working pressure of 1–3 shift fork and 5.7 shift fork; the solenoid regulating valve 2 produces the working pressure of 2–4 shift fork and 6-R shift fork.

5.6 Hydraulic Control System

I. Principle of DCT hydraulic control system

With respect to the electro-hydraulic DCT, the hydraulic control system mainly consists of the hydraulic pump oil supply module, main oil pressure regulating module, shift control module, dual clutch control module and lubrication and cooling module, as shown in Fig. 5.30. Figure 5.31 shows the DCT hydraulic system diagram and Table 5.3 lists the functions of the modules of the DCT hydraulic control system.

1. Hydraulic pump oil supply module

At present, the mechanical fixed-displacement hydraulic pump, with the general displacement of 14–18 mL/r and the basic layout principle as shown in Fig. 5.32, is commonly used in AT and DCT and driven by the engine; or the electrically driven hydraulic pump as shown in Fig. 5.33 is used to provide the required oil pressure for the hydraulic system and meets the oil pressure requirements of the system under various conditions.

2. Main oil pressure regulating module

The main oil pressure regulating module enables the hydraulic control system to control the main pressure of the system, the clutch oil supply pressure, the lubrication and cooling oil supply pressure, and the oil supply pressure of the shift system. Generally, the main pressure of the hydraulic control system changes within a certain range to meet the pressure regulation requirements under different working conditions, as well as the self-protection function under the high system pressure.

The main pressure regulator valve is the core part of the whole DCT hydraulic control system. The hydraulic pump can produce high oil pressure, which is usually limited to between 0.5–2.5 mpa. The variation range of the main oil pressure modulated is determined by the oil pressure and spring of the main pressure controlled solenoid valve. The main pressure regulator valve is controlled by the normally closed VBS (Variable Bleed Solenoid). The basic working principle of the main oil

Table 5.2 Control state of synchronizer in each gear

Action	Solenoid valve S11	Solenoid valve S12	Regulating valve S23	Regulating valve S24	Synchronizer position 5 and 7	Synchronizer position 1 and 3	Synchronizer position 4 and 4	Synchronizer position 6 and R
Engage gear 1	0	0	1	0	0	0 > 1	0	0
Disengage gear 3						4 > 0		
Engage gear 2	1	1	0	1	0	0	0 > 4	0
Disengage gear 4							1 > 0	
Engage gear 3	1	0	1	0	0	0 > 4	0	0
Disengage gear 1						1 > 0		
Engage gear 4	0	1	0	1	0	0	0 > 1	0
Disengage gear 2							4 > 0	

(continued)

Table 5.2 (continued)

Action	Solenoid valve S11	Solenoid valve S12	Regulating valve S23	Regulating valve S24	Synchronizer position 5 and 7	Synchronizer position 1 and 3	Synchronizer position 4 and 4	Synchronizer position 6 and R
Engage gear 5	0	1	1	0	0 > 1	0	0	0
Disengage gear 7					4 > 0			
Engage gear 6	0	0	0	1	0	0	0	0 > 1
Disengage gear R								4 > 0
Engage gear 7	1	1	1	0	0 > 4	0	0	0
Disengage gear 5					1 > 0			
Engage gear R	1	0	0	0	0	0	0	0 > 4
Disengage gear 6								1 > 0

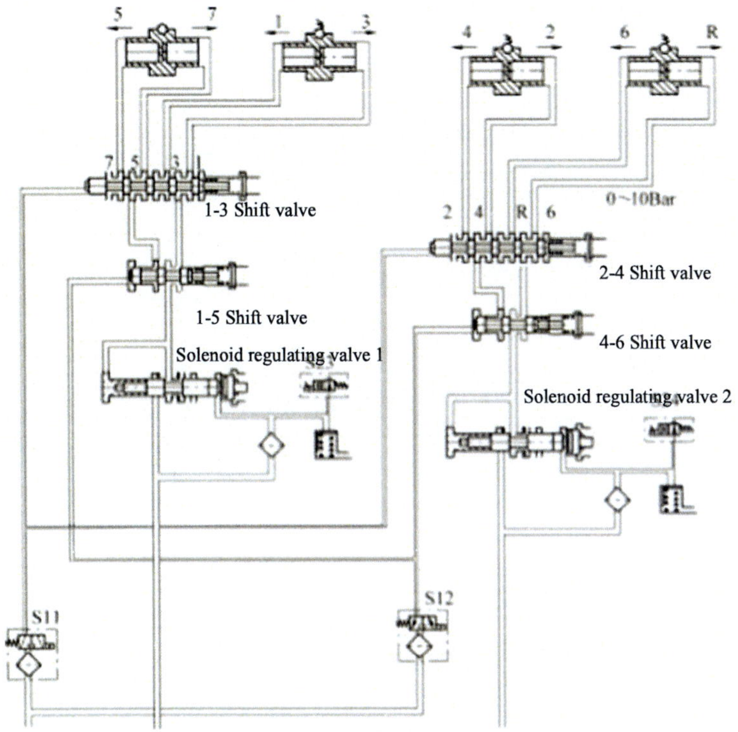

Fig. 5.29 Hydraulic schematic diagram of shift control module

pressure regulating module is shown in Fig. 5.34, and its control principle is shown in Fig. 5.35.

It can be known from Fig. 5.35 that the mechanical equilibrium equation of the main pressure regulator valve is

$$P_L A_1 = P_{VBS} A_2 + F_S \tag{5.7}$$

$$P_L = \frac{A_2}{A_1} P_{VBS} + \frac{F_S}{A_1} \tag{5.8}$$

where

P_L—main oil pressure (Pa);
A_1—main oil pressure action area (m^2);
P_{VBS}—oil pressure of control solenoid valve (Pa);
A_2—oil pressure action area of control solenoid valve (m^2);
F_S—spring force (N).

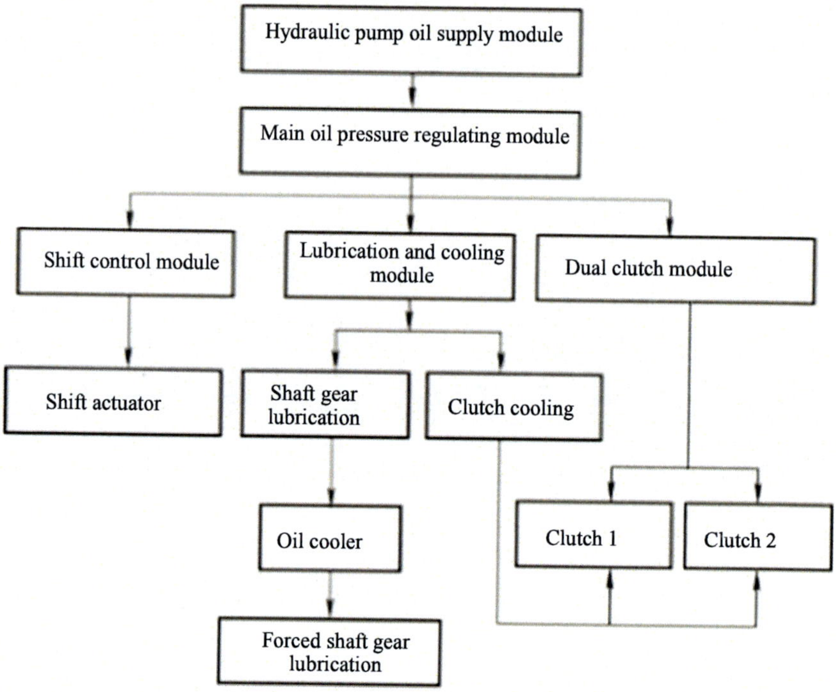

Fig. 5.30 Composition of DCT hydraulic control system

According to formula (5.8), when the spring pretightening force and the main pressure regulator valve action area are determined, the main oil pressure can be obtained only by regulating the oil pressure of the control solenoid valve.

3. **Lubrication and cooling module**

The lubrication and cooling module mainly achieves the clutch lubrication and cooling, shaft gear lubrication and cooling, heat exchange and filtration. Figure 5.36 shows the control principle of the DCT lubrication and cooling system and Table 5.4 lists the requirements and specific implementation methods of the lubrication and cooling module.

II. **Introduction to typical DCT hydraulic system**

The DCT hydraulic system is an actuator achieving the dual clutch alternating control and the shift fork control. Its basic structure is similar to, but different from, the hydraulic control system of the traditional AT and CVT. Figure 5.37 shows the hydraulic system of DCT DQ250 which has been mass produced by Volkswagen. It mainly consists of the oil supply part, dual clutch control part, shift fork control part and auxiliary part.

The oil supply part is composed of the hydraulic pump, reducing valve, main pressure regulating slide valve and pressure regulator valve. The main pressure regulating

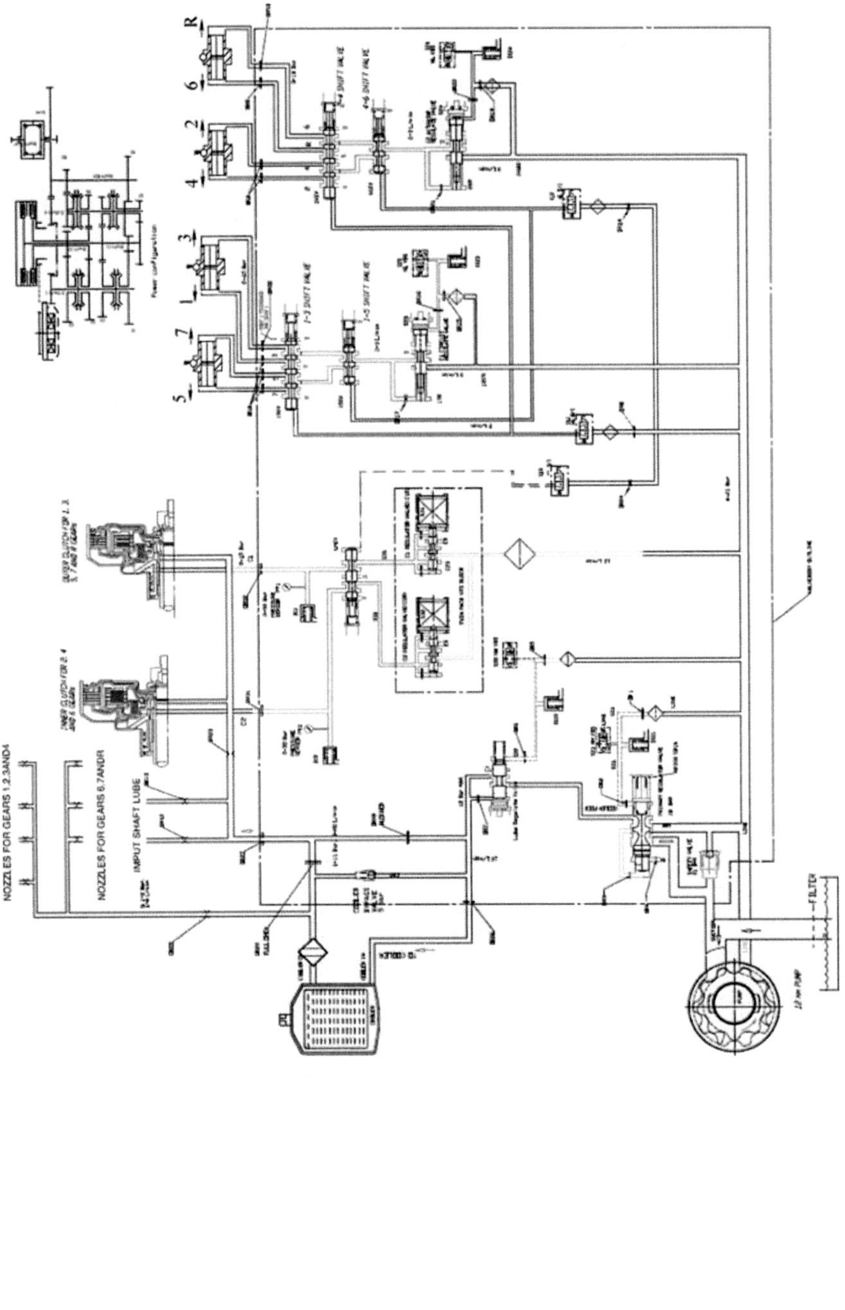

Fig. 5.31 DCT hydraulic system diagram

Table 5.3 Functions of DCT hydraulic control system modules

Module	Function
Hydraulic pump oil supply module	As a power source for hydraulic oil supply, it is driven by the engine or motor
Main oil pressure regulating module	Achieve the regulation of different system oil pressure according to different operating conditions of the transmission
System limit pressure protection module	Ensure the system oil pressure within the designed oil pressure range to prevent the system oil overpressure
Dual clutch module	Adjust the clutch pressure accurately according to different gear shift requirements and driving conditions
Shift control module	Realize accurate engagement and disengagement of gears 1–7 and gear R synchronizer
Lubrication and cooling module	Including the clutch, bearing and gear lubrication and cooling, mainly to accurately control the clutch cooling flow and ensure the reliable operation of the dual clutch
Filter system	Ensure the cleanliness of transmission oil and ensure the cleanliness of the oil into the valve body, so that the solenoid valve and valve element can work normally and reliably

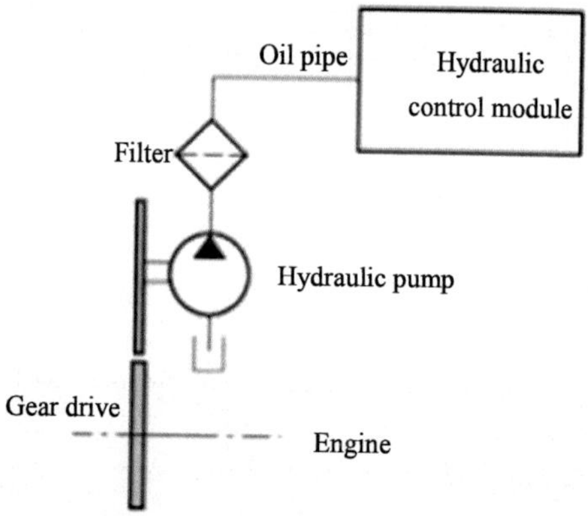

Fig. 5.32 Basic layout principle of mechanical fixed-displacement hydraulic pump

slide valve is controlled through the pressure regulator valve to regulate the main oil pressure of the hydraulic system; when the system fails and the pressure rises to a certain height, the reducing valve is pushed to release the pressure to protect the hydraulic system. The dual clutch control part is mainly composed of two relatively independent oil circuits, controlling clutch C1 and clutch C2 respectively. The two parts control the oil circuit exactly the same, including safety valve, accumulator, pressure sensor and clutch control proportional valve. The safety valve can adjust the oil supply pressure of two clutch control oil circuits and ensure that, in case of

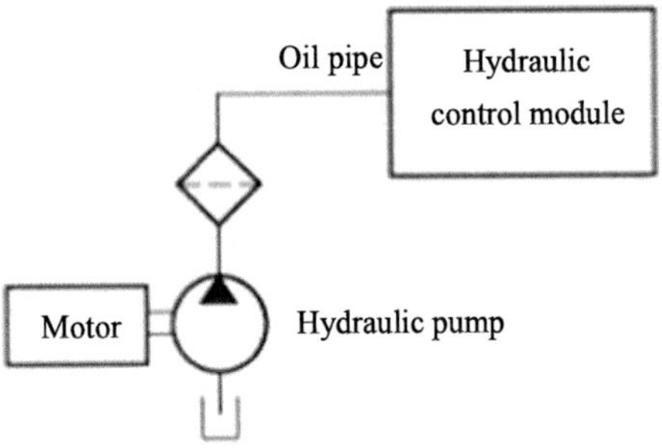

Fig. 5.33 Schematic diagram of electrically driven hydraulic pump

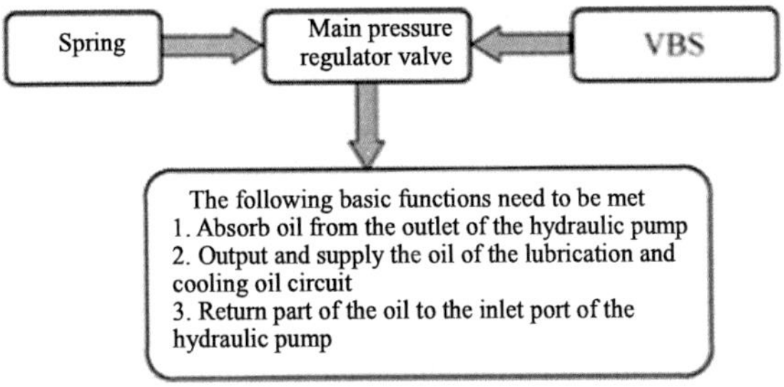

Fig. 5.34 Basic working principle of main oil pressure regulating module

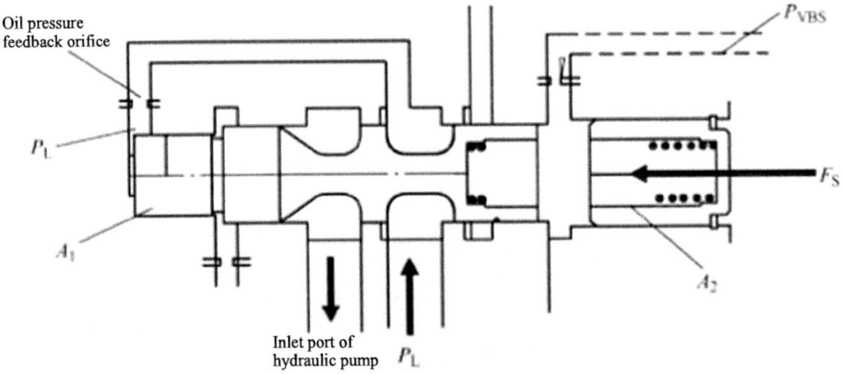

Fig. 5.35 Control principle of main oil pressure regulating module

fault of a clutch, the other clutch can work independently and safely. The clutch valves 1 and 2 are proportional solenoid valves that can achieve accurate control of the clutch working pressure; two pressure sensors provide feedback signal for the accurate control of the clutch pressure. The shift fork control part mainly consists of four on-off valves and a two-position multi-way valve, which controls the change of its operating position through the other on/off valve. The auxiliary part mainly

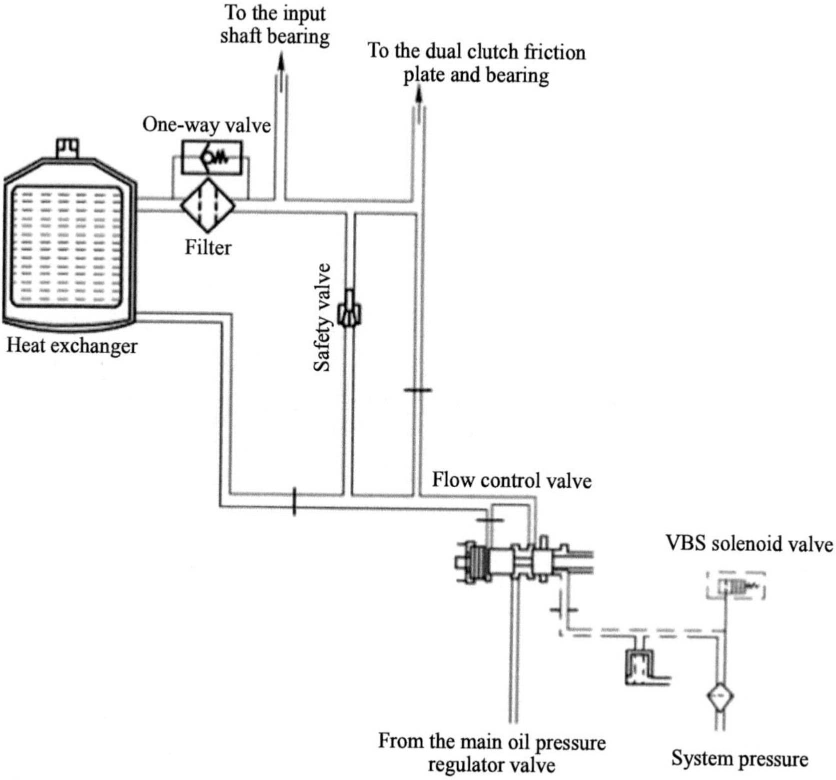

Fig. 5.36 Control principle of DCT lubrication and cooling system

Table 5.4 Requirements and specific implementation methods of lubrication and cooling module

Function	Requirement	Implementation
Clutch lubrication and cooling	The minimum lubrication flow is 2–3L/min, and the maximum lubrication flow is 18L/min, which should be adjusted precisely according to different shift requirements and working conditions	The lubrication and cooling flow of the clutch is regulated by the VBS and flow control valve and controlled by the normally closed solenoid valve to ensure the maximum cooling flow in case of failure of the solenoid valve.
Shaft gear lubrication and cooling	The lubrication flow is 1–4L/min, which may not be controlled accurately and only needs to meet the lubrication need	Obtain sufficient flow by setting the orifice diameter on the lubrication and cooling oil circuit
Achieve heat exchange and efficient filtration	In order to ensure efficient cooling and filtration, a flow of 10–20L/min is required in steady state	Required flow is obtained through the regulation of the flow control valve and the design of the orifice

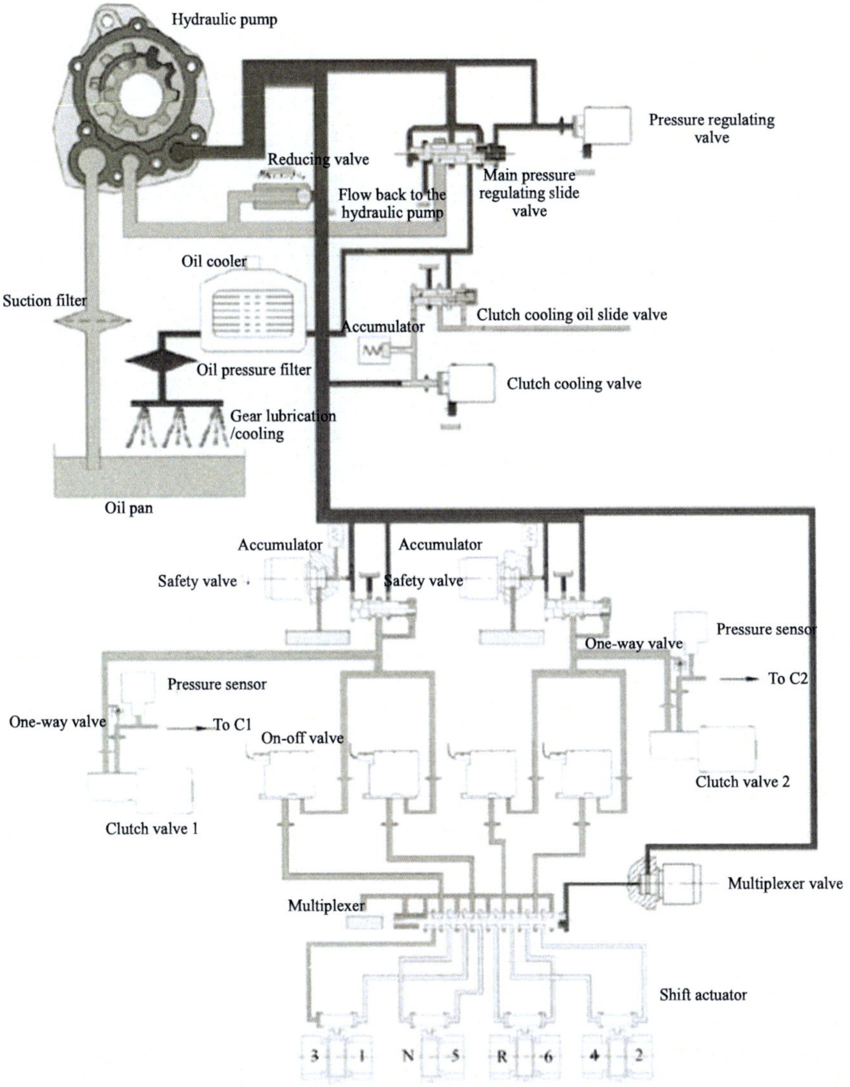

Fig. 5.37 Hydraulic system of DQ250

includes the dual clutch lubrication part, the heat dissipation and filtration part of the hydraulic system.

All kinds of hydraulic valves and solenoid valves are integrated into the hydraulic valve body, as shown in Fig. 5.38. N88 is the gear 1 and gear 3 shift fork control switch solenoid valve; N89 is gear 5 shift fork control switch solenoid valve; N90 is gear 6 and reverse gear shift fork control switch solenoid valve; N91 is gear 2 and gear 4 shift fork control switch solenoid valve; N92 is multi-way valve control switch

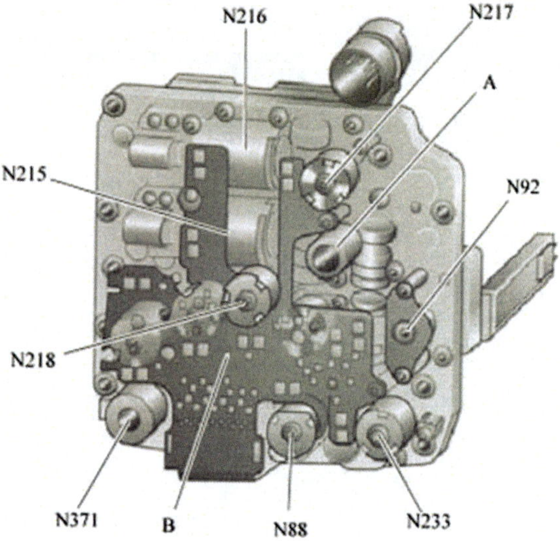

Fig. 5.38 Hydraulic valve body of DQ250

solenoid valve; N215 is the control proportional solenoid valve of clutch C1; N216 is the control proportional solenoid valve of clutch C2; N217 is the control solenoid valve of the main pressure regulating slide valve; N218 is the cooling oil flow control solenoid valve; N233 is the control oil circuit safety valve of the clutch C1; N371 is the control oil circuit safety valve of the clutch C2; A is the main oil circuit reducing valve; B is the solenoid valve power supply connector of the hydraulic valve body. Meanwhile, the hydraulic valve also integrates the pressure sensors of two clutches.

5.7 Control System Hardware Design

The electronic control system is the core device of DCT and the nerve center of the whole transmission. The development of a safe and reliable electronic control system of DCT is an important guarantee for the realization of the vehicle power, economy performance and emission targets.

I. **Composition and principle of DCT electronic control system**

The DCT electronic control system mainly consists of the input/output shaft speed sensor, parking position sensor, oil temperature sensor, TCU, other control unit CAN signal, select-shift actuator, clutch actuator and wire harness, as shown in Fig. 5.39. During the operation of DCT, besides receiving input signals from sensors in the transmission, TCU also needs to obtain input signals from other control units through CAN bus, mainly including the driver gear request signal from the electronic shifter, engine speed signal, engine torque and accelerator pedal position signal from EMS, ABS/ESP status signal and speed signal from ABS/ESP control system. The TCU

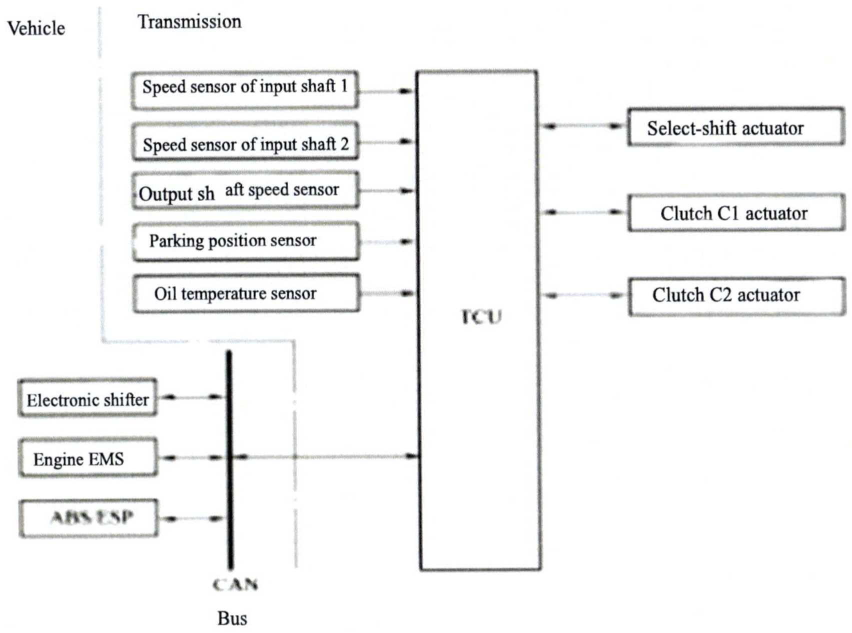

Fig. 5.39 Sketch of DCT electronic control system

analyzes and judges the input signal, makes a decision and then sends a control signal to the select-shift actuator and clutch actuator. The actuator completes the gear selecting and shift and the clutch engagement and disengagement to achieve the vehicle starting and shifting and back feeds the gear selecting and shift status, clutch engagement and disengagement status to the TCU.

II. Monitoring design of DCT electronic control system

The DCT electronic control system components can be divided into safety related components and non-safety related components, among which the non-safety related components account for a large proportion in the whole electronic control system. ISO 26262 has many stringent requirements for the development process and method of safety related components and has no requirements for the development of non-safety related components. If the non-safety related components and safety related components are developed according to the process and method in ISO 26262, the development cost will rise substantially and the development process will be extremely complex. Therefore, the electronic control system of DCT is developed by using the three-layer monitoring concept, as shown in Fig. 5.40. The hardware of the transmission controller consists of two parts: the function controller and the monitoring controller. The software is divided into three layers, of which Layer 1 is non-safety related functional software, used for the basic function control of the transmission; Layer 2 and Layer 3 are the safety related functional software. Layer

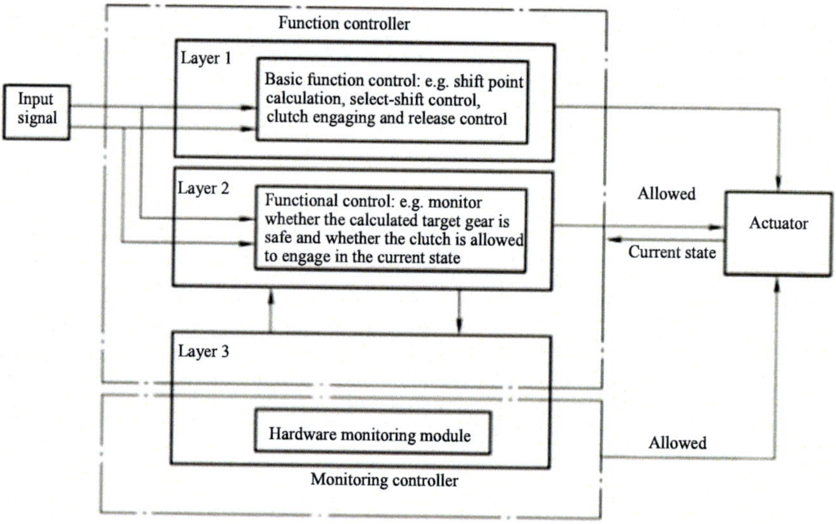

Fig. 5.40 DCT 3-layer monitoring concept

2 is the function monitoring layer, which is used to monitor whether the operation of the functional software on Layer 1 is correct. In case of error in the functional operation of the transmission, Layer 2 enters the failure response mode and restricts the function of Layer 1 by issuing a command of not allowed, so that the vehicle enters the safety status; Layer 3 is the controller monitoring layer to monitor the hardware fault in the function controller by means of the monitoring controller. Layer 2 and Layer 3 of the DCT electronic control system are developed according to the development process and requirements in ISO 26262, while Layer 1 is developed according to the general development process, which can not only meet the requirements of ISO 26262 standard, achieve functional safety, but also minimize the development workload and reduce the development cost.

III. **Safety subsystem design**

According to the functional safety concept of DCT electronic control system obtained from the whole vehicle, all safety related functional modules are listed, and the logical framework of the layer 2 safety subsystem in the DCT3 layer monitoring concept is formulated based on the signal transmission relationship between modules, as shown in Fig. 5.41. According to the DCT running environment and the boundary conditions and combined with the fault tree analysis, failure mode and effects analysis (FMEA), the technical scheme of each module to achieve 7DCT functional safety requirements is formulated, the functional safety requirements are refined into the requirements of each module and the safety mechanism of the interaction between modules and are finally assigned to the corresponding modules and signals in Fig. 5.41 to form the concept of technical safety.

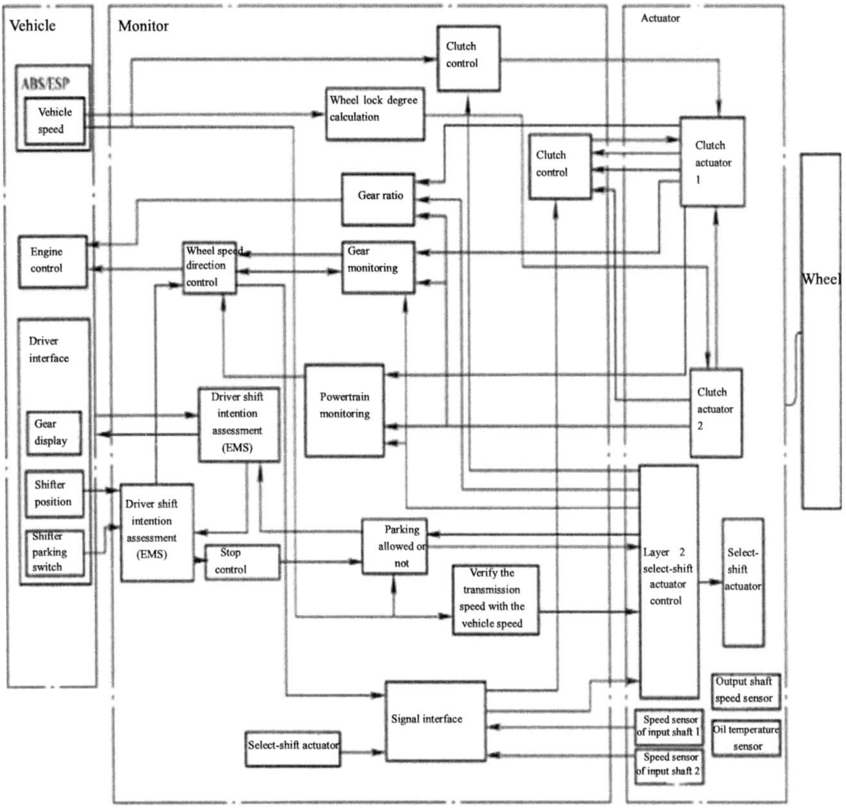

Fig. 5.41 Logical framework of safety subsystem

The safety requirement indexes for the hardware in ISO 26262 are mainly the hardware random failure rate, single-point random failure rate and potential random failure rate. Since the highest ASIL level of the safety goal of the DCT electronic control system ASIL C, according to the requirements of ISO 26262, the hardware indexes related to the safety goal ASIL A are not required and related to the safety goals ASIL B and ASIL C are required as follows: hardware random failure rate PMHF $\leq$ 100FIT, in which 1FIT = 10^{-9} h^{-1}; single-point random failure rate SPFM $\geq$ 97%; potential random failure rate LPFM $\geq$ 80%.

As the electronic control system of DCT is composed of TCU, sensor, actuator and other systems, the above indexes are the total index requirements of the safety goal for the systems. When the safety goal is realized by multiple systems, the sum of the failure rates of these systems cannot be higher than the above indexes.

Due to complex components such as TCU and select-shift actuator in the hardware development process, these components shall be developed in accordance with the ISO26262 process to meet ISO 26262 requirements and corresponding working documents shall be generated. Simple components such as sensors are less developed by companies at home and abroad according to the ISO 26262 process and can be certified according to the requirements of ISO 26262. As long as the performance, failure mode and corresponding failure rate of the components meet the technical safety requirements of DCT, they can be applied in the safety subsystem.

IV. Control system hardware testing

The control system hardware testing is an indispensable part of TCU development and a necessary means to ensure the quality of the TCU control system. The hardware testing for TCU includes hardware unit circuit testing, hardware integration testing and hardware bench testing.

1. Hardware unit circuit testing

The hardware unit circuit testing is the testing for the basic components of the testing object. It should have clear function, performance definition and interface definition, and can be distinguished from other units. Functional testing, fault injection testing and signal integrity testing are needed.

In the functional testing, the drive capability testing is mainly to detect the output capability of the unit circuit outlet, such as whether the voltage and current output by the unit circuit meet the design requirements; the high voltage resistance capability testing is mainly to detect the high voltage resistance of the unit circuit and judge whether the unit circuit meets the high voltage resistance requirements by applying a voltage higher than the rated voltage to the unit circuit and maintaining the voltage for a certain period of time; the fault protection capability testing is to detect whether the unit circuit has the protection function in the case of short circuit, open circuit, power supply reverse connection, etc.

The fault injection testing is to verify whether the protection meets the design requirements under various failure modes of the unit circuit and whether the performance of the unit circuit can be maximized under the fault state.

The signal integrity testing is the testing of signal level, overshoot, timing sequence and other indexes.

2. Hardware integration testing

On the basis of testing the hardware unit circuit, the hardware unit circuits are combined together for testing, mainly to test the accuracy of the integrated whole function and the accuracy of the interface between the units.

The functional testing is to verify the accuracy of the integrated system logic function and verify whether it meets the system requirements according to the relationship between input and output.

Besides unit circuit testing, CAN bus signal testing and clock signal testing are also included in the signal integrity testing.

Robustness testing is a test of the hardware fault tolerance, to check whether the system has sufficient protection under abnormal conditions, and whether the failure cannot be automatically recovered due to some abnormal conditions.

Fig. 5.42 Hardware testing bench

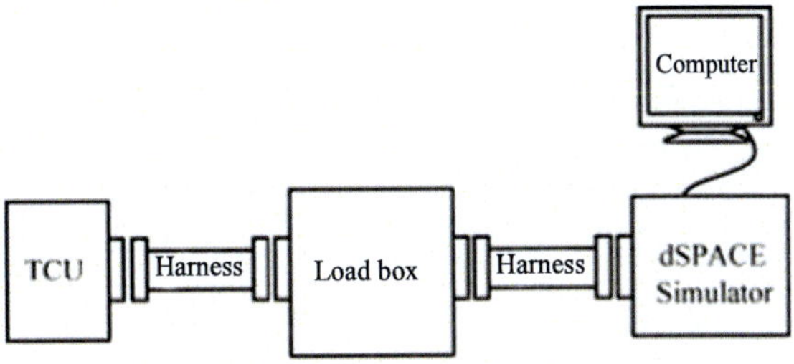

3. **Hardware bench testing**

The hardware bench testing is a comprehensive test of the developed system hardware, consisting of a TCU, a load box, a dSPACE Simulator, and a computer, as shown in Fig. 5.42. TCU is the test object; the load box is an independently designed part, including simulated loads for actuators such as relay, motor and solenoid valve; dSPACE Simulator is used as the operating environment for the transmission drive model, in which different transmission drive models are executed to simulate different types of automatic transmissions; the computer is connected to the dSPACE Simulator via a board for information interaction.

5.8 Control System Software Design

I. **DCT software architecture**

The DCT software fully embodies the developer's design intention, intuitively manifests the intelligence of the control system and mainly performs data acquisition, signal processing, gear decision-making, fault diagnosis and processing, coordination and control of the dual clutch and the engine. As the soul of the TCU, it plays an increasingly important role in the functions and characteristics of the transmission.

The TCU software of the DCT is developed according to the Autosar architecture and is modularized, integrated, interface controllable, extensible and maintainable. Figure 5.43 shows the structure of the Autosar software and Fig. 5.44 shows the interfaces of Autosar software, in which, the application software module can be sensor software module, actuator software module, or other modules.

II. **DCT underlying software**

The underlying driver software that conforms to the Autosar specification is the bridge between the hardware platform and the upper operating system and application software. It is a key software element that affects the quality and development efficiency of embedded systems. The high reliability and high development efficiency of the underlying driver and the loose coupling between the operating system and the application software are the key problems in the driver software development. Figure 5.45 shows the main steps of Autosar underlying software design. According

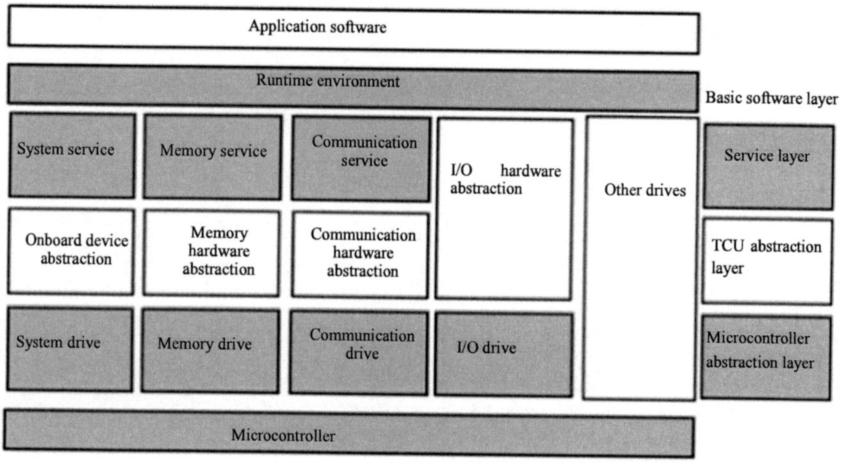

Fig. 5.43 Structure of Autosar software system

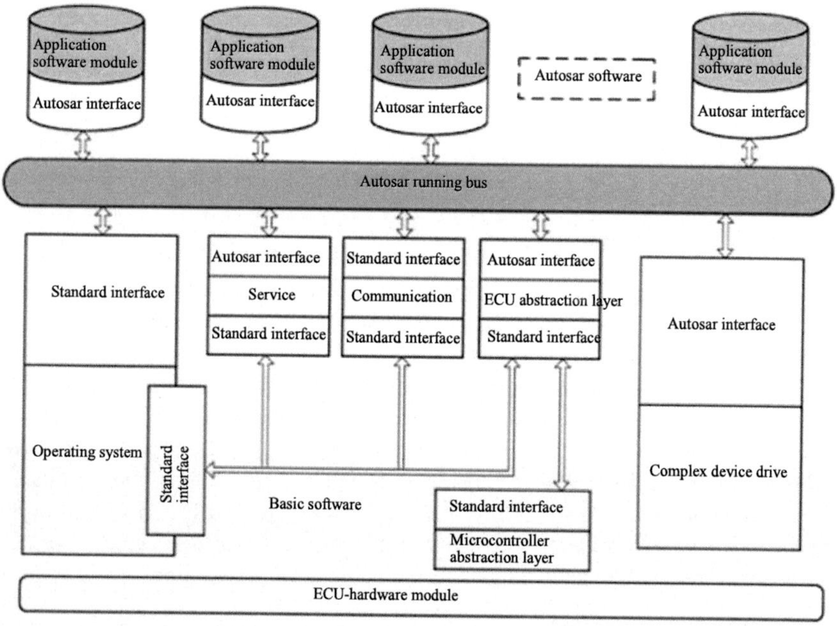

Fig. 5.44 Interfaces of Autosar software

to the Autosar specification, first define the data type and establish the drive engineering file structure system to prepare for the subsequent coding and management. Then define the in-memory images of the registers in the TCU modules, define the driver abstract interface (API) according to the Autosar specification of relevant

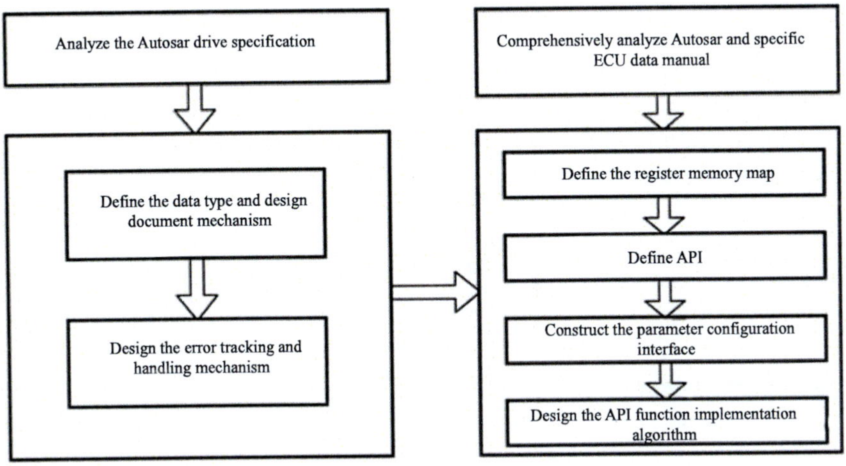

Fig. 5.45 Main steps of Autosar underlying software design

modules and combined with the specific ECU data manual, construct the parameter configuration interface and design the error tracking and handling mechanism according to the functions to be achieved by API and combined with the ECU data manual, design the API function implementation algorithm according to the drive process in the specific ECU and finally carry out the testing.

III. DCT application software

The application software is typically developed using Autosar. The entire application layer is divided into software components (SWC), which are then integrated with the underlying software (BSC) through RTE. Figure 5.46 shows the development process of the Autosar software and Fig. 5.47 shows the DCT control system application software module.

IV. DCT software testing

The DCT software testing is the same as the software testing of other control systems, including modeling specification testing, unit testing, integration testing and system testing.

The modeling specification requires readable modules, clear interfaces, accurate documents, consistent modules, codes and documents; portable model easy to maintain, reuse and modify; fast running speed, small space occupied and easy simulated analysis; easy tracking, verification and validation; easy code generation and high robustness.

The unit testing is the focus of software testing, including functional testing and structural testing of units. In this stage, bugs are found and modified to lay a foundation for the smooth implementation of integrated software.

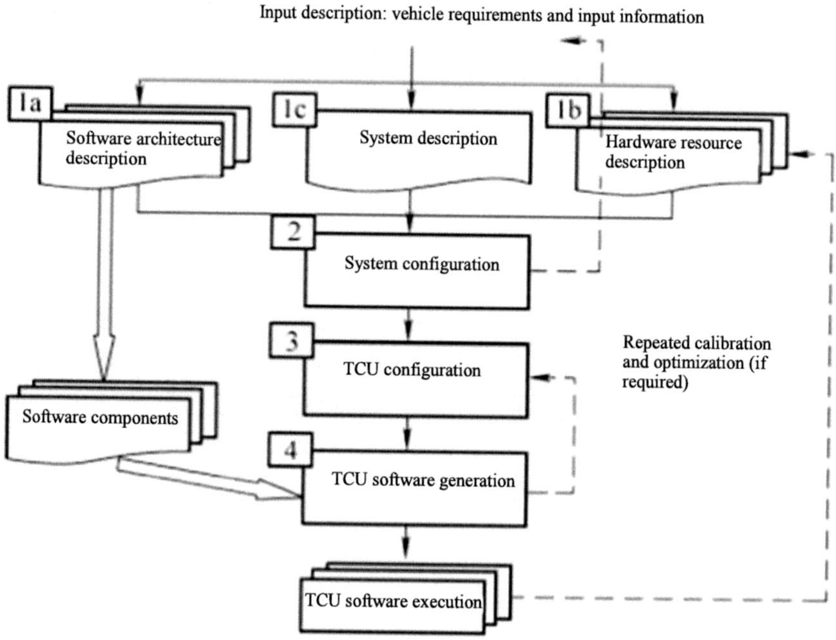

Fig. 5.46 Development process of Autosar software

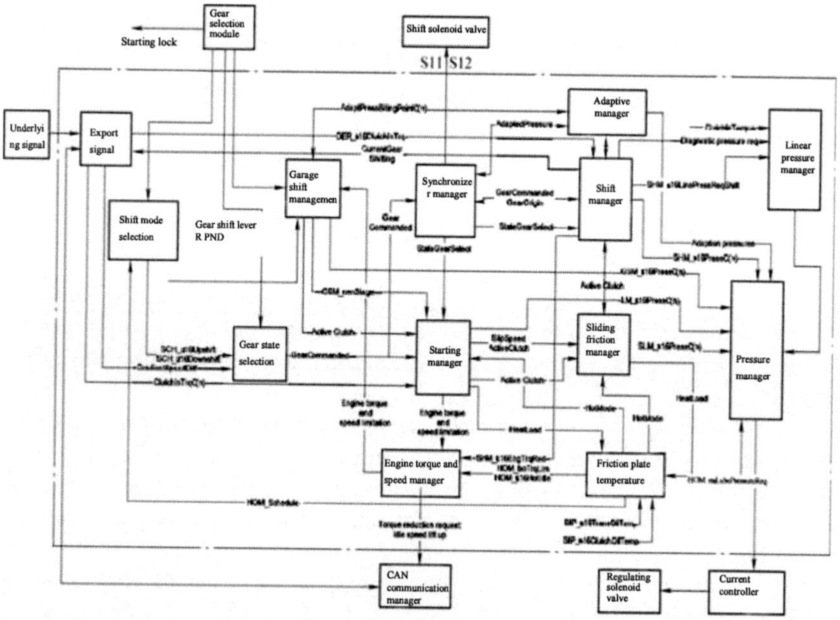

Fig. 5.47 DCT control system application software modules

The integration testing is to test all the software units that have passed the unit testing and find the problems after the integration of all the software. It is the fundamental guarantee of system software quality.

The system testing is the process of using manual or automatic means to run or test system software to verify whether the system under test meets specified system requirements.

For the specific testing of DCT software, please refer to relevant software testing data.

Bibliography

1. Chongbo J, Shihua Y, Xiaolin G (2005) Analysis of dual clutch transmission and its application prospect. Mech Drive 29(3):56–58
2. Berger EJ, Sadeghi F, Krousgrill CM (1997) Torque transmission characteristics of automatic transmission wet clutches: experimental results and numerical comparison. Tribol Trans 40(4):539–548
3. Yang Y, Lam RC (1998) Theoretical and experimental studies on the interface phenomena during the engagement of automatic transmission clutch. Tribol Lett 5(1):57–67
4. Zhenjun L, Datong Q, Ming Y et al (2005) Analysis of double clutch automatic transmission for vehicles. Trans Chin Soc Agric Mach 36(11):161–164
5. Jialin S, Zhenjun L, Datong Q (2009) Anti-interference design of hardware of dual-clutch automatic transmission. Mechatronics 11:29–36
6. Clark KS, Singh T, Buffa RP et al (2015) General motors front wheel drive seven speed dry dual clutch automatic transmission. SAE Int J Eng 8(3):1379–1390
7. Zhixuan J, Yongfu D, Yougang G (2013) Simulating start of wet dual clutch automatic transmission. Mech Sci Technol Aerosp Eng 32(3):410–415
8. Senatore A (2009) Advances in the automotive systems: an overview of dual-clutch transmissions. Recent Pat Mech Eng 2(2):93–101
9. Xi L, Xiusheng C, Wei F (2011) Optimal control of gearshift in wet dual-clucth automatic transmission. Trans CSAE 27(6):152–156
10. Hurley S, Tipton CD, Cook SP (2006) Lubricant technology for dual clutch transmissions. SAE Technical Paper Series, 2006-01-3245
11. Kim N, Lohse-Busch H, Rousseau A (2014) Development of a model of the dual clutch transmission in autonomie and validation with dynamometer. Int J Autom Technol 15(2):263–271
12. Zhenjun L, Fei L, Xiaohong D et al (2011) Analysis of the transmission scheme of a dual clutch transmission. Mech Sci Technol Aerosp Eng 30(2):270–274
13. Cheng Y, Dong P, Yang S et al (2015) Virtual clutch controller for clutch-to-clutch shifts in planetary-type automatic transmission. Math Probl Eng
14. Jianguo Z, Yulong L, Hongbo L et al (2010) Rapid control prototype of dry dual clutch transmission and platform test. J Jilin Univ Eng Technol Edit 40(4):901–905

Chapter 6
Automated Mechanical Transmission

6.1 Overview

Automatic transmission control is an important technology in modern automobile and also the basic function of intelligent vehicles. With the development of the automobile industry, people put forward higher and higher requirements for the convenience and comfort of automotive controllability, and the increase of non-professional drivers has made the automatic transmission that can facilitate driving and reduce the labor intensity of drivers have a broad market. The advantages of automating the powertrain of the automobile: reducing the operating frequency of the driver; easily realizing the best match of powertrain and getting good power and economy performance; reducing exhaust pollution. The automated mechanical transmission (AMT) shown in Fig. 6.1 is developed on the basis of MT. With the advantages of high transmission efficiency, low cost and easy manufacturing, it is very suitable for the development of China's automobile industry and has a good industrialization prospect and a wide range of applications.

By the select-shift actuator and clutch actuator, the AMT is mainly classified into electro-hydraulic AMT and electrically driven AMT. For example, Magneti Marelli electro-hydraulic AMT, as shown in Fig. 6.2, is a set of electro-hydraulic actuator added to the manual transmission. It is more expensive than the popular electromechanical actuators at present, but with its high reliability, it is also welcomed by some OEMs and has been widely used in Chery QQ3, Riich M1, Chevrolet Sail, SAIC MG3 and other models. The electrically driven AMT mainly includes Getrag AMT (Fig. 6.3) and Tsingshan AMT (Fig. 6.4). The models equipped with the Getrag AMT include JOYEAR and Haima Knight, etc. and the characteristic is that the hub shift technology is applied to the AMT; Tsingshan AMT adopts the worm gear electromechanical actuator and is mainly used in Changan mini Benben and JAC Tongyue.

Y. Chen, *Automotive Transmissions*, Springer Tracts in Mechanical Engineering,
https://doi.org/10.1007/978-981-15-6703-2_6

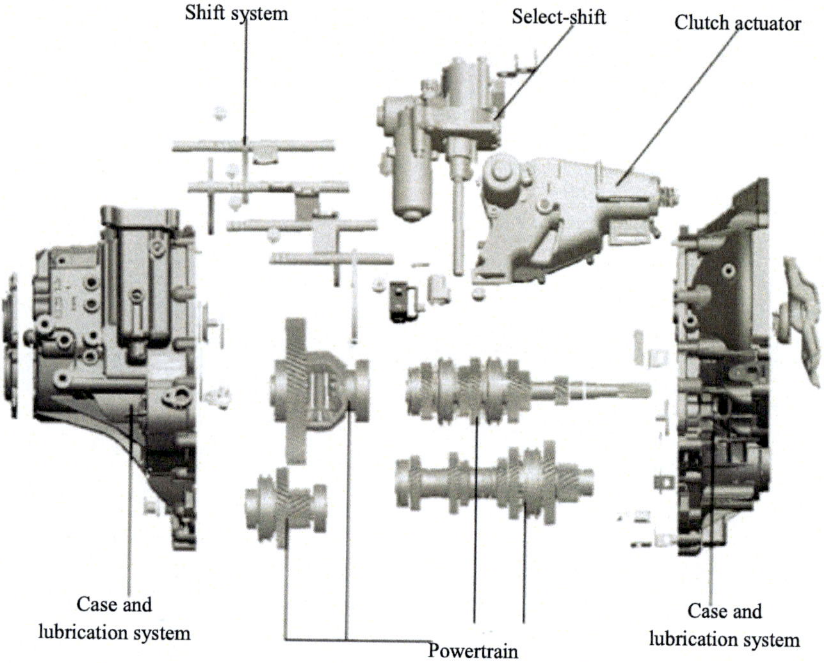

Fig. 6.1 Structure chart of AMT system

Fig. 6.2 Magneti Marelli electro-hydraulic AMT

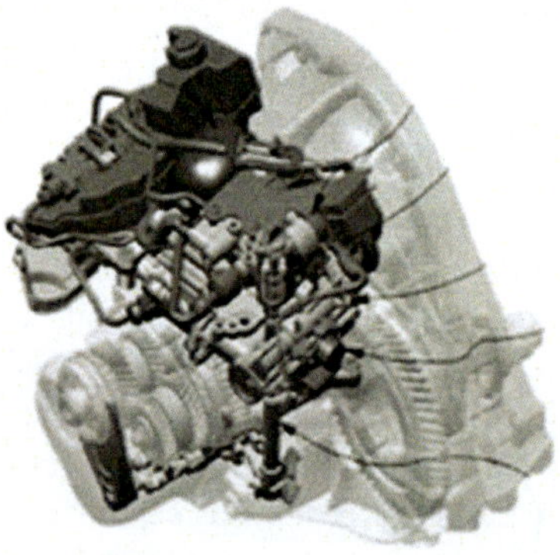

6.2 Composition and Working Principle of AMT Control System

I. Composition of AMT control system

As shown in Fig. 6.5, the AMT control system mainly consists of the software system and hardware system. The hardware system consists of the controlled object,

Fig. 6.3 Getrag AMT

Fig. 6.4 Tsingshan AMT

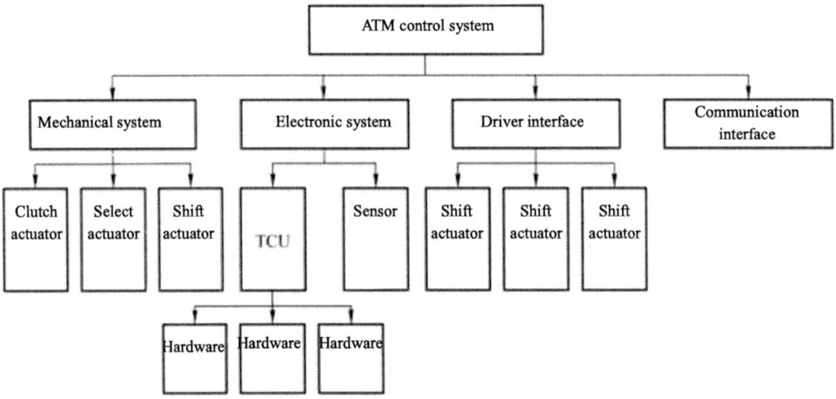

Fig. 6.5 Composition of AMT control system

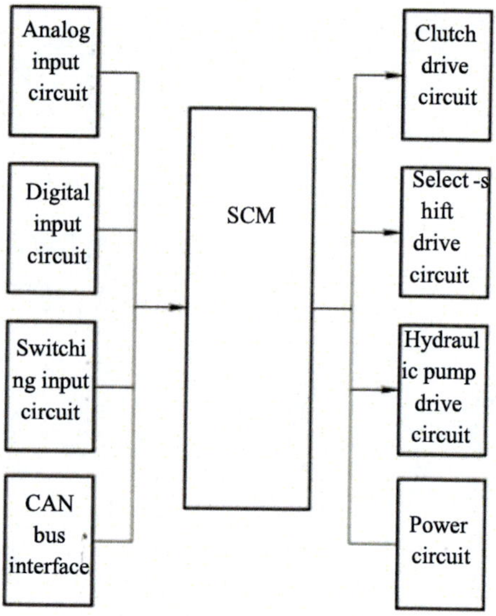

Fig. 6.6 TCU hardware structure

actuator, sensor and electronic control unit (TCU); the software system consists of the application layer software and underlying software.

1. **TCU**

AMT TCU, as the core of its control system, has the functions of program storage, signal collection, processing, judging, making decisions and issuing instructions. The hardware structure of TCU is shown in Fig. 6.6, and its functions are shown in Fig. 6.7. The clutch drive circuit and the select-shift drive circuit are different depending on the actuators used. The solenoid valve is used for the electro-hydraulic control system and the motor drive circuit is used for the full motor drive system; only the electro-hydraulic control system has the hydraulic pump drive circuit. CAN bus interface is used for data exchange and communication between TCU and engine, braking system, instrument, etc. The stored programs mainly include shift law, functions and databases used in starting, shifting and fault diagnosis, data acquisition and processing program, control programs such as gear decision-making, starting, shifting and braking, as well as judgment reasoning programs such as vehicle body mass, road grade and driver type.

2. **Sensor and actuator**

The sensor is to obtain all kinds of information during the vehicle driving, and transmit them to TCU. TCU analyzes the information, identifies the driver intention, judges the operating conditions, and issues corresponding control instructions. With the improvement of control theory and control accuracy, the types of sensors used in AMT control system are increasing. Table 6.1 lists the main sensors used in the current AMT control system and the CAN communication information. The actuator,

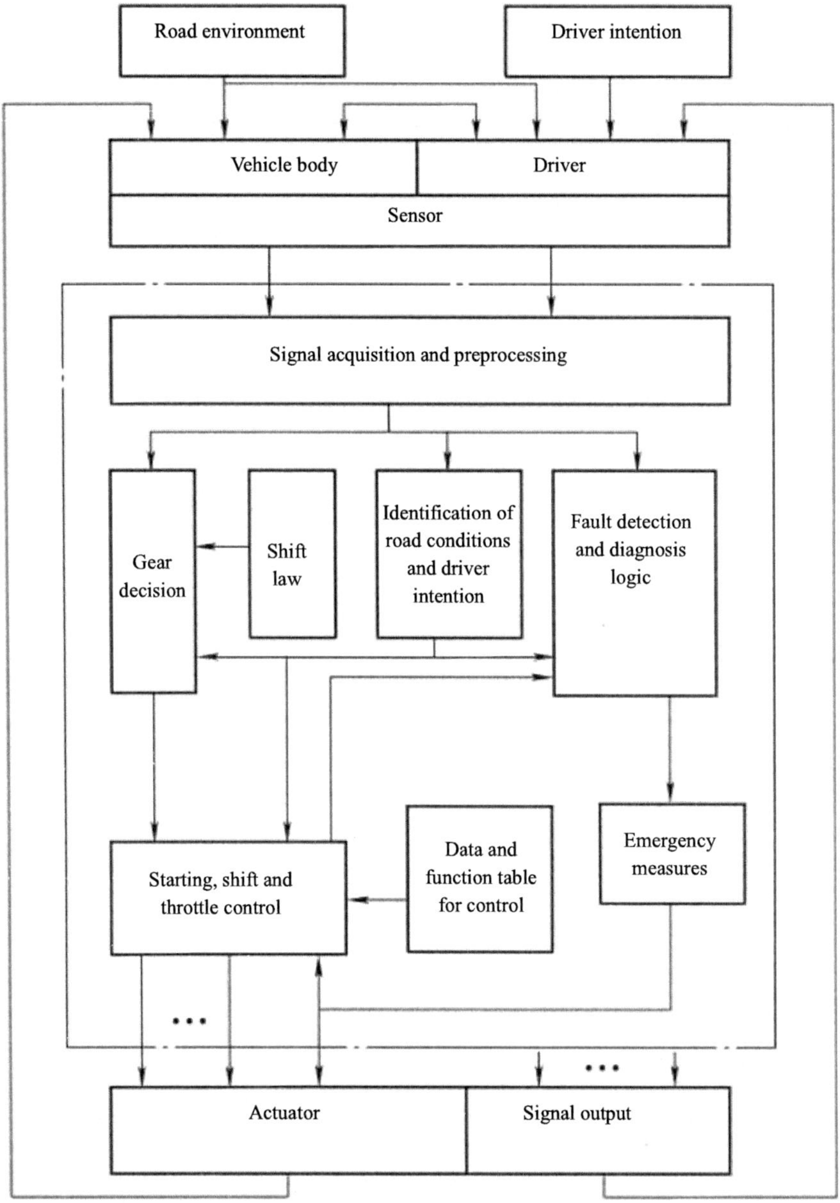

Fig. 6.7 Functions of TCU and its relationship with ambient conditions

Table 6.1 Main sensors used in AMT control system and CAN communication information

Signal feature	Signal name	Remarks
Input shaft speed	Pulse signal	Hall speed sensor
Transmission temperature	0–5 V analog signal	Oil temperature sensor
Gear shift lever position	Consist of four switching signals	R, N, Def, A/M, M+, M−
Clutch position	0–5 V analog signal	Angle sensor
Gear selection position	0–5 V analog signal	Angle sensor
Shift position	0–5 V analog signal	Angle sensor
Brake state	Switching signal 0 and 1	Position sensor/CAN information, sent by EMS
Accelerator pedal position	0–100%	CAN information, sent by EMS
Engine speed	0–16,000 r/min	CAN information, sent by EMS
Real engine torque	0–100%	CAN information, sent by EMS
Torque required by driver	0–100%	CAN information, sent by EMS
Minimum engine torque	0–100%	CAN information, sent by EMS
Friction torque	0–100%	CAN information, sent by EMS
Battery voltage	0–25 V	CAN information, sent by EMS
Vehicle speed	0–360 km/h	CAN information, sent by EMS
Engine throttle position	0–100%	CAN information, sent by EMS

which is classified into hydraulic actuator and motor drive actuator, receives control instructions from the TCU and performs various control actions during starting and shifting. Figure 6.8 shows the schematic diagram of the AMT hydraulic actuator.

II. Basic functions and working principle of AMT control system

The AMT control system has the following basic functions: it can shift gears automatically and smoothly according to the optimal dynamic shift law; it can shift gears automatically and smoothly according to the optimal economic shift law of least fuel consumption; it can start smoothly on various roads; with a certain fault self-diagnosis function and fault display and alarm function, it is easy to maintain and repair; with certain fault-tolerant control measures, it can ensure the basic driving capacity of the vehicle in some fault cases.

In addition to the basic functions mentioned above, the AMT control system shall also meet the following requirements: the shift process is rapid, and the time to

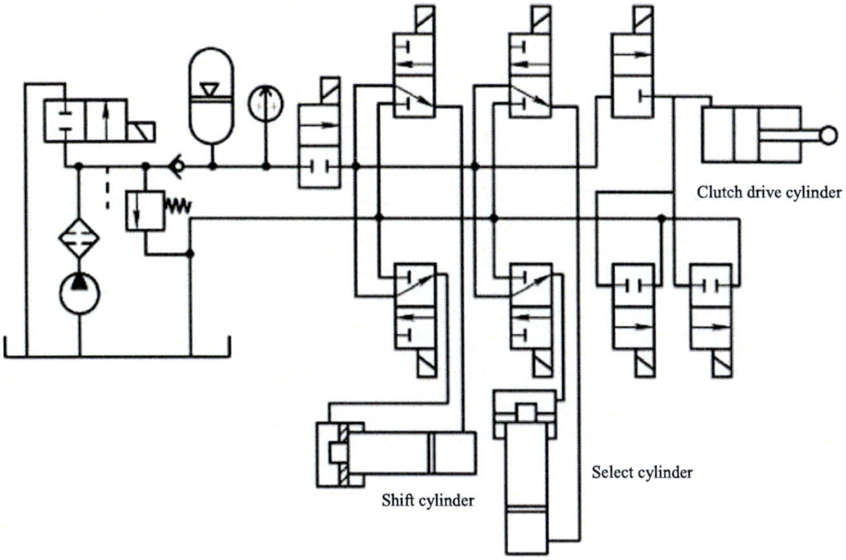

Fig. 6.8 Schematic diagram of AMT hydraulic actuator

complete the whole shift process shall not be too long; during shift, the transmission shall have no obvious impact and noise; the engagement of the clutch shall be smooth and gentle to reduce the sliding friction work, and extend the service life of the clutch friction plate; the control system has strong anti-interference ability and strong robustness.

The working principle of the AMT control system is shown in Fig. 6.9. The TCU controls the corresponding actuators (clutch actuator, select-shift actuator) according to the driver's intention to control the accelerator pedal, brake pedal and joy stick (e.g. start, stop, reverse and forced gear) and the working conditions of the vehicle (engine speed, vehicle speed, gear and clutch status) and appropriate control law (shift law and clutch engagement law) to achieve automatic control of the vehicle powertrain (clutch and transmission).

6.3 Shifter

The shifter is to provide the TCU with the driver's gear requirements. There are many types of shifters. Figure 6.10 shows the appearance of a shifter assembly and its connectors.

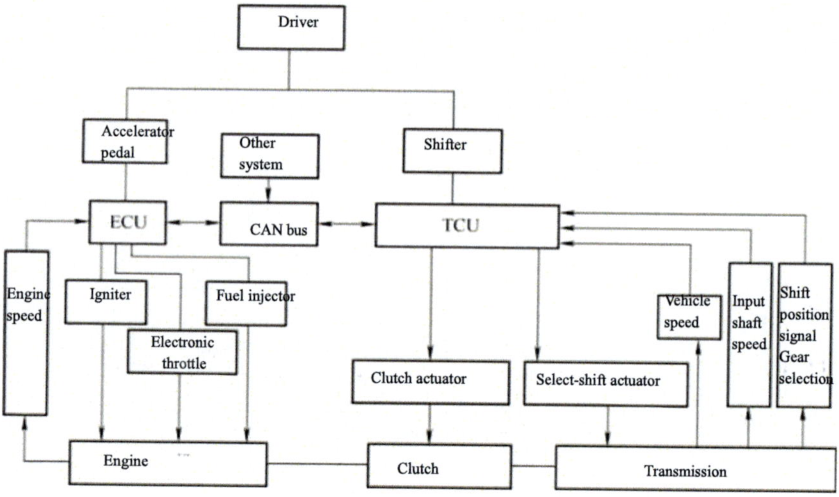

Fig. 6.9 Working principle of AMT control system

Fig. 6.10 Appearance of shifter assembly and its connectors

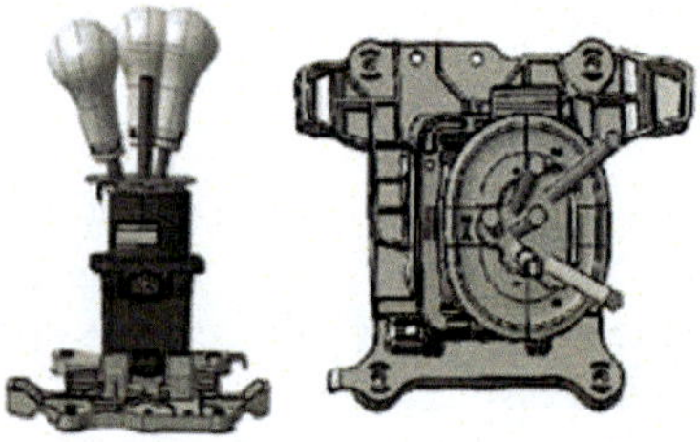

The gear definition and the gear output signals are as follows:

(1) The vehicle can park in D (drive), R (reverse) and N (neutral).
(2) The vehicle cannot park in M/A, + or −. When these three gears are released, the gear shift lever automatically returns to D.
(3) M/A is manual/automatic switching mode, the + and − gears are valid in manual mode, and the current mode can only be judged by the instrument display.

Table 6.2 lists the truth output of the internal circuit when the shifter is at different positions and Fig. 6.11 shows the electronic shifter assembly interface circuit.

When the GSL is at low level (logic 0), $Vad = 1.47/(1.47 + 3.3) \times 5\ V = 1.54\ V$ (transistor break-over voltage not considered).

When the GSL is at high level (logic 1), $Vad = (3.32 + 1.47)/(3.32 + 1.47 + 3.3) \times 5\ V = 2.96\ V$.

Therefore, the voltage signal range at low level is $(1.54 \pm 5\%)$ V and the voltage signal range at high level is $(2.96 \pm 5\%)$ V. The error of the voltage and resistance is not considered in the above calculation formula.

Shift force requirement: the shift force range at each position is 15–20 N.

Table 6.2 Shifter truth table

GSL1	Gear shift lever position	GSL2	GSL0	GSL3
A/M	0	1	1	0
+	0	0	1	1
D	0	1	0	1
–	1	0	1	0
N	1	0	0	1
R	1	1	0	0
Transition state	1	1	1	0

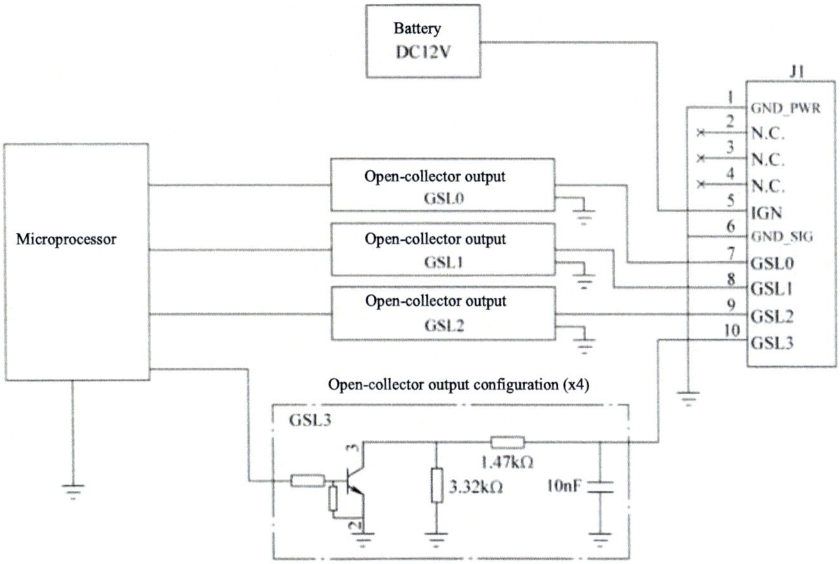

Fig. 6.11 Electronic shifter assembly interface circuit

The shift stroke (stroke at the handle top) is shown in Fig. 6.12.
M/A → D (S1): 32.2 mm;
D → + (S2): 33.1 mm;
D → – (S3): 33.1 mm;
D → right limit position (S4): 57.2 mm;
Right limit position → N (S5): 51.5 mm;
Right limit position → R (S6): 51.5 mm.

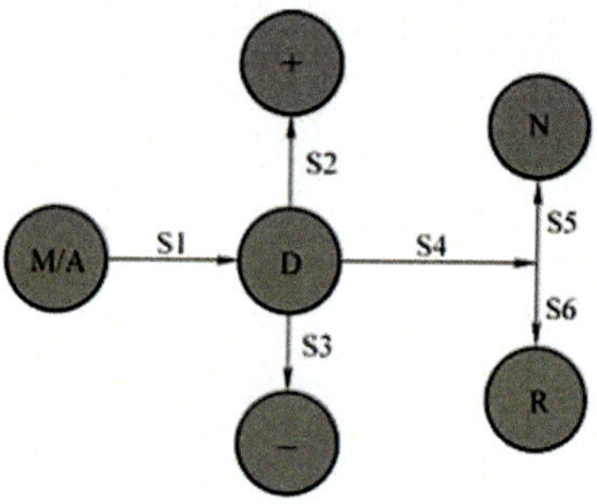

Fig. 6.12 Shift stroke (stroke at the handle top)

6.4 Clutch

The clutch is an important part of the AMT. In the powertrain, the clutch is used to connect the engine and the transmission together to connect and interrupt the power. When starting, the clutch enables the engine and powertrain to engage smoothly, so that the vehicle starts smoothly; when shifting, the clutch disengages the engine from the powertrain, reducing the rotational inertia input by the transmission and facilitating the shift; under large dynamic load impact, the clutch protects the engine and powertrain from large dynamic load impact. The AMT clutch actuator is to complete the automatic operation of the two processes of clutch release and engagement and its performance directly affects the control process of the clutch and the life of the system, thus affecting the starting process and shift quality.

I. **Clutch characteristics**

The clutch used in the AMT automotive powertrain is a diaphragm spring clutch, with the characteristics shown in Fig. 6.13. In the figure, point B is the operating point of the diaphragm spring under the engagement state of the new clutch, point A is

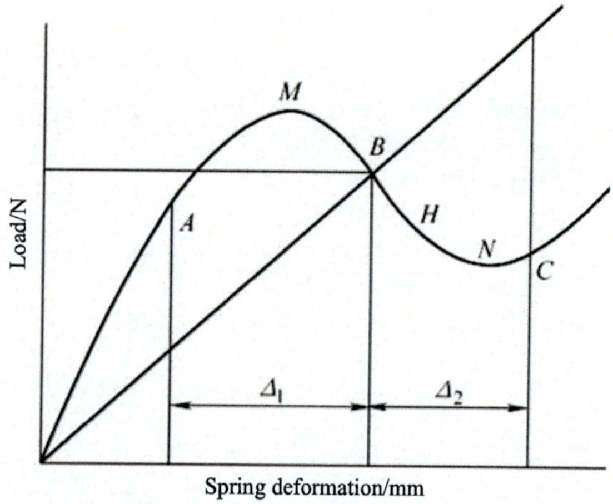

Fig. 6.13 Characteristics of diaphragm spring clutch

the ultimate operating point of the clutch, and section AB is the maximum allowable wear of the clutch friction plate.

The transmitted torque of the diaphragm spring clutch can be expressed as

$$T_c = z\mu_c R_c F_b \tag{6.1}$$

where,

T_c—torque transmitted by clutch;
z—number of friction surfaces of clutch;
μ_c—friction factor;
R_c—equivalent radius of friction;
F_b—friction surface packing force.

The torque transmitted by clutch can be realized by controlling the friction surface packing force. The test shows that the torque transmitted by clutch has the following relationship with the driven plate deformation

$$T_c = a\delta^3 + b\delta^2 + c\delta \tag{6.2}$$

Figure 6.14 shows the torque transmitted by clutch and the OA in the figure corresponds to the total clutch release state; point A corresponds to the state of the clutch just engaged; section AB indicates that with the engagement of the clutch, the torque transmitted by clutch increases gradually, but it cannot balance the torque generated by the sliding force of the vehicle; section BC indicates that the torque transmitted by clutch can balance the torque generated by the sliding force of the vehicle; section CE indicates that with the engagement of the clutch, the transmitted

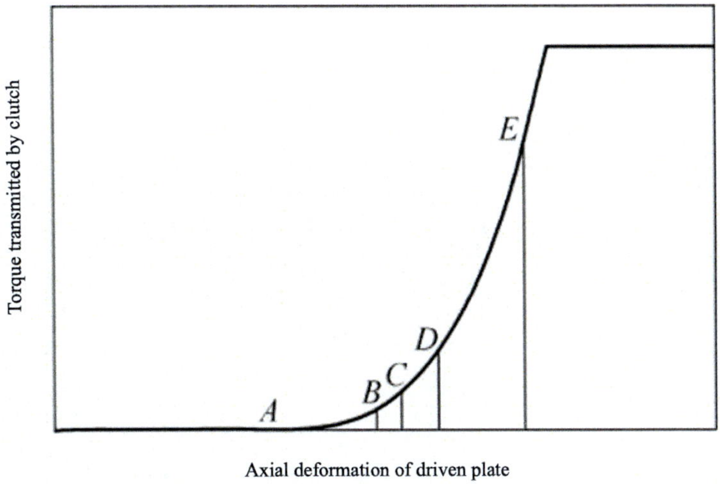

Fig. 6.14 Torque transmitted by clutch

torque increases and the vehicle starts; point E represents the maximum output torque of the engine.

II. Clutch actuator

The AMT clutch actuator mainly has two types: electrically controlled hydraulic type and electrically controlled electric type. Compared with the electronically controlled hydraulic actuator, the electronically controlled electric actuator has the advantages of low cost, high reliability, good system maintenance, low energy consumption, strong environmental adaptability, etc. The electrically controlled AMT clutch actuator is mainly analyzed here.

1. Performance requirements of clutch actuator assembly

The actuator shall be able to make the clutch disengage quickly and completely and engage slowly and flexibly during work and can achieve the automatic compensation for the friction plate wear. The response time of the clutch actuator motor can meet the vehicle requirement for the clutch actuator.

According to the requirements of the clutch actuator assembly, and combined with the vehicle layout, the design requirements of the clutch actuator assembly are as follows: the clutch actuator motor meets the vehicle reliability and durability requirements for the assembly; the clutch actuator meets the layout position and space dimension requirements in the transmission; the clutch actuator shall have a mechanism for fast separation and slow engagement; the clutch actuator shall have the mechanism for automatic compensation of the wear plate wear.

According to the clutch actuator principle and vehicle layout space requirements, the final appearance is shown in Fig. 6.15 and the internal structure is shown in Fig. 6.16.

2. Clutch actuator assembly principle

The clutch actuator is mainly composed of the motor, drive gear assembly, driven gear assembly, housing and housing cover plate, clutch push rod, wear self-adjusting

Fig. 6.15 Appearance of clutch actuator

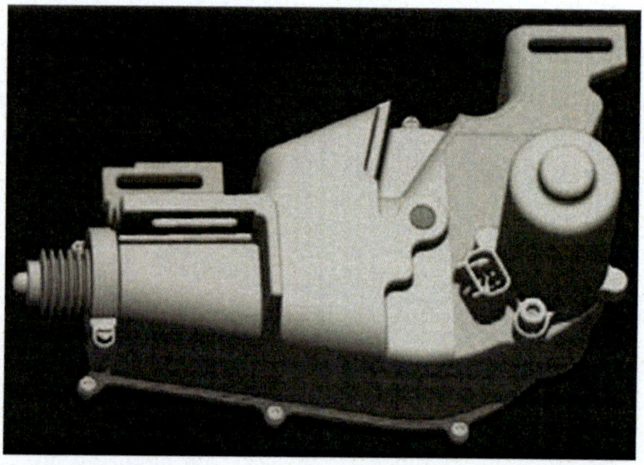

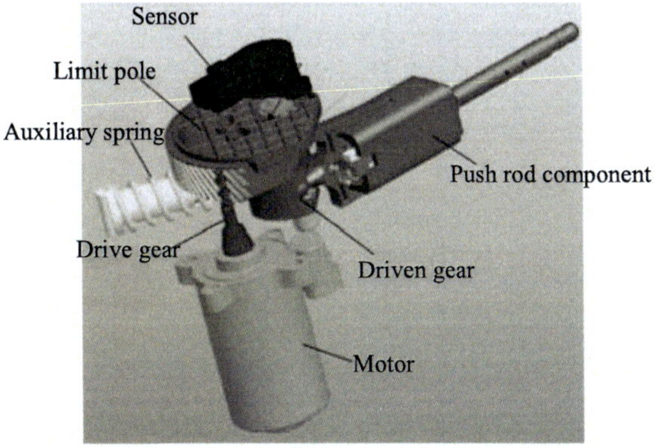

Fig. 6.16 Internal structure of clutch actuator

mechanism, auxiliary spring, damper spring and protective cover. Drive path of the actuator: motor → drive gear assembly → driven gear assembly → clutch push rod → clutch release lever. When turning the axis of the rotating shaft, the auxiliary spring is suddenly extended, pushing the turbine to turn quickly, thus the push rod pushes the clutch release claw, so that the clutch is quickly released.

The clutch actuator is to realize the engagement and release of the clutch, as shown in Fig. 6.17. The fore and aft motion of the push rod of the clutch actuator realizes the release and engagement of the clutch.

According to the design requirements of the clutch actuator in this paper, the principle of the clutch actuator is shown in Fig. 6.18.

The parameter matching calculation process of the clutch actuator is shown in Fig. 6.19.

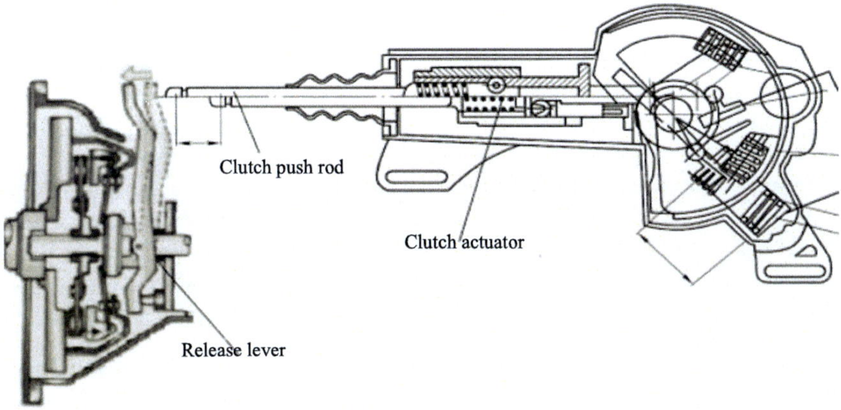

Fig. 6.17 Clutch engagement and disengagement

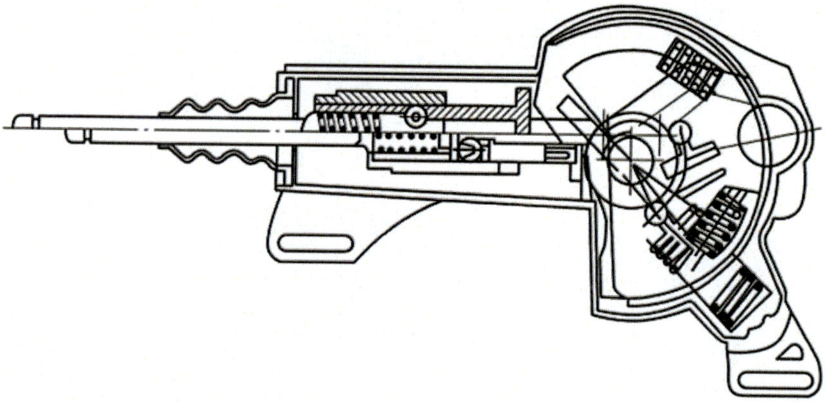

Fig. 6.18 Principle of clutch actuator

6.5 Select-Shift Actuator

The AMT gear shift is accomplished by controlling the select and shift motors. There are strict timing requirements for gear selection and shift control, that is, only after the right gear is selected can the gear shift operation be carried out, and only when the gear is neutral can the gear selection operation be carried out. To achieve a fast and smooth shift, the transmission select-shift control needs to be coordinated with the engine control and clutch control.

I. **Performance and design requirements for select-shift actuator**

The select-shift actuator shall be of high stability and high life and adapt to different conditions; the select-shift shall be completed quickly without shift impact; the locating is required during the shift; two gears cannot be engaged simultaneously; the gear identification function is required.

According to the select-shift actuator requirements and combined with the vehicle layout, the design requirements for the select-shift actuator are as follows: the select-shift motor meets the assembly life requirements; the select-shift actuator meets the layout position and space requirements in the transmission; the select-shift actuator shall be designed with the structure that can achieve fast shift speed and small impact; the select-shift actuator shall have a locating device; the select-shift actuator shall have an interlock plate device; the select-shift actuator shall have an electronic device to identify the gear, such as sensor.

II. **Shift principle**

The select-shift actuator is to achieve the gear disengagement → selection → engagement process of the shift actuator, as shown in Fig. 6.20. The horizontal movement of the select-shift shaft drives the shift finger to move to select the shift; the axial

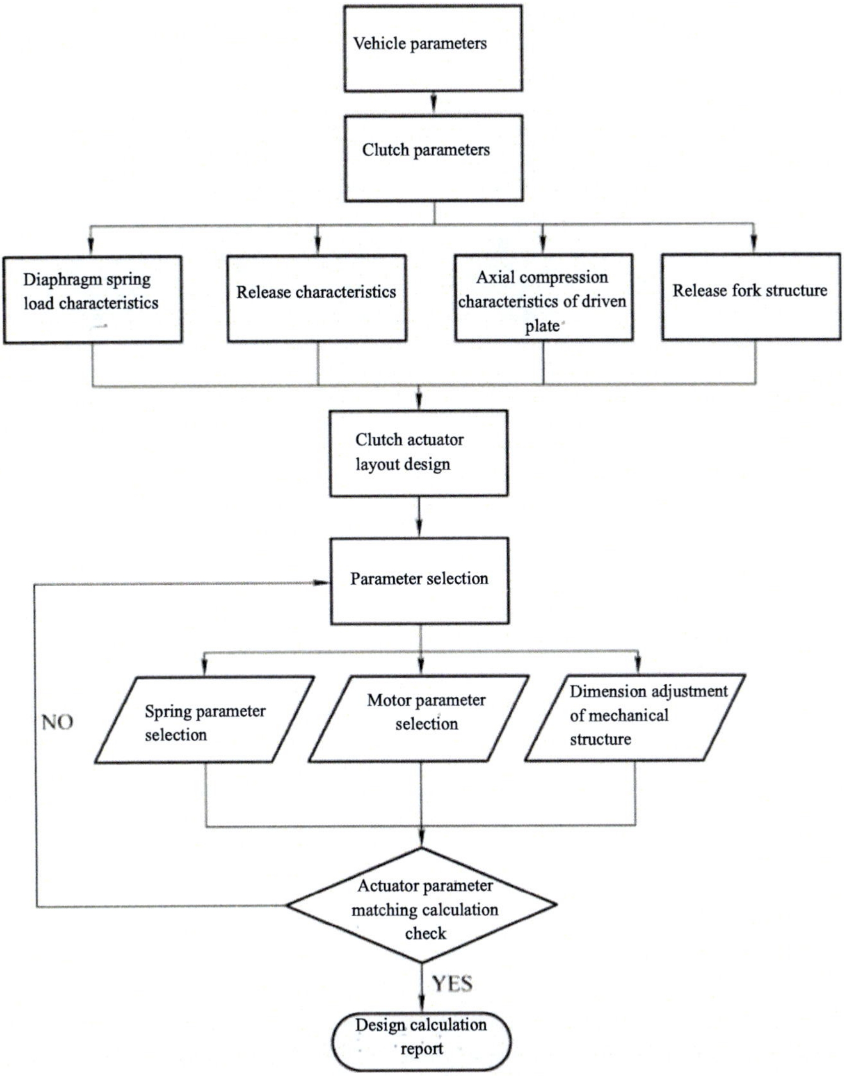

Fig. 6.19 Parameter matching calculation process of clutch actuator

rotation of the select-shift shaft drives the shift finger to rotate and moves the gear shift fork frame to achieve gear disengagement and engagement.

To achieve the horizontal movement and axial rotation of the select-shift shaft, combined with the design requirements of the select-shift actuator, the internal structure of the select-shift actuator adopted is shown in Fig. 6.21.

(1) Shift principle: the axial rotation of the gear shift lever is achieved through the shift system inside the actuator. Mode of motion transmission: shift motor →

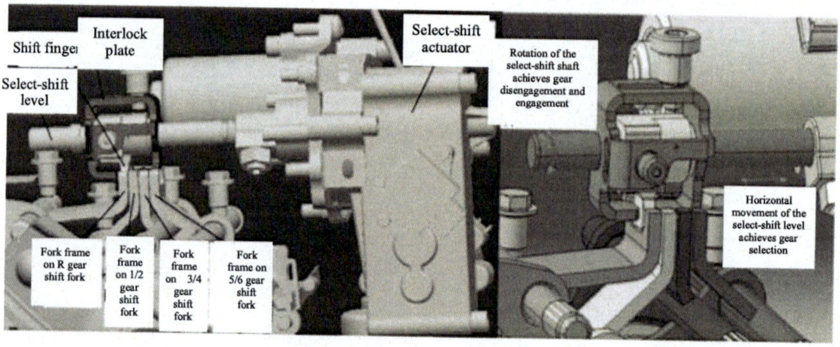

Fig. 6.20 Select and shift principle

shift gear 1 → shift gear 2/3 (the shift gears 2 and 3 are on the same shaft) → shift gear 4. The shift gear 4 is installed in the gear shift lever, so the axial rotation of the select-shift shaft can be achieved.

(2) Select principle: the shift linkage moves up and down through the select system. Mode of motion transmission: select motor → select motor 1 → select gear 2/3(the select gears 2 and 3 are on the same shaft) → select gear 4. The select gear 4 rotates in the chute structure in the gear shift lever, so it can drive the select-shift shaft to move up and down.

(3) Locating principle: the locating device is to keep the shift finger in the expected position to prevent offset at the end of the gear disengagement, selection and engagement. The shift locating device achieves the rotational locating of the gear shift lever through the shift locating slot and shift locating pin; the select locating device achieves the up and down movement and locating of the shift linkage through the select locating slot and select locating pin, as shown in Fig. 6.22.

(4) Interlock principle: the interlock is to prevent two gears from engaging at the same time. In the gear R and gear 1/2, the shift finger engagement is limited by the interlock plate and the shift fork frame will not move, as shown in Fig. 6.23.

(5) Gear identification principle: the angle sensor is used to transmit the current gear information to the TCU, including the shift angle sensor and select angle sensor. The shift angle sensor is installed in the corresponding shaft of the shift gear 4 in the shift system to directly obtain the angle information of the shift gear 4. The select angle sensor is installed in the corresponding shaft of the select gear 2/3 in the select system to directly obtain the angle information of the select gear 2/3.

III. **Select-shift actuator structure parameter matching calculation**

The matching calculation process of the select-shift actuator structure parameters is shown in Fig. 6.24.

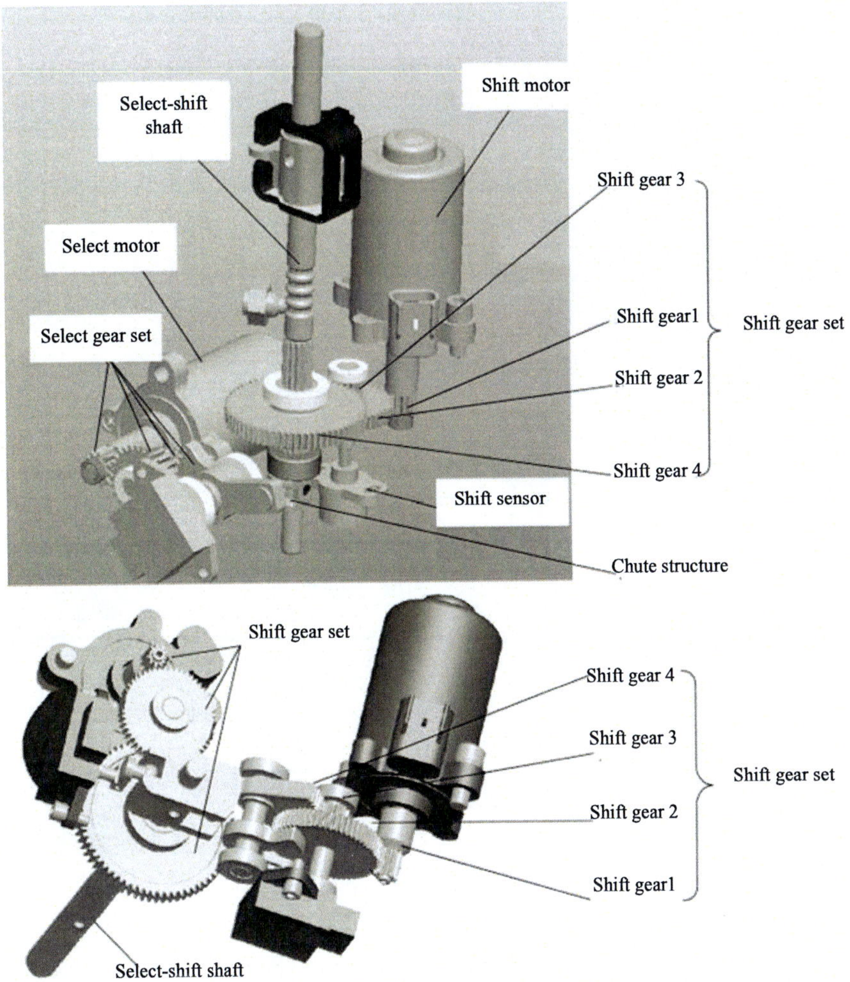

Fig. 6.21 Internal structure of select-shift actuator

6.6 Hydraulic Control System

The hydraulic control system of the AMT is shown in Fig. 6.25. TCU mainly controls the action of the clutch and transmission select-shift actuator through the solenoid valve.

I. **Clutch actuator**

The clutch actuator shall be designed to meet the following basic requirements:

(1) Meet the working stroke of the clutch release and engagement, and have enough driving force and self-locking performance.

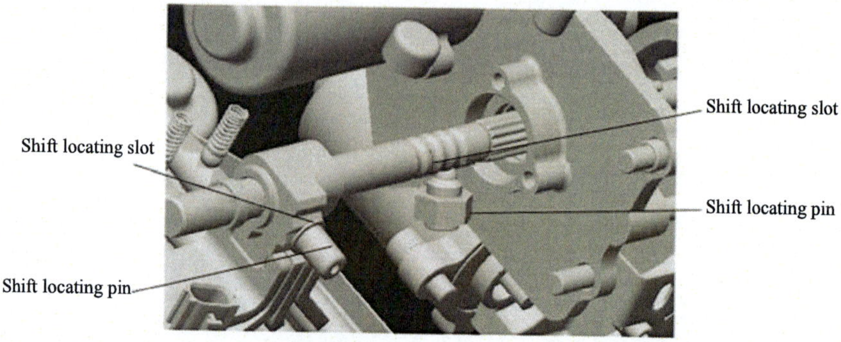

Fig. 6.22 Select-shift locating principle

Fig. 6.23 Interlock principle

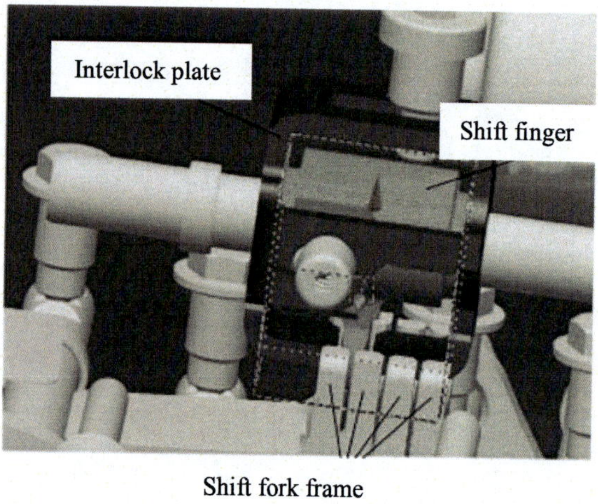

(2) Ensure the maximum release speed and ensure smooth engagement and small impact.
(3) Stable torque output at low speed to avoid velocity fluctuation and obtain accurate position control.
(4) Good frequency response, low lag and low cost.
(5) Reliable mechanism work and compact structure for easy general layout.

Figure 6.26 shows the schematic diagram of the hydraulic clutch actuator. The engine 1 is connected with the flywheel 2; the clutch driven plate 3, clutch pressure plate 4, diaphragm spring 5, release bearing 6, return spring 7, release lever 8 and hydraulic cylinder 9 make up the clutch operating system; the hydraulic pump 14, oil filter 15, one-way valve 16 and accumulator 17 make up the pressure source; the filling valve 10, delivery valve 11 and delivery valve 12 control the supercharge or decompression of the hydraulic cylinder 9. Basic control process of clutch actuator: when the filling valve is open and the delivery valve is closed, the hydraulic oil in the

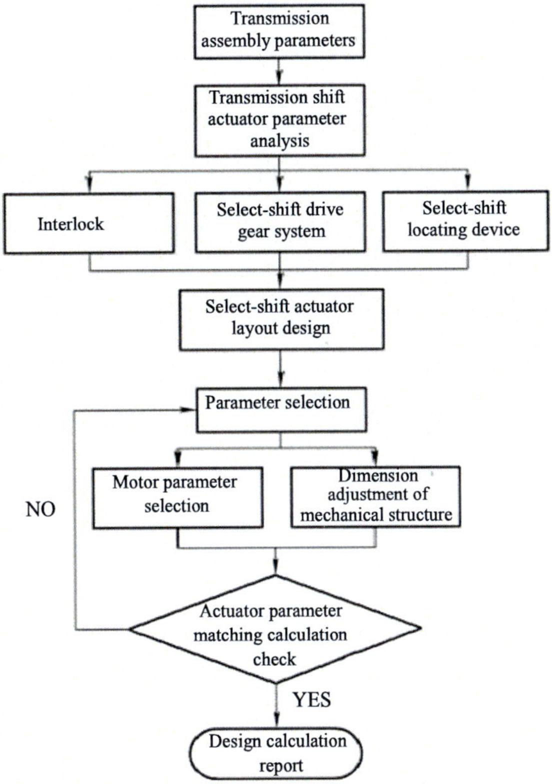

Fig. 6.24 Matching calculation process of select-shift actuator structure parameters

accumulator enters the hydraulic cylinder through the filling valve, and the piston of the hydraulic cylinder moves right under the pressure. At this time, the small end of the clutch diaphragm spring moves left under the action of the release bearing, so that the drive and driven plates of the clutch are released, thus cutting off the power; when the filling valve is closed and the delivery valve is open, the clutch diaphragm spring returns under the action of its own force and the return spring force. The hydraulic cylinder piston moves left under the action of the spring force and the hydraulic oil returns to the tank through the delivery valve. If the filling valve and delivery valve are closed simultaneously, the hydraulic cylinder stops moving and the clutch engagement speed is controlled through different combinations of two delivery valves or the use of PWM for the delivery valve.

II. **Select-shift actuator**

The select-shift actuator must achieve the goals of smooth shift, small impact, fast response speed, accurate locating of middle position and prevention of excessive dynamic load. The specific requirements are as follows:

(1) Requirements for the actuator stroke. If the stroke is too short, the gear may not be engaged or may be disengaged; if the stroke is too long, the synchronizer or

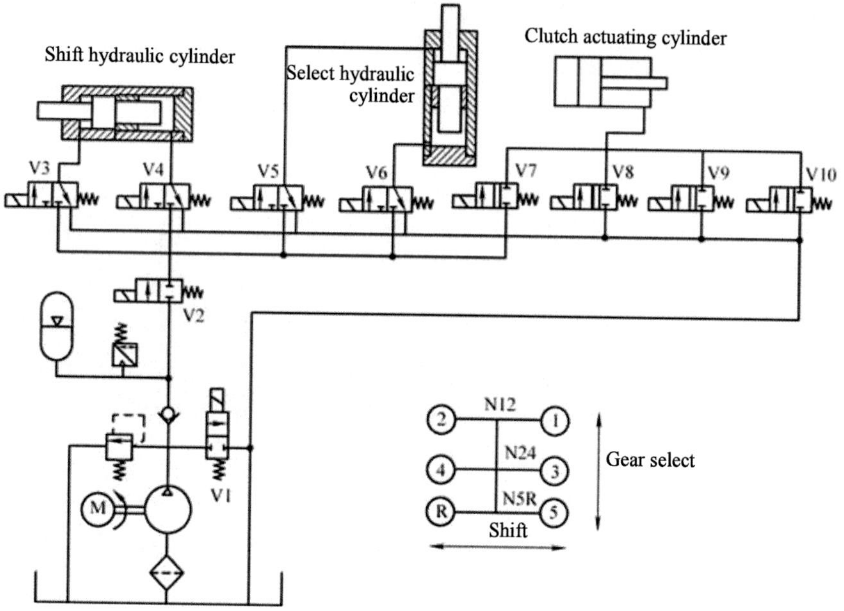

Fig. 6.25 Hydraulic control system of AMT

Fig. 6.26 Schematic diagram of hydraulic clutch actuator. 1—Engine, 2—flywheel, 3—clutch driven plate, 4—clutch pressure plate, 5—diaphragm spring, 6—release bearing, 7—return spring, 8—release lever, 9—hydraulic cylinder, 10—filling valve, 11, 12—delivery valve, 13—fuel tank, 14—hydraulic pump, 15—oil filter, 16—one-way valve, 17—accumulator

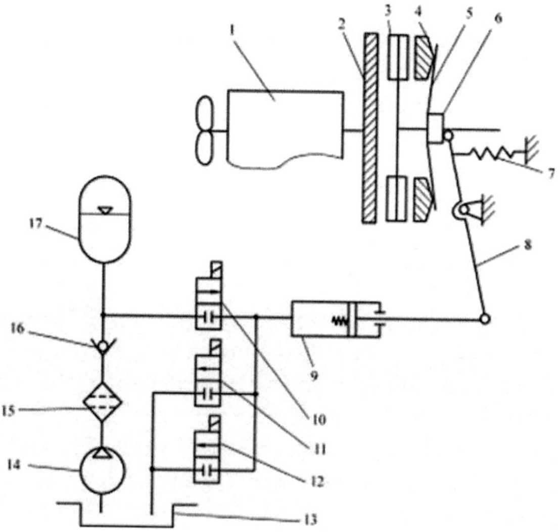

sliding sleeve is still subject to a large axial thrust after gear shift, thus affecting its life.

(2) Requirements for the shift speed. The shift speed affects the shift force. Too fast shift speed will produce large impact force; too slow shift speed will increase the power interruption time, thus affecting the whole engagement time and the vehicle acceleration performance.

(3) Requirements for actuator thrust. The shift reliability and synchronizer life are affected by the thrust. If the thrust is too small, the gear may not be engaged and the slip between the meshing gear and the sliding sleeve results in the early damage of the synchronizer; if the thrust is too large, the life of the synchronizer or sliding sleeve will also be affected. Under normal circumstances, the thrust of the actuator shall be designed according to the maximum shift resistance of the vehicle under dynamic conditions, and the size of the thrust shall be controlled in different ways to ensure the shift quality and the life of the synchronizer. The shift force of the passenger vehicle synchronizer is usually 100–500 N, with a maximum of 800 N.

(4) Requirements for locating. The transmission shift actuator has 3 positions and the inaccurate locating of the middle position will directly lead to the contact between the meshing gear and the sliding sleeve, resulting in meshing gear, synchronizer or sliding sleeve damage and even failure to smoothly select the gear.

The hydraulic drive circuit of the gear shifter shaft is shown in Fig. 6.27. The shift hydraulic cylinder uses two two-position three-way solenoid valves to control two motion directions of the shift shaft. The specific control principle is as follows: in the figure, when the left solenoid valve feeds oil and the right solenoid valve returns oil, the piston rod moves right; when the left solenoid valve returns oil and the right solenoid valve feeds oil, the piston rod moves left; when two solenoid valves feed oil, the piston rod moves to the middle position; when two solenoid valves return oil, the piston rod stops moving. The hydraulic drive circuit is suitable for both gear select control and shift control.

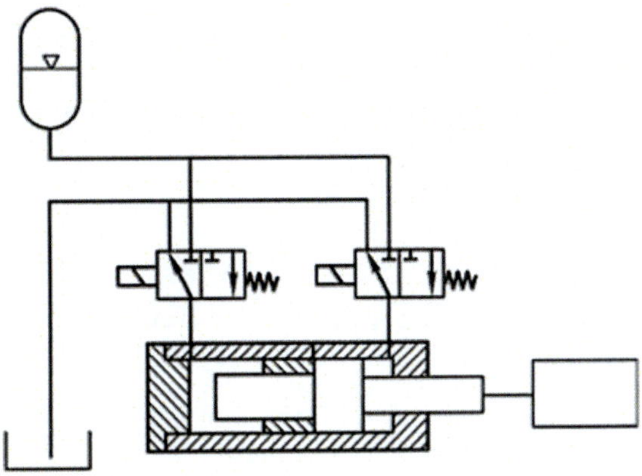

Fig. 6.27 Hydraulic drive circuit of gear shifter shaft

6.7 AMT Control Strategy

The main framework of the AMT control strategy is shown in Fig. 6.28.

I. Derived signal module

The derived signal module is to amplify the hard wire signal, the underlying software processing signal, the CAN bus signal and the signal sent by the control strategy, and return them to the control strategy for use. The derived signals to be calculated include accelerator pedal change rate, input shaft speed change rate, output shaft speed change rate, engine speed change rate, vehicle speed change rate, slip, output shaft speed, clutch input torque, inertia moment, dynamic detection, Roll-down, speed ratio, air density, vehicle load and estimated output shaft speed.

II. Shift law selection module

The shift law includes normal mode, winter mode, performance mode, cold mode, economic mode, warm up mode, engine brake mode, catalyst heating request mode, uphill mode, downhill mode, hot mode and plateau mode.

The shift law selection module finds the required mode and shift point from the above shift law.

III. Gear selection module

The gear selection module determines the current gear status according to the position of the gear shift lever, including manual, automatic, neutral and reverse. The target gear is determined as follows:

(1) Manual mode: change the target gear by adding and subtracting manually. If the target gear selected manually is not reasonable, the program will filter out and correct the unreasonable gear.

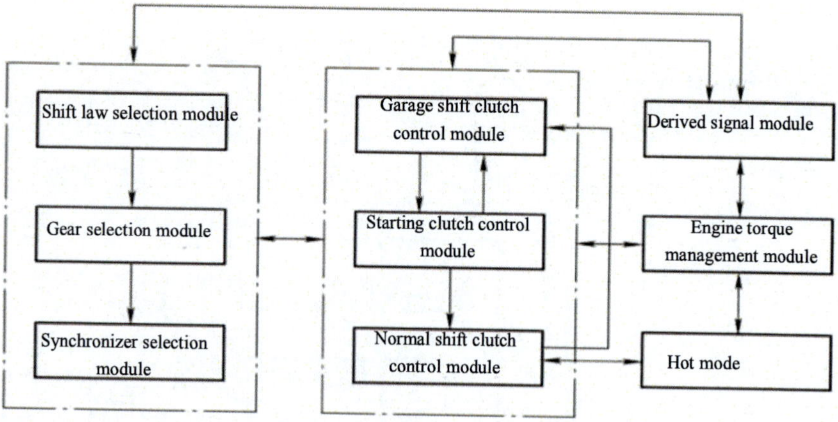

Fig. 6.28 Main frame of AMT control strategy

(2) Automatic mode: determine the target gear in the automatic shift mode according to the shift point given by the shift law selection module. If the target gear selected by the shift law selection module is not reasonable, the program will filter out and correct the unreasonable gear.
(3) Neutral mode: enter this mode when the gear shift lever is in position N and achieve this function through the clutch release at the high speed.
(4) Reverse mode: enter this mode when the gear shift lever is in position R. The reverse gear will not engage immediately in the forward driving at high speed.

IV. Garage shift clutch control module

The garage shift clutch control module controls the clutch release before the vehicle starting. After the clutch release, the transmission is engaged and then the clutch is engaged in the release point for waiting. The module functions include the clutch self-learning and control correction. The control principle of the garage shift clutch is shown in Fig. 6.29.

V. Starting clutch control module

The starting clutch control module is mainly to make coordination control of the clutch engagement speed and engine request torque and guarantee the smooth and quick vehicle starting, timely engine response and low engine speed. The clutch control in the starting process is shown in Fig. 6.30.

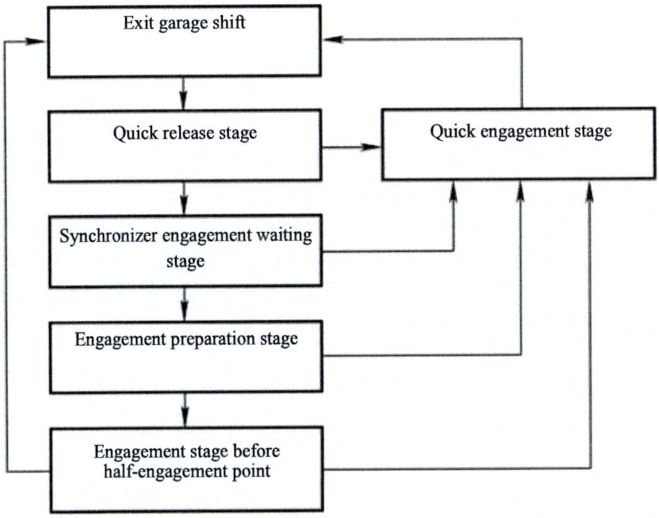

Fig. 6.29 Control schematic diagram of garage shift clutch

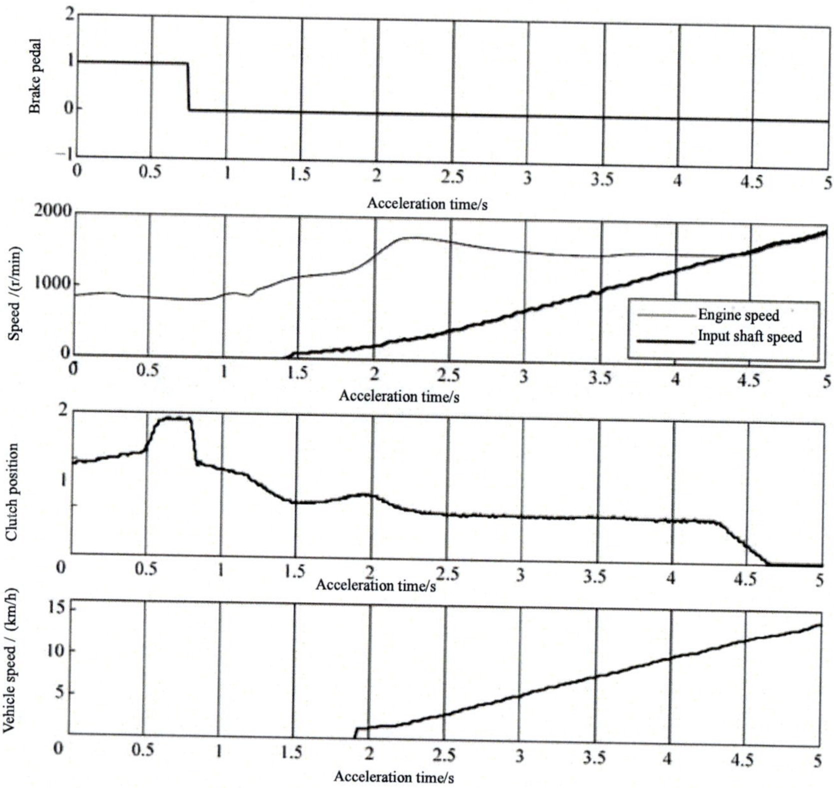

Fig. 6.30 Clutch control in the starting process

VI. **Normal shift clutch control module**

The normal shift clutch control module is used for the clutch control in the garage shift, starting and normal shift. The normal shift clutch control principle is shown in Fig. 6.31 and the normal shift clutch control is shown in Fig. 6.32.

VII. **Synchronizer control module**

The synchronizer control module controls the synchronizer engagement and release according to the target gear generally after total release of the clutch, including self-learning and fault recovery. The synchronizer control principle is shown in Fig. 6.33.

VIII. **Engine torque management module**

To prevent the engine flame-out and limit the engine speed in the starting and normal shift, requests need to be made to the engine control system, including the required torque of the transmission, target idle, and maximum engine speed, with the emphasis on the engine torque reduction control in the process of clutch release and engagement and the engine speed control in the process of total release. The evaluation criteria are

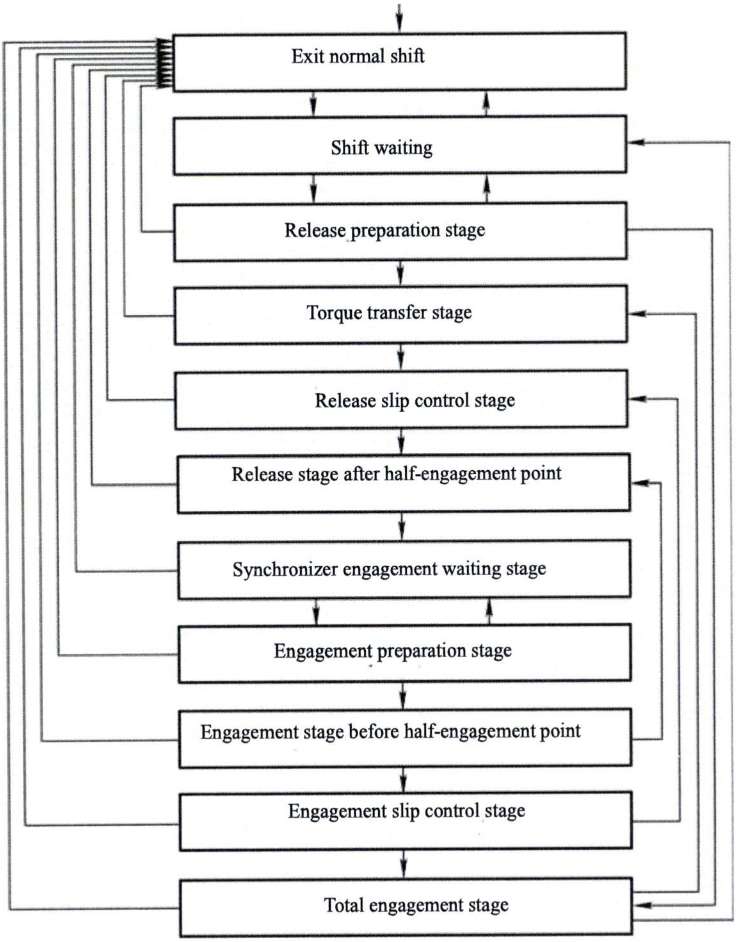

Fig. 6.31 Normal shift clutch control principle

that the release process and engagement process are stable, fast and good in tracking, and the engine speed cannot increase in the case of total release. The torque control in the starting process is shown in Fig. 6.34 and in the normal shift process is shown in Fig. 6.35.

XI. **Hot mode**

The hot mode is mainly to protect the powertrain from overheating by reducing the excessive load and the shift times.

X. **EMS and TCU coordination control technology**

TCU can establish communication with the engine ECU through CAN bus for data interaction. TCU can control the engine torque and speed through the CAN bus and

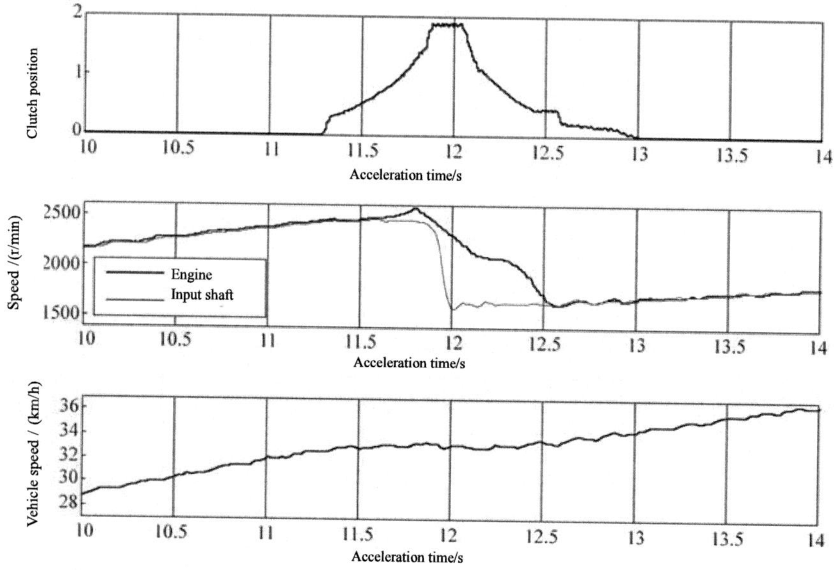

Fig. 6.32 Clutch control in normal shift process

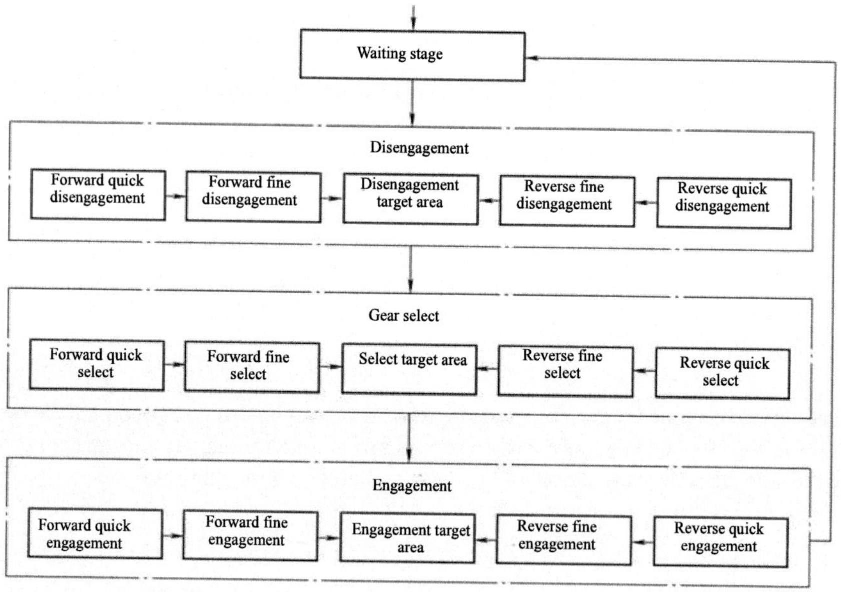

Fig. 6.33 Synchronizer control principle

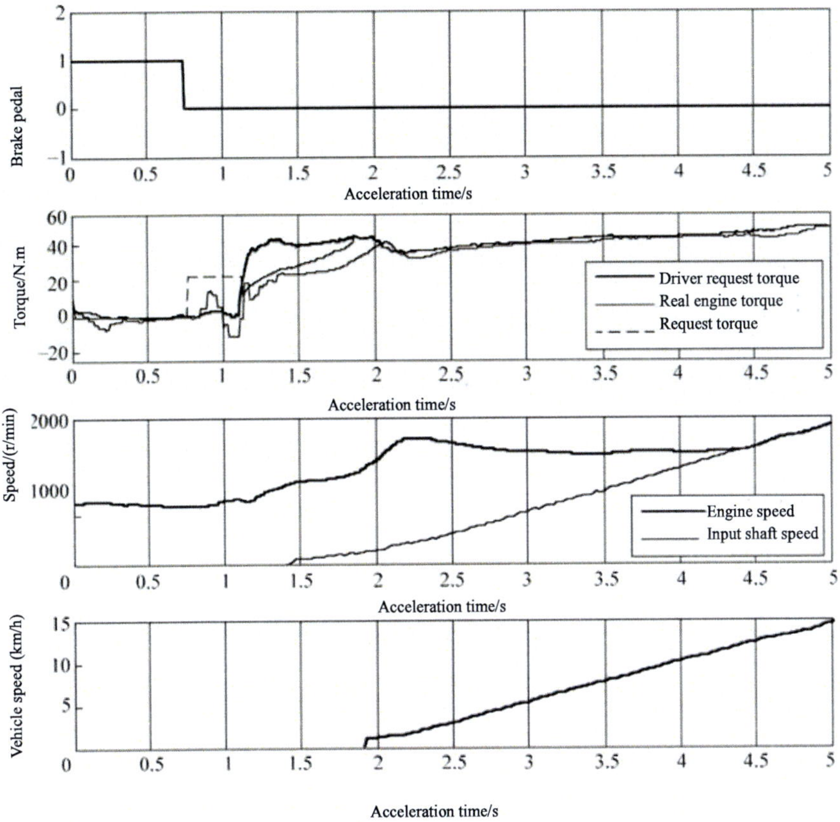

Fig. 6.34 Starting process torque control

obtain the actual engine output torque, throttle percentage and engine speed signal sent by EMS. Figure 6.36 shows the coordination control communication mode among the engine, transmission and clutch. Figure 6.37 shows the AMT torque control in the shift process.

6.8 AMT Performance Evaluation Indexes

I. Functional indexes

(1) Tip in
Tip in 45%: the function of downshift when the constant throttle percentage (45%) is increased to 100% suddenly in gear 4.
Tip in 60%: the function of downshift when the constant throttle percentage (60%) is increased to 100% suddenly in gear 5.

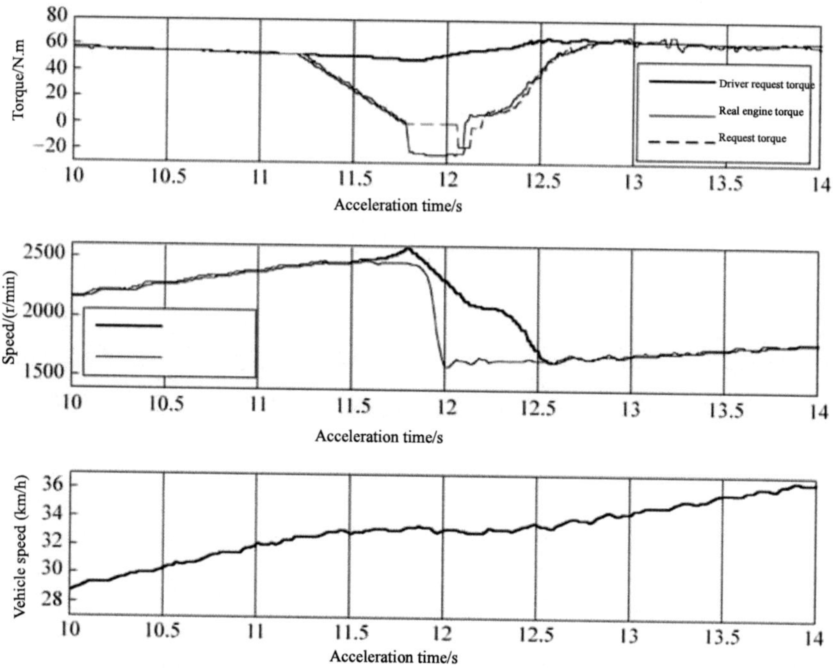

Fig. 6.35 Torque control in normal shift process

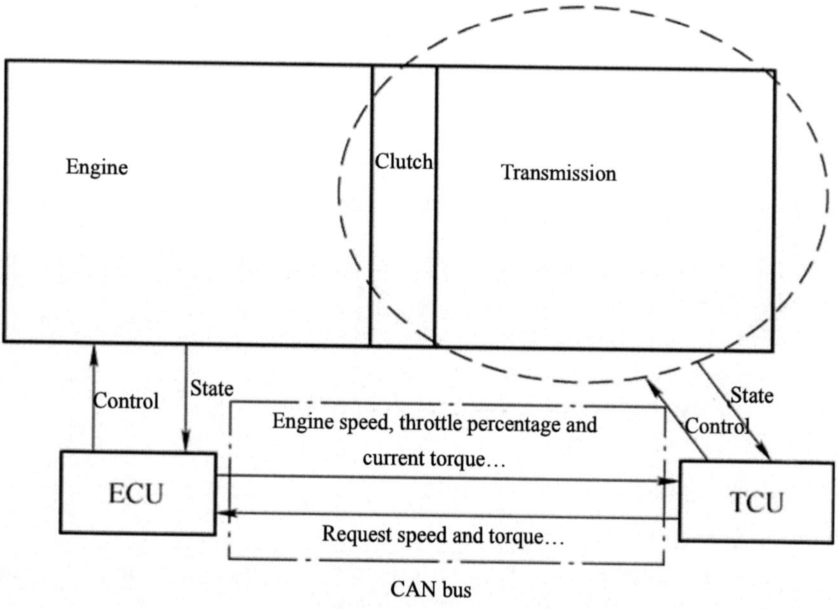

Fig. 6.36 Coordination control communication mode among the engine, transmission and clutch

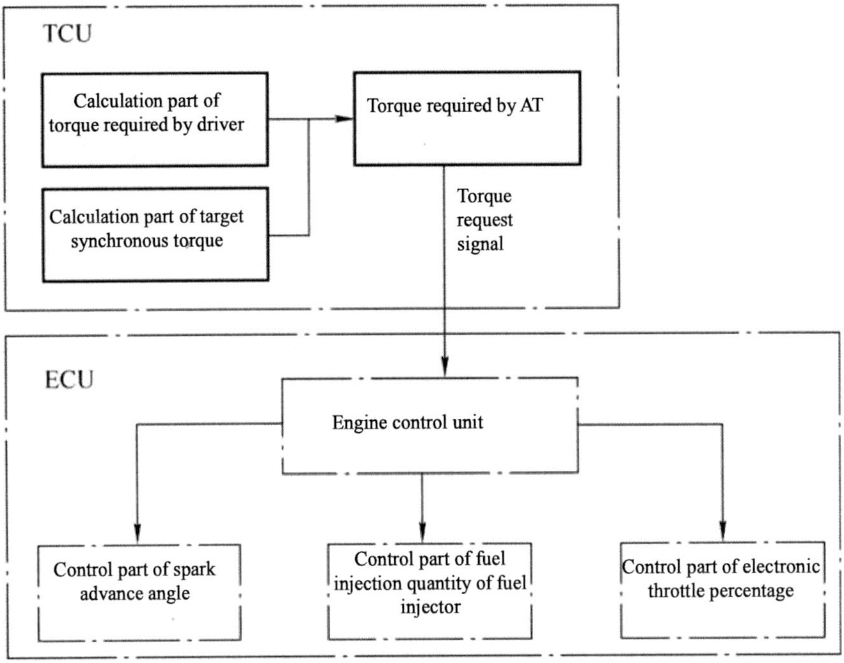

Fig. 6.37 AMT torque control in the shift process

(2) Tip out
Tip out 30%: in the driving process at 30 km/h in gear 2, the driver generally releases the accelerator pedal quickly from the throttle percentage 30% to the throttle percentage 0 to delay upshift.
Tip out 75%: in the driving process at 65 km/h in gear 3, the driver generally releases the accelerator pedal quickly from the throttle percentage 60% to the throttle percentage 0 to delay upshift.

(3) Brake control logic: in normal braking, gear 6, vehicle speed 100 km/h: in the moderate braking, the vehicle speed is gradually decreased and the acceleration changes smoothly; in the emergency braking, the vehicle speed is quickly reduced to 0 and the gear quickly decreases to gear 1.

(4) Controllability: frequent shift is not allowed in the continuous steering and the driving at 50 km/h.

(5) Starting on level road

1) Idle creeping start.
No load: starting time 6 s and maximum acceleration 0.49 m/s^2 at the starting.
Full load: starting time 6.5 s and maximum acceleration 0.49 m/s^2 at the starting.
2) Starting from small throttle percentage (25% ~40%) to gear 2.

No load: starting time 2.69 s and maximum acceleration 2.49 m/s^2 at the starting.
Full load: starting time 2.71 s and maximum acceleration 2.53 m/s^2 at the starting.

3) Starting from middle throttle percentage (50–65%) to gear 2.
No load: starting time 2.21 s and maximum acceleration 3.76 m/s^2 at the starting.
Full load: starting time 2.50 s and maximum acceleration 3.0 m/s^2 at the starting.
4) Starting from large throttle percentage (>65%) to gear 2.
No load: starting time 2.34 s and maximum acceleration 3.96 m/s^2 at the starting.
Full load: starting time 2.74 s and maximum acceleration 3.48 m/s^2 at the starting.

(6) Acceleration performance (throttle percentage 100%): acceleration time 6.56 s from 0 to 60 km/h; 14.86 s from 0 to 100 km/h; 38.66 s from 0 to maximum speed; 8.29 s from 60 km/h to 100 km/h.
(7) Uphill starting (gradient 10–15%): starting at 0% throttle percentage from gear 1 to gear 2, and the backward sliding distance within 100 mm; starting at 30% throttle percentage from gear 1 to gear 2, and the backward sliding distance within 50 mm.
(8) Downhill protection
Automatic mode: press the accelerator pedal down 0%, 0% or lightly touch the brake pedal to maintain gear 1.
Manual mode: press the accelerator pedal down 0%, 0% or lightly touch the brake pedal to achieve manual downshift.
(9) Ramp correction function: including ramp identification function to achieve uphill on different ramps.
(10) Vehicle mode state: the gear shift lever signals include N, R, D, A-M, M+, M– and transition signal NAA and the gear shift lever modes include:

1) Neutral mode: enter this mode when the gear shift lever is in position N.
2) Reverse mode: enter this mode when the gear shift lever is in position R.
3) Automatic mode: enter the automatic mode after the gear shift lever is from N to D, or from R to D; enter the automatic mode after from D to A-M if it is originally manual mode.
4) Manual mode: enter the manual mode after the gear shift lever is from N to A-M, or from R to A-M;

Enter the manual mode after from D to A-M, or from D to M+and M– if it is originally automatic mode.

II. AMT shift quality indexes

1. Shift impact

The index impact j is used to evaluate the smoothness of the clutch engagement process during shift and is the change rate of the vehicle longitudinal acceleration a, i.e.

$$j = \frac{da}{dt} \tag{6.3}$$

The transmission output torque T_0 is variable, so the shift impact is controlled when the transmitted torque of the clutch is controlled during shift.

2. Clutch sliding friction work during shift

The index sliding friction work affects the clutch service life during the shift and is the work of the sliding fiction between the drive and driven frictions in the clutch engagement process, i.e.

$$W = \int_{t_0}^{t_s} T_c[\omega_e(t) - \omega_c(t)]dt \tag{6.4}$$

where,

i—clutch torque (N m);
ω_e—angular velocity of drive member (rad/s);
ω_c—angular velocity of driven member (rad/s);
t_0—start engagement;
t_s—total engagement.

With respect to the clutch, the greater the sliding friction work, the higher the temperature rise, the shorter the clutch life. Therefore, in order to extend the clutch life, the sliding friction work should be minimized, which will inevitably increase the shift impact. Reasonable clutch control can give consideration to both impact and sliding friction work.

3. Shift time

In terms of the clutch, the shift time t includes the clutch release time t_1, release holding time t_2 and engagement time t_3. The release holding time t_2 includes the disengagement time, select time and engagement time. The shift time is

$$t = t_1 + t_2 \tag{6.5}$$

It can be seen from the evaluation indexes of the gear shift that there are contradictions among impact, sliding friction work and shift time in shift quality, but they affect each other. Therefore, to achieve satisfactory shift quality, the clutch and engine must be comprehensively controlled.

Bibliography

1. Changwan H, Ruixiang S, Xin J (2007) Automated mechanical transmission. Int Mechatron Technol 12:23–24
2. Yulong L, Yongjun L, Anlin G (2000) Starting process control for automated mechanical transmission. Chin J Mech Eng 36(5):69–71
3. Hongjun Z, Anlin G (2000) Automated mechanical transmission. Auto Electric Parts 1:11–13
4. Jun L, Jianwu Z, Jinzhi F et al (2000) Development, current situation and forecast of automated mechanical transmission. Autom Technol 3:1–3
5. Chenglin L, Zhang J, Qingchun L (2005) Coordinated powertrain control method for shifting process of automated mechanical transmission in the hybrid electric vehicle. Chin J Mech Eng 41(12):37–41
6. Xuexun G, Tao W (1999) Development status and prospect of automotive automatic transmission in China. Chin J Autom Eng 6:7–10
7. Qingmei C, Zhili Z, Mingzhu Z (2005) Shifting qualities control on HMT of vehicle. J Henan Univ Sci Technol (Natl Sci) 26(1):18–21
8. Guangqiang W, Weibin Y, Datong Q (2007) Key technique of dual clutch transmission control system. Chin J Mech Eng 43(2):13–21
9. Kim D, Peng H, Bai S et al (2007) Control of integrated powertrain with electronic throttle and automatic transmission. IEEE Trans Control Syst Technol 15(3):474–482
10. Mingkui N, Xiusheng C, Bingzhao G et al (2004) A study on shifting characteristics of dual clutch transmission. Autom Eng 26(4):453–457
11. Zhong Z, Kong G, Yu Z et al (2012) Shifting control of an automated mechanical transmission without using the clutch. Int J Autom Technol 13(3):487–496
12. Ronghui Yu, Dongye S, Datong Q (2006) Study on dynamic gear-shift law of automated mechanical transmission. Trans Chin Soc Agric Mach 37(4):1–4
13. Haj-Fraj A, Pfeiffer F (2001) Optimal control of gear shift operations in automatic transmissions. J Franklin Inst 338(2–3):371–390
14. Yongjun L, Shuxing C, Yong C et al (2003) Integrated control of the starting process of automated mechanical transmission. Autom Eng 25(2):178–181

Chapter 7
Transmission for New Energy Vehicle

7.1 Overview

The hybrid electric vehicle is a hybrid vehicle powered by two or more energy converters. Depending on the hybrid energy, the hybrid power can be divided into engine (gasoline or diesel) + motor, engine (gasoline or diesel) + hydraulic system, engine (gasoline or diesel) + mechanical system and so on. At present, the most common hybrid electric vehicle is of the "gasoline engine + motor" form.

1. **Classification by power and transmission structure**

HEV has a variety of classification methods, among which, the common classification method is by power and transmission structure and the HEV can be divided into SHEV, PHEV and SPHEV.

The SHEV features a motor that drives the vehicle alone, with the engine only acting as a power source for the generator to charge the battery. The engine works in the most efficient area, with stable speed and torque and simple control. However, since energy needs to be transformed in multiple steps, the overall efficiency of the engine, generator, battery charge, battery discharge and motor in the whole process is relatively low, so the improvement of fuel economy is limited. Meanwhile, since the vehicle is driven by motor only, the power performance of the vehicle is not good. Therefore, SHEV is rare in the current HEV mass production models.

The two power sources (engine and motor) of a PHEV can drive the vehicle jointly or independently. Therefore, PHEV has a more flexible drive mode, and has been greatly improved in terms of the power performance and fuel economy performance compared with the SHEV.

The SPHEV combines the advantages of both SHEV and PHEV configurations, making it the easiest to optimize for energy, but also the most complex.

Y. Chen, *Automotive Transmissions*, Springer Tracts in Mechanical Engineering,
https://doi.org/10.1007/978-981-15-6703-2_7

2. **Classification by hybridization rate**

The HEV can also be classified by hybridization rate. Hybridization rate refers to the proportion of motor power in the total power. Depending on the hybridization rate, the vehicle can be classified into mild HEV, medium HEV and full HEV.

For the mild HEV, a belt driven motor (BSG motor) with the power generally 2–3 kW is generally added, which is mainly used for quickly starting the engine. 5–7% fuel can be saved by achieving the idle downtime function of the engine. LaCrosse HEV produced by SGM is a typical mild HEV.

The typical structure of medium HEV is to install an ISG motor with the power generally about 10 kW at the power output end of the engine crankshaft to quickly start the engine, provide assistance for the high power demand, recover the braking energy at the deceleration and braking and generate power actively when necessary. About 20% fuel can be saved through idle downtime, engine displacement reduction (load rate increase), engine operating point optimization, braking energy recovery and other basic functions. Honda Civic is a typical medium HEV.

The full HEV is typically characterized by electric drive with a motor power of more than 20 kW. The EV driving at low speed (low power demand) can significantly reduce the fuel consumption at low speed. With other fuel saving functions, the total fuel saving rate can reach 40%. Toyota PRIUS and Lexus CT200h are typical full HEVs.

3. **Classification by external charge**

Depending on the external charge, the HEV can be classified into non-off-vehicle-chargeable hybrid electric vehicle and plug-in hybrid electric vehicle. The above HEVs are non-off-vehicle-chargeable HEVs. The PHEV (Plug-in Hybrid Electric Vehicle) is characterized by off-vehicle charge and long electric range. In case of low battery, the fuel can be used to achieve long driving range. The PHEV can be divided into three modes. Chevrolet VOLT and JAC RS PHEV are typical series PHEV, also called series extended range HEV; BYD F3DM and Volvo V60 are parallel PHEV; the plug-in hybrid system of Toyota Prius is typical SPHEV.

4. **Application of AMT, AT, CVT and DCT in hybrid electric vehicle**

Among the NEW transmissions, the AMT, AT, CVT and DCT have been tried and applied on the HEV and have different structural features, transmission principles and control methods and have advantages and disadvantages on HEV.

The AMT, as an HEV transmission, is characterized by low cost, reliable transmitted torque, high transmission efficiency and coordination with a variety of dynamic coupling modes; the shift control strategy of the AMT can be incorporated into the vehicle control strategy to give the HEV better fuel economy, and the motor compensates for the engine drive torque during the shift to improve the vehicle power performance. The disadvantage is that the motor in this scheme cannot be involved in speed change.

As an HEV transmission, AT is usually used to replace the hydraulic torque converter with the motor. AT is used as the transmission in Honda Accord hybrid electric car and GM U-Model hybrid concept vehicle. Through the motor control, the AT achieves the steady starting and smooth shift of the engine, completing the drive and energy recovery and improving the vehicle fuel economy.

As an HEV transmission, the CVT can overcome the shortcomings of traditional CVT vehicles and give full play to the advantages of HEV in energy saving and emission reduction. Its integration with CVT technology will certainly give full play to the HEV advantages of energy saving and environmental protection.

The DCT is used as an HEV transmission. In the Honda transmission scheme, the motor is integrated in the odd input shaft and shall perform synchronous tracking during shift to reduce the speed difference between input and output of synchronizer, and the control strategy is complicated; in BYD model, the motor is integrated in the output shaft and the motor output does not take part in the shift. Its biggest disadvantage is that it does not have the function of parking charge.

7.2 Power and Economy Performance of HEV

Depending on the drive mode, the HEV can be classified into hybrid drive mode, engine drive mode and electric drive mode. In the hybrid drive mode, the vehicle is driven by the engine, motor and all other onboard powertrain according to the management logic (vehicle control strategy); in the engine drive mode, the vehicle is driven by the engine; in the electric drive mode, the vehicle is driven by the motor.

The HEV has two or more power sources, and Table 7.1 lists the characteristics of HEV power sources. As for the engine and motor power source, the overall performance is optimized by full use of the advantages of the engine drive and motor drive through the transmission and corresponding control strategy.

Table 7.1 Characteristics of HEV power sources

Type of power source	Advantages	Disadvantages	Efficient operation mode
Engine	(1) Large output torque (2) Compact structure (3) High energy density	(1) Low efficiency at low output torque (2) Engine working at stop	(1) High load (2) Long distance operation
Motor	(1) Achieve starting at zero speed (2) Energy recovery (3) High efficiency at low output torque (4) Reverse rotation	(1) Low energy density (2) High high-voltage safety requirements	(1) At stop (2) Starting (3) Low load (4) Deceleration and reverse

I. Power performance

Power performance is one of the main evaluation indexes of the HEV. The power indexes of the hybrid drive mode include maximum speed, acceleration time from 0 to 100 km/h or from 0 to 50 km/h, 30 min maximum speed, climbing speed, hill starting ability and maximum gradeability. The power indexes of the electric drive mode include maximum speed, 30 min maximum speed, acceleration time from 0 to 50 km/h, climbing speed and hill starting ability.

1. Power index of hybrid drive mode

Maximum speed: average of the maximum speed reached to maintain the hybrid driving for more than 1 km according to the procedures stipulated in GB/T19752—2005 *Hybrid Electric Vehicles—Power Performance—Test Method.*

Acceleration performance from 0 to 100 km/h (0 to 50 km/h): minimum time required to accelerate from 0 to 100 km/h in the hybrid driving according to the procedures stipulated in GB/T19752—2005 *Hybrid Electric Vehicles—Power Performance—Test Method* (test the acceleration performance from 0 to 50 km/h when the maximum speed of the HEV is less than 110 km/h).

30 min maximum speed: average of the maximum speed reached to maintain the hybrid driving for 30 min according to the procedures stipulated in GB/T19752—2005 *Hybrid Electric Vehicles—Power Performance—Test Method.*

Climbing speed: maximum average speed reached to maintain hybrid driving for more than 1 km on roads with gradients of 4 and 12% according to the procedures stipulated in GB/T19752—2005 *Hybrid Electric Vehicles—Power Performance—Test Method.*

Hill starting ability: maximum gradient in the starting and hybrid driving for more than 10 m according to the procedures stipulated in GB/T19752—2005 *Hybrid Electric Vehicles—Power Performance—Test Method.*

Maximum gradeability: maximum gradeability in the hybrid mode according to the procedures stipulated in GB/T19752—2005 *Hybrid Electric Vehicles—Power Performance—Test Method.*

2. Power indexes of electric drive mode

Maximum speed: average of the maximum speed at which an electric vehicle can drive continuously for a distance of more than 1 km round trip.

30 min maximum speed: maximum average speed at which the electric vehicle can drive continuously for more than 30 min.

Acceleration time from 0 to 50 km/h: minimum time required for the electric vehicle to accelerate from 0 to 50 km/h.

Climbing speed: maximum average speed at which the electric vehicle can drive continuously more than 1 km on a ramp with a given gradient.

Hill starting ability: maximum gradient at which the electric vehicle can start on a ramp and drive upwards at least 10 m within 1 min.

II. Economy

The ability to drive economically with as little energy consumption as possible on the premise of ensuring power performance is called the energy economy, or economy for short. The energy economy is usually measured by the energy consumption for a certain distance driven by a vehicle under a certain operating condition or by the mileage driven by a certain amount of energy.

In China and Europe, fuel consumption or power consumption of 100 km under certain operating conditions is used to measure the vehicle energy economy, in units of L/100 km and kW h/100 km, respectively. The higher the value, the worse the economy. In the United States, the energy economy of a vehicle is measured by the mileage driven by a certain amount of fuel or power under certain operating conditions, in units of mile/USgal and mile/(kW h), respectively. The higher the value, the better the economy.

HEV consumes both power and fuel, so fuel consumption and power consumption need to be taken into account for the energy consumption evaluation. Ordinary HEVs cannot be charged externally. The power consumption during operation accounts for a small proportion. Moreover, the power consumed still comes from engine fuel consumption, and the power battery SOC fluctuates within a small range. Under the same test conditions, there is a certain linear relationship between energy consumption and fuel consumption. It is easy to obtain accurate equivalent fuel consumption evaluation results by using this linear relationship.

With respect to the PHEV, the relationship between the power consumption and fuel consumption is uncertain because the power battery SOC varies widely and most of the power consumption comes from the external grid. In view of this situation, the United States Environmental Protection Agency (EPA), for direct comparison between traditional vehicles and NEVs, adopts the equivalent fuel consumption measurement method, that is, the electric energy of 33.7 kW h is equivalent to the energy of 1 USgal (3.785 L) gasoline and the equivalent fuel economy is used for evaluation.

The driving cycles used to evaluate the vehicle energy economy include: New European Driving Cycle (NEDU) adopted in Europe and China, Urban Dynamometer Driving Schedule (UDDS) and High Way Fuel Economy Driving Schedule (HWFET) stipulated by EPA and JC09 in Japan.

Under the drive cycle, the vehicle energy economy test is generally carried out on the chassis dynamometer. The vehicle fuel consumption can be measured by carbon balance method, while the power consumption can be calculated by measuring the amount of electricity required to charge the battery to the specified SOC value.

With the aggravation of the energy crisis, the governments of various countries have set more stringent energy economy targets for automobiles. The implementation of the *Notice of the State Council on Printing and Issuing the Planning for the Development of the Energy-Saving and New Energy Automobile Industry (2012–2020)* puts forward requirements on the average fuel consumption of passenger vehicle enterprises, and defines the targets of reducing the average fuel consumption of passenger vehicle products in China to 6.9 L/100 km and 5.0 L/100 km in 2015 (Phase III) and

Table 7.2 Limits of fuel consumption for passenger vehicles in China (unit: L)

Kerb mass/kg	Phase III MT	Phase III AT	Phase IV	Kerb mass/kg	Phase III MT	Phase III AT	Phase IV
(0, 750)	5.2	5.6	3.9	(1540, 1660)	8. 1	8.4	5.5
(750, 865)	5.5	5.9	4.1	(1660, 1770)	8.5	8.8	5.7
(865, 980)	5.8	6.2	4.3	(1770, 1880)	8.9	9.2	5.9
(980, 1090)	6.1	6.5	4.5	(1880, 2000)	9.3	9.6	6.2
(1090, 1205)	6.5	6.8	4.7	(2000, 2110)	9.7	10.1	6.4
(1205, 1320)	6.9	7.2	4.9	(2110, 2280)	10.1	10.6	6.6
(1320, 1430)	7.3	7.6	5.1	(2280, 2510)	10.8	11.2	7.0
(1430, 1540)	7.7	8.0	5.3	Above 2510	11.5	11.9	7.3

2020 (Phase IV), respectively. In addition, China also restricts the fuel consumption of passenger vehicles with different kerb mass, as shown in Table 7.2, and China adopts the national standard of *Fuel Consumption for Passenger Vehicles.*

The calculation of the vehicle energy economy in the design stage needs to go through the following stages:

(1) Discretization of driving conditions: for a determined driving cycle, the vehicle driving process is discretized to calculate the total power demand in each period

$$P = \left(mgf\cos\alpha + mgf\sin\alpha + \frac{C_D A v^2}{21.15} + \delta m\frac{dv}{dt}\right)\frac{v}{3600} \quad (7.1)$$

where,

α—ramp angle in vehicle driving (°);
C_D—drag coefficient;
A—frontal area of vehicle (m^2).

(2) Determination of engine operating point and calculation of fuel consumption: the operating points of all system parts, including the engine, motor, battery and transmission, are determined according to the system structure and energy control strategy. According to the vehicle speed, transmission gear information and the engine power in this period, the engine speed can be obtained as

$$n_e = \frac{v i_g i_0}{0.377r} \quad (7.2)$$

If the engine stops, the fuel consumption per unit time is $Q_t = 0$; if the engine is idle, the fuel consumption per unit time is $Q_t = Q_d$; if the engine is working, the specific fuel consumption b is determined from the universal characteristics of the engine by means of the engine power P_e and speed n_e and the fuel consumption per unit time is further determined as

$$Q_t = \frac{P_e b}{3.671 \times 10^5 \rho g} \quad (7.3)$$

where,

b—specific fuel consumption (G/kW h);
ρ—fuel density (kg/L).

In this period, the fuel consumption Q_i is

$$Q_i = Q_t t_i \quad (7.4)$$

(3) Determination and efficiency of motor operating point: assuming that the total motor power is P_m in this period, the motor speed and torque can be obtained according to the vehicle speed and gear information. Then the motor efficiency under this condition is determined according to the motor efficiency characteristics.
(4) Determination of battery operating point and calculation of power consumption: according to the motor charge and discharge and the motor efficiency η_m, the battery input and output power is determined as

$$P_b = \begin{cases} \dfrac{P_m}{\eta_m} & \text{The motor is in drive state} \\ P_m \eta_m & \text{The motor is in charging state} \end{cases} \quad (7.5)$$

Power consumption or storage E_i (unit kW h, + for charge and – for discharge):

$$E_i = \begin{cases} \dfrac{P_b t_i}{3600 \eta_b} & \text{The motor is in drive state} \\ \dfrac{P_b \eta_b t_i}{3600} & \text{The motor is in charging state} \end{cases} \quad (7.6)$$

The relationship among the charge and discharge current I_b (+ for charge and – for discharge), charge and discharge time t_i, battery voltage U and charge and discharge capacity E_i is

$$I_b t_i = \frac{E_i}{U} \quad (7.7)$$

If the battery charge quantity is C_b, the new battery SOC state is

$$SOC_i = SOC_{i-1} + \frac{I_b t_i}{C_b} \quad (7.8)$$

(5) Calculation of energy consumption in the whole cycle: the fuel consumption and power consumption during all periods are superimposed to obtain the fuel

consumption Q_f and power consumption E in the whole cycle. If the whole cycle is S (km) and it is converted to a distance of 100 km, then the fuel consumption and power consumption of 100 km can be obtained.

$$Q_f = \frac{100 \sum Q_i}{S} \tag{7.9}$$

$$E_f = \frac{100 \sum E_i}{S} \tag{7.10}$$

7.3 AMT Hybrid Transmission

With the characteristics of simple structure, high transmission efficiency and good manufacturability, the AMT has been applied in the orthodox cars and HEVs. Using the AMT for the HEVs achieves all the advantages of the AMT while also compensating for power outages during gear shifts. As a result, the HEV AMT will stand out in future hybrid technology developments with lower fuel consumption and improved comfort. Figure 7.1 shows several common HEV AMT arrangement forms, and their advantages and disadvantages are shown in Table 7.3.

Figure 7.2 shows Hongqi AMT hybrid power system, which can realize different drive modes such as idle downtime of engine, engine driving the vehicle alone, drive motor driving the vehicle alone and engine and drive motor driving the vehicle in combination. The motor A is used to drive the vehicle alone or in combination; the motor B mainly starts the engine and charges the battery, or drives the vehicle with the engine.

Figure 7.3 shows Tsingshan AMT hybrid power system, which adopts the hub range shift for shifting, achieves the separate engine drive, separate motor drive and combined drive and has the parking charge function. There are two gear ratios for the separate motor drive. The biggest disadvantage of this system is that there is only motor driven reverse rather than engine driven reverse.

Figure 7.4 shows the AMT hybrid power system of FEV. The AMT can be used to achieve motor drive, engine drive, combined drive, braking capacity recovery, separate engine or motor drive of air conditioning and other functions. The realization of the gears is shown in Table 7.4. The air conditioning may be driven by the engine or motor and is driven by the motor when the engine does not work.

Figure 7.5 shows a US patented AMT hybrid power system, of which, the engine drive may realize 6 forward gears and the motor drive may realize 3 forward gears. The sliding friction of the clutch is used to assist the drive torque and reduce the shift impact during the shift.

The most typical application of the AMT hybrid power system is Magneti Marelli AMT hybrid transmission, as shown in Fig. 7.6, which has hybrid drive, low speed electric drive, shift torque compensation, energy recovery and other functions.

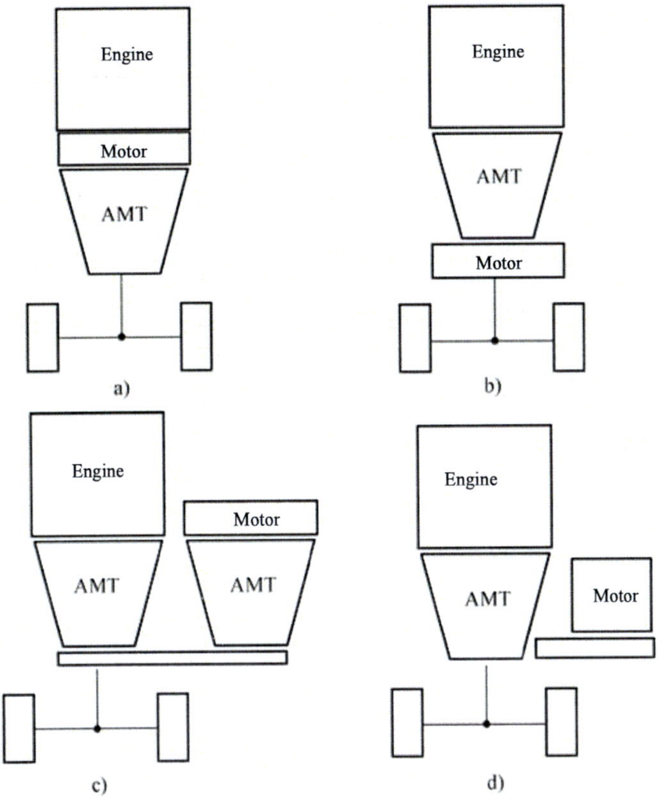

Fig. 7.1 Common HEV AMT arrangement forms

Figure 7.7 shows Magneti Marelli AMT hybrid power control scheme. In addition to identifying the driver intention, the vehicle control unit also makes integrated control of the engine ECU, transmission TCU, motor controller INV and battery capacity to achieve global optimization of the control system. In case of small torque required by the driver, the motor provides the full required torque within the torque range that can be provided by the motor; in case of large torque required by the driver, the engine is started and the motor drive is changed to the engine drive; in case of engine drive and sudden increase in the torque required by the driver, the motor provides the power to avoid downshift; in case of low battery and low engine load, the engine charges the battery in addition to driving the vehicle.

Figure 7.8 shows the torque compensation in the AMT hybrid power shift. During the shift, the clutch interrupts the engine power. At this time, the motor output torque is increased and the torque of the output shaft changes little or remains unchanged to improve the shift comfort.

Table 7.3 Advantages and disadvantages of the HEV AMT arrangement forms

Detail drawing number in Fig. 7.1	(a)	(b)	(c)	(d)
Arrangement form	The motor is directly connected to the engine	The motor is connected to the transmission output	The motor and the engine use their respective AMTs, and the power converges at the output	The motor uses some gears of the AMT
Advantages	The motor uses the AMT to change the gear ratio and operates in a high efficiency area at the same speed as the engine	The motor generates uninterrupted power during shift and achieves high comfort	The motor uses the AMT to change the gear ratio and generates uninterrupted power during shift and achieves high comfort	The motor uses part of the AMT to change the gear ratio, generates uninterrupted power during shift and has the same structure as the traditional AMT
Disadvantages	Power failure during shift	The motor cannot change the gear ratio and has low efficiency at high speed	Complex structure, high cost and large mass	Complex control system

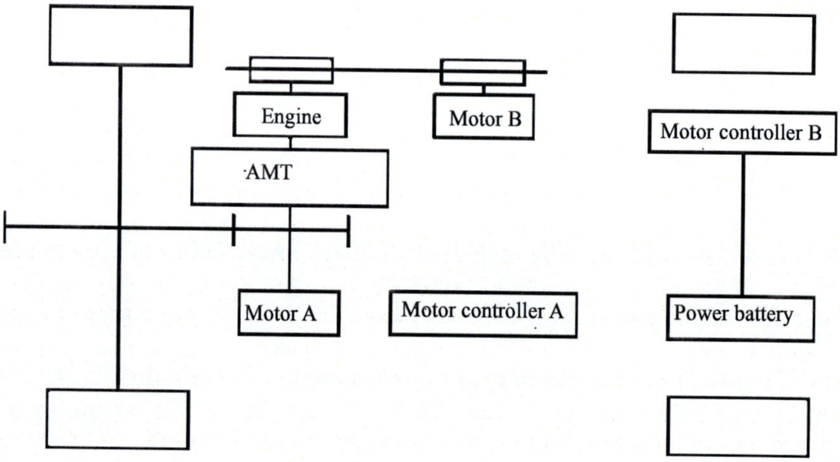

Fig. 7.2 Hongqi AMT hybrid power system

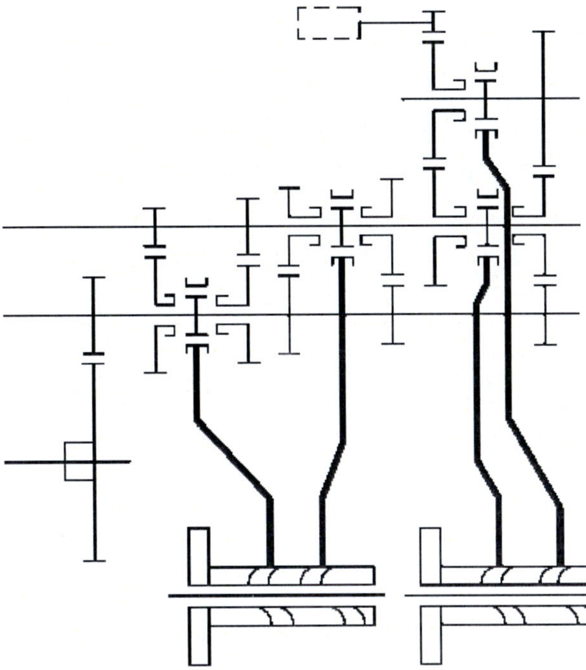

Fig. 7.3 Tsingshan AMT hybrid power system

7.4 AT Hybrid Transmission

There are two typical ways to form hybrid transmission on the basis of AT. One is to add a small motor between the engine and the hydraulic torque converter, the small motor drives the vehicle through the transmission together with the engine and the vehicle is a medium HEV; the other is to replace the hydraulic torque converter in AT with a large motor, torque damper and engine disengaging clutch. The vehicle is driven electrically at low speed. The engine disengaging clutch is engaged and the motor drives the transmission together with the engine at the medium and high speed. At this time, the motor can also act as a brake and recover the energy in the braking to charge the battery. Figure 7.9 shows the AT-based hybrid drive scheme of ZF.

The hybrid drive is primarily designed to maximize the compatibility of the core elements of the basic transmission and avoid additional parts and secondly to minimize the installation space. AT mild hybrid system, AT medium hybrid system, AT full hybrid system and AT plug-in full hybrid system are obtained by adding different parts.

AT mild hybrid system achieves the start-stop function by adding a hydraulic impulse storage (HIS) shown in Fig. 7.10 on the basis of traditional AT and can save additional 5% fuel in NEDC. The energy stored in the HIS is used after the stopping

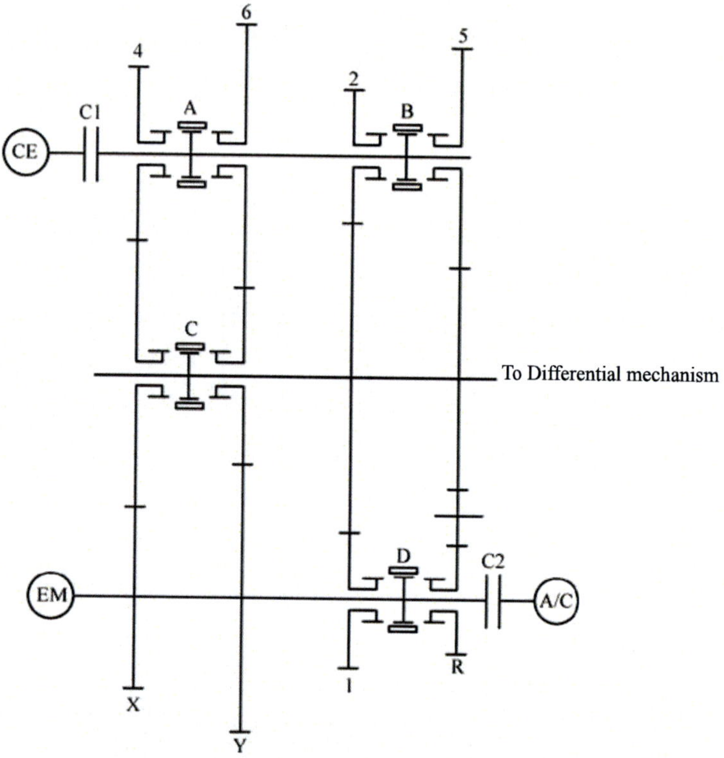

Fig. 7.4 AMT hybrid power system of FEV

phase. The auxiliary hydraulic pump accelerates the oil to fill the shift clutch (the generation of system pressure), reducing the response time from the release of the brake to the operation of the power system. The HIS is operated as follows: when a vehicle is driving on the road, the plunger chamber is filled with a certain amount of transmission oil. Once the filling process is complete, the piston is held at the position shown in Fig. 7.10b through the electromechanical locking device; in the stopping phase, the driver releases the brake pedal and releases HIS piston while starting the engine, which injects the system with the transmission oil stored in HIS to generate the system pressure normally generated by the transmission hydraulic pump. With this amount of additional oil, the time it takes the oil to fill the clutch is significantly reduced, while the driver is allowed to continue driving undetected.

The AT medium hybrid system achieves energy recovery, parking charge, engine start, motor power and other functions by integrating an ISG motor at the output end of the engine on the basis of the traditional AT.

Figure 7.11 shows AT full hybrid transmission, which achieves the separate motor drive, separate engine drive, hybrid drive, energy recovery, parking charge, engine start, motor power and other functions by integrating the clutch and drive motor at

Table 7.4 Gear realization of AMT hybrid power system of FEV

Engine drive					
Gear	A	B	C	D	C1
1	Left shift	Middle	Left shift	Left shift	Engaged
1L	Right shift	Middle	Right shift	Left shift	Engaged
2	Middle	Left shift	Middle	Middle	Engaged
3	Left shift	Middle	Right shift	Middle	Engaged
4	Left shift	Middle	Left shift	Middle	Engaged
5	Middle	Right shift	Middle	Middle	Engaged
6	Right shift	Middle	Right shift	Middle	Engaged
7	Right shift	Middle	Left shift	Middle	Engaged
R	Right shift	Middle	Right shift	Right shift	Engaged
RL	Left shift	Middle	Left shift	Right shift	Engaged
1	Middle	Middle	Middle	Left shift	Released
2	Left shift	Left shift	Middle	Middle	Released
3	Right shift	Left shift	Middle	Middle	Released
4	Middle	Middle	Left shift	Middle	Released
5	Left shift	Right shift	Middle	Middle	Released
6	Middle	Middle	Right shift	Middle	Released
7	Right shift	Right shift	Middle	Middle	Released
R	Middle	Middle	Middle	Right shift	Released

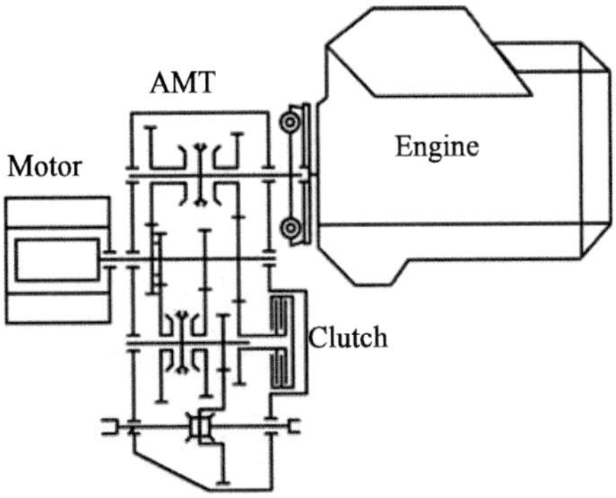

Fig. 7.5 US patented AMT hybrid power system

the output end of the engine and adding an electronic hydraulic pump on the basis of the traditional AT. The AT plug-in full hybrid transmission is structurally the same as the full hybrid transmission, except that it has a larger power battery capacity than the traditional full hybrid system and is equipped with a charging motor to charge the power battery from an external power source during parking.

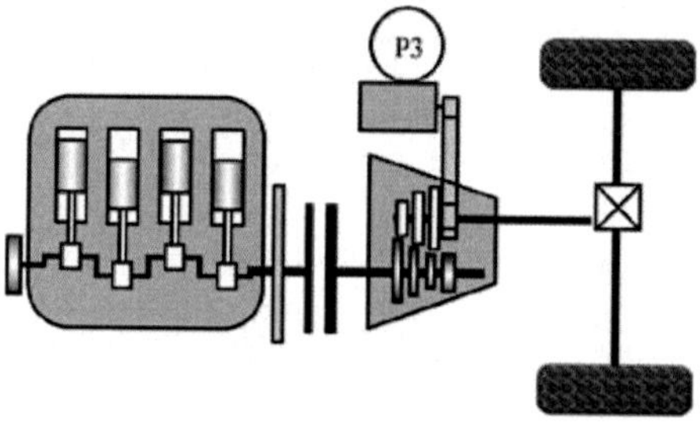

Fig. 7.6 Magneti Marelli A MT hybrid transmission

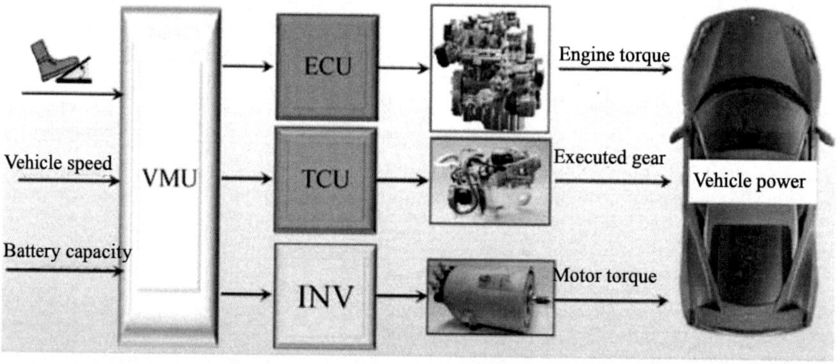

Fig. 7.7 Magneti Marelli AMT hybrid power control scheme

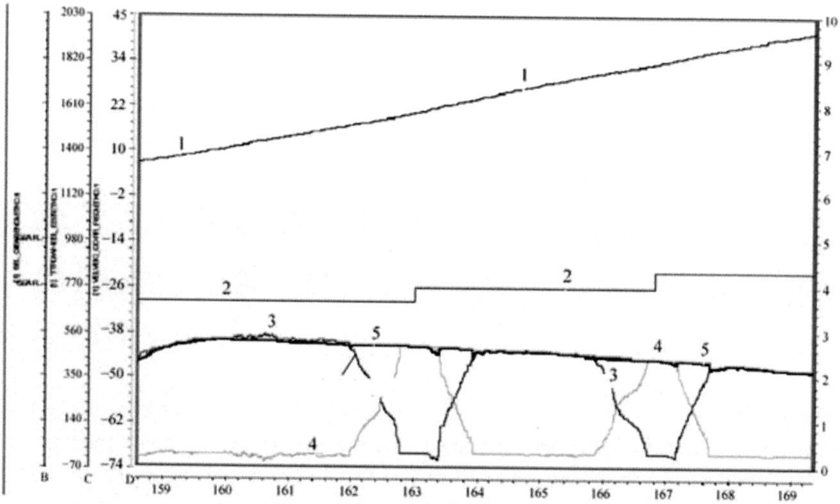

Fig. 7.8 Torque compensation in AMT hybrid power shift

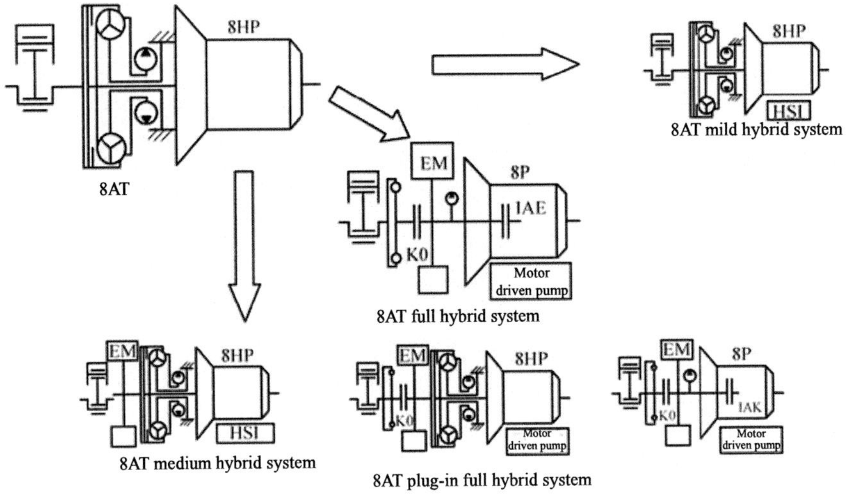

Fig. 7.9 AT-based hybrid drive scheme of ZF

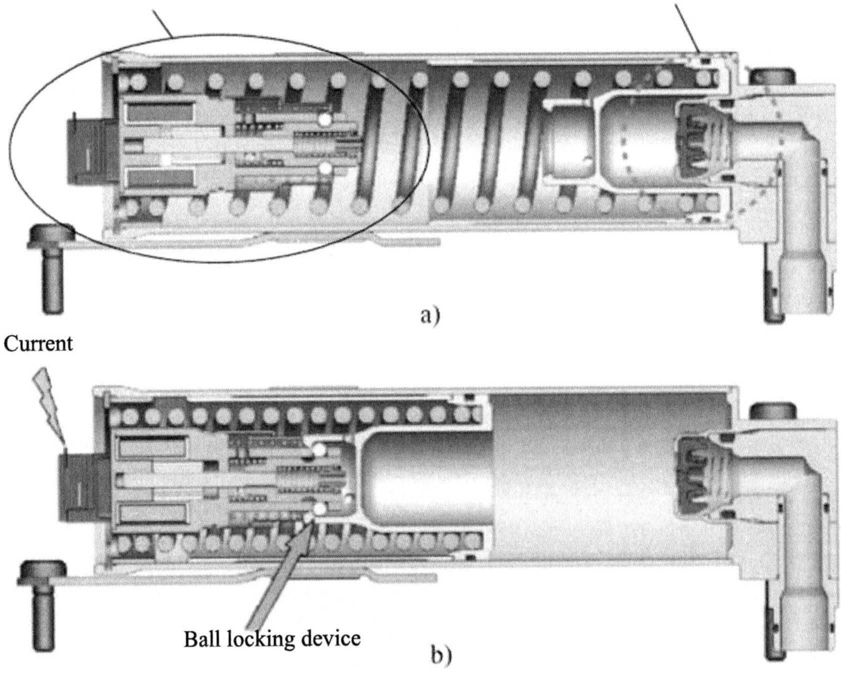

Fig. 7.10 Hydraulic impulse storage (HIS)

Fig. 7.11 AT full hybrid transmission

7.5 CVT Hybrid Transmission

I. Configuration scheme

There are three common configuration schemes of CVT based hybrid power system. The first is the four-wheel drive scheme, with the engine driving the front axle (or rear axle) and the motor driving the other axle. The advantage of this scheme is that the mechanical part of CVT is changed less and the change is mainly made to the CVT speed ratio control clutch control, brake control and other electronic control technology. The scheme is simple and practicable. Its disadvantage is that the motor torque is not through the transmission and the motor working range is large, not conducive to the full play of the motor performance. The second is the longitudinal rear drive scheme, with the engine placed longitudinally and connecting ISG motor coaxially in parallel and driving the rear axle through CVT. The advantage of this scheme is that mechanical part of CVT is changed less (the changes are mainly made to the front part of CVT, the mechanical connection between the front case and the ISG motor shell, and the mechanical connection between the hydraulic torque converter and the ISG motor stator) and the change is mainly made to the electronic control technology. The scheme is relatively simple. The idle downtime, engine displacement reduction, braking energy recovery and other functions can be achieved. The third is the horizontal front drive scheme, which is the powertrain layout of most cars. Due to the limitation of installation space, CVT needs to cancel the hydraulic torque converter. The original hydraulic torque converter space is used to install the drive motor.

The advantage of the CVT hybrid transmission is that the fuel economy of the vehicle is greatly improved by controlling the engine to work in the economic area and the motor to work in the non-economic area. The fuel economy of the HEV can be further improved if the engine can be controlled to work on the "optimal curve" using the CVT speed ratio continuous regulation characteristics within the engine operating range. The HEV motor drive mode and regenerative braking mode can also make use of the speed ratio regulation function of CVT to make the motor work on the “optimal curve”, ensuring that less power is consumed in the motor drive and more energy is recovered in the regenerative braking, further reducing the energy consumption of the vehicle. On the other hand, the use of the electric hydraulic pump instead of the ordinary hydraulic pump and the integration of the energy conservation and emission reduction advantages of the HEV with CVT technology will certainly give full play to the HEV advantages of energy saving and environmental protection.

II. Main forms of CVT based HEV

1. Hybrid transmission consisting of engine and ISG motor/generator

Figure 7.12 shows a hybrid transmission consisting of engine and ISG motor/generator. The ISG motor/generator acts as the starter of the engine at the engine start to start the engine; is driven by the engine in the normal working of the engine and charges the power battery as a generator; provides auxiliary power for the vehicle in the acceleration and climbing; recovers the braking energy as the generator in the vehicle braking.

2. HONDA CIVIC HEV transmission

Figure 7.13 shows HONDA CIVIC HEV transmission, which combines the ISG motor and engine as auxiliary power to the same shaft. The torque of the moor and engine drives the vehicle in parallel. The powertrain of this model adopts the horizontal front drive scheme and the hydraulic torque converter is canceled from the CVT, saving a certain space. It mainly has the following working conditions:

(1) Hybrid drive: a large power, if required for the vehicle driving, is jointly driven by the engine and motor.
(2) Separate engine drive: when the vehicle requires appropriate power and the engine works in the high efficiency interval, the engine drives the vehicle separately. The engine provides additional generating power for the motor to charge the battery while providing the driving power of the vehicle: (1) to maintain the relative balance of the battery capacity in case of low battery; (2) to improve the efficiency of the vehicle powertrain in case of low power required by the vehicle and the engine working in the area with low fuel efficiency.

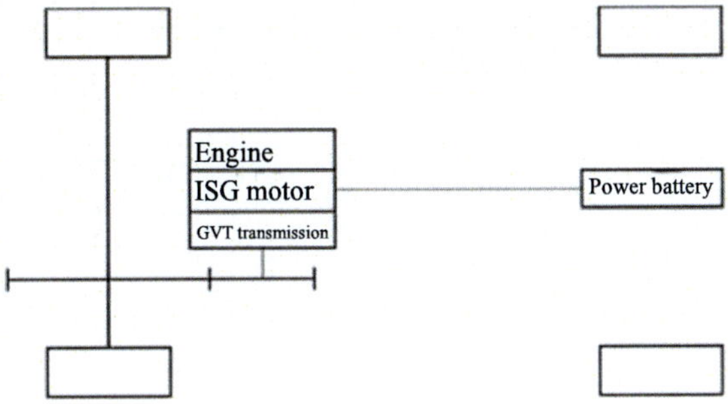

Fig. 7.12 Hybrid transmission consisting of engine and ISG motor/generator

Fig. 7.13 HONDA CIVIC hybrid transmission

(3) Checking braking: the vehicle braking system brakes the vehicle and the motor does not generate power in case of high battery capacity; if the battery SOC does not exceed the upper limit, the motor will run as a generator and recovers partial kinetic energy of the vehicle to charge the power battery in the checking braking process. At this time, the insufficient braking strength is compensated by the traditional braking system.

3. **Hybrid transmission consisting of generator and two motors**

Figure 7.14 shows a hybrid transmission consisting of generator and two motors. The motor A may drive the vehicle alone through CVT or with the engine when the clutch is engaged. It acts as the generator in braking to charge the power battery; the motor B charges the power battery in parking and acts as the starter in the engine start.

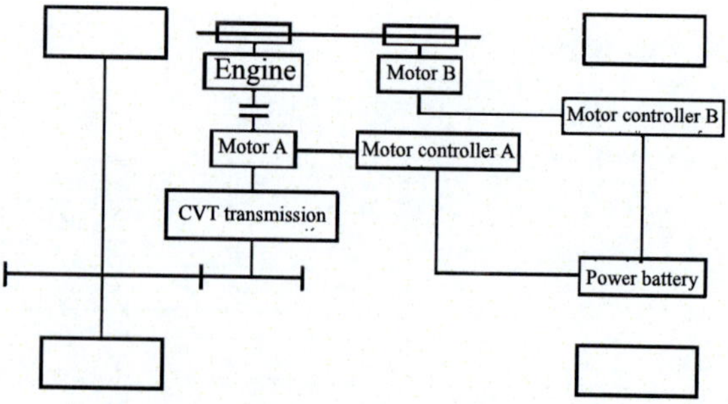

Fig. 7.14 Hybrid transmission consisting of generator and two motors

This scheme is adopted for the Tino powertrain of Nissan. The engine is connected to a 13 kW starter/generator via a drive belt and can be used as a starter when the engine needs to be started, and can be used as a generator to charge the battery pack at ordinary times. The engine is connected by an electromagnetic clutch to a 17 kW drive motor, whose rotor is connected to the CVT. The vehicle operating mode is switched through the electromagnetic clutch. When the clutch is released, the vehicle can achieve electric and regenerative braking mode; when the clutch is engaged, the vehicle can achieve the engine drive, motor power and driving power generation mode.

Due to the removal of the hydraulic torque converter and reversing gear, Tino achieves more smooth speed change process and lower fuel consumption by use of the speed ratio regulation of the CVT. The fuel consumption of 100 km reaches 4.3 L/100 km. Compared with cars of the same class, its fuel consumption under urban conditions is reduced by 50% and CO_2 emission is reduced by 50%. However, due to control technology and production costs, only 100 units of this model have been produced before production stop.

4. **Four-wheel drive hybrid transmission**

Figure 7.15 shows a four-wheel drive hybrid transmission, which realizes the coupling between the engine and motor A by means of the planetary gear train and clutch, drives the front wheel through CVT and realizes the conditions such as motor A starting the engine, motor A charging the power battery in the engine stop, engine and motor A driving the front wheel through different combinations of the planetary gear train and clutch; the motor B is used to drive the rear wheel or recover the braking energy in the braking. Under the coordination of the control system, the engine, motor A and motor B can achieve electric two-wheel drive, electric four-wheel drive, separate engine drive, front wheel hybrid drive, four-wheel hybrid drive and braking energy recovery.

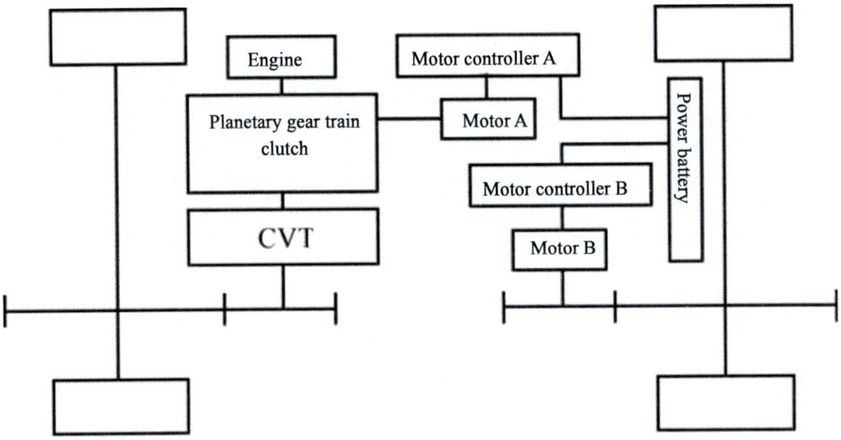

Fig. 7.15 Four-wheel drive hybrid transmission

Estima and Alphard of Toyota, as early representative products of THS-C (Toyota Hybrid System-CVT), can achieve four-wheel drive. The front wheels are driven by a 96 kW engine and 13 kW permanent magnet synchronous motor. Their power is coupled by the planetary gear train, and metal belt CVT is used to realize the infinitely variable speed; the rear wheels are driven by an 18 kW AC synchronous motor, which is integrated with the main reducer and differential mechanism. When the ground adhesion of the front wheel is insufficient and "slip" is about to occur, the engine divides the power into two parts: one part drives the front wheels through CVT; the other party drags the front motor to generate power through the planetary gear train and supplies the electricity generated by the front motor to the rear motor to drive the rear wheels, realizing the composition of the engine power and effectively using the ground adhesion of four wheels. In case of insufficient vehicle driving force or regenerative braking, the rear motor can also be involved. The CVT hydraulic pump of the THS-C is driven by motor and the hydraulic torque converter is canceled.

5. **Hybrid transmission based on CVT**

The hybrid transmission based on CVT is shown in Fig. 7.16. The motor is responsible for driving and generating electricity, and the clutch CL1 connects or disconnects the engine and motor, using a multi-disk clutch dry clutch; the clutch CL2 connects or disconnects the CVT and the motor, using a multi-disk wet clutch. Through the control of clutch CL1 and clutch CL2, various modes as shown in Figs. 7.17, 7.18, 7.19, 7.20, 7.21 and 7.22 are realized.

The powertrain torque demand (PTD) is used in the HEV to achieve integrated control of the engine, motor and CVT, as shown in Fig. 7.23. The target driving force is first set and then the corresponding torque and gear ratio of the target driving force are established. The energy management module needs to consider charging state and other factors, and output the optimal action point instructions for engine, motor torque control and CVT variable speed control, and each unit executes the action according to the instructions. In addition, the adaptive control system can modify the above control output according to the driver's intention.

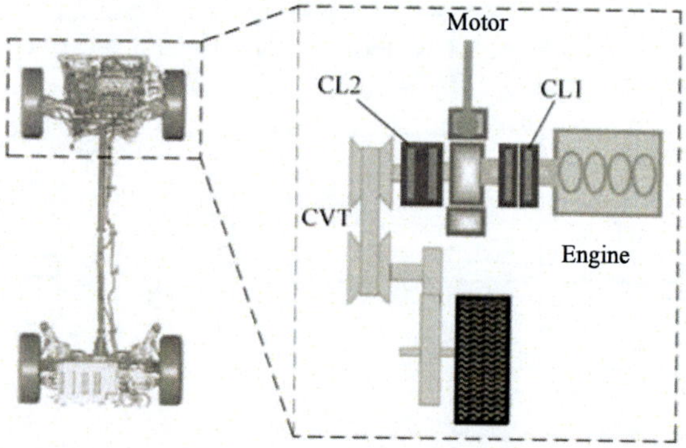

Fig. 7.16 Hybrid transmission based on CVT

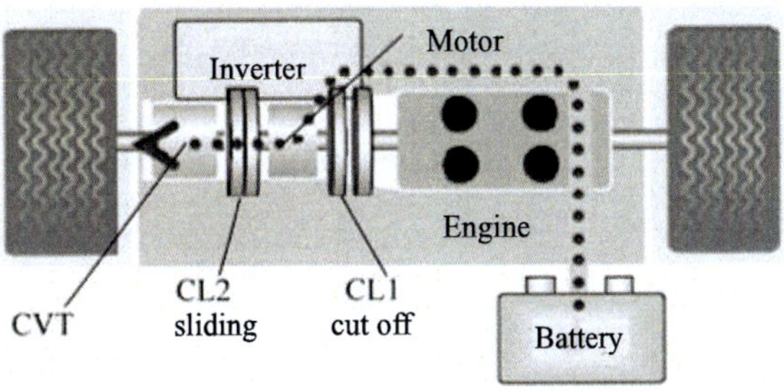

Fig. 7.17 Motor drive for starting at the engine stop

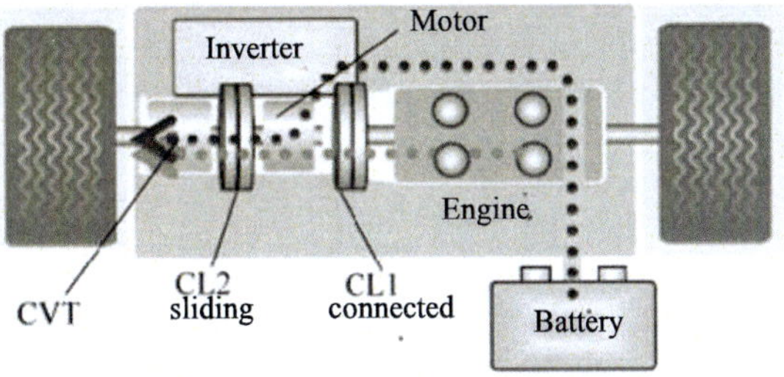

Fig. 7.18 Hybrid drive of engine and motor for starting

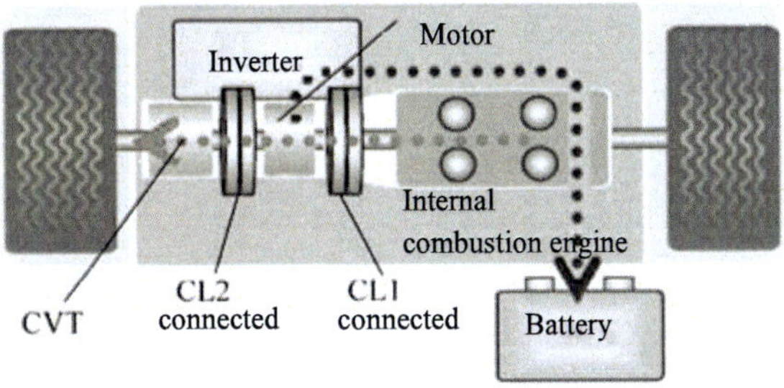

Fig. 7.19 Engine drive while charging

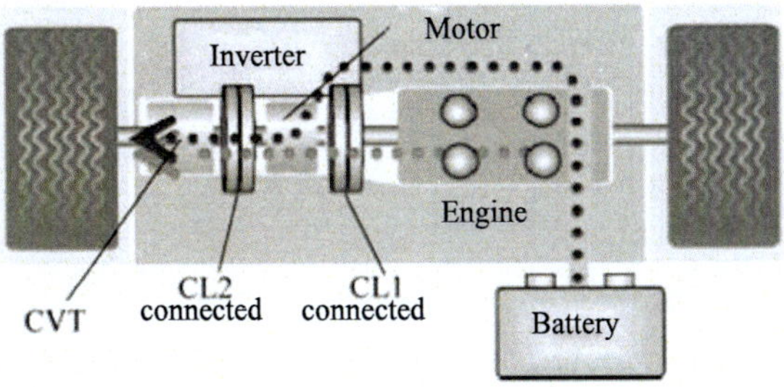

Fig. 7.20 Hybrid drive of engine and motor

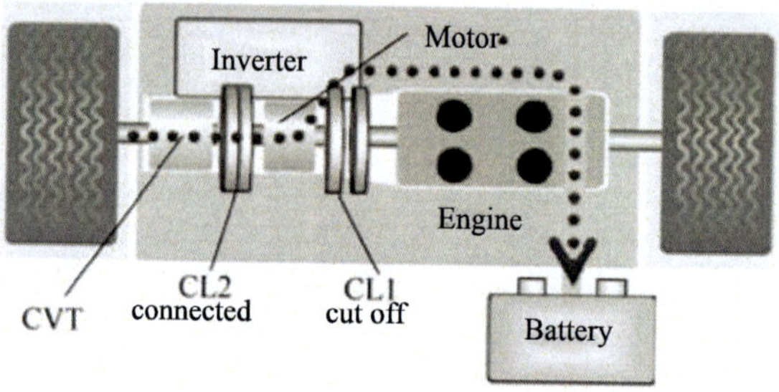

Fig. 7.21 Energy recovery in sliding condition

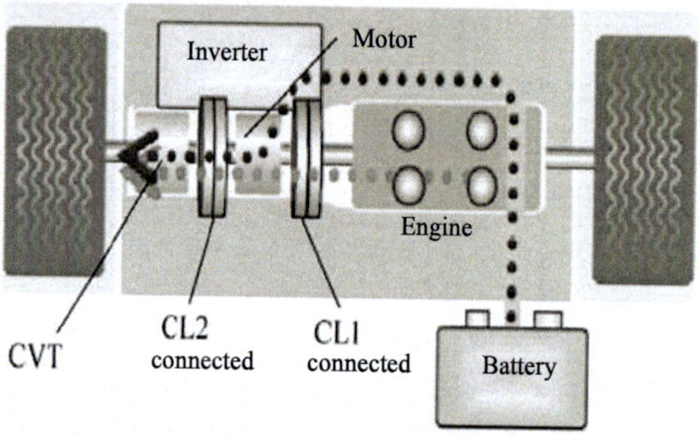

Fig. 7.22 Exhaust reduction at low temperature

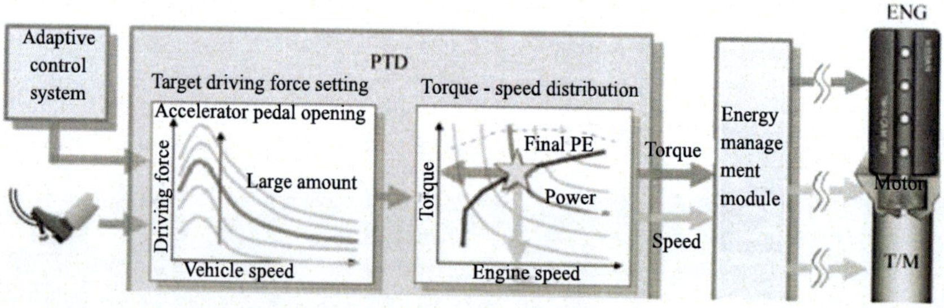

Fig. 7.23 HEV integrated control flow

7.6 DCT Hybrid Transmission

A typical application abroad of the DCT based hybrid transmission system is Honda Fit dry 7DCT hybrid power system, as shown in Fig. 7.24. It has the electric drive mode (EV), hybrid drive mode, engine drive mode and energy recovery mode.

In the electric drive mode (EV) shown in Fig. 7.25, the motor drives the vehicle alone; in the 7DCT hybrid drive mode shown in Fig. 7.26, the engine and motor

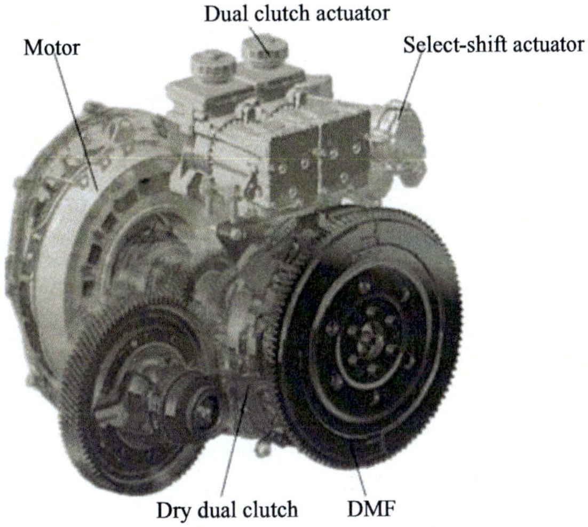

Fig. 7.24 Honda Fit dry 7DCT hybrid power system

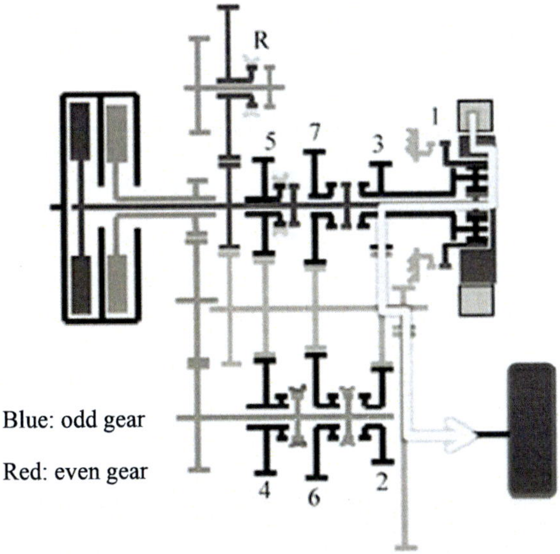

Fig. 7.25 7DCT electric drive mode (EV)

drive the vehicle; in the 7DCT engine drive mode shown in Fig. 7.27, the engine drives the vehicle alone; in the 7DCT energy recovery mode shown in Fig. 7.28, the vehicle power is used to drive the motor and the motor, in the power generating mode, charges the battery.

A typical application at home of the DCT based hybrid transmission system is DCT hybrid transmission system of BYD Qin, as shown in Fig. 7.29. It has the electric drive mode (EV), hybrid drive mode, engine drive mode and energy recovery mode.

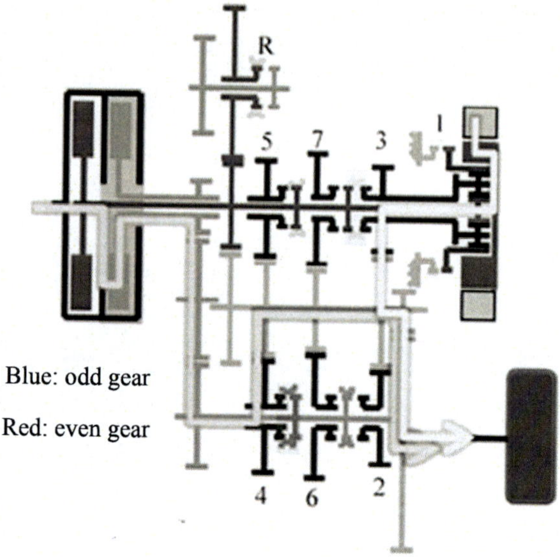

Fig. 7.26 7DCT hybrid drive mode

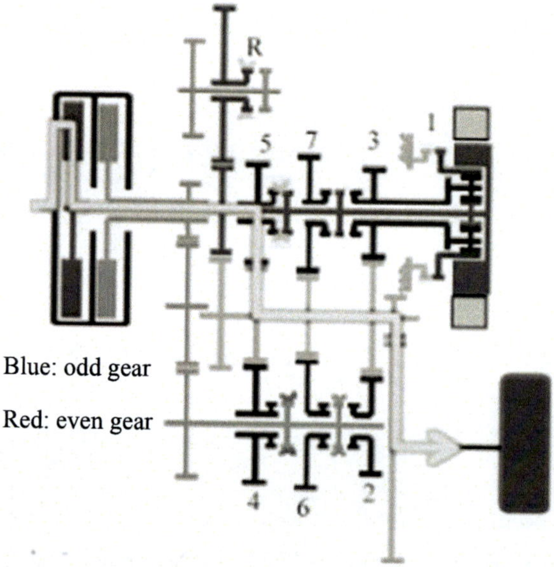

Fig. 7.27 7DCT engine drive mode

In case of high battery SOC and low vehicle speed, the motor is used to drive the vehicle, i.e. electric drive mode, as shown in Fig. 7.30; in case of high vehicle power demand, the motor and engine drive the vehicle, i.e. hybrid drive mode, as shown in Fig. 7.31; in case of low vehicle power demand, the engine drives the vehicle alone and the motor can charge the battery, i.e. engine drive mode, as shown in Fig. 7.32. The power transmission route in the braking energy recovery mode is the same as

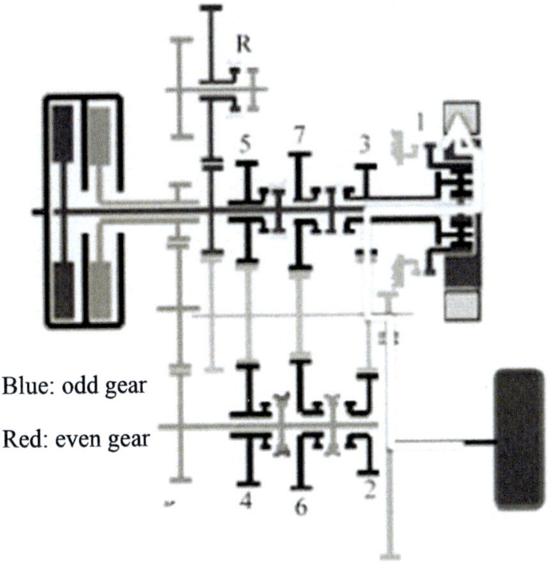

Fig. 7.28 7DCT energy recovery mode

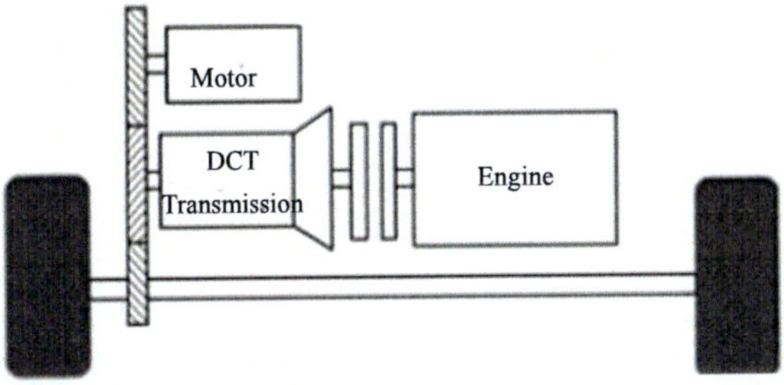

Fig. 7.29 DCT hybrid transmission system of BYD Qin

in Fig. 7.32, but the engine does not output power at this time. The disadvantages of the DCT hybrid power system are that it cannot implement the power generation at shutdown, the motor cannot be involved in the speed change and cannot always ensure the optimization of energy.

7.7 Planetary Gear Hybrid Transmission

The PSHEV is a special hybrid power system and PRIUS hybrid power system is a typical example of PSHEV, with the structure as shown in Fig. 7.33. The PSHEV is mainly composed of motor 1, motor 2, engine and planetary gear train, among which motor 1 is connected to the center gear, motor 2 is connected to the annular gear, and the engine is connected to the planetary carrier. Its biggest feature is the use of a

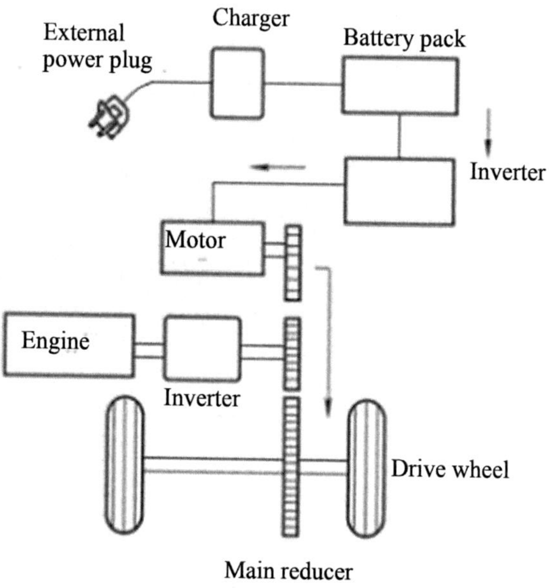

Fig. 7.30 Electric drive mode

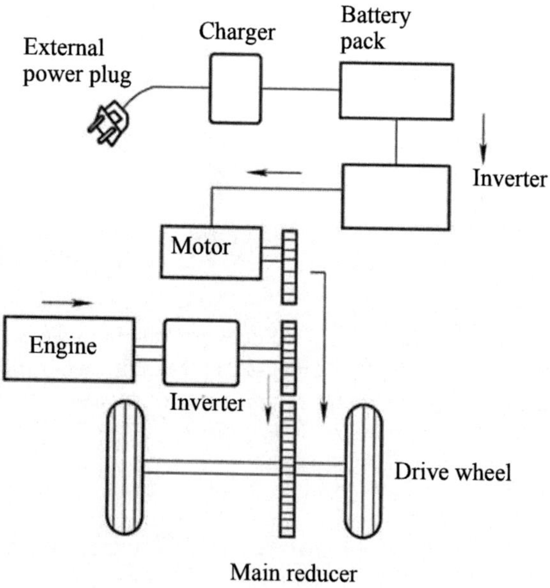

Fig. 7.31 Hybrid drive mode

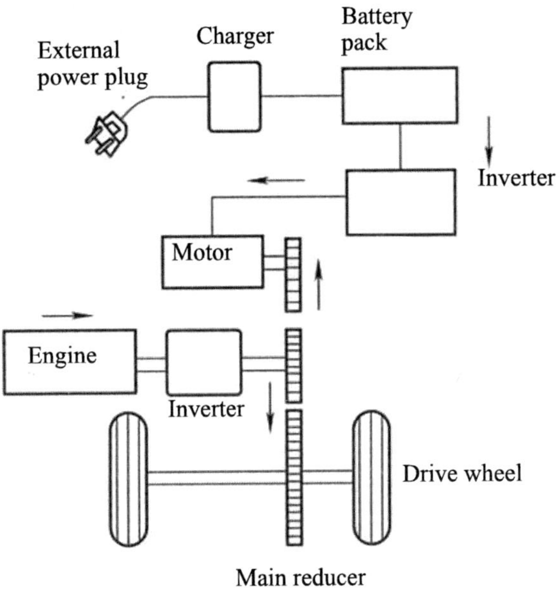

Fig. 7.32 Engine drive mode

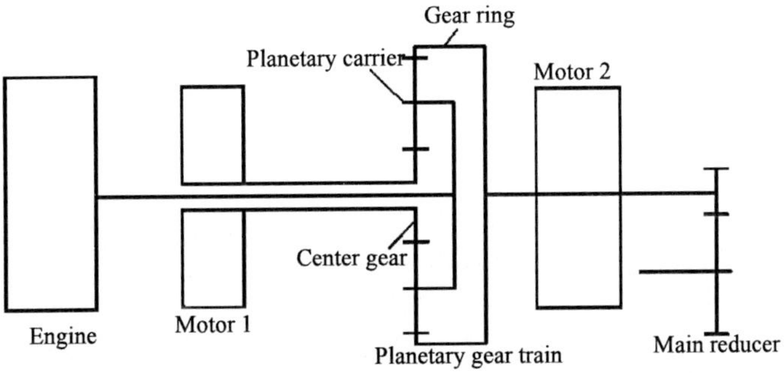

Fig. 7.33 Prius PSHEV

planetary gear set to couple two motors and the engine together. The planetary gear set can realize the function of infinitely variable speed, making the whole system more efficient. Its biggest disadvantage is that it can only carry out constant torque distribution, and the system efficiency is low when cruising at high speed.

In the planetary gear train, the center gear speed, annular gear speed, planetary carrier speed and the number of gear teeth have the following relationship

$$n_s + an_r = (1 + a)n_c \tag{7.11}$$

where,

n_s—center gear speed (r/min);
n_r—annular gear speed (r/min);
n_c—planetary carrier speed (r/min);
a—characteristic coefficient, a = $\frac{Z_r}{Z_s}$, Z_s is the number of the center gear teeth and Z_r is the number of annular gear teeth.

In the planetary gear train, the center gear torque, annular gear torque, planetary carrier torque and characteristic coefficient a have the following relationship

$$T_r = aT_s \tag{7.12}$$

where,

T_x—center gear torque (N m);
T_r—annular gear torque (N m).

$$T_c = T_r + T_s \tag{7.13}$$

式中 T_c—planetary carrier torque.

$$T_s = \frac{T_c}{1+a} \tag{7.14}$$

$$T_r = \frac{aT_c}{1+a} \tag{7.15}$$

As for the hybrid power system in Fig. 7.21, when the planetary gear train is balanced, the engine torque is equal to the planetary carrier torque, the torque of motor 1 is equal to the center gear torque, and the torque of motor 2 is not equal to the annular gear torque. The planetary gear train may be used to achieve switching of the following modes:

(1) In case of vehicle starting and small load without large power output to drive the vehicle, the power provided by the motor 2 can meet the vehicle driving requirements when the power battery permits. At this time, the engine does not rotate, the motor 1 reverses and is in the electric drive condition.
(2) In the acceleration or climbing, the vehicle requires a large power and the torque output by the engine is assigned to the annular gear according to the calculation results of formula (7.15) to drive the vehicle; the torque output by the engine is assigned to the center gear according to the calculation results of formula (7.14) to generate electricity for motor 1; the motor 2 provides power torque for the annular gear using the electricity provided by motor 1, and the torque assigned by the engine to the annular gear and the torque provided by motor 2 drive the

vehicle. The deeper the accelerator pedal goes, the more drive torque provided for the vehicle.

(3) When the vehicle is in the cruising condition, the engine only needs to provide power to overcome the air drag and rolling resistance rather than excessive power output. At this time, the engine operates under economic conditions, with motor 1 switching from power generation to drive state, and motor 2 switching from drive state to power generation, entering the peculiar mode of the PRIUS hybrid power system.

(4) When the accelerator pedal is released, the vehicle enters the sliding mode, the engine is in the down state and the motor 2 is in the power generation state to charge the power battery. According to different control strategies, the inertia of the vehicle may be fully used instead of power generation in these conditions, thereby further improving the energy economy of the system.

(5) When the vehicle is braking, motor 2 increases the generated power and generates greater resistance to reduce the speed. The insufficient demand for braking is partly compensated by the traditional braking system, and the energy recovered from braking is stored in the power battery.

(6) When the battery is low, the system will control motor 1 or motor 2 to generate electricity in real time according to the running state of the power system and charge the power battery.

7.8 Electric Vehicle Transmission

Since the electric vehicle has a variety of powertrain architectures, the connections between the motor and the transmission or other drive mechanisms are diverse. The common electric vehicle structure is shown in Figs. 7.34, 7.35, 7.36, 7.37, 7.38, 7.39, 7.40 and 7.41, in which C is the clutch, D is the differential mechanism, FG is the reducer with fixed speed ratio, GB is the transmission, M is the motor, and VCU is the vehicle control unit.

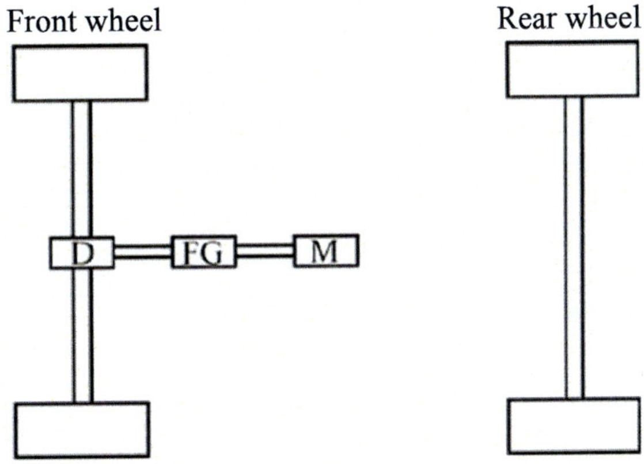

Fig. 7.34 Clutchless single-gear drive electric vehicle

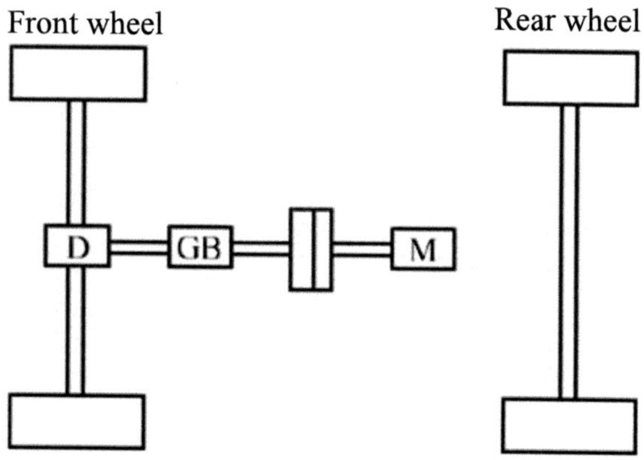

Fig. 7.35 Multi-speed electric vehicle

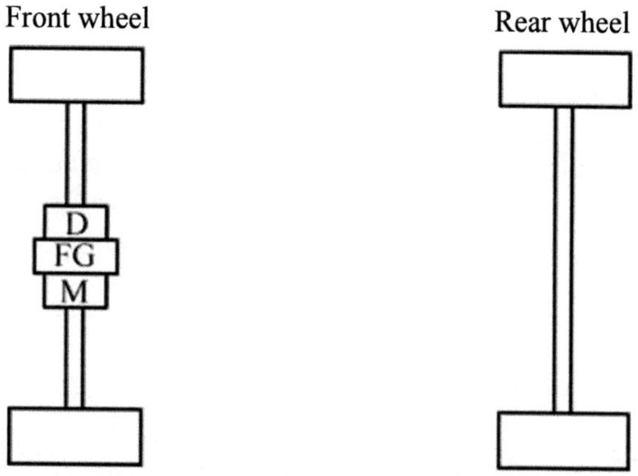

Fig. 7.36 Drive system integrated electric vehicle

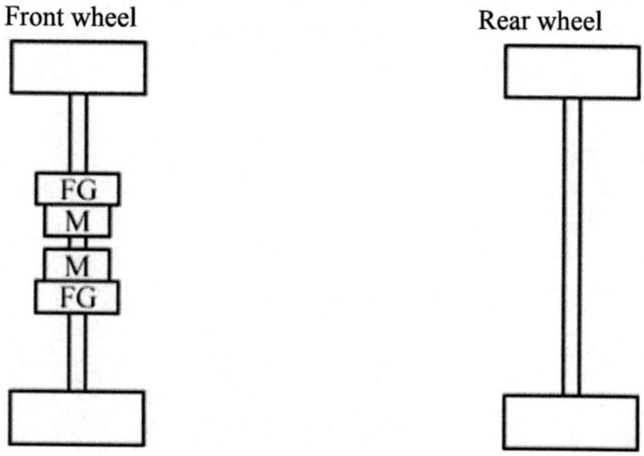

Fig. 7.37 Two-wheel motor drive system with fixed gear ratio

Fig. 7.38 Dual-motor and fixed-gear direct drive system

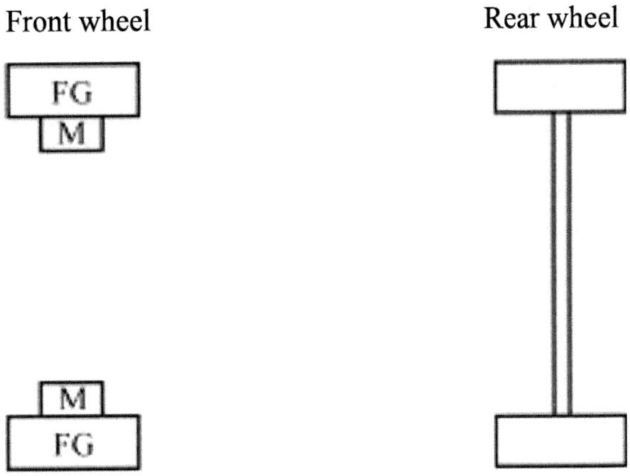

Fig. 7.39 Dual-hub motor drive system

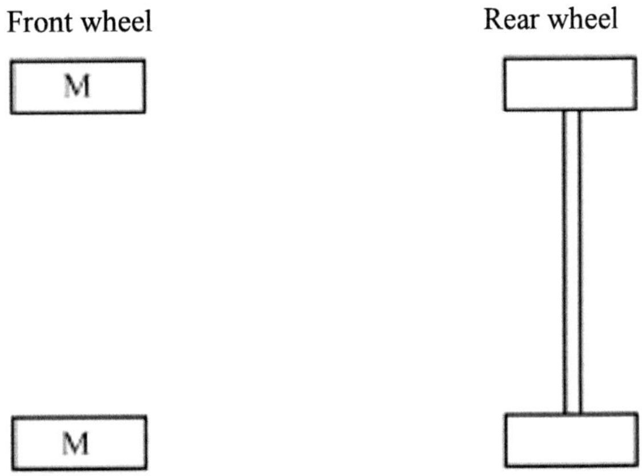

Fig. 7.40 Dual-motor four-wheel drive system

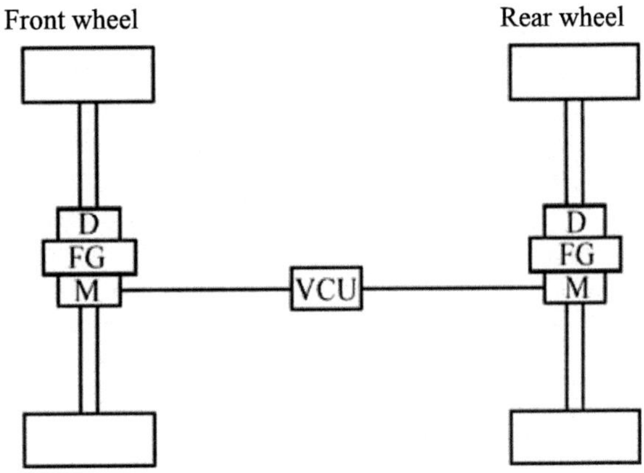

Fig. 7.41 Four-hub motor drive system

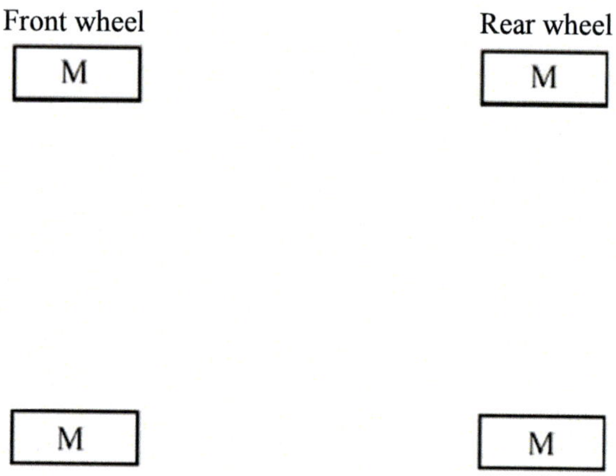

In the clutchless single-gear drive electric vehicle shown in Fig. 7.34, the motor, the transmission with a fixed speed ratio and the differential mechanism constitute the power system of the electric vehicle. The structure of the power system takes advantage of the constant torque at the low motor speed and the constant power characteristics in the large range of speed change. The reducer with fixed speed ratio is used to replace the reducer with multiple speed ratios. Based on this replacement, the requirement for the clutch is reduced and thus the clutch can be canceled, so as to reduce the volume and mass of the mechanical drive and simplify the drive system; the disadvantage of this system is that it is unable to optimize the operating point efficiency of the motor under variable working conditions. At the same time, in order to meet the requirements of vehicle acceleration/climbing and high-speed working conditions, it is usually necessary to choose the motor with larger power.

In the multi-speed electric vehicle shown in Fig. 7.35, the electric motor replaces the internal combustion engine in the traditional ICE vehicle and, together with the clutch, transmission and differential mechanism, constitutes the power drive system similar to the traditional vehicle. The motor outputs the driving force instead of the internal combustion machine and the clutch can realize the motor driving force and driving wheel connection or disconnection. The transmission provides different gear ratios and matches the load demand by changing the speed - power (torque) curve. The differential mechanism is used to realize that the wheels on both sides of the vehicle are driven at different speeds when turning.

Figure 7.36 shows a drive system integrated electric vehicle, in which, the motor, the reducer with fixed speed ratio and the differential mechanism are further integrated and even combined into individual components. The axle shaft connected with the wheel is directly connected with the composite unit, and the drive system is further simplified and miniaturized. It is the most common drive form of battery electric vehicles currently.

Figure 7.37 shows two-wheel motor drive system with fixed gear ratio. The mechanical differential mechanism is canceled and two motors drive the wheels on the respective sides through the reducer with fixed speed ratio. In the vehicle turning, the electronic differential mechanism controls the motor to run at different speeds, so as to achieve normal turning of the vehicle.

Figure 7.38 shows a dual-motor and fixed-gear direct drive system. The drive motor and the planetary gear reducer with fixed speed ratio are installed in the wheel, so this drive system is also called in-wheel driven system, thus further simplifying the drive system. In the drive system, the planetary gear reducer is mainly to reduce the motor speed and increase the motor torque.

Figure 7.39 shows a dual-hub motor drive system, which completely abandons the mechanical connection device between the motor and the driving wheel. The motor drives the wheels directly, and the motor speed control is equivalent to the wheel speed control, i.e. vehicle speed control. Such a drive system structure puts forward special requirements for the motor, such as high torque characteristics in the vehicle acceleration or deceleration. The external rotator motor at low speed is generally used.

Figure 7.40 shows a dual-motor four-wheel drive system. The front and rear wheels are driven by the motor through differential mechanism. Different motors can be used to drive the vehicle under different working conditions, or two motors can be used to drive the vehicle according to a certain proportion of torque distribution, so as to maximize the drive system efficiency.

Figure 7.41 shows a four-hub motor drive system, which completely abandons the mechanical connection device between the motor and the driving wheel and directly drives four wheels using four motors.

The motor drive systems mentioned above are all single reduction systems, so it is difficult for the motor to work in the high efficiency area all the time, and the NVH performance of the motor becomes poor at high speed. For this reason, 2- or 3-speed motor transmission is developed. Compared to the single-speed motor transmission, the two-speed motor transmission has the following advantages: the high speed ratio of single speed can achieve better acceleration performance and gradeability; the low speed ratio of two-speed can improve the maximum speed and widen the speed range; the lower motor running speed and better NVH can be obtained; the motor can be miniaturized; a wider range of high efficiency area and higher overall operating efficiency can be achieved, with smaller battery capacity or improved electric endurance mileage.

Figure 7.42 shows a 2-speed motor transmission, which consists of the double-row planetary gear train, rear axle differential mechanism, motor, shift actuator and power device, so that the motor can transmit the power according to two different gear ratios, as shown in Fig. 7.43. Figure 7.44 shows the comparison of efficiency of two motor transmissions. As can be seen from the figure, the high efficiency area range of the 2-speed motor transmission is obviously larger than that of the single-speed motor transmission, thus improving the motor efficiency and increasing the driving range.

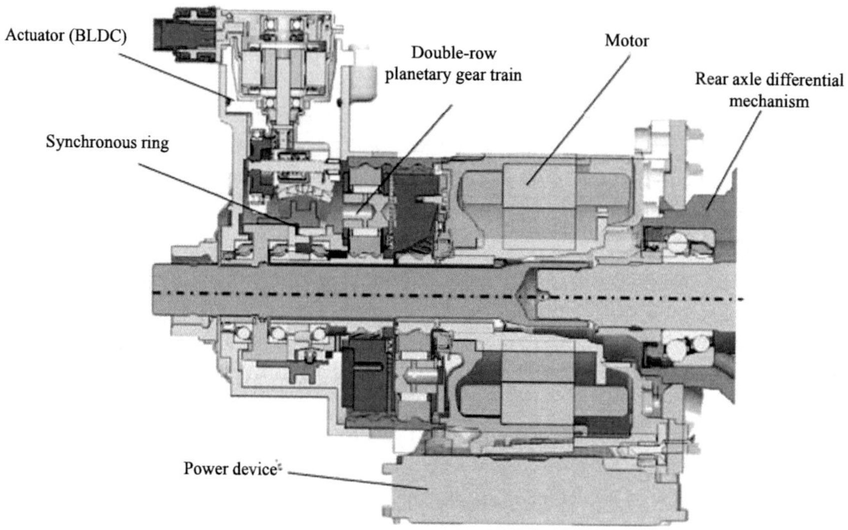

Fig. 7.42 2-speed motor transmission

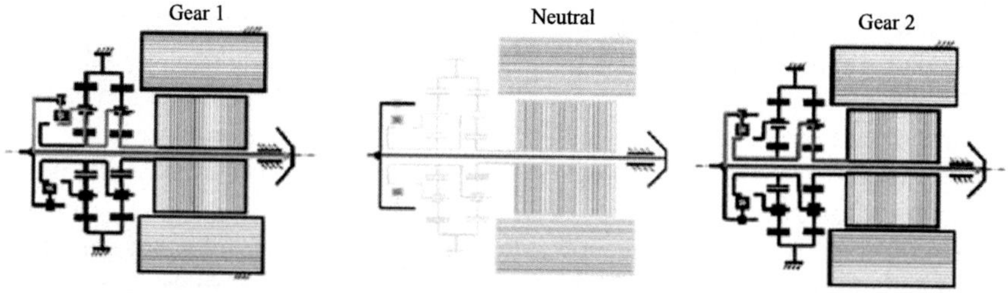

Fig. 7.43 Transmission route of 2-speed motor transmission

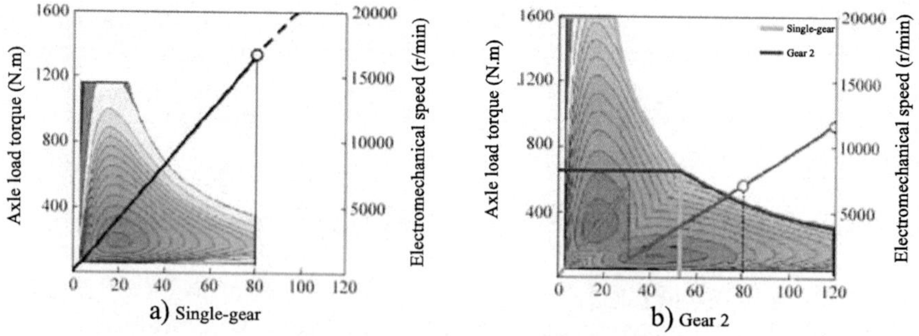

Fig. 7.44 Comparison of efficiency of two motor transmissions

7.9 Other Hybrid Power Plants

The previous sections focused on transmissions for motor and engine-powered HEVs. It is important to note that non-electric HEVs may be more economical than electric HEVs. Mechanical HEVs and hydraulic HEVs are described here.

I. Mechanical HEV

On the basis of powertrain, the mechanical HEV integrates the mechanical hybrid power plant as shown in Fig. 7.45, which is mainly composed of flywheel assembly, coupling device, clutch, transmission and control clutch. The flywheel assembly is used for energy storage and operates in a vacuum container. The transmission is used to change the gear ratio between powertrain and flywheel assembly. The clutch is used to control the process of engagement, release and sliding friction of mechanical hybrid power plant and automobile powertrain. When braking, the clutch is engaged gradually and the flywheel is accelerated. When the flywheel accelerates to a certain degree, the clutch is disconnected and the flywheel rotates at high speed in the vacuum chamber. The flywheel is used to store the kinetic energy of the vehicle and the clutch is gradually engaged when starting or accelerating. The flywheel kinetic energy accelerates the vehicle and provides power together with the engine. The system is characterized by high energy density, compact structure, low mass, low cost and low maintenance cost compared with the electric HEV. The mechanical hybrid power plant has been successfully applied in racing cars, and has broad application prospects in buses, trucks and engineering vehicles. It is still in the trial sub stage in the passenger vehicles.

The biggest problem with the mechanical hybrid power plant is that the flywheel assembly rotates at speeds up to tens of thousands of revolutions per min, close to

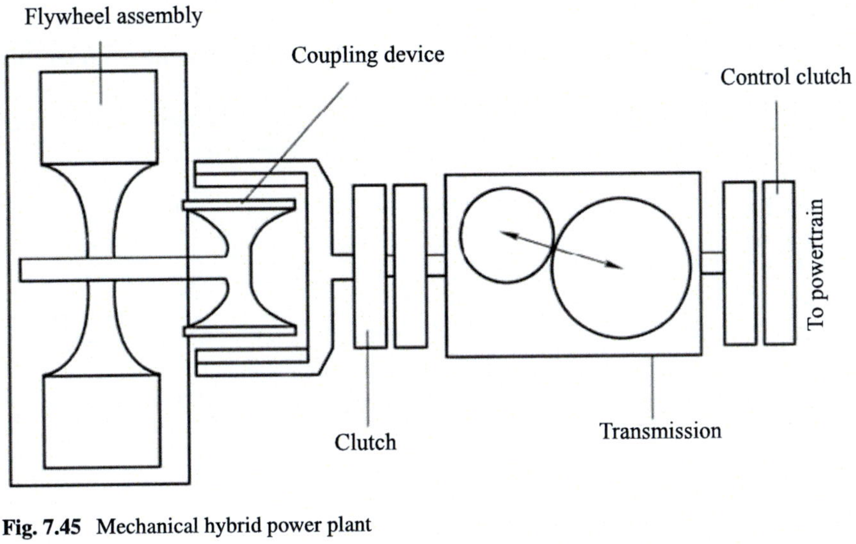

Fig. 7.45 Mechanical hybrid power plant

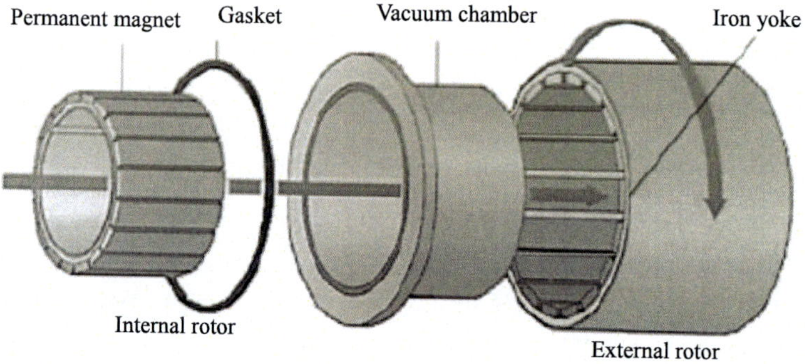

Fig. 7.46 Magnetic coupling

the sound velocity externally. In order to reduce the loss of air drag, the flywheel assembly needs to work in a vacuum tight container. Structurally, the shaft of the flywheel transmitting power must pass through the vacuum tight container. After long-term use, air will inevitably enter the container and the solution is to vacuumize with a vacuum pump.

In order to facilitate the sealing of the coupling device in the mechanical hybrid power plant, a magnetic coupling is also adopted, as shown in Fig. 7.46. In addition to transmitting power through magnetic force, the magnetic coupling can also obtain the speed change function through different number of pole pairs of internal and external rotors. The magnetic coupling can be used to separate the power transmission shaft from the vacuum seal without the need for vacuum pumping.

II. **Hydraulic HEV**

In addition to using the internal combustion engine to generate the driving force, the hydraulic system can also provide the power to drive. In the hydraulic system, energy is stored in the accumulator in the form of compressed liquid or similar. To supercharge the liquid, the internal combustion engine is needed to generate power and energy to drive a hydraulic pump, and a hydraulic motor is required to extract the energy. In other words, a hydraulic pump is like a generator, a hydraulic motor is like a motor, and the pressurized liquid in an accumulator is like a battery. It can be seen that there is a one-to-one equivalent relationship between hydraulic system and electrical system.

In the HEV (Hydraulic Hybrid Vehicle) system architecture shown in Fig. 7.47, the hydraulic pump is used instead of the AC motor, the hydraulic motor instead of motor, HIS instead of battery, HHV controller instead of HEV controller and hydraulic valve system instead of the power electronic system.

Figure 7.48 shows the physical architecture of a complete HHV truck, in which, HIS includes a high pressure accumulator filled with a safe gas such as nitrogen. The pressure inside the cylinder can be as high as 21–35 MPa (3000–5000 psi), while the pressure inside the low pressure cylinder can be as low as a few hundred pa. The internal combustion engine drives the hydraulic pump, which pulls liquid

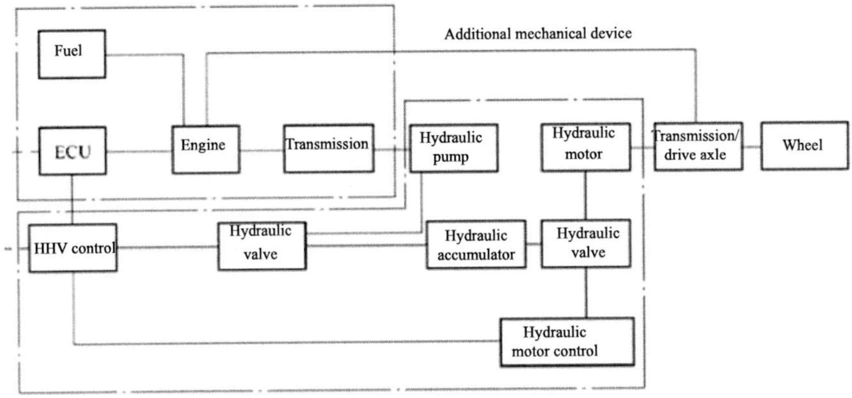

Fig. 7.47 HHV system architecture

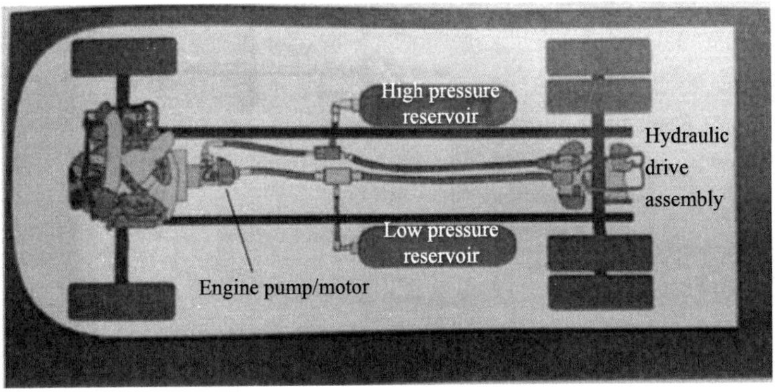

Fig. 7.48 Physical architecture of HHV truck

out of the low pressure cylinder and pumps it to a very high pressure into the high pressure cylinder, where the mechanical energy is eventually stored as high pressure gas. When driving the wheel, the high pressure fluid from the high pressure cylinder flows through the hydraulic motor to drive the powertrain. The hydraulic motor sucks in the high pressure liquid and converts its stored energy into mechanical energy at the wheel. As the liquid flows through the hydraulic motor, the pressure plummets and the liquid is sent to the low pressure cylinder. At this point, a hydraulic cycle is complete.

It should be noted that the energy storage in the hydraulic energy storage system is very low in terms of energy stored per kilogram of mass. For example, the energy storage density of the HIS is about 1.9 W h/kg, while the energy density of the battery can reach 30–120 W h/kg. However, the power density of the hydraulic system can reach 2500 W/kg, while the power density of the electric system is only 650 W/kg. Therefore, it is obvious that the hydraulic hybrid power system is very suitable for systems with large power demand and relatively low energy demand, especially for the conditions of rapid acceleration and deceleration at high power.

One of the biggest benefits of using the hydraulic system for driving is that it can avoid the demand for heavy conventional powertrain. Meanwhile, the liquid can be moved from one point to another more easily, without the need for sophisticated transmissions or mechanical connection device. However, in order for the internal combustion engine to operate at the optimal operating point, some form of energy storage system is required, which can be achieved in the hydraulic hybrid power system by means of a gas-filled accumulator.

As can be seen from Fig. 7.49, for regenerative braking of hydraulic HEV, the efficiency of the hydraulic pump and hydraulic motor is slightly over 90%, and the efficiency of the accumulator (including high pressure and low pressure cylinders) is about 98%. The efficiency of the whole regenerative braking process is about 82%.

Figure 7.50 shows bladder type hydropneumatic accumulator and diaphragm accumulator. Figure 7.50a shows a bladder type hydropneumatic accumulator, in which, the bladder B is filled with compressible gas (volume V). When the liquid flows into the cylinder from the bottom, it can expand or contract so that energy can be stored in the gas during compression and released from the gas during expansion. Therefore, energy is transferred by the liquid. Figure 7.50b shows a diaphragm accumulator. The liquid flowing from the bottom drives the diaphragm D and the deformation of the diaphragm D changes the volume V of the gas above it, which is filled between the cylinder and the diaphragm.

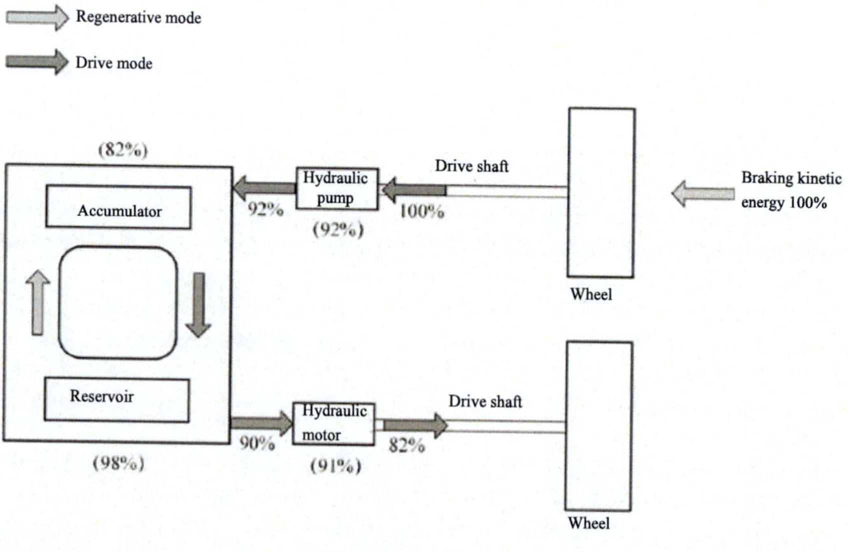

Fig. 7.49 Distribution of regenerative braking efficiency in HHV

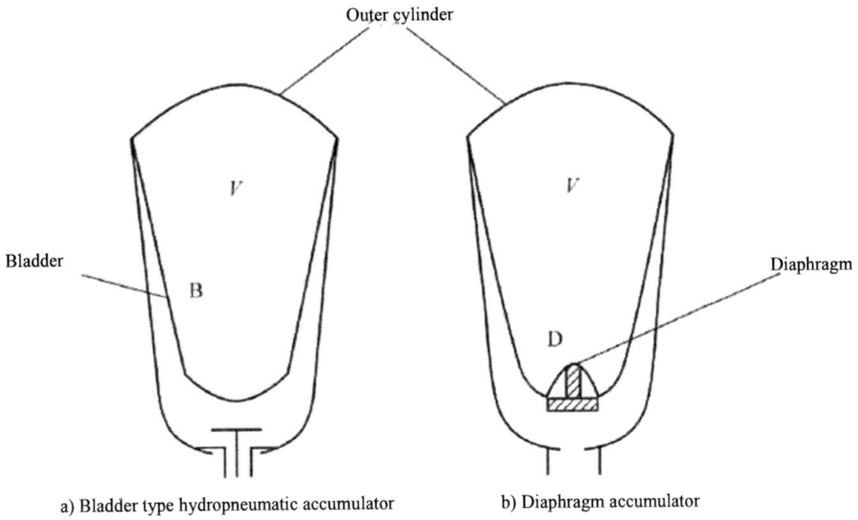

Fig. 7.50 Bladder type hydropneumatic accumulator and diaphragm accumulator

Bibliography

1. Qishan F, Chengliang Y, Zhu Yu (2005) Road test for dynamic and economy performance of hybrid electric vehicle. Chin J Mech Eng 41(12):19–24
2. Youliang Ma, Quanshi C (2001) The development of hybrid electric vehicle (HEV). J Highw Trans Res Dev 18(1):78–80
3. Yutao L, Sijia Z, Kegang Z (2007) Analysis on AMT shifting process of hybrid electric vehicle. J South China Univ Technol (Natl Sci Edn) 35(2):33–37
4. Yanchun G (2006) Gear shift schedule and gearshift control for AMT in hybrid electric vehicle. Shanghai Jiao Tong University, Shanghai
5. Tao D, Dongye S, Datong Q et al (2009) Regenerative braking simulation for hybrid electric vehicle with continuously variable transmission. J Mech Eng 45(9):214–220
6. Zhonghao B, Libo C, Yaonan W (2007) Modeling and simulation of hybrid electric vehicle based on CVT. Comput Simul 24(6):235–238
7. Ying F, Yongge L (1999) Development survey of continuously variable transmission. J Hubei Univ Autom Technol 13(4):15–18
8. Yuqiang Z (2008) Research on power-shift model and hydraulic control characteristics based on DCT structure. Chongqing University of Technology, Chongqing
9. Zhixuan L (2013) Configuration analysis, parameters matching and research on the optimal control for the HEV equipped with DCT. Jilin University, Changchun
10. Bu Xi D, Aimin XF (2006) Theoretical study and simulation on planetary gear set for hybrid electric vehicle. Autom Eng 28(9):834–838
11. Yuan Z, Guangyu T, Quanshi C et al (2004) A research on the system efficiency of a hybrid electric vehicle with power-split powertrain. Autom Eng 26(3):260–265
12. Zeng H, Juhua H (2010) The matching of EV between the motor and the transmission. Equip Manuf Technol 2:40–42
13. Xiumin Yu, Shan C, Jun L et al (2006) Present study situation and developing trend of control strategies for hybrid electric vehicle. Chin J Mech Eng 42(11):10–16
14. Weigang Z, Tan Yu, Xiaolin Z (2006) Application of hydraulic transmission in motor vehicle hybrid power brake energy conservation. Mach Tool Hydraul 6:144–146

15. Jiuyu D, Shihua Y, Chao W et al (2009) The application and development of hydraulic hybrid powertrain of vehicle. Mach Tool Hydraul 37(2):181–184
16. Demirdöven N, Deutch J (2004) Hybrid cars now, fuel cell cars later. Science 305(5686):974–976
17. Bitsche O, Gutmann G (2004) Systems for hybrid cars. J Power Sour 127(1):8–15
18. Anderson CD, Anderson J (2004) Electric and hybrid cars: a history. McFarland, North Carolina
19. Pourbafarani Z, Montazeri-Gh M (2010) Optimisation of the gearshift strategy for a hybrid electric vehicle equipped with AMT. M. S. thesis, School of Mechanical Engineering, Iran University of Science & Technology
20. Miller JM (2006) Hybrid electric vehicle propulsion system architectures of the E-CVT type. IEEE Trans Power Electron 21(3):756–767
21. Baraszu RC, Cikanek SR (2002) Torque fill-in for an automated shift manual transmission in a parallel hybrid electric vehicle. In: Proceedings of the 2002 American control conference (IEEE Cat. No. CH37301). IEEE, no 2, pp 1431–1436
22. Hofman T, Dai CH (2010) Energy efficiency analysis and comparison of transmission technologies for an electric vehicle. In: 2010 IEEE vehicle power and propulsion conference. IEEE, pp 1–6

Chapter 8
Transmission Design

Transmission is an important part of powertrain. In the transmission design, in addition to the transmission type, basic performance characteristics and purpose, this chapter mainly introduces the gear design, bearing selection and design, case design, parking mechanism design, synchronizer design, seal selection and transmission ventilation design and transmission test. The design of hydraulic torque converter and planetary gear train will be introduced in corresponding sections.

8.1 Gear Design

With the increasing requirements for energy saving and environmental protection, the automobile engine technology has made great progress, its powertrain has also made great progress in small lightweight, high performance and high output power, thus greatly increasing the high strength requirements for the gears in the automobile powertrain system and their materials. At present, how to further improve the fatigue strength of automobile gears has become an important technical research topic for automobile companies and alloy steel material companies.

The gear strength mainly refers to the gear bending fatigue resistance strength and tooth surface contact fatigue strength. The high strength gear technology includes gear design, processing, material selection, heat treatment and surface hardening treatment, gear lubrication, friction and wear, etc. In this section, combined with years of research and practice, the author summarily expounds the achievements and application status of researches on high strength automobile gear materials and material surface hardening in recent years in terms of the gear material technology, discusses the development trend of high strength gear material technology and analyzes the service life of gear.

Y. Chen, *Automotive Transmissions*, Springer Tracts in Mechanical Engineering,
https://doi.org/10.1007/978-981-15-6703-2_8

I. Gear strength calculation

With respect to the automobile transmission gear, the common damage forms are bending fatigue failure and tooth surface fatigue failure, in which the bending fatigue failure is judged by the tooth root bending stress. The tooth root bending stress is determined by the following parameters

$$\sigma_w = \frac{F_t}{bm_n} Y_{Fa} Y_{Sa} Y_\varepsilon Y_\beta K_A K_V K_{F\beta} K_{F\alpha} \tag{8.1}$$

where, σ_w—tooth root bending stress (N);

F_t—nominal circumferential force on the reference circle (N);
b—tooth width (mm);
m_n—standard modulus (mm);
Y_{Fa}—shape factor;
Y_{Sa}—notch stress concentration factor;
Y_ε—contact ratio;
Y_β—helical overlap coefficient;
K_A—application factor;
K_V—dynamic factor;
$K_{F\beta}$—cumulative frequency distribution coefficient of longitudinal load;
$K_{F\alpha}$—lateral coefficient.

The permissible tooth root bending stress is $[\sigma_w]$ and the safety factor is S_w. The tooth root bending stress must meet the following formula to meet the design requirements

$$\sigma_w \le \frac{[\sigma_w]}{S_w} \tag{8.2}$$

The tooth surface fatigue failure is judged according to the tooth surface contact stress, which is determined by the following parameters

$$\sigma_j = Z_{B/D} Z_H Z_E Z_\varepsilon Z_\beta \sqrt{\frac{F_t(u+1)}{d_1 bu}} \sqrt{K_A K_V K_{H\beta} K_{H\alpha}} \tag{8.3}$$

where, σ_j—tooth surface contact stress (N);

$Z_{B/D}$—gear contact coefficient;
Z_H—partition coefficient;
Z_E—elastic coefficient;
Z_ε—contact ratio;
Z_β—helical overlap coefficient;
d_1—pinion base diameter;
u—gear ratio;
$K_{H\beta}$—cumulative frequency distribution coefficient of longitudinal load;
$K_{H\alpha}$—lateral coefficient.

The permissible tooth surface contact stress is $[\sigma_j]$ and the safety factor is S_j. The tooth surface contact stress must meet the following formula to meet the design requirements

$$\sigma_j \leq \frac{[\sigma_j]}{S_j} \quad (8.4)$$

Gear design is carried out based on the tooth root bending stress and tooth surface contact stress. In order to realize the lightweight and high strength of gear, it is necessary to explore the countermeasures of high-strength gear material and its processing technology from the research of gear materials.

II. Gear material technology with high bending fatigue strength

1. **Analysis of the causes for gear bending fatigue failure and material countermeasures**

The main direction of gear strength research is to solve the problem of gear tooth fracture, namely, insufficient gear bending fatigue strength. The requirement for automobile gear materials is good machinability and heat treatment carburizing and quenching performance. Generally, the carbon with the mass fraction of 0.2% is selected and added carburizing alloy steels with Ni, Cr, Mn, Mo and other alloying elements individually or in combination. In Japan, SCr420H or SCM420H carburizing alloy steel is generally used and there are strict requirements on the stability of quenching distance on the top of the steel. Table 8.1 shows the composition of above materials.

The gear tooth fracture and breakage is mainly caused by the gradual expansion of the cracks caused by the repeated action of the overload concentrated stress on the gear root. Figure 8.1 shows an example of gear failure caused by insufficient gear bending fatigue strength in a car transmission endurance test. The electron microscopy observation photos shown in Fig. 8.2 are a metallographic analysis of the fracture section of a broken gear tooth. It can be seen from the metallographic analysis that the bending fatigue failure crack of the gear is generated from the electron microscopy observation layer of the grain boundary oxygen failure on the surface part of the gear, and spreads along the austenitic grain boundary below the surface layer to the deep area of the hardened case, thus causing the grain boundary failure. As shown in Fig. 8.2c, the results of metallographic observation and analysis of the impact failure easily caused by the bevel gears on automobiles are similar.

Table 8.1 Composition of common carburizing alloy steel gear materials in Japan (mass fraction, %)

Material trademark	C	Si	Mn	P, S	Cr	MO
SCr420H	0.18–0.23	0.15–0.35	0.60–0.85	Below 0.03	0.9–0.12	–
SCM420H	Ditto	Ditto	Ditto	Ditto	Ditto	0.15–0.30

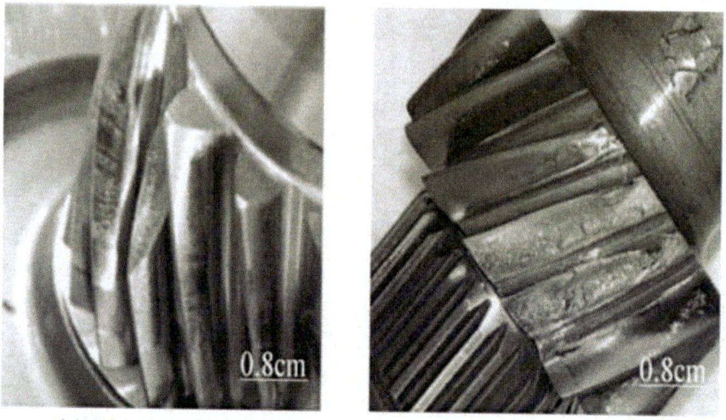

(a) Drive gear break (b) Input shaft gear break

Fig. 8.1 Example of automobile gear bending fatigue failure

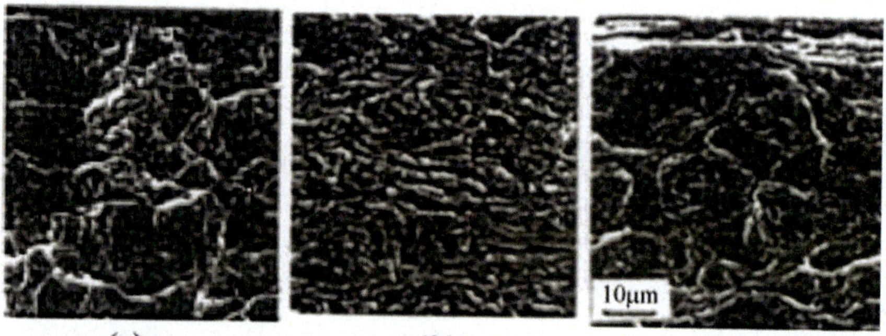

(a) (b) (c)

Fig. 8.2 Gear bending fatigue failure and impact

Because the gear bending fatigue failure is caused by the grain boundary oxide layer on the gear surface, reducing the thickness of the grain boundary oxide layer and the abnormal carburizing layer, improving the crystal boundary and strengthening and refining the grains are important measures to improve the bending fatigue strength of the gear material.

The grain boundary oxide layer on the material surface is mainly generated by Si, Mn, Cr and other alloying elements. Meanwhile, Si, Mn and Cr are the alloying elements that improve the quenching performance and are prone to produce incomplete quenching near the grain boundary, and the abnormal layer of incomplete carburizing is composed of troostite and bainite. The electron microscopy observation photos shown in Fig. 8.3 are the grain boundary oxidation structure of SCr420H. The black part in Fig. 8.3b is the oxide of Si, Mn and Cr. The countermeasures to reduce the thickness of the abnormal carburizing layer are to increase the quenching speed or increase the content of alloying elements beneficial to improve the quenching

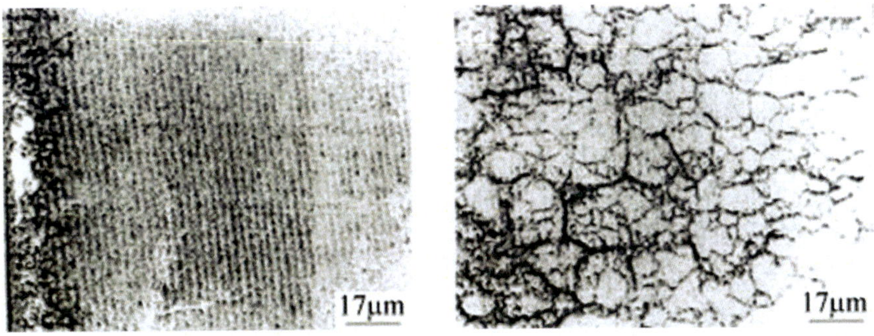

(a) Cross-section diagram (b) Oblique section

Fig. 8.3 Grain boundary oxidation structure on gear surface

performance. The former method is easy to produce large tooth surface deformation; the latter method, while reducing the contents of Si, Mn and Cr, is to ensure the quenching performance, inhibit the generation of the abnormal layer of incomplete carburizing, improve crystal boundary, strengthen and refine grains, increase the contents of Ni, Mo and other alloying elements, and adjust and appropriately increase the content of retained austenite. Another method is to use nitrocarburizing heat treatment method to reduce the degree of internal oxidation of alloying elements. This method is helpful to adjust the content of retained austenite and restrain the development of initial fatigue cracks to depth. The effect of reducing the content of Si, Mn and Cr or of nitrocarburizing on reducing the thickness of the abnormal carburizing layer is shown in Fig. 8.4. The abnormal carburizing layer thickness of the common carburized and quenched steel (SCr420H) is about 20 μm and of the newly developed carburized and quenched steel or the carburized and quenched steel subject to nitrocarburizing is only about 10 μm.

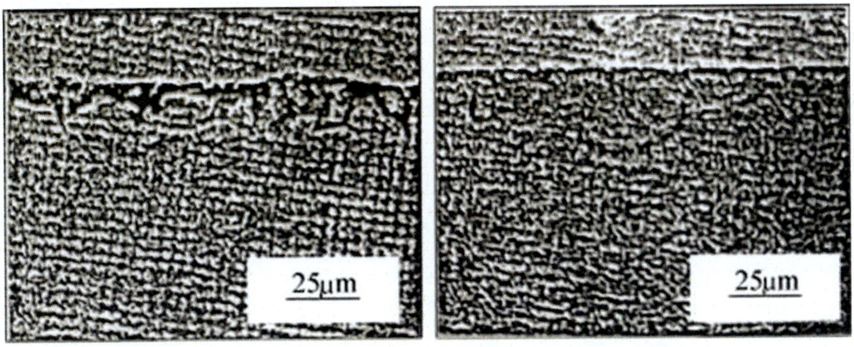

(a) Common carburizing alloy steel (b) Nitrocarburizing heat treatment

Fig. 8.4 Electron microscopy observation photos of abnormal carburizing layer of gear

2. **Influence of effective case depth on bending fatigue strength**

The effective case depth generally refers to the thickness of the hardened case from the hardened surface of the gear to the hardness of 550 HV. As shown in Fig. 8.5, under the bending fatigue load of high-order rotating cyclic stress, the fatigue life of the specimen gear with a hardened case thickness of 0.2–0.25 mm is better than that of the specimen gear with a hardened case thickness of 0.4–0.5 mm. The optimum effective case depth of the gear shall be determined according to the gear modulus, stress state and purpose. As shown in Fig. 8.6, the shallow hardened case should be adopted at the tooth root and the deep hardened case should be adopted on the

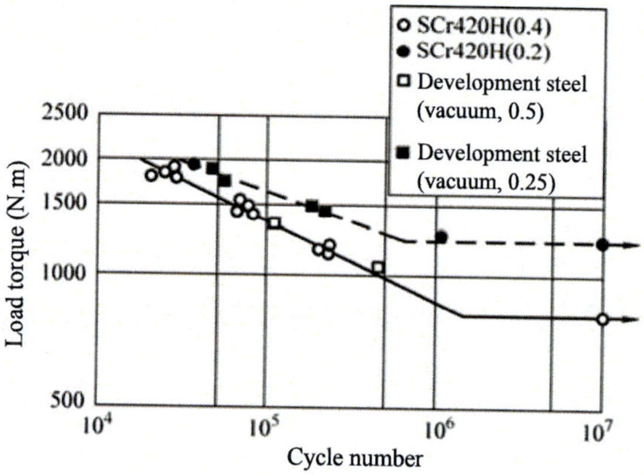

Fig. 8.5 Relationship between hardened case thickness and gear bending fatigue life

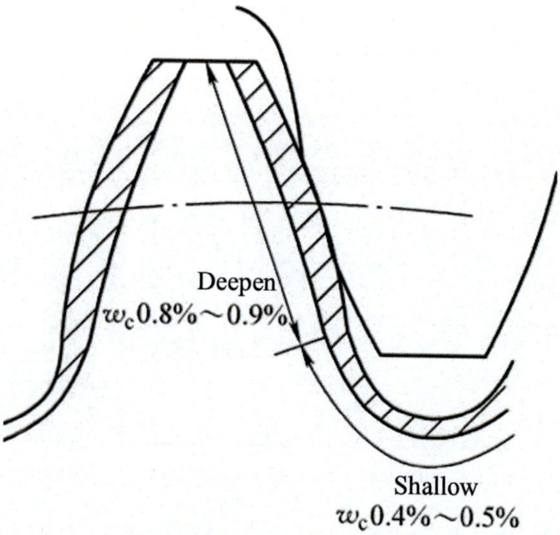

Fig. 8.6 Ideal hardened case thickness and carbon mass fraction

involute meshing tooth surface to adapt to high bending and high surface pressure bending fatigue load.

3. **Shot peening strengthening and residual compressive stress on tooth surface**

The gear bending fatigue strength mainly depends on such factors as the tooth surface hardness, residual compressive stress near the surface layer, the relaxation of stress concentration, the tooth surface roughness and the thickness of the abnormal layer of incomplete carburizing, with the following relationship

$$\sigma_w = 396 + \left(\frac{583}{R_{max}} + D_{in}\right) - 0.94\sigma_{rm} \tag{8.5}$$

where, σ_w—rotary bending fatigue strength (MPa);

Ra_{max}—minimum surface roughness value (mm);

D_{in}—thickness of abnormal layer of incomplete carburizing (mm);

σ_{rm}—maximum residual compressive stress near surface layer (MPa).

In order to improve the gear bending fatigue strength, the research focus of Japanese automobile company is to develop and popularize various forms of tooth surface shot peening strengthening and related material technology. The parts surface shot peening strengthening was mainly used for cleaning and deburring before. Mazda first developed the technology of surface shot peening strengthening for high-pressure nozzle-type gear teeth in 1987. That is, a nozzle-type peening machine with excellent controllability was used to collide with the gear tooth surface of the high-speed cast steel shots (Φ0.4–0.6 mm) of fatigue strength subject to the gear surface shot peening strengthening in Fig. 8.7 at high pressure (3.5–4.5 kg/cm^2), high speed and room temperature, so that the tooth surface was strengthened at the moment of repeated heating and quenching and the residual compressive stress near

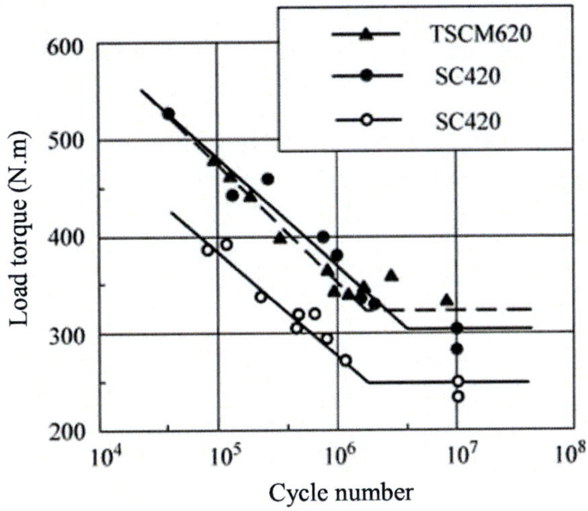

Fig. 8.7 Fatigue strength of gear surface after shot peening strengthening

Table 8.2 Processing conditions affecting the effect of shot peening strengthening

Projection conditions	Projection velocity, projection time, projection density, projection quantity, projection angle, projection mode, stress state and temperature
Projection shot peening	Material, density, hardness, particle size and particle size distribution
Projection object	Material heat treatment structure, material mechanical properties, gear modulus and processing state before heat treatment

the surface layer of gear teeth was significantly increased, so as to inhibit the spread of the fatigue cracks. The effect is shown in Fig. 8.7. After the surface shot peening strengthening, the gear bending fatigue strength could be increased by more than 1.5 times. In view of the modulus size and the use of the gear, two researches on the shot peening strengthening and small diameter shot peening strengthening technology recently have also achieved good results. The optimal effect of the shot peening strengthening is mainly affected by the following processing conditions, as shown in Table 8.2.

III. Material technology to improve the tooth surface fatigue strength

1. **Research background of material technology with high tooth surface fatigue strength**

The continuous improvement of fuel consumption standards for passenger vehicles has promoted the development of the combined automatic control technology of engine and transmission, as well as the small lightweight and high output power technology of CVT and MT. In particular, a new change has taken place in the structure composition of lubricating oil used for AT and CVT, that is, from the main lubrication performance to improvement of the control performance, resulting in increase in the friction factor between metals and deterioration of lubricated friction conditions. Therefore, it is very important to improve the durable fatigue strength of the gear with high surface pressure.

Figure 8.8 shows the gear load and failure model. It can be seen from the figure that in the past, the bending failure strength of gear teeth was lower than the pitting strength of the tooth surface, but in recent years, through the research and development of gear materials and surface peening treatment technology, the bending fatigue strength of the gear has been greatly improved and exceeded the pitting strength of the tooth surface. At present, how to further improve the fatigue strength life of the tooth surface and reduce the tooth surface abrasion has become an important and urgent technical research topic for major automobile companies and steel materials companies.

2. **Analysis of the causes for tooth surface fatigue failure and material countermeasures**

The tooth surface fatigue failure is caused by the repeated action of tensile stress caused by different tooth surface contact stress and relative sliding speed of tooth

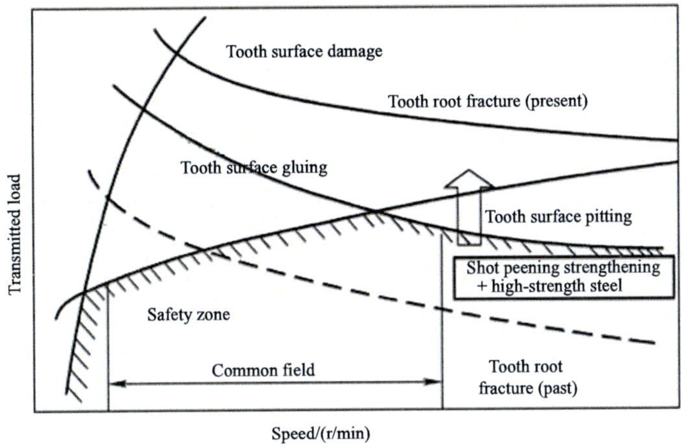

Fig. 8.8 Gear load and failure model

surface meshing in the gear pair. The failure forms are shown in Fig. 8.9, mainly including spalling and destructive pitting corrosion (pitting). Due to different stress states, the failure forms are different from the tooth root bending fatigue failure forms. As shown in Fig. 8.9a, firstly, a crack (grain boundary failure) at a certain acute angle appears in the normal direction of the tooth surface near the meshing part of the gear, which gradually extends to the bottom of the tooth root and makes the surface layer peel off.

In order to prevent the failure of tooth spalling, the following measures are generally taken:

(1) Increase the thickness of the hardened case.
(2) Increase the hardness of the hardened core.
(3) Appropriately increase the hardness at tens of microns away from the surface layer according to formula (8.6)

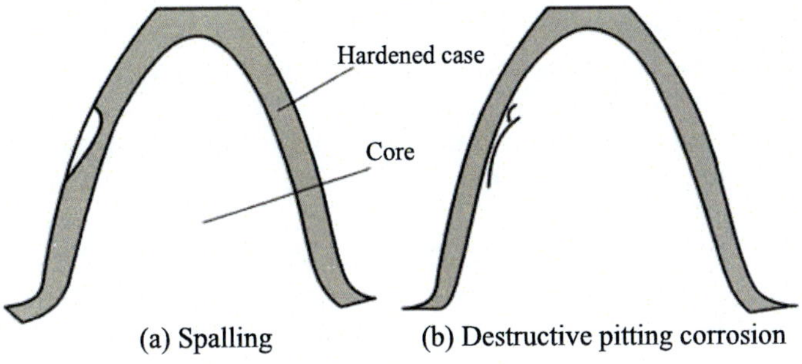

Fig. 8.9 Main failure form of gear fatigue failure

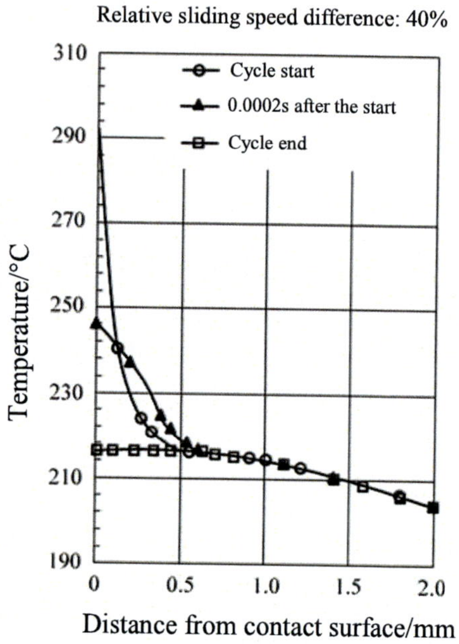

Fig. 8.10 Surface temperature distribution of cylindrical rollers during pitting fatigue test

$$\frac{\tau_{45}}{0.55} \leq \tau_y = HV/6 \tag{8.6}$$

where, τ_{45}—shear stress (MPa);

τ_y—shear yield stress (MPa);

HV—Vickers hardness.

The failure form of destructive pitting corrosion is shown in Fig. 8.9b. The crack is at a small angle (25°–30°) from the shallow surface of the hardened case and gradually extends to the surface of the gear in a diamond shape. In addition, pitting is directly related to the surface temperature of gear meshing. Figure 8.10 shows the surface temperature distribution of cylindrical rollers during pitting fatigue test using finite element analysis. As can be seen from the figure, when the relative sliding speed difference between rollers is 40% and the contact stress is about 200 kg/mm^2, the instantaneous rising temperature of the surface of the material is 250–300 °C, which is equivalent to the carburizing tempering temperature of the gear. In general, the relative sliding speed difference of automobile gear pair is about 20%, and when the tooth surface contact stress is about 160 kg/mm^2, the instantaneous rising temperature of material surface is about 160 °C. As shown in Fig. 8.11, the fatigue life of the tooth surface can be effectively improved by improving the tempering hardness and tempering resistance of the material. In the figure, steel A is commonly used chrome-molybdenum alloy steel (1.0 Cr–0.4 Mo), while steel B is chrome-molybdenum alloy steel with increased Cr and Si alloy content. It can be seen that increasing the content of Si, Cr and other alloying elements or adopting high concentration carburizing method can generate hard carbides on the material surface in order to inhibit the

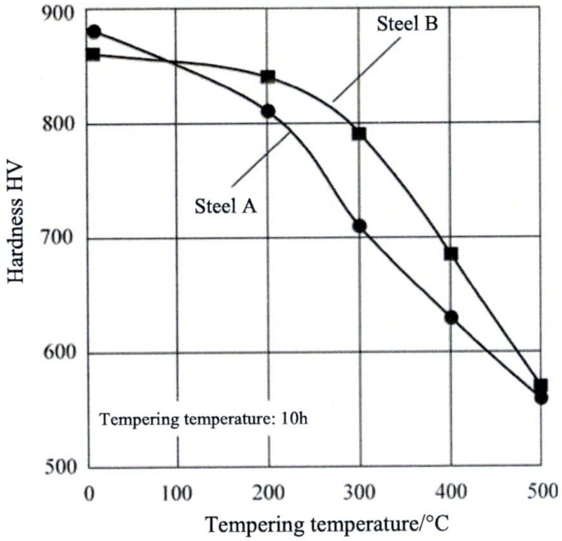

Fig. 8.11 Comparison of tempering resistanc

high-temperature softening of the material surface. Recently, the application and development of the carbo nitriding heat treatment method has also achieved a very good effect in improving the gear tooth surface fatigue life. Its main function is to reduce the solid solution chromium generated by internal acidification on the material surface, so as to minimize the production of chromium nitride, promote the refinement of carbides, and thus improve the surface hardness and tempering resistance.

3. **Research and development of material surface strengthening technology**

According to the research and practice of Japanese automobile company, it is not enough to greatly improve the tooth surface fatigue strength only by improving the alloy composition of the material and adopting the heat treatment method. The research and development of the material surface strengthening technology must also be combined. In recent years, Nissan and Mazda have developed and applied chemical treatment methods for tooth surface coating of molybdenum disulfide (MoS_2) and phosphoric acid coating, which have achieved good results and increased the pitting fatigue strength of tooth surface by more than 2 times, as shown in Fig. 8.12. The main principle is that after coating, a soft layer of 2–3 m thick is generated on the tooth surface, which flattens out most of the concave and convex cutting ripples on the tooth surface, reduces the local maximum meshing contact stress and metal surface friction factor, and improves the oil film thickness and lubrication condition in gear engagement.

Figure 8.13a, b show the electron microscopy observation photos of the tooth surfaces of the gears with and without molybdenum disulfide (MoS_2) coating after the initial gear engagement and operation. It is shown that the smoothness of the

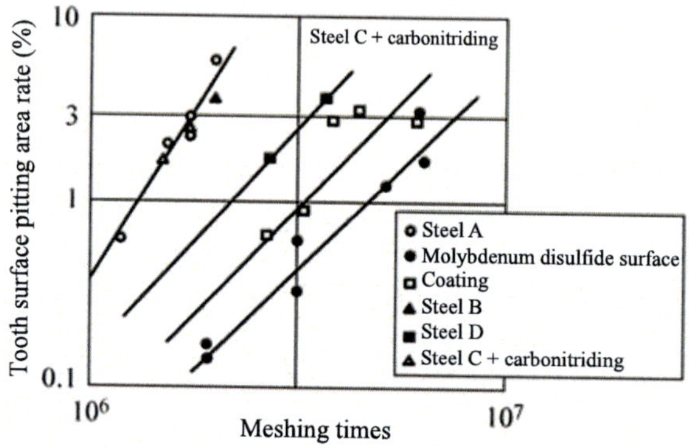

Fig. 8.12 Change of tooth surface pitting area rate

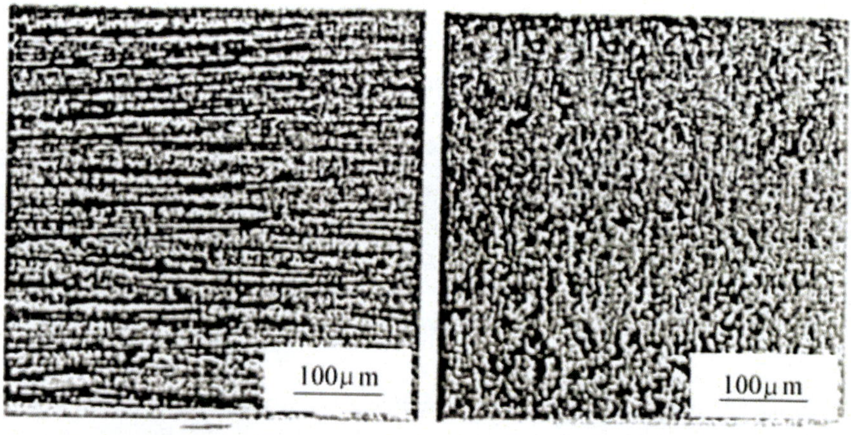

(a) Without coating (b) With coating

Fig. 8.13 Tooth surface condition after initial gear engagement and operation

coated contact surface is obviously improved during initial operation. However, the cost of these two chemical treatment methods is high.

In recent years, the Barreling method, which was jointly studied by Mr. Hoyashita of Saga University and Sumitomo Heavy Industries, has achieved very good experimental results and attracted the attention of experts in the United States and Japan. The most important characteristic of the Barreling method is that it greatly improves the tooth surface fatigue strength with low cost and has extensive application prospect. The sketch of the Barreling method is shown in Fig. 8.14. The mixture of grinding stones and grinding powder with a diameter of several millimeters rotates (100 r/min) in the same direction as the Barreling tank, while the gear shaft rotates in the opposite direction, with a treatment time of 15–30 min. The effect of the Barreling method on improving the tooth surface fatigue strength is shown in Fig. 8.15. The main

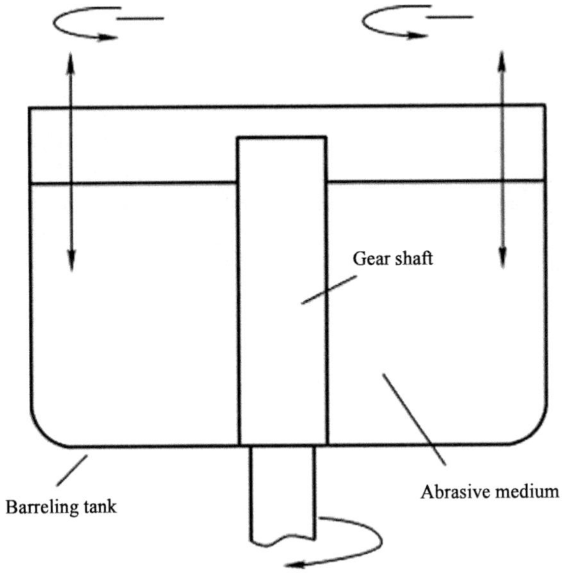

Fig. 8.14 Sketch of Barreling method

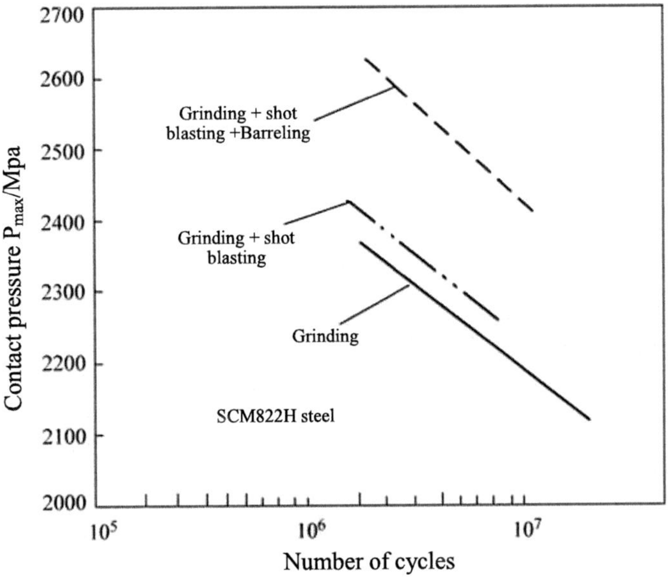

Fig. 8.15 Effect of Barreling on improving the tooth surface fatigue strength

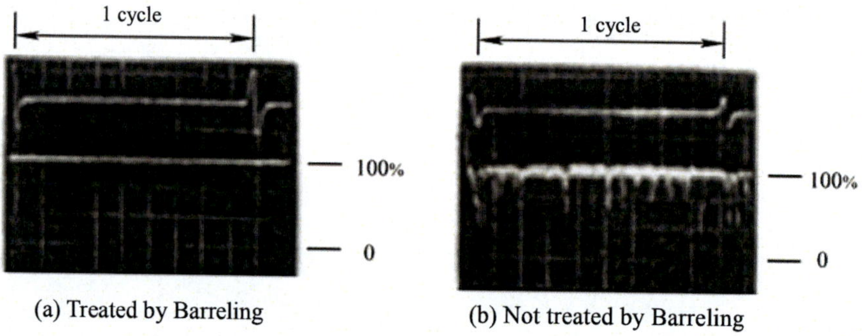

Fig. 8.16 Formation of oil film between gear pairs after 10^7 load rotations

principle is that the Barreling makes the concave and convex cutting ripples on the tooth surface gentle and uniform, thus reducing the surface roughness value of the tooth surface and the meshing friction factor of the tooth surface, and improving the grain boundary lubrication condition in gear engagement. As shown in Fig. 8.16, compared with the gear pair without Barreling, the formation of oil film of the gear pair treated by Barreling is greatly improved.

(1) Application practice of gear manganese phosphate conversion coating: in the research of gear antifriction coating, the friction factor can be effectively reduced by preparing the functional antifriction and antiwear coating on the tooth surface. The manganese phosphate conversion coating on the alloy steel gear surface after phosphating can effectively reduce the friction factor of the friction pair surface and, with good corrosion resistance, can prevent occlusion or scratches. After the preparation of the manganese phosphate conversion coating, a soft layer of several microns thick is formed on the gear surface, which flattens out most of the concave and convex cutting ripples on the gear surface, reduces the local maximum meshing contact stress and metal surface friction factor, and improves the oil film and lubrication condition in gear engagement. A superfine manganese phosphate conversion coating can be obtained by controlling the process parameters of the manganese phosphate conversion coating to affect the coating density and grain size. After treatment, a soft layer of 3–5 μm thick is formed on the tooth surface and the density of the formed coating surface is about 2.2 g/m^2. The ordinary manganese phosphate conversion coating treatment and superfine manganese phosphate conversion coating treatment shall be selected combined with the gear processing technology and actual working conditions. The comparison of the surface topography after treatment by both methods is shown in Fig. 8.17.

The manufacturing processes of the manganese phosphate conversion coating mainly include pretreatment of the gear surface with degreasing agent in a 70–95 °C degreasing bath, washing with water and then surface conditioning at the

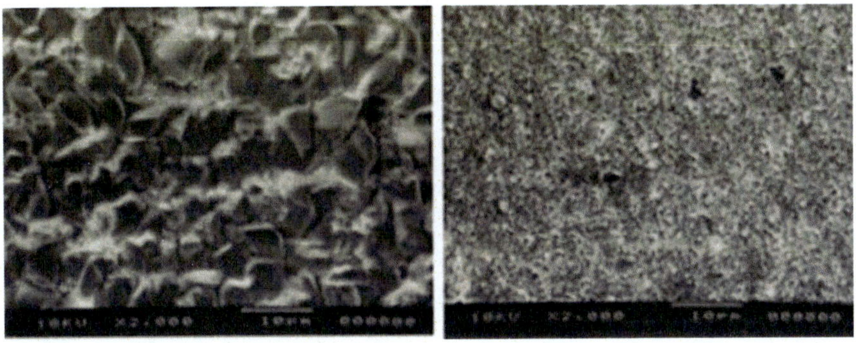

Fig. 8.17 Surface topography after ordinary manganese phosphate conversion coating treatment and superfine manganese phosphate conversion coating treatment

treatment temperature 40–80 °C, with the phosphating temperature range of 80–100 °C, acid ratio of 5.6–6.2 and treatment time of 10–15 min. The manganese phosphate conversion coating is formed by the following chemical reaction:

$$Mn(H_2PO_4)_2 \rightleftarrows MnHPO_4 + H_3PO_4$$

$$Fe(H_2PO_4)_2 \rightleftarrows FeHPO_4 + H_3PO_4$$

$$3(Mn, Fe)\,(H_2PO_4)_2 \rightleftarrows (Mn, Fe)_3(PO_4)_2 + 4\,H_3PO_4$$

Figure 8.18 shows the formation of oil film by cylindrical roller. The formation of the oil film by cylindrical roller is observed and tested with the separation voltage resistance measurement method, with complete contact at 0 V and complete separation at 0.1 V. The separation voltage of the cylindrical roller after the manganese phosphate conversion coating treatment begins to rise after 30 min, and its oil film formation ability is significantly better than that of the untreated cylindrical roller pair.

The antifriction and antiwear characteristics of manganese phosphate conversion coating under certain working conditions are evaluated by friction and wear test. As shown in Fig. 8.19, the friction and wear test of the manganese phosphate conversion coating specimen is conducted according to the gear working load and lubrication conditions. It can be seen that the superfine manganese phosphate conversion coating has good continuous antifriction and antiwear characteristics under low and high loads.

Figure 8.20 shows the fatigue pitting of two groups of AT gear pairs of the automatic transmission provided to Volkswagen by JATCO automatic transmission company in accordance with the requirements of endurance test after the test under the same test conditions. The two gear pairs have undergone the same processing

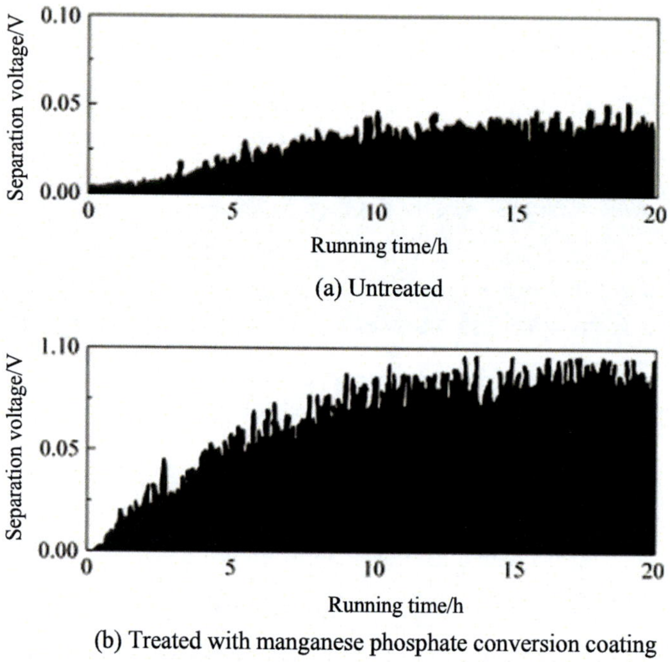

Fig. 8.18 Formation of oil film by cylindrical roller

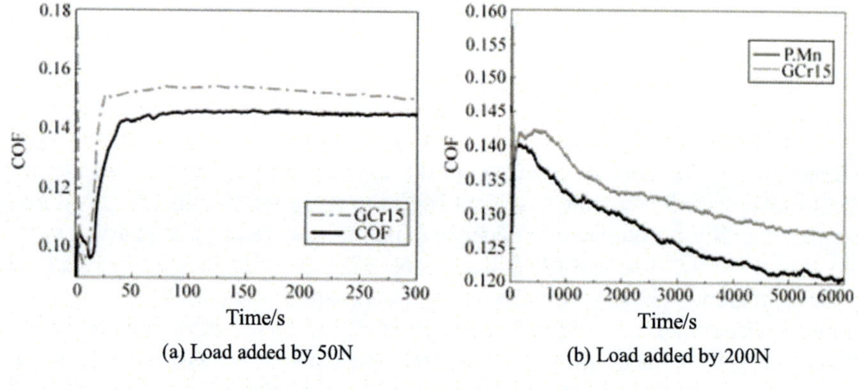

Fig. 8.19 Change curve of friction factor with load before and after treatment

technology. The difference is that the gear pair in group B is treated with manganese phosphate conversion coating, while the gear pair in group A is not.

As shown in Fig. 8.21, the test is conducted under the maximum pressure at the tooth surface contact point of 2000 MPa and the maximum sliding speed at the tooth root of 7.8 m/s. The abscissa axis represents the number of meshing cycles, and the ordinate axis represents the pitting area rate. The pitting failure occurs in the gear

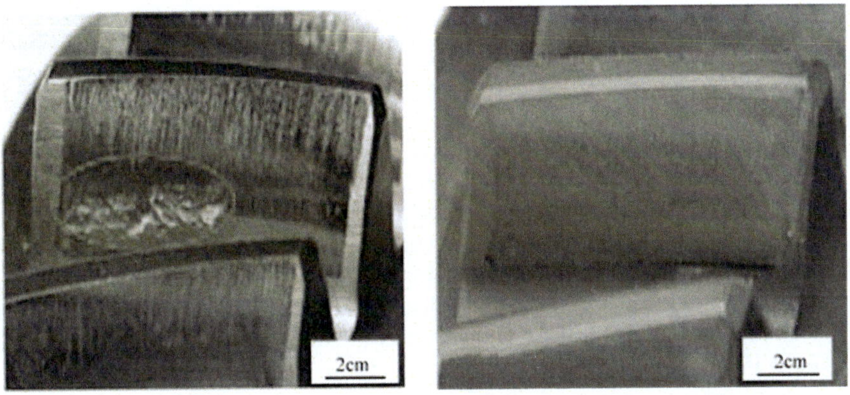

(a) Untreated

(b) Treated with manganese phosphate conversion coating

Fig. 8.20 Gear fatigue pitting before and after treatment

Fig. 8.21 Change in gear fatigue life and pitting area rate

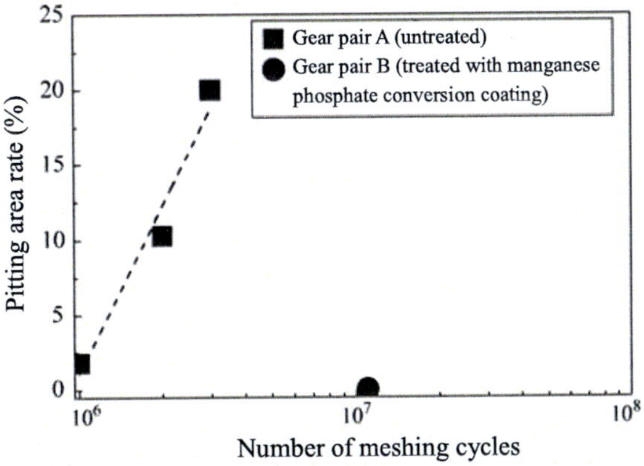

pair of group A at 3×10^6 cycles, while the pitting failure does not occur in the gear pair of group B at 1.1×10^7 cycles. The test shows that the manganese phosphate conversion coating increases the pitting fatigue strength of the transmission gear by 3–4 times.

(2) Superfine composite material metal spraying: in recent years, some scholars have studied and applied the spraying technology using manganese sulfide (MoS_2) and superfine composite material, as a new engineering method in the field of pressure spraying, which has achieved a good practical effect in the field of high strength gear surface strengthening. Its technological process is shown in Fig. 8.22.

Under high temperature and high pressure, the composite material impinges and penetrates into the base material at a depth of 1–20 μm, which changes the phase

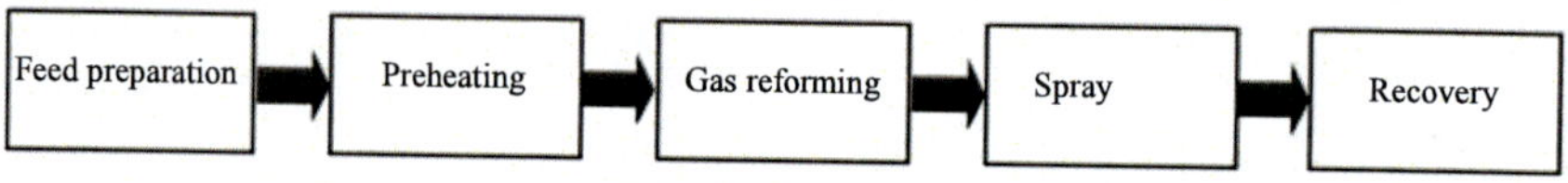

Fig. 8.22 Process flow chart of gear composite spraying

structure at a depth of 20 μm below the metal surface. Solid lubricant is added in the processing process, and after the solid lubricant penetrates into the metal surface, the surface becomes self-lubricating. In this technology, the high-pressure inert gas drives the grinding ball and solid lubricant and strikes the workpiece to be machined with the mixture injection. Due to the impact of spherical particles on the surface of the workpiece, a number of tiny holes are formed, so that the surface structure is compressed and refined, resulting in the reduction of external stress. Moreover, the solid lubricant is attached to the surface of the workpiece to improve the self-lubrication of the workpiece. Figure 8.23 shows that the inner edge of the needle bearing of a transmission gear has been sprayed with composite material to improve the smoothness of the surface indentation, and numerous fine pits have been formed on the surface, which is conducive to the formation of the oil film, thus improving the oil film adhesion on the friction surface and improving the fatigue strength.

IV. Gear design based on service life

In the past, the calculation of gear life in transmission design is based on the traditional allowable stress method. That is, according to a certain calculation formula, the working stress of gear fatigue bending and fatigue contact is calculated, and then compared with the corresponding allowable stress to predict whether the designed gear life is enough. The main problem with this method is that the calculated load cannot reflect the actual cyclic load on the gear under the operating conditions. Thus, the resulting calculation of gear life is only rough and conditional. With the

(a) Before treatment (b) After treatment

Fig. 8.23 Surface morphology before and after spray treatment with superfine composite material development of the research on the load spectrum of automobile transmission, it is possible to predict the gear service life accurately.

The load spectrum is used to describe the load-time history signal. For example, Fig. 8.24 shows the typical load spectrum of a certain period of time and describes the engine torque, engine speed and gear. In order to shorten the development time and reduce the test cost, the rain flow extrapolation method and time domain extrapolation method of the load spectrum are usually used to obtain the data of the load spectrum during the whole service life. The data can also be collected by real vehicle or bench test. For a part of the transmission, such as a gear, the drive torque and speed in the whole service life of the gear are divided into several sections according to the gear ratio, forming a load spectrum time distribution table of a gear as listed in Table 8.3.

The allowable stress of the gear is related to the bending strength and surface bearing capacity of the gear. The fatigue strength of gears of different materials and shapes can be obtained by testing. After fitting, the number of gear revolutions (life) N and input torque (stress) T have the following relation within the limited life span

$$T^k N = \mathrm{C} \tag{8.7}$$

where, K and C—test fitting results, both are constants, related to the stress properties and materials.

In Table 8.3, the gear speed is divided into 7 data segments, and the torque is divided into 10 data segments. A total of 70 cumulative running times are calculated.

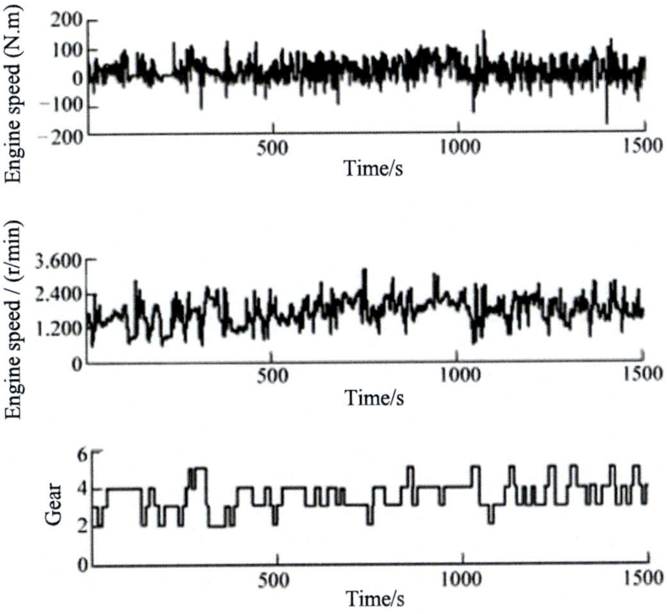

Fig. 8.24 Typical load spectrum

Table 8.3 Load spectrum time distribution table of a gear

Torque T (Nm)	Gear speed n/(r/min)						
	800	1200	1600	2000	2400	2800	3200
15	11317	54971	192	164	28	13	0
30	6519	112	147	66	13	8	0
45	80	401	1014	35	47	34	0
60	723	1489	2669	104	42	21	0
75	1637	4614	3816	2145	22	17	0
90	3139	2330	1157	678	27	14	0
105	1595	1910	2645	1360	31	12	0
120	2059	4277	3613	2085	1126	13	0
135	1030	3400	3222	2085	1144	22	0
150	104	876	212	1889	3432	213	0

Using these data, the cumulative damage C^1 of the bending fatigue strength and surface bearing capacity of the gear can be calculated respectively, and the calculation formula is

$$C^1 = \sum_{j=1}^{70} T_j^k \left(\frac{\text{nt}}{60}\right)_j \tag{8.8}$$

With respect to a specific gear, the allowable torque under the allowable stress σ_D is T_D and the allowable number of cycles is N_D. Then the gear life N_{th} is

$$N_{\text{th}} = \frac{T_D^k \sum_{j=1}^{70} \left(\frac{nt}{60}\right)_j}{C^1} \times N_D \tag{8.9}$$

8.2 Shaft Design

I. Design criteria

The design requirements for the transmission shaft are to avoid notches, reduce bending moment and increase the critical speed. To meet these requirements, the following design criteria shall be followed:

(1) Reduce the distance between bearings by compact design.
(2) Arrange the gear bearing a large force near the bearing to reduce the deflection and bending moment of the shaft and improve the critical speed. The deformation of the transmission shaft will lead to the gear tipping and uneven

Table 8.4 Reference values of allowable deflection and allowable bending angle of transmission shaft

Shaft	Deflection	Bending angle
General rules for gear	$\leq 0.01 m_n$ (m_n is standard modulus)	$\leq \frac{2d_w}{10^4 b}$ (d_w is pitch diameter and b is tooth width)
Reference value	0.02–0.06 mm	≤0.005 (cylinder); ≤0.001 (cone)

bearing stress. Table 8.4 lists the reference values of allowable deflection and allowable bending angle of the transmission shaft.

(3) There should be a transition between the stepped shafts, and it is better to use tapered design or a large radius of curvature for the transition rather than the shaft shoulder.
(4) The shaft and gear are connected by splines or interference fit rather than flat keys.
(5) The snap ring is used only at the shaft end and the spacer ring is used in the middle of the shaft for axial constraint.
(6) Make the stress near the giant ring groove smooth by using a unload groove or correcting the inner edge of the ring groove.
(7) The method to reduce the notch effect at the shaft shoulder is to set up an unload groove in the transition section of the shaft shoulder, adopt a large knuckle, adopt a radial stress unload groove, and add an additional unload groove in the transition area.
(8) For the journal that fits with the hub, the diameter of the shaft fitting with the gear shall be increased and the hub thickness decreases gradually in the direction from large to small shaft diameter.
(9) Unload the cross dill in the shaft with radial hole by increasing the hole diameter or using the transition radius.
(10) Reduce the rotational inertia of the parts on the shaft to reduce the deflection of the shaft and increase the critical speed.
(11) The stress trough is used to realize the gradual transition of stress.

II. Shaft design method

The design calculation of shaft mainly includes strength calculation, stiffness calculation and structural analysis of shaft. In the design, the axial dimension of the transmission should be tried to be as small as possible, which involves the calculation of the journal size, shaft strength, shaft stiffness and spline on the shaft.

1. **Shaft strength calculation**

Simplify the shaft stress, analyze the stress conditions of the shaft in all gears under the maximum engine input torque and calculate the radial force, axial force and tangential force on the shaft gear and bearing in the corresponding gear; plot the shear force, bending moment and torque of the shaft in all gears; find the dangerous section and calculate the reference torque at the dangerous section, depending on the bending moment, torque and load correction factor; preliminarily determine the minimum diameter of the shaft based on the reference torque; finally calculate the dangerous section strength of the shaft. First calculate the bending stress and torsion stress on the dangerous section, then calculate the resultant stress and judge whether the shaft strength meets the requirements under a certain safety factor. For the specific steps and methods, refer to relevant technical data.

2. **Shaft stiffness calculation**

Insufficient shaft stiffness will cause a large deflection and bending angle of the shaft, resulting in the failure of the gear to engage properly and causing vibration and noise, thus reducing the service life of the gear. In the shaft stiffness calculation, the shaft deflection and bending angle are mainly calculated and shall meet the shaft design criteria and relevant technical requirements.

3. **Shaft structural analysis**

After completion of the shaft strength and stiffness calculation, the specific structural analysis of the shaft has been completed basically. Finally, the shaft structure shall be analyzed, and the shaft life, stiffness and fatigue process can be analyzed using FEM software.

Firstly, establish a finite element mesh model as shown in Fig. 8.25 for the shaft and the elements in the shaft; then set the accurate material information of each component on the shaft, including modulus of elasticity, Poisson's ratio, density, yield strength and tensile strength, as well as the radial stiffness and tilt stiffness of the bearing shown in Fig. 8.26; then set the boundary conditions and define the axial, radial clearance and sliding friction factor of the bearing and synchronizer. Generally, the axial clearance is 0.1 mm, the radial clearance is 30 μm and the sliding friction factor is 0.1. Then set the load conditions, load the shaft load on the spline and the corresponding gear of the analysis gear conditions, input the torque in the spline and input the engagement pressure on the gear tooth surface. The load position is as follows: connect the node in the bearing area of the tooth surface bearing the load by

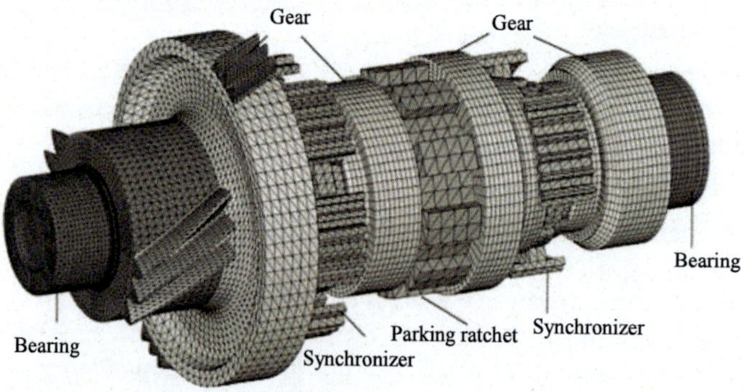

Fig. 8.25 Finite element mesh model of shaft

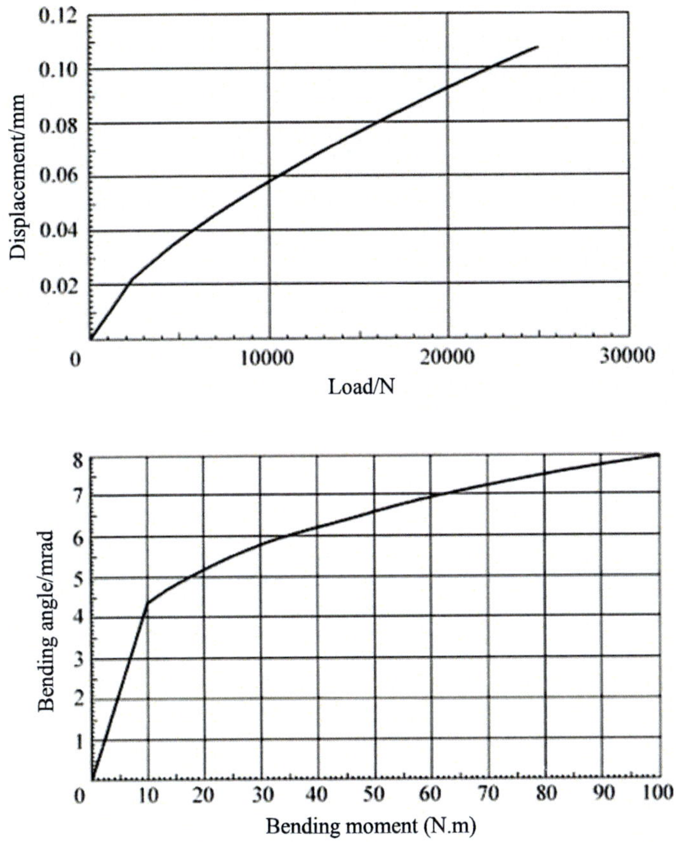

Fig. 8.26 Radial stiffness and tilt stiffness of bearing

means of distributed coupling, and input the gear engagement pressure at this node. The tooth surface bearing position is about 70% in the direction of tooth height and 80% in the direction of tooth width; create a rigid connection to the nodes of the spline tooth surface on the bearing side by means of distributed coupling and input the torque in the center node. Finally make the static strength analysis and stiffness analysis. Figure 8.27 shows the static strength analysis results of the shaft, with the maximum mean stress of 527 MPa; Fig. 8.28 shows the static stiffness analysis results of the shaft, with the maximum deflection of 130 μm.

For the shaft fatigue process analysis, it is necessary to input the tensile strength, yield strength, S–N curve and linear static parameters of the material; set the surface roughness, technical size, dispersion and temperature; set the influence factors; set the analysis type and survival rate; check σ_m = CONST or R = CONST (σ_m = CONST and R = CONST are two different calculation methods). The evaluation criterion is that the minimum fatigue safety factor of the shaft non-contact area is greater than the design requirement and the design index is usually as follows: when

Fig. 8.27 Static strength analysis results of shaft

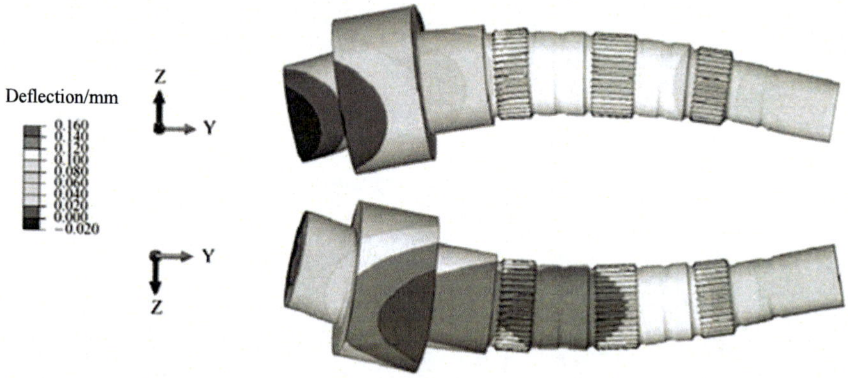

Fig. 8.28 Static stiffness analysis results of shaft

σ_m = CONST, the fatigue safety factor is greater than 1.5; when R = CONST, the fatigue safety factor is greater than 1.25. Figure 8.29 shows the distribution cloud diagram of the safety factor (SF_{min}) of the shaft under the maximum working load.

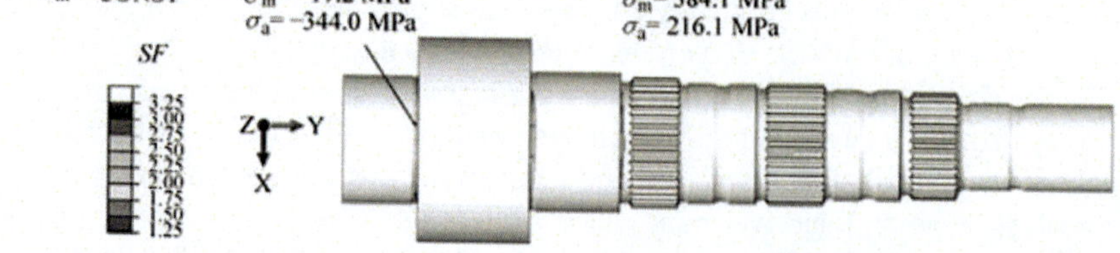

Fig. 8.29 Distribution cloud diagram of the safety factor of the shaft under the maximum working load

8.3 Bearing Selection and Design

The function of the transmission bearings is to support and guide the relative moving parts, absorb the acting force and pass the force to the case. According to the type of relative motion, the bearings may be classified into sliding bearing and rolling bearing. The rolling bearing is mostly used in the automobile transmission. Typical requirements for the rolling bearing for transmission are:

(1) Even if the transmission shaft is inclined due to force, the bearing capacity shall also be guaranteed.
(2) The bearing is able to withstand high operating temperature and low lubricant viscosity.
(3) High radial stiffness and axial stiffness.
(4) Strong anti-pollution capacity.

I. Transmission bearing type

The automobile transmission bearings mainly include support bearings, thrust bearings, needle bearings, linear bearings and release bearings, in which, the support bearings include deep groove ball bearing, cylindrical roller bearing, tapered roller bearing and angular contact ball bearing. Table 8.5 lists the types and properties of various transmission bearings.

Table 8.5 Types and properties of various transmission bearings

Characteristics	Deep groove ball bearing	Angular contact ball bearing	Cylindrical roller bearing	Tapered roller bearing	Needle bearing	Thrust ball bearing
Radial carrying capacity	Satisfactory	Satisfactory	Satisfactory	Satisfactory	Satisfactory	Unsatisfactory
Axial carrying capacity	Bidirectional	Unidirectional	Unsatisfactory	Unidirectional	Unsatisfactory	Unidirectional
High-speed rotation performance	4★	4★	4★	3★	3★	1★
High-speed rotation accuracy	3★	3★	3★	3★		1★
Good NVH performance	4★	3★	1★		1★	1★
Low friction moment	4★	3★	1★			
Good stiffness			2★	2★	2★	
Vibration impact resistance			2★	2★	2★	
Ultimate inclination of inner and outer rings	1★		1★	1★		
Axial positioning	Bidirectional	Unidirectional		Unidirectional	Unidirectional	
Axial movement	On mating surface		On raceway		On raceway	On mating surface

(1) Support bearing: the transmission support bearings are usually installed on both ends of the shaft to jointly bear the axial and radial forces of the shaft assembly. The front and rear bearings of different specifications and types may form different composite structures, generally including the combination of deep groove ball bearing and cylindrical roller bearing, combination of tapered roller bearing and tapered roller bearing and combination of angular contact ball bearing and angular contact ball bearing. In the case of identical or equivalent mounting installations (inner and outer diameters and width of different bearings), the combination of the tapered roller bearing and tapered roller bearing has the highest capacity to withstand the load and longest life and the ball bearing can withstand higher limit speed than other bearings; in terms of the transmission efficiency of the whole transmission, the combination of the deep groove ball bearing and cylindrical roller bearing has the highest efficiency, while the combination of the tapered roller bearing and tapered roller bearing has the lowest efficiency; in terms of the assembly, the pre-tightening is required for the use of the tapered roller bearing and angular contact ball bearing in pairs to adjust the internal play, so the adjustable pads shall be added generally during assembly, thus increasing the assembly difficulty; in terms of stiffness, the combinations other than the combination of the deep groove ball bearing and cylindrical roller bearing have forward and reverse installation methods, which require different support stiffness of the supported shaft. In the forward installation, the span between the front and rear supporting points is smaller and the bearings are more rigid, so the bearings are generally subject to forward installation if used in pairs in the transmission. In addition, the accuracy class 6 is enough for the support bearing of the light-load car transmission. If the accuracy class is too high, the bearing price will increase and the cost will increase. If the accuracy class is too low, the bearing performance will decrease, resulting in the decrease in the overall performance of the transmission.
(2) Thrust bearing: the thrust bearings are generally installed at both ends of the part that only bears axial force and rotates relative to its connected part, and are widely used in planetary gear train of the AT. The thrust bearing structure may be selected according to GB/T273.2-2006 *Rolling Bearings—Thrust Bearings—Boundary Dimensions, General Plan.* Generally, the load borne by the transmission thrust bearing is a dynamic load and the bearing life may be calculated according to GB/T6391-2010 *Rolling Bearings—Dynamic Load Ratings and Rating Life.*
(3) Needle bearing: the needle bearing used to connect the transmission gear to the shaft generally bears the static load only. In special cases, it is possible to bear the dual load with different dynamic and static loads (not simultaneously, but under different working conditions). For example, that is the case with the needle bearing of the gear 1 driven and reverse drive dual gear in a 6MT transmission. The needle bearing is under static load in the gear 1 of the transmission and is under dynamic load in the reverse gear of the transmission. Therefore, in the calculation of the bearing life, whether the equivalent static and dynamic loads meet the working condition requirements shall be calculated. The needle

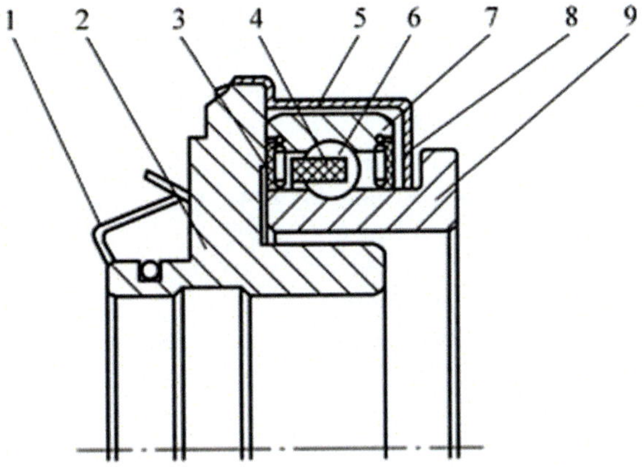

Fig. 8.30 Clutch release bearing. 1—Clamp spring, 2—spring retainer, 3—seal ring, 4—retainer, 5—outer cover, 6—Steel ball, 7—outer ring, 8—spring washer, 9—inner ring

bearing life may be calculated according to GB/T6391-2010 *Rolling Bearings—Dynamic Load Ratings and Rating Life* and GB/T4662-2012 *Rolling Bearings—Static Load Ratings.*

(4) Linear bearing: the linear bearing is generally used in the shift fork of the transmission shift actuator, its boundary dimensions and tolerances may be designed according to GB/T16940-2012 *Rolling Bearings—Sleeve Type Linear Ball Bearings—Boundary Dimensions and Tolerances* and its technical conditions can be referred to JB/T5388.2010 *Rolling Bearings—Sleeve Type Linear Ball Bearings—Specifications.*

(5) Release bearing: the clutch release bearing, with the general structure shown in Fig. 8.30, consists of the ball bearing, bearing block and connectors. The clutch release bearing selected shall meet the overall layout size requirements and dynamic load stress requirements. According to the specific needs of the vehicle type, the bearing connecting parts (mainly the size of the outer ring 7 in Fig. 8.30) matching with the clutch release fork are designed and the overall structure of the release bearing is determined.

II. Transmission bearing selection

(1) Confirm the application conditions and environmental conditions, including the transmission functions and structure, bearing application site, bearing load size and direction, speed, vibration and shock, bearing temperature (environment temperature and temperature rise) and ambient environment (corrosiveness, cleanliness and lubricity).

(2) Select and determine the bearing type and structure, related to the allowable space of bearing, bearing load (size, direction, vibration and shock), speed, rotation accuracy, stiffness, inclination of inner and outer rings, friction moment, bearing arrangement (free end and fixed end), installation and disassembly, economy and marketability.

(3) Select and determine the bearing dimension, related to the design life, equivalent static load and equivalent dynamic load, safety factor, limit speed, maximum axial load and allowable space of the transmission.
(4) Select the bearing accuracy, related to the changes in the run-out accuracy, speed and friction moment of the driver shaft.
(5) Select and determine the bearing fit tolerance and internal play, related to the material, shape and fit of the shaft and bearing, temperature difference of inner and outer rings, inclination of inner and outer rings, load (size and nature), pretension and speed.
(6) Select the type and material of the retainer, related to the speed, noise, vibration and shock, torque load and lubrication method.
(7) Select the lubrication method and sealing method, related to the application temperature, speed, lubrication method, sealing method, maintenance and inspection.
(8) Determine the special requirements for the bearings, including the application environment (high temperature, low temperature, vacuum and drugs) and high reliability.
(9) Determine the bearing assembly and disassembly methods, including mounting dimensions, installation and disassembly sequence.

III. Transmission bearing design

The design flow of the transmission bearing is shown in Fig. 8.31.

8.4 Case Design

As the basic part of the transmission, the case assembles gears, shafts, bearings, forks and other parts of the transmission into an integral whole and must ensure accurate relative position of the parts and absorb the applied force and applied moment during working; with good heat transfer performance, stiffness and strength and light weight, it can isolate and attenuate the transmission noise and is easy to assemble and disassemble.

In terms of the case machining accuracy, the size, geometry, mutual position and surface roughness of the key positions must be strictly required in the case design, so as to guarantee the normal transmission operation and meet the overall transmission design requirements. The tolerance and surface roughness of the key positions of the case are generally selected according to Tables 8.6 and 8.7.

In terms of the case strength and stiffness, the case is under complex load during the transmission working, so different positions will bear the bending and distortion. To ensure the gear engagement accuracy, the case must have enough stiffness to avoid excessive deformation; to ensure reliable bearing working, the bearing block must have enough stiffness; to ensure reliable transmission seal, the sealing plane must have enough stiffness, otherwise in the bolt pre-tightening, the plane seal will

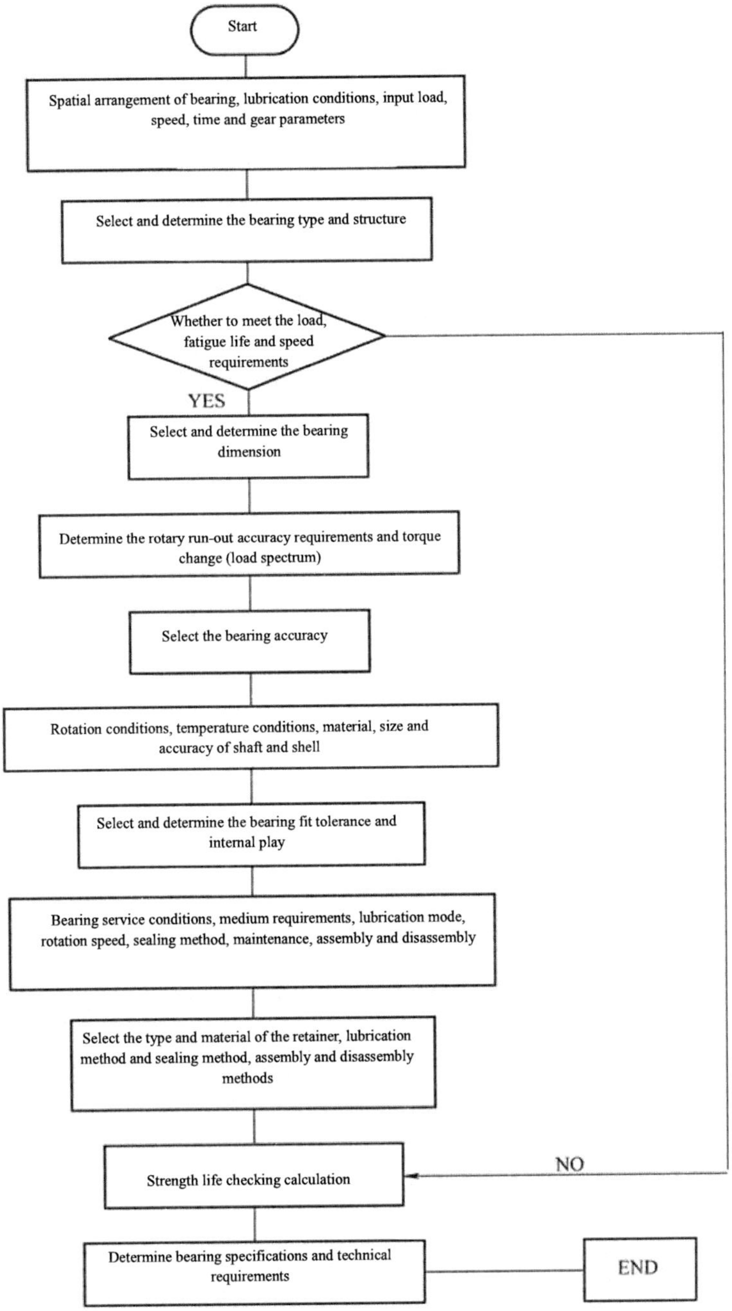

Fig. 8.31 Design flow of transmission bearing

Table 8.6 Dimensional tolerance and geometric tolerance requirements of key hole diameters and mounting surfaces

Item	Dimensional tolerance	Form tolerance	Position tolerance
Pin hole	IT6–IT8	IT6–IT7	IT6–IT7
Bearing hole	IT6–IT8	IT6–IT7	IT6–IT7
Shift fork shaft mounting hole	IT7–IT8	IT7–IT8	IT6–IT8
Parking brake device mounting hole	IT8–IT10	IT8–IT10	IT8–IT10
General hole	IT10–IT12	IT10–IT12	IT10–IT12

Table 8.7 Machined surface roughness requirements for some positions of case (unit: μm)

Item	Key hole diameter	Tapping hole and via hole	Large assembly plane	Other
Surface roughness requirement	*Ra*1.6	*Ra*6.3	*Ra*1.6	*Ra*12.5

be deformed and leaky, thus affecting the normal operation of the transmission. The case structure can generally meet the strength requirements, but if the structure is not designed reasonably, it cannot meet the stiffness requirements. The stiffness of case is far more important than the strength. Therefore, it is the first guiding principle of the case design to increase the stiffness of case as much as possible under the condition that the weight is as light as possible.

In terms of the case material selection and lightweight, the case has a complex structure and a large size, and its weight is roughly 30% of that of the transmission assembly. Therefore, the case lightweight is the key to the transmission assembly lightweight. The transmission cases of the cars subject to batch production are made of die casting aluminum alloy or magnesium alloy. The reduction of the case weight is limited by the material strength and the minimum wall thickness allowed by the casting process.

Case is one of the parts with the highest requirements in the transmission design. The following is an analysis of transmission case structure design, case wall thickness design, case stiffener arrangement, case lightweight design and other aspects.

I. Structural design of transmission case

The transmission case has integral and split structures. The integral case is characterized by easily guaranteed coaxiality of the front and rear bearing holes of the transmission and convenient assembly and inspection. It is generally a casting part with low productivity. The integral structure is mostly used for heavy goods vehicles. The split case (Fig. 8.32) is further divided into front-rear split and left-right split, which are characterized by the difficulty in ensuring the coaxiality of the front and rear bearing holes, high requirements for machining accuracy. It is mostly aluminum alloy die casting with high productivity and mostly used for cars and light vehicles.

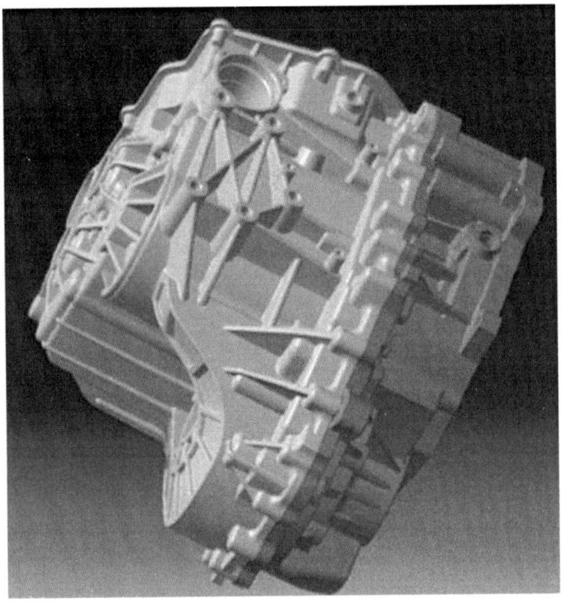

Fig. 8.32 Split case

The basic dimensions of the transmission case depend on the need for the gear rotation motion and shift track, and shall enable the gears to move freely within the case without interference. The layout needs of the valve body, hydraulic pump and hydraulic torque converter shall also be considered for the transmission case with a hydraulic system. Moreover, the case shall be as small and light as possible and have enough stiffness to ensure the gear engagement accuracy and service life. The inherent frequency shall be within the range of 500–1000 Hz. Adequate clearance shall be left for the transmission case at the beginning of design according to the previous design experience, the analysis of competitive products as well as the influence factors such as the manufacturing tolerance, deformation and wear of all parts to prevent interference of the moving parts in the case with the case in any case. The basic dimensions of the case shall be determined according to the above factors.

Most of the car transmission cases are of split structure. Generally speaking, the front and rear cases are bolted together. The layout of bolt holes shall be determined according to the size of the outer contour of the case, with relatively uniform spacing of 50–80 mm. The bolt holes are better arranged symmetrically on the case wall, so that the joint surfaces can be evenly stressed to avoid deflection or stress concentration in some parts. In case of failure to arrange symmetrically, the offset distance shall be as small as possible, and the stiffeners shall be used to prevent bolt deflection and case deformation under pretightening force. To prevent the lubricating oil from leaking from the front and rear case joint surfaces, the joint surfaces shall be provided with a rubber tank. Generally, the inner edge of the case is rounded R1–R2 mm fillets to ensure that the effective width of the sealant at the joint surface is 2–3 mm.

The transmission case NVH design is an effective way to reduce the transmission noise. In the case design stage, a computer is used for analysis, and during the sample trial production, instruments are used to measure and analyze the vibration and inherent frequency of the trial-produced case, so as to find out the weak links and optimize the structure.

The case shape shall meet the stiffness requirements and the vibroacoustic radiation area should be small to avoid large plane transition. The surfaces shall be designed to be connected by facets in different normal directions to reduce the case wall vibration and radiation energy. Most transmission cases at home and abroad adopt a stepped design scheme, as shown in Fig. 8.33.

II. Case wall thickness design

The transmission case is of complex space structure, like a thick wall box structure by shape and also like a spatial plate-beam composite structure by the stress state. The case strength and stiffness mainly depend on the distribution of metal materials.

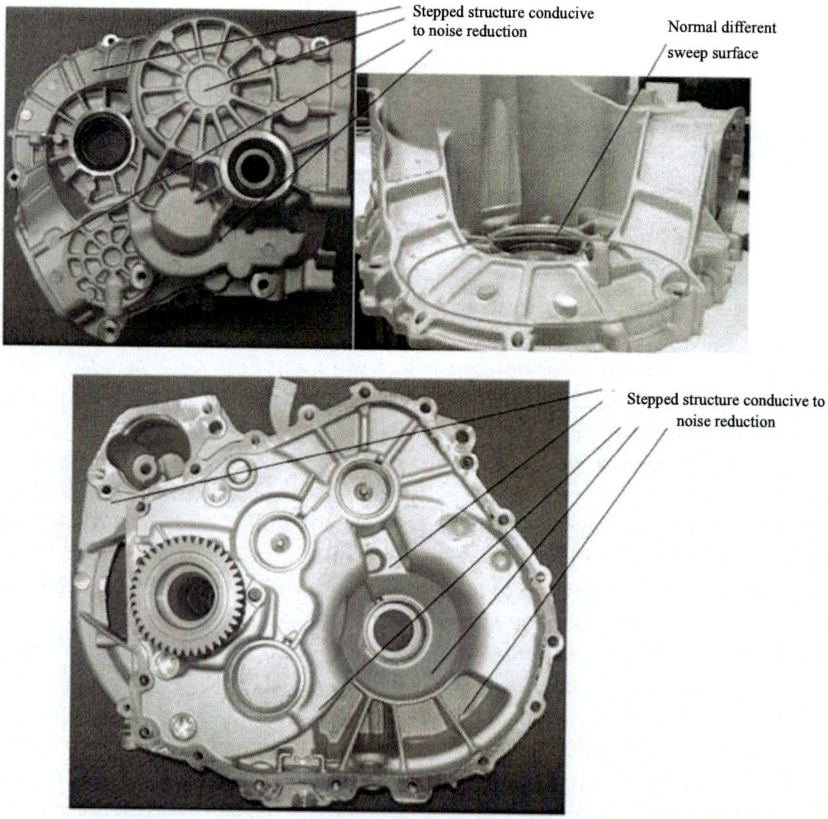

Fig. 8.33 Stepped structural case

Therefore, in the case design, a variety of stiffeners are usually arranged on the basis of the box case structure according to its stress condition and the requirements of stiffness and strength, and the wall thickness of local areas is increased as required, so that the case can meet the design requirements of the same strength under the condition of the lightest weight.

The case wall thickness is closely related to casting process, case stiffness and strength and increases with the increase of the casting size. However, the case wall thickness shall not be designed too large. If it exceeds the critical wall thickness, the grains in the central part will be coarse, and defects such as shrinkage and porosity will often occur, which will lead to the reduction of mechanical properties.

The appropriate wall thickness of the transmission case (which is convenient for die-casting and can give full play to the mechanical properties of materials) is generally 3.5–5.0 mm; for a medium-torque transmission (drive torque of 200–350 N m), the appropriate wall thickness should be 4.0 mm. To ensure the uniformity of the case wall thickness and prevent the local wall thickness from being too large or too small, Materialise Magics 9.0 software is commonly used to detect the case wall thickness, as shown in Fig. 8.34.

In order to strengthen the local stiffness of the case and prevent excessive deformation in the local area, it is necessary to increase the wall thickness in the local

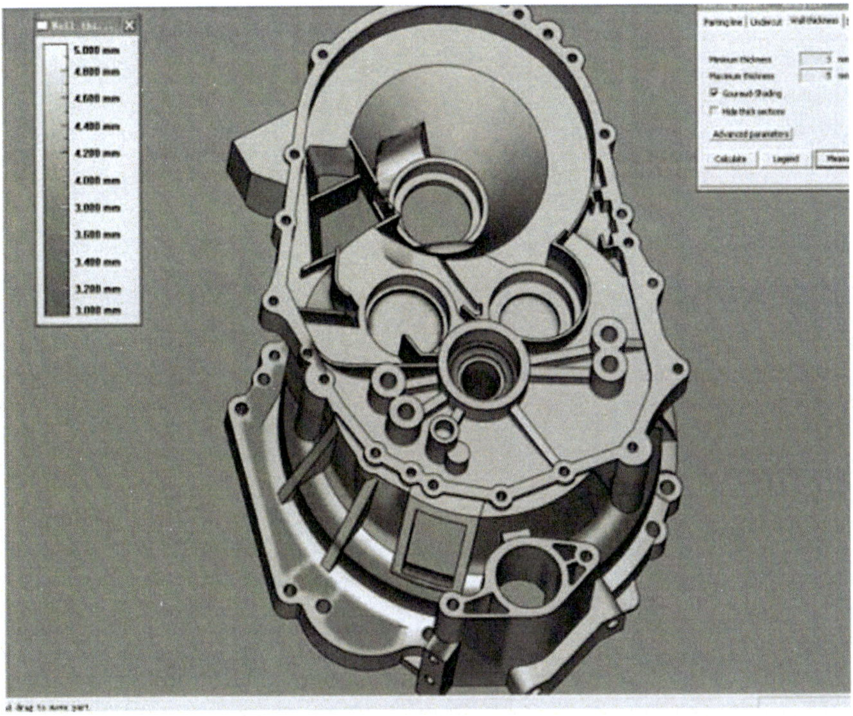

Fig. 8.34 Detection of common case wall thickness range

stress concentration area while adopting a thinner wall thickness on the basic wall. Due to the concentrated force and torque on the bearing block, the wall thickness of this part needs to be increased locally generally to 5.5–7.5 mm to prevent excessive local deformation. However, it should be noted that the actual wall thickness shall not exceed the critical wall thickness (generally 3 times of the minimum wall thickness), as shown in Table 8.8.

The front and rear case joint surfaces are subject to a large shear force. In order to prevent the case from oil leakage due to deformation caused by excessive stress, the thickness of the joint surfaces should be increased to strengthen the strength and stiffness of the case. According to previous design experience, the wall thickness at the joint surface should be 10–12 mm.

III. Case stiffener arrangement

The stiffness and strength of the case mainly depend on the number, height and distribution of stiffeners, while the case wall thickness has relatively little influence on the overall stiffness. As for the transmission, the main stress areas are input, output and reverse bearing block. Figure 8.35 shows the distribution diagram of

Table 8.8 Minimum allowable wall thickness for sand mold casting (unit: mm)

Alloy type	Casting profile size					
	<50	50–100	100–200	200–400	400–600	600–800
Aluminum alloy	3	3	4–5	5–6	6–8	8–10
Magnesium alloy	4	4	5	6	8	10
Zinc alloy	5	4	–	–	–	–

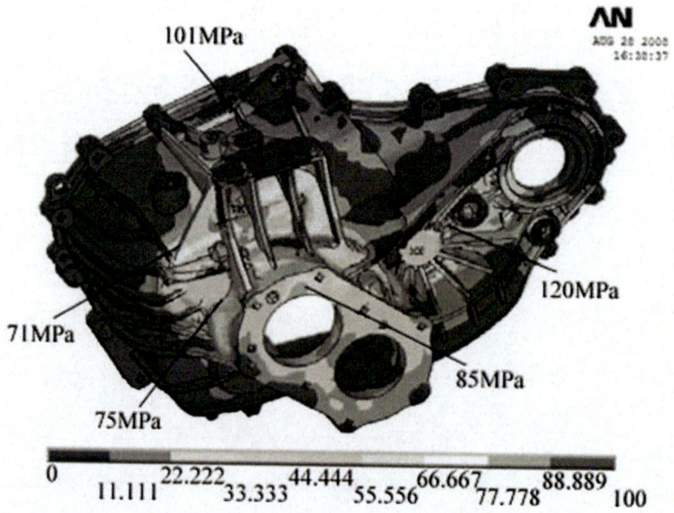

Fig. 8.35 Distribution diagram of stress concentration sites of case

stress concentration sites in the case. In order to strengthen the local stiffness of the case and prevent excessive deformation in the local area, the stiffeners shall be set in these areas to meet the stiffness and strength requirements of the case. Generally speaking, the width of the stiffener is roughly equal to the wall thickness, and its height shall not be less than the wall thickness, otherwise the increase of the bearing section will not be enough to offset the uneven stress distribution caused by the stiffener, and the expected effect will not be achieved; if the stiffener is too high, it will lead to local stress increase and casting difficulties, and the height is generally not more than the bolt boss diameter or the main bearing width.

The stress on the bearing block in the transmission case is most concentrated, so the stiffeners shall be set here to increase the case stiffness and strength. The stiffeners in the bearing block are generally arranged in a radial shape uniformly, so as to meet the requirements of increasing the strength and stiffness in the bearing block with as few stiffeners as possible, so as to meet the design requirements of lightweight. The number and width of stiffeners are related to the stress on the bearing block and the stress on the reverse shaft is more concentrated, so the width and number of stiffeners are greater than that in other parts. Figure 8.36 shows the distribution of transmission case stiffeners. The transmission case is transitioned with a large surface inside and outside. In order to increase its strength and stiffness, large stiffeners are arranged. Figure 8.37 shows the distribution of the outer wall stiffeners of transmission case.

IV. Case lightweight design

The case weight accounts for 30% of the overall transmission weight. Therefore, the case lightweight design plays a decisive role in the whole transmission lightweight. The case lightweight may be achieved in terms of material, wall thickness and structure.

The case material shall have sufficient strength and good casting properties, and shall meet the lightweight requirements. The aluminum alloy is commonly used for

Fig. 8.36 Distribution of transmission case stiffeners

Fig. 8.37 Distribution of the outer wall stiffeners of transmission case

the transmission case of the passenger vehicles. To meet the design requirements of case lightweight, the overall wall thickness should be as small as possible under the condition of meeting the requirements of strength and casting process design. In addition, it is also a common method to set a pocket locally in the case. Figure 8.38 shows a transmission case pocket.

V. Case design flow

The reducer case design is a series of processes and the overall design flow is shown in Fig. 8.39.

8.5 Parking Mechanism Design

Parking mechanism is a kind of safety device in AT to prevent the vehicle from sliding in parking state, which is used to make the vehicle stop in a certain position or even on the slope reliably without time limit. The parking mechanism generally uses a reliable mechanical driving mechanism rather than hydraulic or pneumatic to avoid faults.

The AT parking mechanism can be classified into manual, electro-hydraulic and electronic parking mechanisms. The early AT and CVT adopted the manual parking mechanism, and the driver manipulated the shift knob to drive the control cam, parking pawl and parking ratchet through the cable to realize the transmission parking function, as shown in Fig. 8.40. With respect to the electro-hydraulic parking mechanism, there is no mechanical connection between the parking pawl of the transmission and the shift-by-wire shift knob. The engagement or disengagement of the parking pawl is fully determined by the electro-hydraulic device, that is, by the position of the

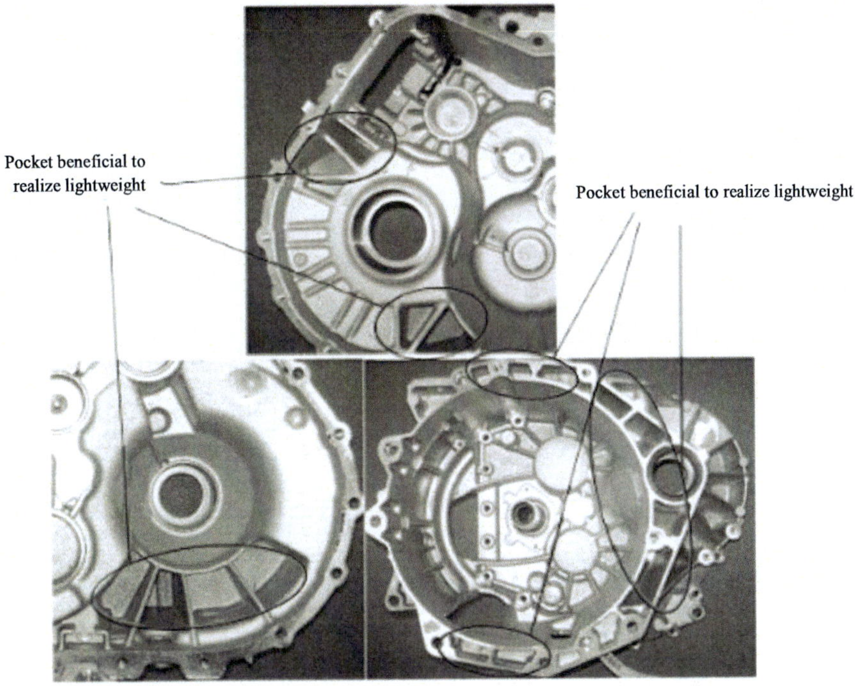

Fig. 8.38 Transmission case pocket

shift knob or other safety factors, such as opening the driver's side door, powertrain working, or pulling out the ignition key. Figure 8.41 shows the hydraulic control unit of Benz 9G-TRONIC AT parking mechanism. The electronic parking mechanism is shown in Fig. 8.42. The gear selection, shift and parking are implemented by the select-shift motor. In the above three drive modes, only the actuators acting on the control push rod are different, and the position relations of the control push rod, parking pawl and parking ratchet are the same.

I. Basic design principles of parking mechanism

(I) Basic input

(1) Curb weight and tire radius.
(2) Maximum parking slope (30%).
(3) Number of gears (depending on the gear switch requirements and layout position).
(4) Angle between gears (depending on the matching shifter).
(5) Cable stroke (depending on the execution mode).
(6) General layout space and layout position (position of the output shaft or differential mechanism).

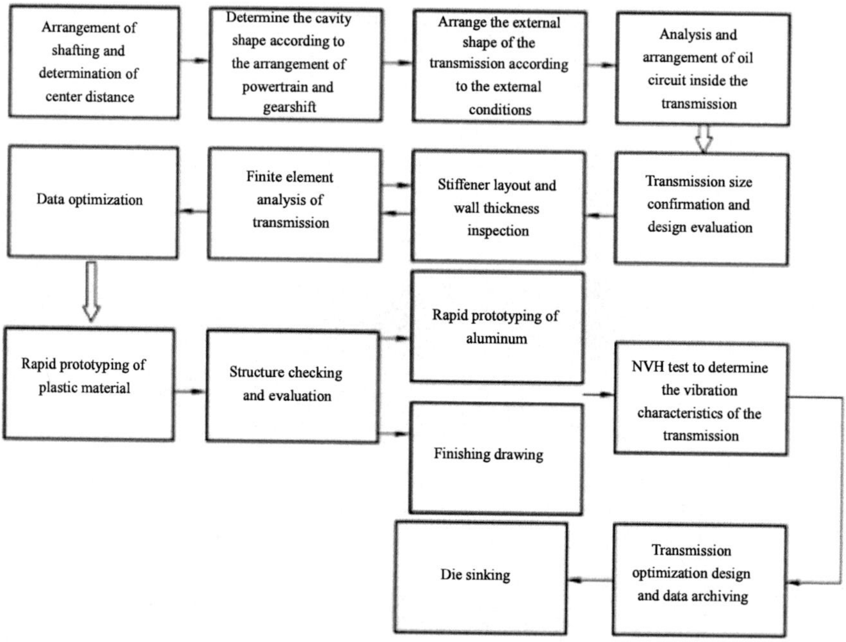

Fig. 8.39 Case design flow

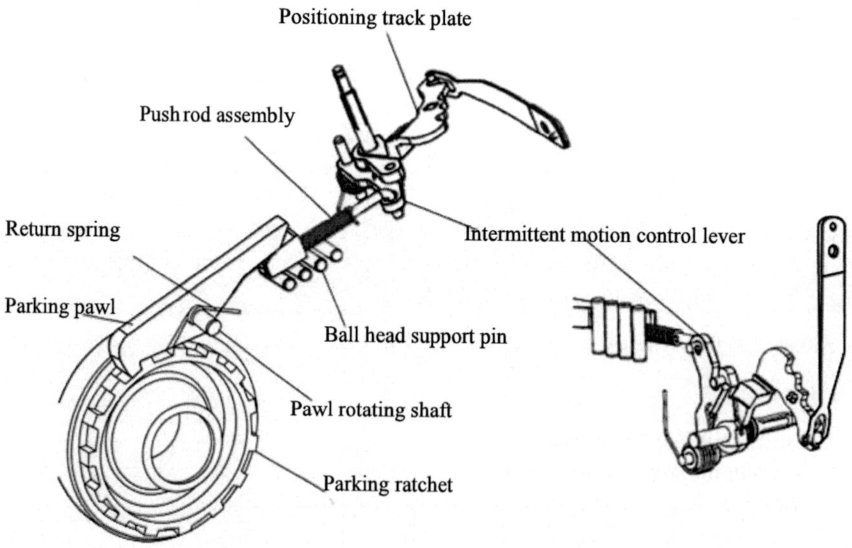

Fig. 8.40 AT parking mechanism

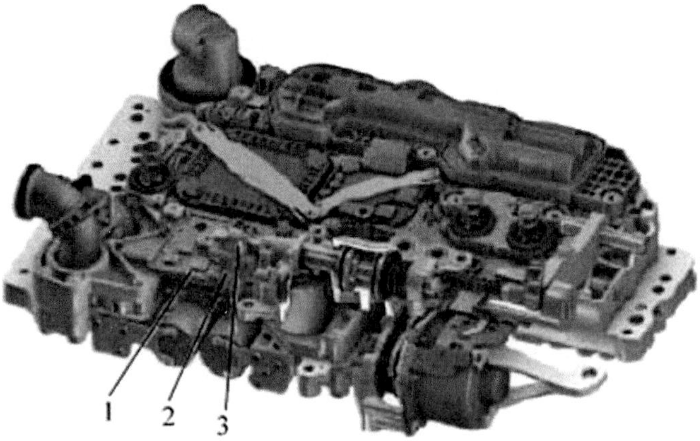

Fig. 8.41 Hydraulic control unit of Benz 9G-TRONIC AT parking mechanism. 1—Fully integrated transmission control unit, 2—parking pawl position sensor, 3—permanent magnet

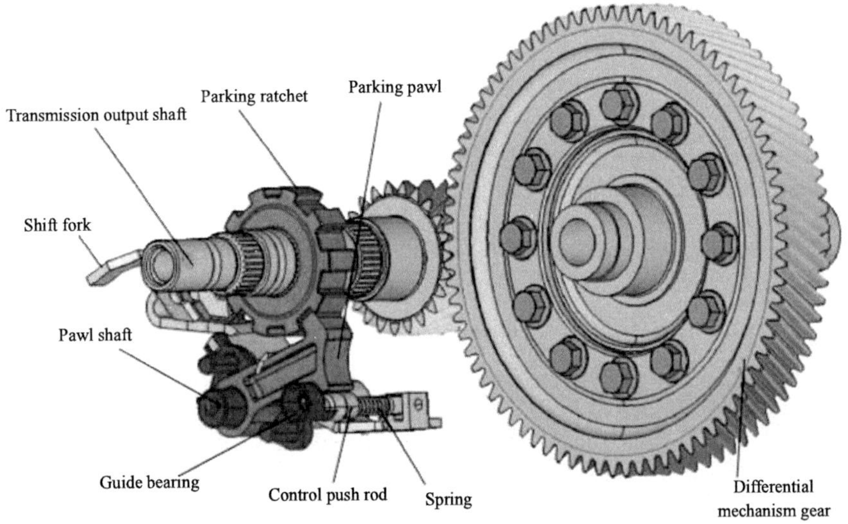

Fig. 8.42 Electronic parking mechanism

(II) Basic functional requirements

(1) When the vehicle is driving at a speed not higher than 5 km/h, the parking mechanism can realize safe parking.
(2) When the vehicle is in the non-parking condition, the parking mechanism cannot automatically park the vehicle regardless of any abnormality.
(3) When the vehicle is parking, the parking mechanism cannot be automatically out of gear.

Fig. 8.43 Load of vehicle on the ramp

(4) When the vehicle needs driving, the parking mechanism can make the vehicle smoothly out of parking gear.

(III) Basic design calculation

Figure 8.43 shows the load of the vehicle on the ramp. The maximum slope of the vehicle parking on the ramp is affected by the vehicle structure parameters and load distribution. When the maximum parking slope is determined, the ground braking force provided by the parking mechanism is

$$F_z = mg\sin\alpha \tag{8.10}$$

where, F_z—ground braking force;
m—vehicle mass.

Figure 8.44 shows the stress of the parking mechanism (Fig. 8.40) when the vehicle is on the ramp, and the main release torque M_1 is

$$M_1 = Fa \tag{8.11}$$

where, F—normal force of ratchet (N);
a—distance from pawl spindle to the line of action of normal force of ratchet (m).

Friction torque M_2 when parking mechanism is released

$$M_2 = F\mu c \tag{8.12}$$

where, μ—friction factor between pawl and ratchet;
c—distance from pawl spindle to the line of action of ratchet friction (m).

Radial force F_r acting on the control lever

$$F_r = \frac{M_1 - M_2}{b} \tag{8.13}$$

where, b—distance from pawl spindle to the point of action of control lever (m).

Axial holding force F_{hold} acting on the control lever

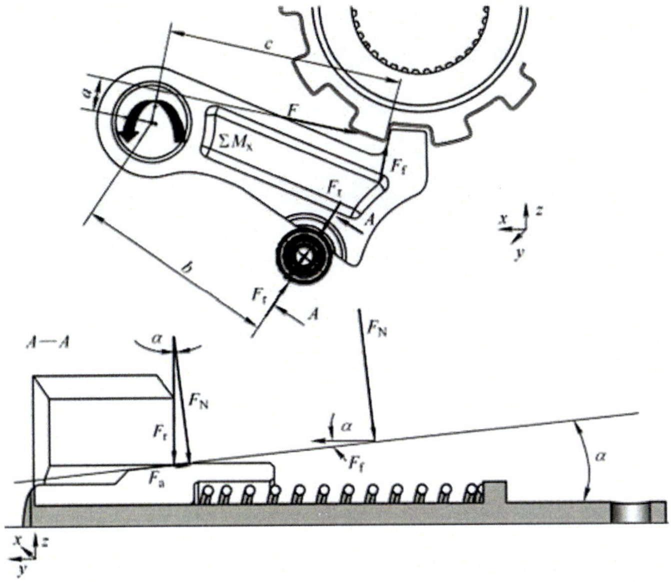

Fig. 8.44 Stress of the parking mechanism when the vehicle is on the ramp

$$F_{hold} = 2F_r \tan\alpha - 2F_r\mu \tag{8.14}$$

where, α—cone angle of control lever.

Figure 8.45 shows the dynamic stress analysis of pawl and Fig. 8.46 shows the axial force estimation of the control lever. Parameters involved in the figure: β is cone angle; b is the distance from pawl spindle to the point of action of control lever; M_a is the torque on the pawl due to the action of the actuator; I is rotational inertia; α is the angle of pawl; t is engagement time; ω_g is the angular velocity of ratchet; Δφ is the notch angle when the pawl is engaged with ratchet; v is vehicle speed; r is wheel radius; i is main gear ratio; Mr is the torque of pawl release spring.

For the pawl, the following is true

$$M_a + M_r = I\ddot{\alpha} \tag{8.15}$$

$$\alpha = \frac{1}{2}\ddot{\alpha}t^2 \tag{8.16}$$

For the ratchet, the following is true

$$t = \frac{\Delta\varphi}{\omega_g} \tag{8.17}$$

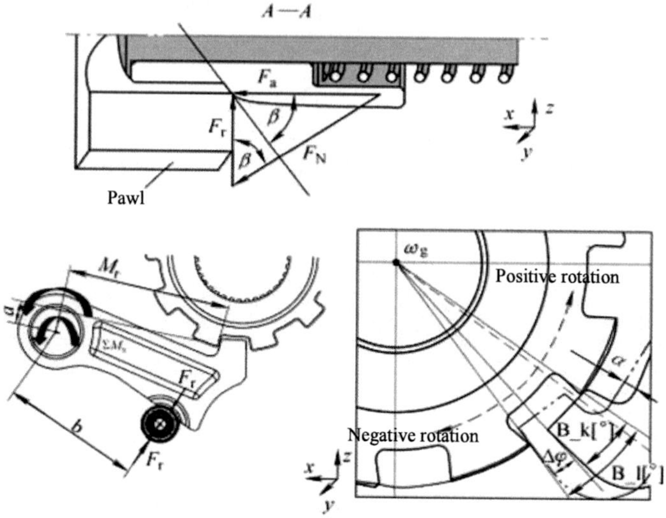

Fig. 8.45 Dynamic stress analysis of pawl

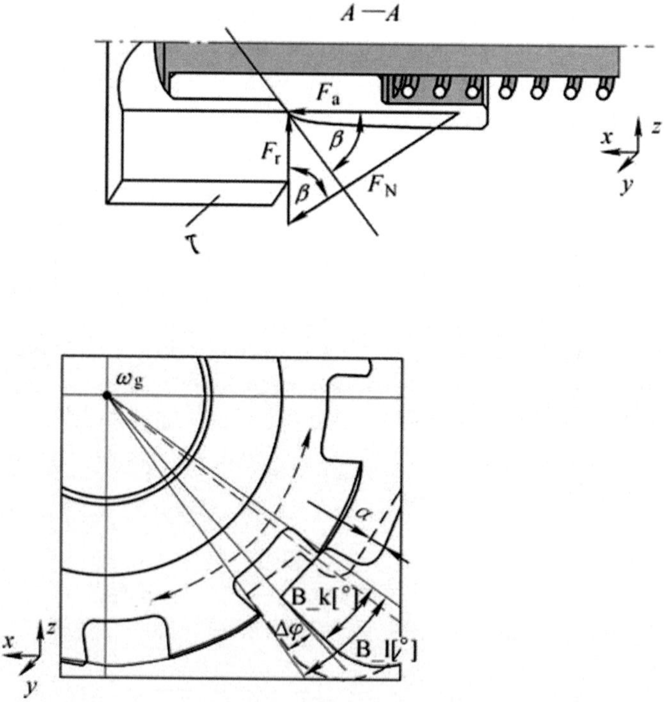

Fig. 8.46 Axial force estimation of control lever

The angular velocity of ratchet, the angular velocity of wheel and the vehicle speed have the following relationship

$$\omega_g = \omega_w \times i_0 \tag{8.18}$$

$$\omega_w = \frac{V}{r} \tag{8.19}$$

$$M_a = \frac{F_a b}{\tan\beta} \tag{8.20}$$

$$F_a = \frac{M_a \tan\beta}{b} \tag{8.21}$$

where, i_0—gear ratio.

Substitute Eqs. (8.15) and (8.16) into Eq. (8.21) to obtain

$$F_a = \frac{-M_r + I\frac{2\alpha}{t^2}}{b} tan\beta \tag{8.22}$$

Substitute Eqs. (8.17)–(8.19) into Eq. (8.22) to obtain:

$$F_a = \frac{2I\alpha}{b}\left(\frac{V i_0}{\Delta\varphi r}\right)^2 \tan\beta - \frac{M_r}{b}\tan\beta \tag{8.23}$$

$$V = \frac{r\Delta\varphi}{i_0}\sqrt{\frac{F_a b\text{ctan}\beta + M_r}{2\alpha I}} \tag{8.24}$$

(IV) Simulation analysis check

1. **Multi-body dynamic model building**

The three-dimensional modeling software Proe/Creo, CATIA and SolidWorks are used to build the 3D entity model of the vehicle parking mechanism, which is imported into the many-body dynamics software Adams. In the environment of Adams View, model material or density parameters are given, connection relationship between the parts is added, and contact conditions between the parts are determined to drive the moving parts. The multi-body dynamic model is shown in Fig. 8.47.

(1) Static strength simulation of parking mechanism: set the analysis conditions: in the gear P, gradually increase the output shaft torque until system failure to obtain the maximum torque that the parking mechanism can bear and determine the safety reserve factor. The simulation results are shown in Fig. 8.48. Taking downhill as an example, at 0.85 s, the system fails and the maximum bearing torque of the mechanism is 6.0×10^6 N mm.

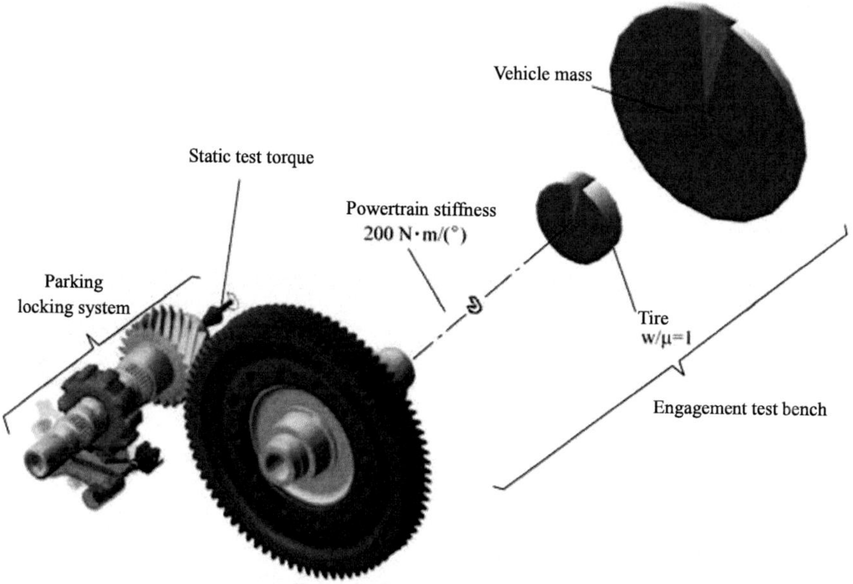

Fig. 8.47 Multi-body dynamic model

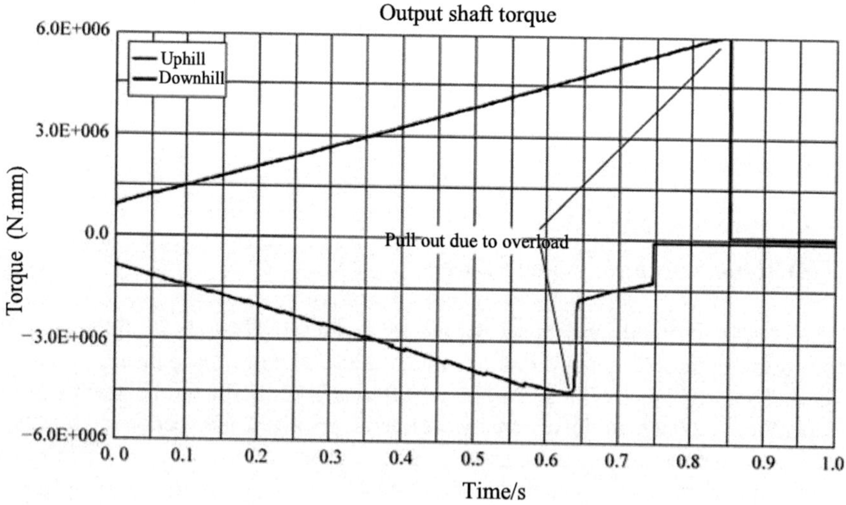

Fig. 8.48 Static strength simulation torque

(2) Simulation of system pull out force: set the analysis conditions: in the full load condition of the vehicle, simulate the gear P out situation in the parking on 30% ramp. The simulation results are shown in Fig. 8.49, and the change process of the pull out force will be different for different structures.
(3) Critical parking speed simulation: set the analysis conditions: the vehicle starts sliding at 10 km/h and puts in gear P. The critical speed at the parking in gear P is simulated. The simulation results are shown in Fig. 8.50 and the vehicle completes parking in gear P at 0.48 s.

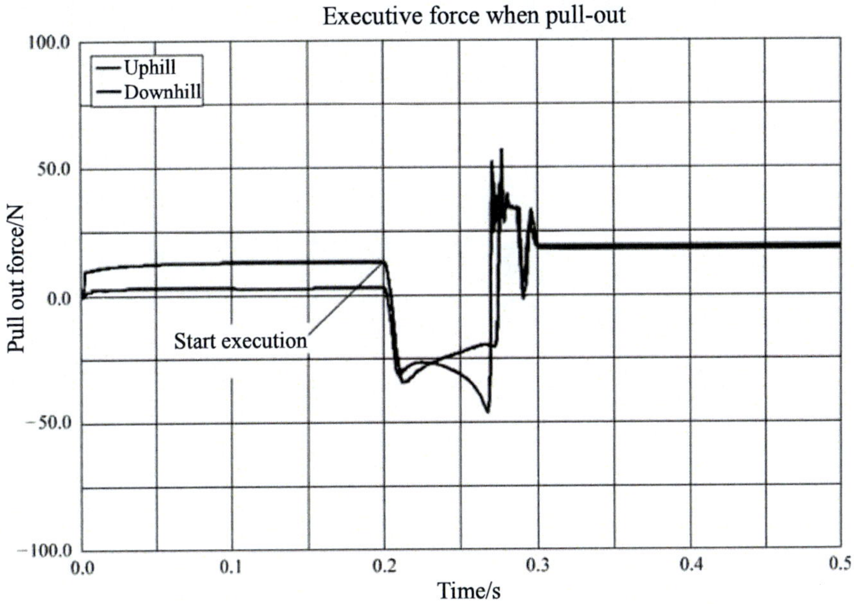

Fig. 8.49 Pull out force simulation

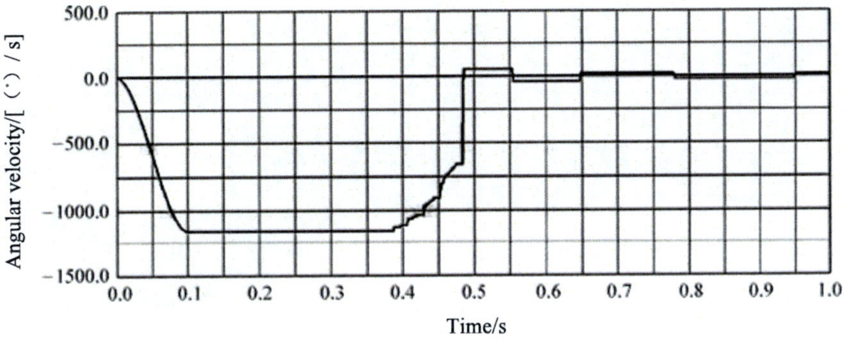

Fig. 8.50 Critical parking speed simulation

2. **Finite element analysis (FEA)**

The completed 3D model is imported to the finite element software Hypermesh and meshed. After pretreatment, it is imported to LS_DYNA to complete the strength analysis. The analysis results are shown in Figs. 8.51, 8.52, and 8.53.

Analysis of the static strength of the tapered pressure head using the CAE software can clearly and intuitively reflect that the stress state at each site of the parts meets

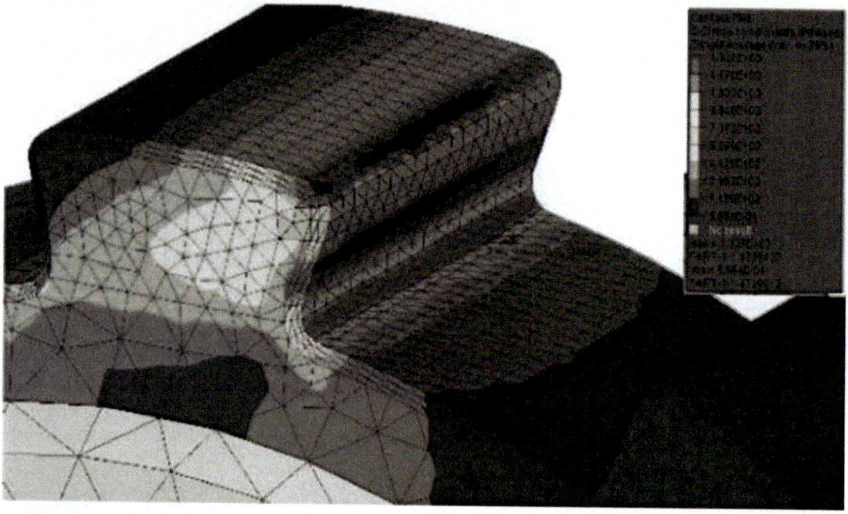

Fig. 8.51 Ratchet strength analysis

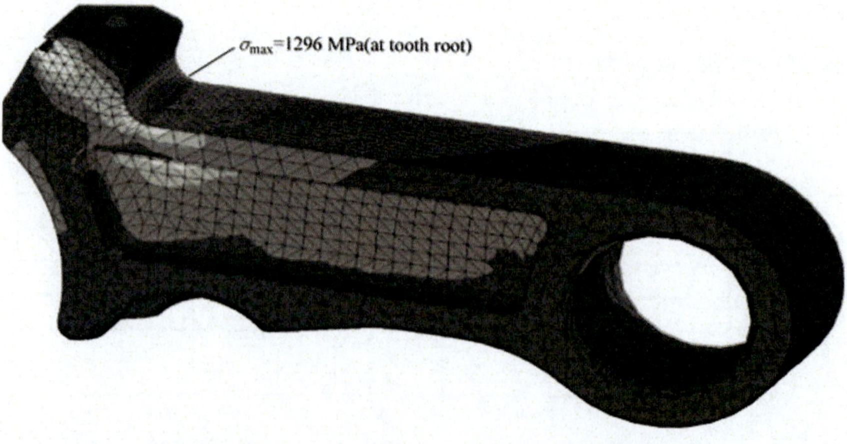

Fig. 8.52 Pawl strength analysis

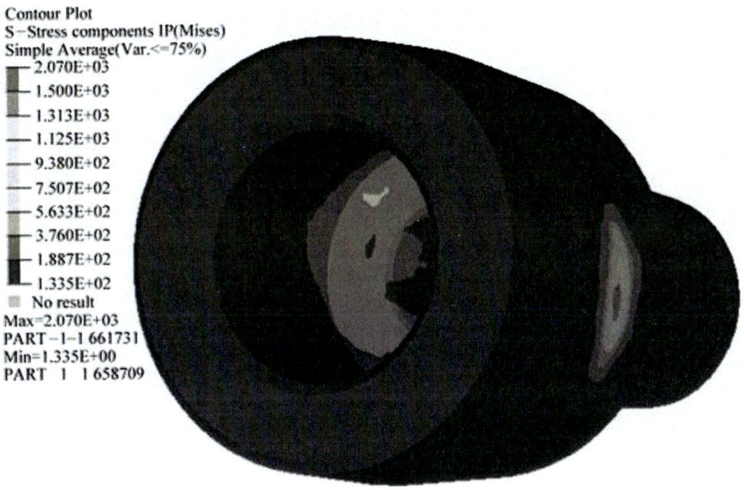

Fig. 8.53 Tapered pressure head strength analysis

the design requirements, and the criterion is that the maximum stress analyzed by CAE is less than the yield strength of the material.

II. Partial design calculation processes of another parking mechanism

As shown in Fig. 8.54, the parking mechanism consists of the ratchet (Fig. 8.55), pawl, torsion spring, pawl rotating shaft, actuator motor, motor fork, concave block, push rod, compression spring washer, compression spring, guide column and trapezoidal head.

(I) Design of parking mechanism actuator level spring

1. **Ramp parking speed calculation** (Fig. 8.56)

$$v_{wl} = \sqrt{2e_1 S_r} \tag{8.25}$$

where, v_{wl}—heel speed (maximum speed that can be achieved in parking on a ramp with a gradient of 30%) (m/s);

e_1—wheel acceleration (m/s^2), $e_1 = g\sin\alpha$;
g—acceleration of gravity;
α—slope (°);
S_r—wheel rolling distance, $S_r = \frac{2\pi r}{n_t i_d}$
r—wheel radius (m);
n_t—number of ratchet teeth;
i_d—final ratio.

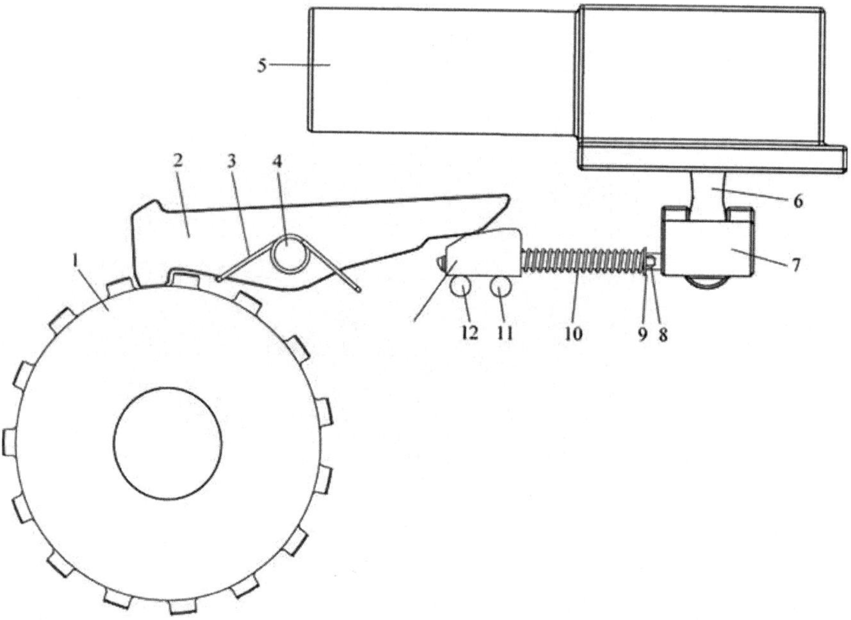

Fig. 8.54 Electric parking mechanism. 1—Ratchet, 2—pawl, 3—torsion spring, 4—pawl rotating shaft, 5—actuator motor, 6—motor fork, 7—Concave block, 8—push rod, 9—compression spring washer, 10—compression spring, 11—guide column, 12—trapezoidal

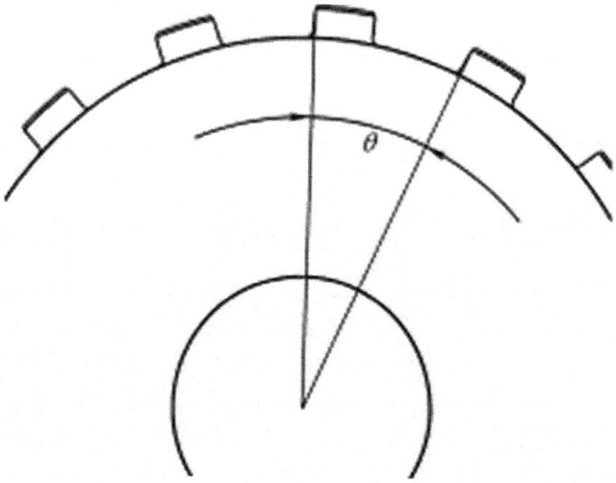

Fig. 8.55 Ratchet rotation angle

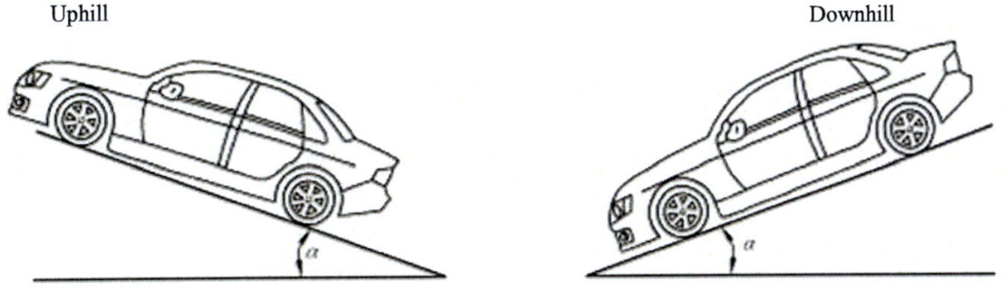

Fig. 8.56 Ramp packing

2. **Calculation of actuator level spring force**

(1) Calculation of ratchet rotation time in parking

$$t_1 = \frac{\Delta\phi}{\omega_g} \tag{8.26}$$

where, t_1—ratchet rotation time (s);

$F_{\max} = 2F_{gear}$—ratchet rotation angle (rad);

ω_g—rotational angular velocity of ratchet (rad/s), $omega_g = 2\pi n_p$

n_p—ratchet speed in parking (r/min), $n_p = n_o = n_w i_{oi} i_d$.

n_o—output shaft speed (r/min);

i_{oi}—gear ratio from the output shaft to the countershaft;

n_w—wheel speed, $n_w = \frac{v_w}{2\pi r \times 3.6}$.

v_w—critical parking speed (m/s), as shown in Fig. 8.57;

r—wheel radius (m);

i_d—final ratio.

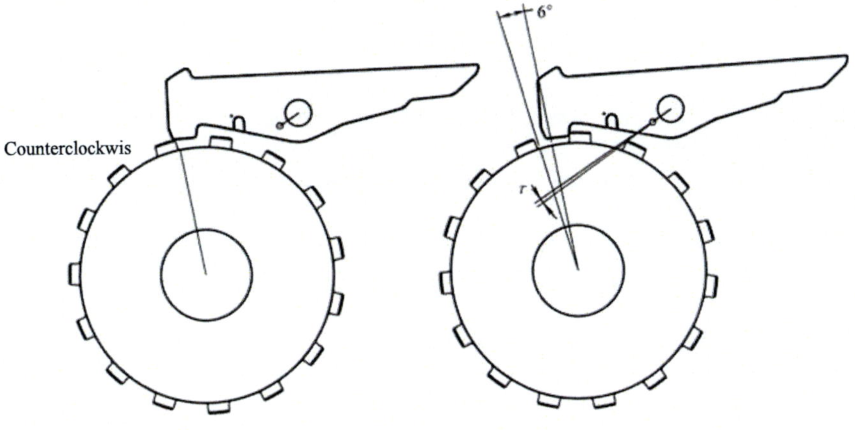

Fig. 8.57 Parking critical state

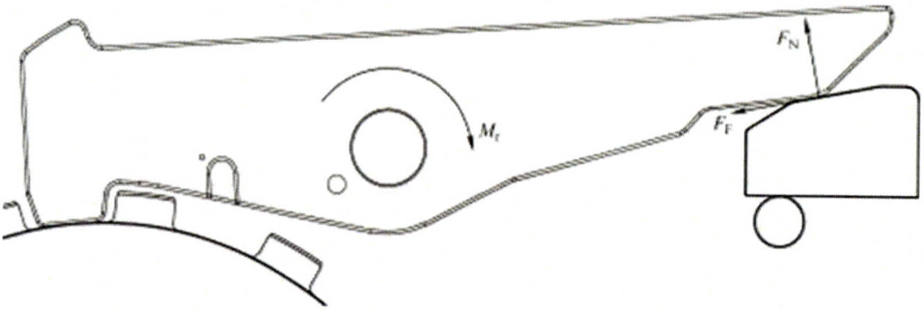

Fig. 8.58 Stress analysis of pawl in parking position

(2) Calculation of required rotation torque of pawl in parking at critical speed

$$t_2 = \sqrt{\frac{2\gamma J}{T_1}} \tag{8.27}$$

where t_2—pawl rotation time (m);
J—rotational inertia of pawl (kg m^2);
γ—pawl deflection angle (rad);
T_1—pawl rotation torque (N m).
The pawl rotation torque T_1 can be calculated from $t_1 = t_2$.

(3) Calculation of actuator level spring force: with the pawl as the research object, as shown in Fig. 8.58, according to the stress balance analysis of the pawl, the torque provided by the actuator level spring shall overcome the pawl rotation torque, return torsion spring torque and friction torque, then

$$F_N R_1 - F_F R_2 - M_r = T_1 \tag{8.28}$$

where, F_N—positive pressure of the trapezoidal head on pawl (N);
R_1—arm of force of F_N (m);
F_F—friction of trapezoidal head against pawl (N), $F_F = \mu_1 F_N$, in which, μ_1 is sliding friction factor;
R_2—arm of force of F_F (m);
M_r—return torsion spring torque (N m);
T_1—pawl rotation torque.
As shown in Fig. 8.59, the stress balance equation is listed with the trapezoidal head as the research object

$$F_a = F_{N1} \sin\beta + F_{F1} \cos\beta + F_{F2} + F_z \tag{8.29}$$

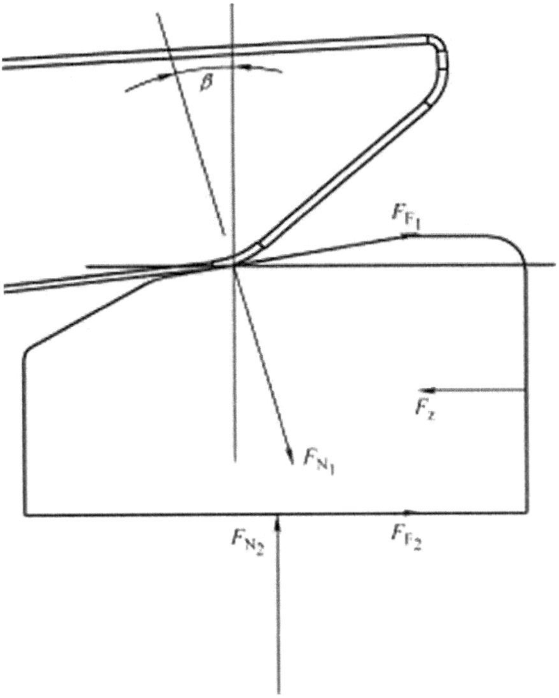

Fig. 8.59 Stress analysis of trapezoidal head

where, F_a—actuator level spring force (N);

F_{N1}—positive pressure of pawl on the trapezoidal head (N);

β—included angle between the contact surface and the horizontal plane of the trapezoidal head and pawl (°);

F_{F1}—friction of pawl against trapezoidal head (N), $F_{F1} = \mu_1 F_{N1}$, μ_1 is sliding friction factor;

F_{F2}—friction of guide column against trapezoidal head (N), $F_{F2} = \mu_1 F_{N2}$;

where, F_{N2}—positive pressure of the guide column on trapezoidal head (N), with the formula of

$$F_{N2} = F_{N1} \cos\beta - F_{F1} \sin\beta;$$

F_Z—force generated by the trapezoidal head acceleration (N), $F_z = m_z e_z$

m_Z—mass of trapezoidal head (kg);

S_1 —trapezoidal head stroke (m).

e_Z—trapezoidal head acceleration (m/s^2), $e_z = \frac{2\times S_1}{t_1^2}$.

The calculation process of ratchet negative rotation is the same as above.

(II) Design calculation of parking mechanism ratchet

1. **Maximum load calculation of parking ratchet**

$$F_g = \frac{M_g}{r_g} = \frac{mg\sin\alpha r}{i_d r_g} \tag{8.30}$$

where, m—vehicle full mass (m);
g—acceleration of gravity;
α—included angle between the ramp and the horizontal plane (°);
r—wheel radius (m);
i_d—gear ratio of the differential mechanism to the output shaft of the ratchet;
r_g—contact radius of parking ratchet (m).

2. **Calculation of ratchet load in overused parking condition**

$$F_{max} = 2F_g \tag{8.31}$$

3. **Calculation of contact stress on ratchet tooth surface** (Fig. 8.60)

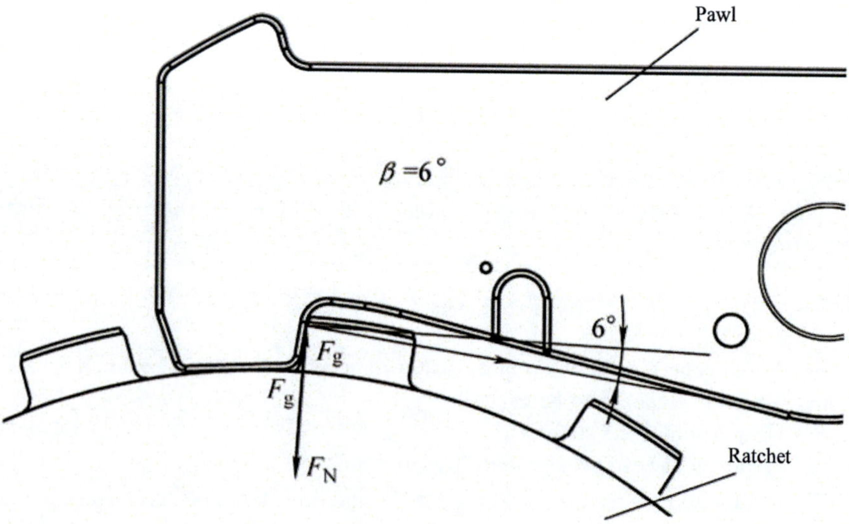

Fig. 8.60 Force of pawl on ratchet

$$\sigma_H = 189.8\sqrt{\frac{F_n}{b\rho}} \tag{8.32}$$

where, σ_H—contact stress on ratchet tooth surface;

F_n—positive pressure of pawl on ratchet (N), $F_n = F_g\cos\beta$;

b—length of contact line between ratchet and pawl (m);

ρ—contact radius of curvature (m).

The contact stress on the tooth surface of the parking mechanism is usually not more than 6000 MPa under the overuse condition, and not more than 4000 MPa under the normal parking condition.

4. **Ramp rollback calculation**

$$S_r = \frac{2\pi r}{n_t \times i_d}$$

The ramp rollback S_r is the same with the wheel (maximum ramp parking) rolling distance in the "Design of parking mechanism actuator level spring" and shall be less than the stipulated ramp rollback.

5. **Internal spline strength calculation**

The spline strength check is mainly to check the bending strength and contact strength.

(1) Bending strength

$$\sigma_F = \frac{T_{max}}{D_b e L z_e s_1} \tag{8.33}$$

where, T_{max}—maximum torque borne by the spline (N m), $T_{max} = s_{11}T$, s_{11} is safety factor, with the recommended value of 2, T is the torque actually borne by the spline (N m), $T = M_g$;

D_b—reference diameter (m);

e—reference circle tooth thickness (m);

L—contact length (m);

z_e—actual number of active teeth;

S_1—meshing load factor, gene—rally 0.25.

(2) Contact strength

$$\sigma_H = \frac{F_i}{S s_1 s_2} \tag{8.34}$$

where, F_i—pressure on each spline (N), $F_i = \frac{2T}{D_b Z_e}$;

S—contact area (m^2), equal to the product of the contact length L and the contact height;

S_2—area load factor, indicating the actual contact area, recommended 1.

The calculation process of ratchet negative rotation is the same as above.

(III) Design of parking mechanism pawl

1. Calculation of pawl load in maximum ramp parking

$$F_{n1} = F_n \tag{8.35}$$

where, F_{n1}—positive pressure of ratchet on pawl (N);

F_n—positive pressure of pawl on ratchet (N), $F_n = F_g \cos\alpha$, as shown in formula (8.32).

2. Calculation of pawl load in overused parking condition

$$F_{max} = 2F_{n1} \tag{8.36}$$

3. Calculation of contact stress on pawl tooth surface

$$\sigma_H = 189.8\sqrt{\frac{F_{n1}}{b\rho}} \tag{8.37}$$

where, F_{n1}—positive pressure of ratchet on pawl (N);

b—length of contact line between ratchet and pawl (m);

ρ—radius of curvature of contact arc (m).

The contact stress on the tooth surface of the parking mechanism is usually not more than 6000 MPa under the overuse condition, and not more than 4000 MPa under the normal parking condition.

The calculation process of ratchet negative rotation is the same as above.

(IV) Calculation of self-locking of parking mechanism tapered pressure head

1. With the pawl as the research object

As shown in Fig. 8.61, according to the stress balance of the pawl, the torque provided by the actuator level assembly is equal to the sum of the return torsion spring torque, torque of ratchet on pawl and torque of trapezoidal head on pawl, i.e.

$$F_{N3}R_3 + F_{N3}RF_4 = F_{n1}R_1 - F_{f1}R_2 + M_r \tag{8.38}$$

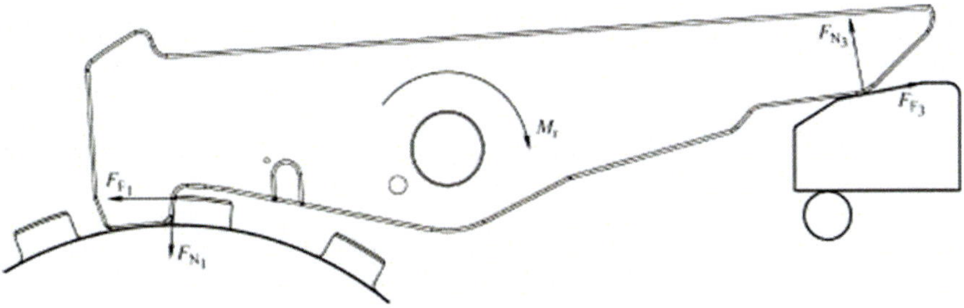

Fig. 8.61 Stress analysis of paw

where, F_{N3}—positive pressure of the trapezoidal head on pawl (N);

R_3—arm of force of F_{N3} (m);

F_{F3}—friction of trapezoidal head against pawl (N), $F_{F3} = \mu_2 F_{N3}$, μ_2 is static friction factor;

R_4—arm of force of F_{F3} (m);

F_{N1}—positive pressure of ratchet on pawl (N);

R_1—arm of force of F_{N1} (m);

F_{F1}—friction of ratchet against pawl (N), $F_{F1} = \mu_2 F_{N1}$, μ_2 is static friction factor;

R_2—arm of force of F_{F1} (m);

M_r—return torsion spring torque (N m).

2. **With the trapezoidal head as the research object** (Fig. 8.62)

$$F_h = F'_{F3}\cos\beta - F'_{N3}\sin\beta + F_S + F_{F4} \tag{8.39}$$

where: F_h—pull out force of trapezoidal head (N);

F'_{F3}—positive pressure of pawl on the trapezoidal head (N), $F'_{F3} = F_{F3}$;

F'_{F3}—friction of pawl against trapezoidal head (N), $F'_{F3} = \mu_2 F_{F3}$;

β—included angle between the contact surface and the horizontal plane of the trapezoidal head and pawl (°);

F_S—pretightening force of actuator level spring (N);

F_{F4}—friction of guide column against trapezoidal head (N), $F_{F4} = \mu_2 F_{N4}$.

F_{N4}—positive pressure of guide column on trapezoidal head (N), $F_{N4} = \mu_2 F'_{N3}\sin\beta + F'_{N3}\cos\beta$.

When $F_h \geq 0$, the tapered pressure head may be kept self-locking.

The calculation process of ratchet negative rotation is the same as above.

(V) Design of parking mechanism pawl return torsion spring

When the gear P is not engaged, the pawl shall not park in the ratchet and shall meet the following equation

$$M_P abk - M_r \leq 0 \tag{8.40}$$

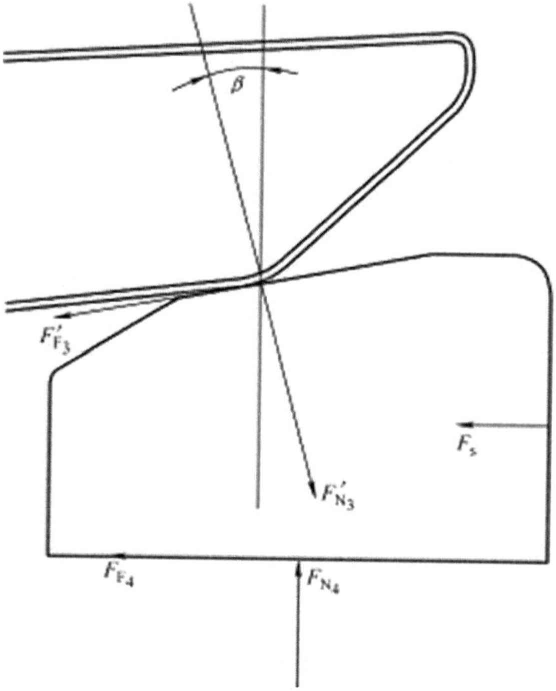

Fig. 8.62 Stress analysis of trapezoidal head

where, M_P—pawl mass (kg);

a—pawl vibration acceleration (m/s^2);

B—distance between the center of gravity of the pawl and the center of the pawl rotating shaft (m);

k—safety factor;

M_r—torque of the pawl return torsion spring (N m).

8.6 Synchronizer Design

I. Classification and construction of synchronizers

The synchronizer is classified into constant pressure, inertial, inertial boost and elastic meshing types, in which, the constant pressure synchronizer does not work reliably and has been eliminated basically. The inertial synchronizer is widely used in automobiles, which can be divided into two types: lock ring type and lock pin type. The lock ring type inertial synchronizer is mostly used in the car transmission and generally includes cone type, lever and boost types. The lock pin type inertial synchronizer is mostly used in the large vehicle transmissions. With different structures, they consist of the friction elements, locking elements and elastic elements. The structure of the lock ring type inertial synchronizer is mainly introduced below.

As shown in Fig. 8.63, the lock ring type inertial synchronizer mainly consists of the synchronizer hub, gear sleeve, slider and synchronous ring and can be classified into single-cone, double-cone and multi-cone types depending on the quantity of the cones of friction. Its structure is characterized by the friction moment generated by the interaction between the cone of friction on the synchronous ring and the cone of friction on the soldered tooth, thus achieving the effect of synchronization. The specific functions of each component of the lock ring type inertial synchronizer are as follows.

(1) Synchronizer hub: used for axial fixation and with the same spline tooth as the soldered tooth ring and synchronous ring.
(2) Gear sleeve: used to connect the synchronizer hub, synchronous ring and soldered tooth ring and with the same spline tooth as the soldered tooth ring and synchronous ring.
(3) Slider: the slider is mounted in the axial groove of the synchronizer hub with a locating pin (steel ball) for positioning in neutral, and the two ends extend into the notch of the synchronous ring.

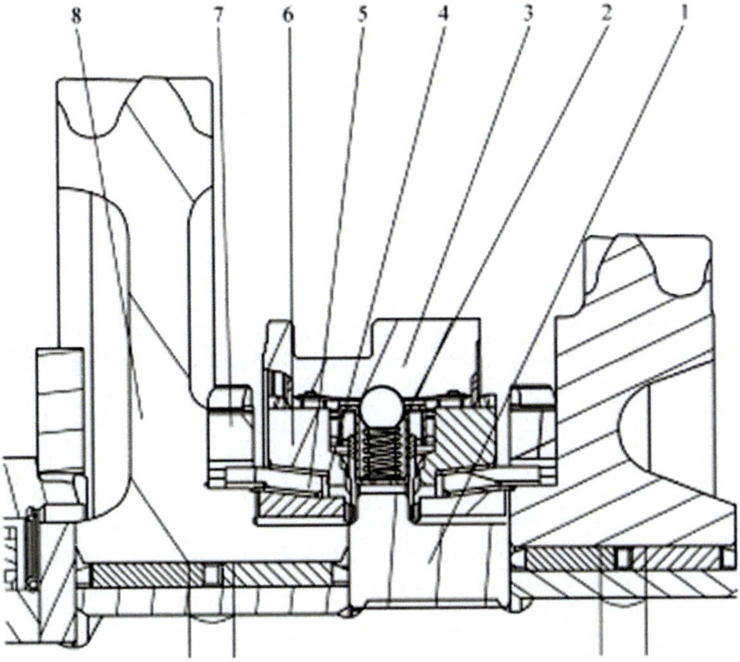

Fig. 8.63 Double-cone lock ring type inertial synchronizer. 1—Synchronizer hub, 2—slider, 3—gear sleeve, 4—inner ring of synchronous ring, 5—middle ring of synchronous ring, 6—outer ring of synchronous ring, 7—soldered tooth ring, 8—gear

(4) Synchronous ring: the chamfer (lock angle) of the synchronous ring is the same as that of the gear sleeve. The single-cone synchronous ring has an inner cone with a coating on it. Generally, the multi-cone synchronous ring has a coating on the inner cone of the outer ring, a coating on the inner core of the middle ring or a coating on the outer cone of the inner ring; some synchronous rings have spiral grooves on the inner core, so as to destroy the oil film after the two cones contact, and increase the friction between the cones.
(5) Soldered tooth: the soldered tooth fits the gear sleeve spline to transmit the motion and torque.

II. Functions of synchronizer

During transmission shift, the main functions of the synchronizer are as follows

(1) Ensure that the gear engagement is not impacted during gear shift to extend gear life.
(2) Reduce the powertrain noise when shifting gears.
(3) Make shift fast, smooth and light.
(4) Improve the power performance, economy and comfort of the vehicle.

III. Shift process and principle of synchronizer

The shift force applied by the driver causes the gear sleeve to move axially through the shift actuator and is applied on the synchronous ring through the slider until the synchronous ring cone is in contact with the cone on the soldered tooth ring. The angular velocity difference between two cones generates a friction moment that causes the synchronous ring to rotate by an angle relative to the gear sleeve and the slider, and to be positioned by the slider. Then, the gear end of the gear sleeve is in contact with the locking surface of the gear end of the synchronous ring, so that the movement of the gear sleeve is blocked and the synchronizer is locked, as shown in Fig. 8.64.

If the driver continues to apply the shift force, the shift force will continue to press the synchronous ring against the cone and increase the friction moment. At the same time, the ring toggle moment in the opposite direction to the friction moment acts on the locking surface. The angular velocities of the soldered tooth ring and the synchronous ring gradually approach, and at the instant the angular velocities are equal, the synchronization process ends, as shown in Fig. 8.65.

After realization of synchronization, the friction moment disappears and the ring toggle moment is greater than the friction moment. When the synchronous ring is pushed side, two locking surfaces are separated, the synchronizer is unlocked and the gear sleeve is engaged with the soldered tooth ring through the synchronous ring under the action of the shift force to complete synchronized shift, as shown in Fig. 8.66.

IV. Design flow of synchronizer

The main input parameters of the synchronizer are designed as general layout of synchronizer, center distance between the shaft, speed at maximum engine power,

Fig. 8.64 Locking state

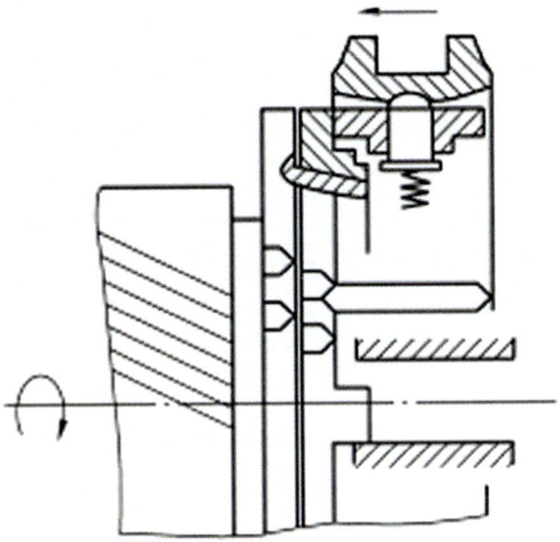

Fig. 8.65 End of the synchronization

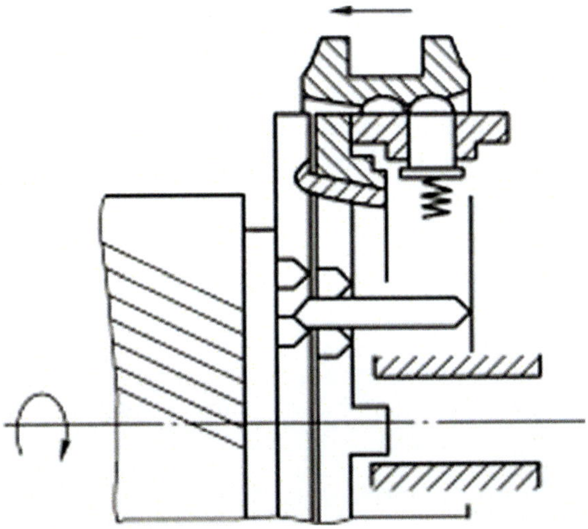

maximum engine torque, resistance moment of clutch driven plate at maximum power/speed, tire rolling radius, gear ratio, rotational inertia of clutch driven plate, rotational inertia at input end, rotational inertia on output shaft, allowable synchronization time, shift strategy and shift points, dynamic friction factor of the cone of friction of synchronizer, static friction factor of the cone of friction of synchronizer and wear characteristics of friction materials.

The general flow to determine the synchronizer design scheme is shown in Fig. 8.67.

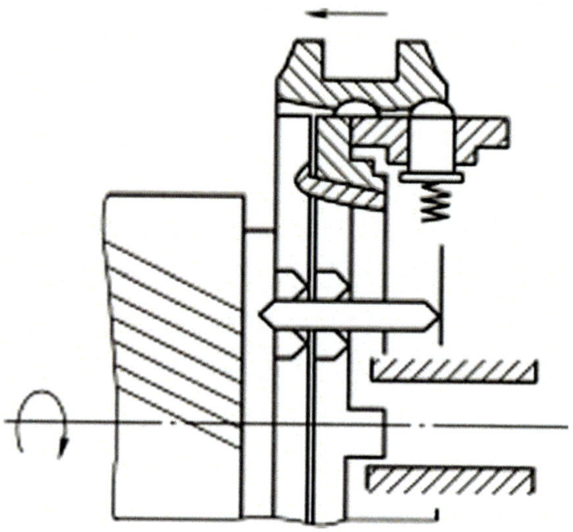

Fig. 8.66 Completion of shift

1. **Synchronous ring spline design**

The synchronous ring spline has the same principal parameters as the synchronizer hub spline and the spline slot width may be slightly greater than the synchronizer hub width; the synchronous ring spline modulus is the same with the soldered tooth modulus. The small modulus design can improve the matching width of soldered tooth.

2. **Synchronous ring taper hole diameter design**

The large end diameter of the synchronous ring taper hole shall be determined according to the reference diameter of the spline tooth. For a car transmission, the difference between the two is generally 12–15 mm, and for a multi-cone synchronizer, the middle ring can be chosen to limit the diameter of each cone of friction inside and outside the synchronous ring.

3. **Synchronous ring cone angle design**

It can be known from the friction moment formula $T_R = F\frac{d}{2}\frac{\mu_C}{\sin\alpha}$ that, the smaller the cone angle α, the greater the friction moment. When α is too small, the synchronous ring is easy to lock. At this time,$N\sin\alpha > \mu_C N\cos\alpha$, so $\tan\alpha > \mu_C$ shall be satisfied. As shown in Fig. 8.68, the synchronous ring cone angle is recommended to be 6.5°–7° for single-cone and 7°–12° for multi-cone. As for the fit between the multi-cone synchronous rings, the tolerance design of the inner and outer rings needs to ensure the contact of the large ends. Usually the negative tolerance is used for the outer cone and the positive tolerance for the inner cone.

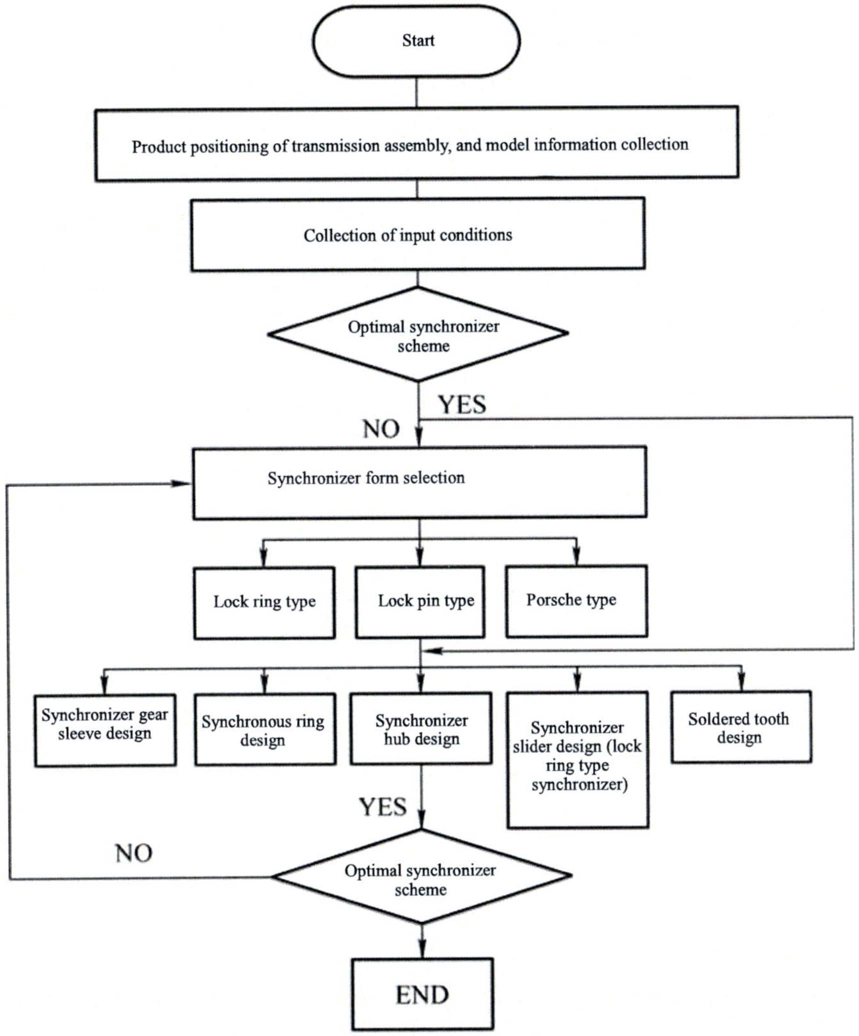

Fig. 8.67 Determine flow of synchronizer design scheme

4. **Design of synchronous ring lock angle and soldered tooth reverse cone angle**

In theory, the lock angle of the gear sleeve should be the same as that of the synchronous ring, but for processing reasons, the two angles are exactly the same and difficult to control, so the two angles are machined differently, as shown in Fig. 8.69. Since the gear sleeve material is usually harder than the synchronous ring material, the lock angle of the synchronous ring should be 3°–5° less than that of the gear sleeve used with it. The reverse cone angle is to lock the gear and prevent it from slipping out of gear.

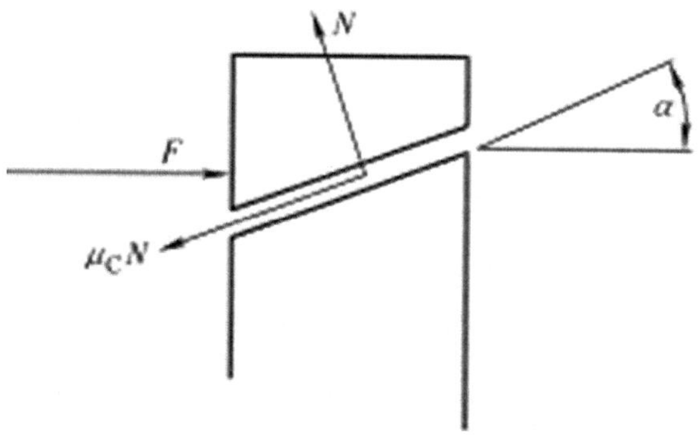

Fig. 8.68 Synchronous ring cone angle

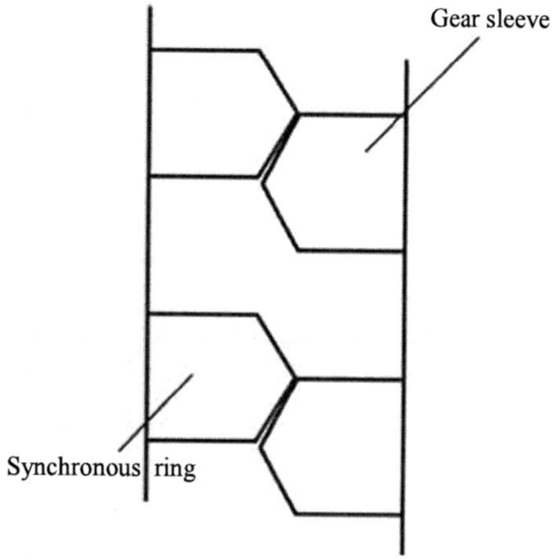

Fig. 8.69 Lock angle

5. **Synchronous ring face width design**

The synchronous ring face width is limited by the axial space size of the structure. Within the allowable range, the larger the face width, the larger the strength and stiffness of the synchronous ring, the better the heat dissipation and wear resistance. The synchronous ring face width is generally 6–14 mm.

6. **Synchronous ring thread and drain design**

For a multi-cone synchronous ring, the synchronous ring cone will bond the friction material or be sprayed with molybdenum, so there is no need to design the thread and drain. For a single-cone synchronous ring, it is necessary to design the thread and drain. Generally, the lead angle is 40° or 60°, the crest width 0.12–0.2, the thread pitch 0.6 mm, the threaded hole depth 0.25–0.5 mm; the drain width 3–4 mm, the depth greater than the threaded hole depth with the minimum of 0.8 mm and the quantity 6 or 9.

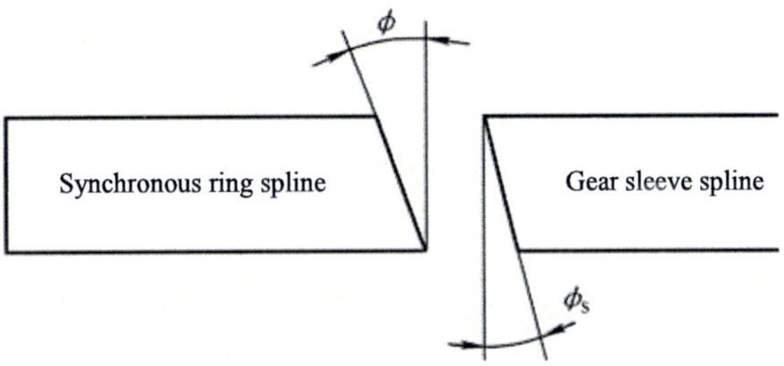

Fig. 8.70 Locking surface

7. **Gear sleeve fork groove design**

There is a circle of fork groove on the gear sleeve, which is designed symmetrically as far as possible. It is matched with the clearance of shift fork, and the clearance on both sides is generally 0.3–0.5 mm. The surface roughness of the fork groove wall is guaranteed to reduce wear. The depth of the fork groove shall not be less than 0.4 mm, and the thinnest part shall not be less than 2 mm with the large diameter of the spline to ensure that it is not easy to be hardened.

8. **Gear sleeve width design**

The gear sleeve must be more than twice as wide as the shift stroke to prevent the slider from coming off. The width of the gear sleeve shall be slightly larger than the length of the spline, and the length of the gear sleeve spline shall ensure the effective engagement length of about 2 mm after each gear is engaged.

9. **Gear end ridge angle design**

The ridge angle Φ_S at the gear end of the gear sleeve spline is smaller than the ridge angle Φ of the synchronous ring, as shown in Fig. 8.70; otherwise, it is easy to make local contact on the top of the locking surface of the synchronous ring and increase wear. Recommended value: $\Phi = 8°$, $\Phi_S = 6° \pm 1.5°$.

10. **Synchronizer hub spline design**

The synchronizer hub spline works with the gear sleeve spline, with the same principal parameters as the synchronous ring and soldered tooth splines. The spline tooth thickness shall be controlled reasonably. Excessive thickness will cause the gear sleeve to slide unsmoothly. Too small thickness can cause the gear sleeve to swing too much. The design of a small modulus can improve the length of fit between the soldered tooth and gear sleeve. However, if the modulus is too small, it will be easily quenched during heat treatment, so the tooth thickness is generally greater than 2 mm. The runout of the mating part between the gear sleeve and synchronizer hub is usually used to control the spline parameter and is generally not greater than 1 mm/Φ 100 mm.

11. **Synchronizer hub width design**

The synchronous ring positioning, axial layout space, web width, slider length, clearance and other factors shall be comprehensively considered for the synchronizer hub width design and the idle stroke of the synchronous ring shall be generally 0.3–1 mm.

12. **Slider chute width design**

The peripheral movement on each side of the synchronous ring soldered tooth is mπ/4, which is realized by the peripheral clearance of the synchronous ring bump in the hub slider chute. The exradius r_0 of the bump is less than the reference radius r of the spine soldered tooth. In order to ensure the clearance $C = \frac{r_0}{r}\frac{m\pi}{4}$ of each side of the bump in the chute, the slider chute width in the synchronizer hub is $H = h + 2C$.

V. Synchronizer performance calculation

1. **Calculation of positioning force in neutral position**

$$F_{\text{positioning force}} = \frac{3 \times F_R(\mu + \tan\theta)}{1 - \mu \tan\theta} \tag{8.42}$$

where, F_R—slider spring force (N);

μ—friction factor between steel ball and gear sleeve V-groove contact surface;

θ—supplementary angle of chamfer at the bottom of gear sleeve V-groove (°).

2. **Shift force calculation**

$$F_{\text{Shift force}} = \left(\frac{J_r \Delta\omega}{t_R} \pm T_V\right) \times \frac{\sin\alpha}{\mu_C R j} \tag{8.42}$$

where, $F_{\text{shift force}}$—shift force;

J_r—synchronous inertia (kg m^2);

T_V—trailing moment ($-T_V$ in upshift and $+T_V$ in downshift) (N m);

Δω—relative speed difference (angular velocity) (rad/s);

j—number of friction surfaces (1 for single-cone, 2 for double-cone and 3 for three-cone);

μ_C—friction factor of synchronous ring friction material;

α—cone angle of synchronous ring cone (°);

R—mean friction radius of synchronous ring (m);

t_R—synchronization time (s).

3. **Calculation of trailing moment**

$$T_V = T_{is} i \tag{8.43}$$

where, T_{is}—trailing moment at input shaft end (N m);
i—gear ratio.

4. **Calculation of mean friction diameter of synchronous ring**

$$d = \frac{2}{3} \times \frac{d_1^3 - d_2^3}{d_1^2 - d_2^2} \tag{8.44}$$

where, d_1—small end diameter of synchronous ring (m);
d_2—large end diameter of synchronous ring (m).

5. **Synchronous inertia calculation**

A transmission usually has 1–2 output shaft. When calculating the synchronous inertia, it is required to convert the inertia of all relevant parts to the same input shaft, which requires the equivalent conversion of inertia.

For example, the inertia J2 of the rotating member with radius R_2 on shaft 2 is converted to the rotating member with radius R_1 on shaft 1, and is combined with the inertia J1 of the rotating member on shaft 1 to obtain the total equivalent inertia J_{eq} on shaft 1.

The inertia of J_2 converted to shaft 1 is

$$J_{2->1} = \left(\frac{R_1}{R_2}\right)^2 J_2 \tag{8.45}$$

Therefore, the total equivalent inertia on shaft 1 is

$$J_{eq} = J_1 + \left(\frac{R_1}{R_2}\right)^2 J_2 \tag{8.46}$$

Figure 8.71 is the layout of a 5-speed MT. Assuming engagement from gear 2 to gear 1, the formula of the synchronous inertia is

$$J_{r,i} = J_i + \sum_{k=1}^{i} J_k \frac{1}{i_k^2} \tag{8.47}$$

where, J_i—rotational inertia of the gear in the current gear (kg m^2);

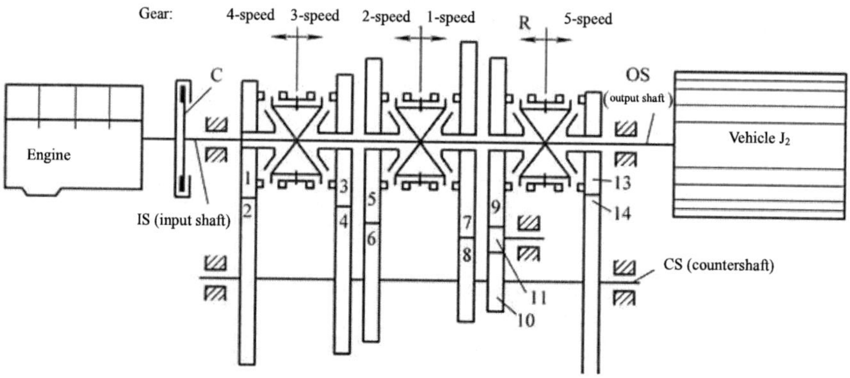

Fig. 8.71 Layout of a 5-speed MT

J_k—rotational inertia of the gear in the target gear (kg m^2);
i_k—speed ratio in each gear.
According to the above formula, the following results can be obtained

$$J_{r,7} = J_7 + (J_C + J_{is} + J_1)\left(\frac{z_7}{z_8}\right)^2\left(\frac{z_2}{z_1}\right)^2 + (J_{CS} + J_2 + J_4 + J_6 + J_8 + J_{10} + J_{14})\left(\frac{z_7}{z_8}\right)^2$$
$$+ \left[J_3\left(\frac{z_4}{z_3}\right)^2 + J_5\left(\frac{z_6}{z_5}\right)^2 + J_9\left(\frac{z_{10}}{z_9}\right)^2 + J_{11}\left(\frac{z_{10}}{z_{11}}\right)^2 + J_{13}\left(\frac{z_{14}{}^2}{z_{13}}\right)\right]\left(\frac{z_7}{z_8}\right)^2 \quad (8.48)$$

The above formula is the exact algorithm for calculating the inertia. However, in the actual calculation, the following simplified calculation method is often used

$$J_r = J_{is} i_k^2 \quad (8.49)$$

where, J_{is}—sum of the inertia of all relevant parts converted to the input part of the shaft of the synchronizer (kg m^2);
i_k—speed ratio in each gear.

6. **Shift impulse calculation**

$$I = F t_R \quad (8.50)$$

where, F—shift force acting on the gear sleeve (N);
t_R—upshift synchronization time (s).

7. **Friction moment calculation**

$$T_R = jF\frac{d}{2}\frac{\mu_C}{\sin\alpha}\frac{1}{1000} \tag{8.51}$$

where, j—number of friction surfaces;
F—shift force acting on the gear sleeve (N);
μ_C—friction factor of synchronous ring friction material;
α—cone angle of synchronous ring cone (°);
d—mean friction diameter of synchronous ring (m).

8. **Synchronous ring pressure calculation**

$$2P_{2R} = \frac{2F}{2A_{2R}\sin 2\alpha} \tag{8.52}$$

where, P_R—synchronous ring pressure (Pa);
A_R—total friction area of synchronous ring (m^2).

9. **Calculation of total friction area of synchronous ring**

$$A_R = \frac{\pi DBb}{p} \tag{8.53}$$

where, D—cone working diameter (m);
B—synchronous ring cone width (m);
b—thread crest width of synchronous ring cone (m);
p—synchronous ring cone thread pitch (m).

10. **Calculation of relative linear velocity in synchronization of synchronous ring**

$$v = \frac{\Delta\omega\pi R}{30} \tag{8.54}$$

where, v—relative linear velocity in synchronization;
Δω—relative speed difference (angular velocity) (rad/s);
R—mean friction radius of synchronous ring (m).

11. **Calculation of friction work carried by synchronous ring**

$$W = \frac{J_r \Delta\omega^2 \pm T_V \Delta\omega t_R}{2} (-, \text{n upshift and} + \text{in downshift}) \tag{8.55}$$

Where, W—friction work of synchronous ring (J).

12. **Calculation of friction work per unit area of synchronous ring**

The above formula calculates the total work done by synchronous ring at one time. The heat will inevitably be produced in the work. To evaluate whether the work per unit area of the synchronous ring exceeds the standard, it is also necessary to divide the total work by the total working area of the synchronous ring to get the work per unit area of the synchronous ring.

$$W_A = \frac{W}{A_R} \tag{8.56}$$

where, W_A—friction work per unit area of synchronous ring (J).

13. **Calculation of friction power per unit area of synchronous ring**

$$\boldsymbol{P_A} = \frac{|\boldsymbol{W_A}|}{\boldsymbol{t_R}} \tag{8.57}$$

where, P_A—friction power per unit area of synchronous ring (W).

14. **Calculation of ring toggle moment**

$$\boldsymbol{T_Z} = \boldsymbol{FR}\frac{\cos\boldsymbol{\beta} - \boldsymbol{\mu_D}\sin\boldsymbol{\beta}}{\sin\boldsymbol{\beta} + \boldsymbol{\mu_D}\cos\boldsymbol{\beta}} \tag{8.58}$$

where, T_Z—ring toggle moment (N m);
F—shift force acting on the gear sleeve (N);
R—pitch radius of synchronous ring (m);
β—gear sleeve lock angle (half-angle) (°);
μ_D—friction factor between the gear sleeve locking surface and synchronous ring locking surface.

15. **Calculation of locking safety factor of synchronous ring**

After calculation of the above performance indexes, the locking safety of the synchronous ring shall also be confirmed. As for the locking safety, it is required to ensure that the gear sleeve will not slide to the soldered tooth through the synchronous ring before synchronization of the synchronous ring. To ensure this requirement, it is required to ensure that the ring toggle moment T_Z applied on the gear sleeve is less than the friction moment T_R of the synchronous ring, namely $T_Z < T_R$

$$K = \frac{T_R}{T_Z} \tag{8.59}$$

where, K—locking safety factor of synchronous ring.

VI. Synchronizer performance evaluation indexes

1. **Maximum allowable shift force** (Table 8.9)
2. **Allowable synchronization time** (Table 8.10)
3. **Reference trailing moment**

The trailing moment is usually caused by the friction of the bearing and the friction between the clutch driven plate and the transmission fluid. In the actual test, the trailing moment at the input shaft end is different at different speed and oil temperature. However, in theoretical calculation, to avoid too complicated calculation, the

Table 8.9 Maximum allowable shift force

Item	Gear	Passenger vehicle		Commercial vehicle		
				Basic transmission	Front-mounted auxiliary transmission	Rear-mounted auxiliary transmission
Shift force/N	Gear 1–top gear	MT	<120	<250	Pneumatic	Pneumatic
		AMT, DCT	<1400 (gear sleeve)			

Table 8.10 Allowable synchronization time

Item	Gear	Passenger vehicle	Commercial vehicle		
			Basic transmission	Front-mounted auxiliary transmission	Rear-mounted auxiliary transmission
Synchronization time/s	Gear 1–top gear	0.15–0.25	<0.14	0.15	0.2

Table 8.11 Reference trailing moment

Item	Passenger vehicle	Commercial vehicle	Commercial vehicle with range gear
Trailing moment/N m	2–5.5	4–8	10–14

Table 8.12 Locking safety factor of synchronous ring

Item	Single-cone	Double-cone	Three-cone
Locking safety factor	≈1.2	1.3–1.5	≥1.8

Table 8.13 Design reference values of different friction materials

Material of synchronous ring	Friction factor	Maximum line velocity/(m/s)	Friction work per unit area/(J/mm^2)	Friction power per unit area/(W/mm^2)	Compressive stress per unit area/(N/mm^2)
Special brass	0.08–0.12	5	0.09	0.45	3
Mo-sprayed	0.08–0.12	7	0.53	0.83	6
Sintered material	0.08–0.12	9	1.00	1.50	7
Carbon material	0.10–0.13	16	1.50	10	12

trailing moment can usually be selected as the fixed value generally based on the experience of engineers at the initial stage of design, as shown in Table 8.11.

4. **Allowable locking safety factor** (Table 8.12)

5. **Design reference values of different friction materials** (Table 8.13)

The parameters of carbon materials vary greatly. For example, due to the addition of various microelements, the friction factor of carbon particle friction materials can range from 0.08 to 0.2 and even higher, while the friction factor of woven fiber carbon friction materials is usually 0.08 to 0.14. Therefore, regarding the design reference values of carbon materials, the supplier shall be consulted to confirm the performance limit value of materials to ensure the accuracy of theoretical calculation.

VII. Specification for strength and fatigue analysis of synchronizer

1. **Software requirements**

The pre-processing modules of HYPERWORKS, ANSYS, ANSA and other software are generally used as the pre-processing software; generally, the ABAQUS and ANSYS are used to solve the strength, the FEM-FAT, Fe-safe and NCODE are used to solve the fatigue and the HYPERWORKS, ANSYS and ABAQUS are used to view the analysis results.

2. **Input conditions**

The input conditions for the strength CAE analysis of the synchronizer system mainly include:

(1) Digital models of synchronizer gear sleeve, synchronizer hub, soldered tooth and transmission shaft, assembled according to the drawing requirements without penetration in the models; the finite element mesh quality meets the solver requirements (minimum angle greater than 20° and minimum mesh greater than 0.1 mm).
(2) Material information of each component of synchronizer, including modulus of elasticity, Poisson's ratio and density.
(3) Friction factor between components of the synchronizer system.
(4) Torque borne by the transmission shaft and soldered tooth.

The input conditions for the fatigue CAE analysis of the synchronizer system are mainly the results of strength calculation, ODB file and material S–N curve.

3. **Boundary conditions**

According to the operating conditions of the synchronizer system, the boundary conditions are set as follows:

(1) The distributed coupling (rbe3) element is used to rigidly connect the two constrained surface nodes with the shaft center nodes.
(2) The two shaft center nodes are constrained to constrain the degrees of freedom of the transmission shaft rotating in 5 directions except x axis, and the degrees of freedom of the soldered tooth rotating in 5 directions except x axis.
(3) The shaft is in full contact with the synchronizer hub, the synchronizer hub is in full contact with the gear sleeve, and the gear sleeve is in 50% contact with the soldered tooth (as practical). The synchronizer coupling element and contact form are shown in Fig. 8.72.

4. **Loading conditions**

The loads of the synchronizer system is applied at the transmission shaft and soldered tooth, which are equal in size and opposite in direction, at the two shaft center nodes that have been created.

5. **Strength output**

The analysis name, parameters, types and output results are defined. The synchronizer hub stress and displacement cloud diagram under maximum working load is shown in Fig. 8.73. The stress of the synchronizer hub shall be less than the yield strength of the synchronizer hub material.

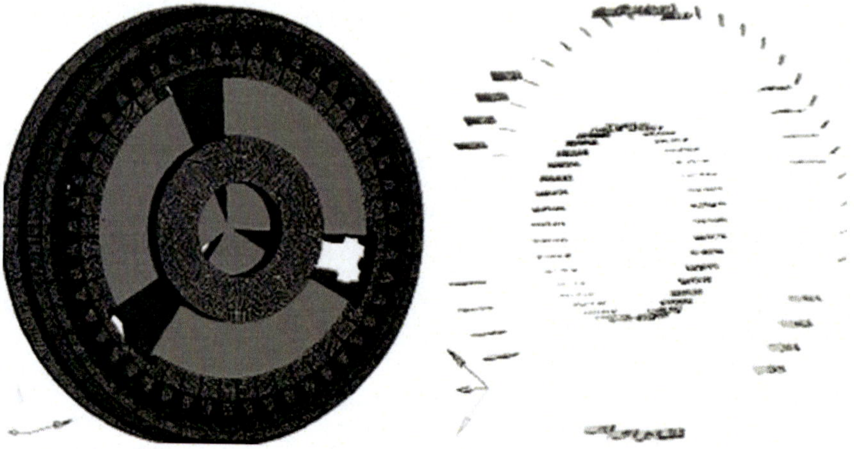

(a) Coupling element (b) Contact form

Fig. 8.72 Synchronizer coupling element and contact form

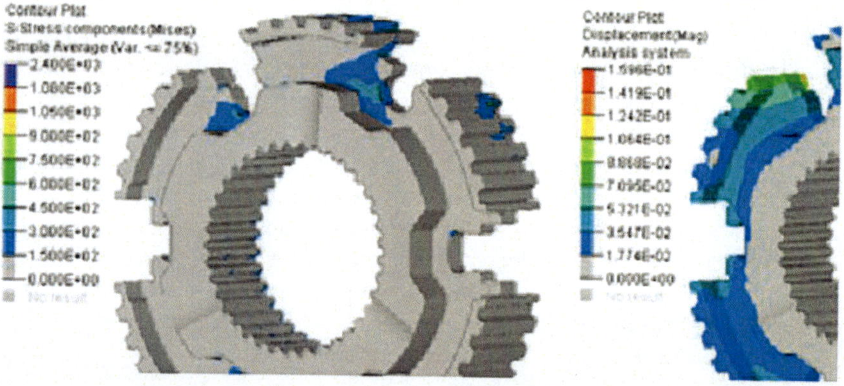

Fig. 8.73 Stress and displacement cloud diagram of synchronizer hub

6. **Fatigue analysis**

Import the strength analysis results. ODB file into the FEMFAT software and analyze the setting conditions:

(1) Input the tensile strength, yield strength, S–N curve and linear static parameters of the material.
(2) Set the surface roughness, technical size, dispersion and temperature.
(3) Set the influence factors, analysis type and survival rate.

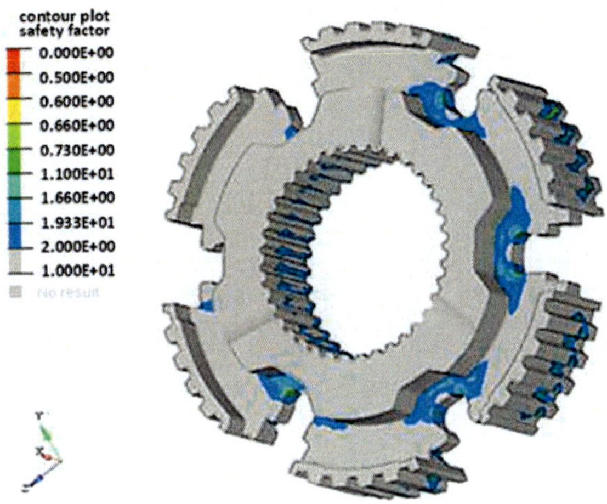

Fig. 8.74 Cloud diagram of synchronizer hub safety factor distribution

7. **Fatigue output**

Taking synchronizer hub as an example, the cloud diagram of its safety factor distribution under the maximum working load is shown in Fig. 8.74. The minimum fatigue safety factor in the contact area of the synchronizer parts shall be greater than the design requirements.

8.7 Selection of Seals

Requirements for transmission service life and environmental considerations have made reliable, durable transmission seals an important issue. If the seal fails, repairs can cost many times as much. On the automobile transmission, there are many places need to seal, such as the input and output ends of the shaft, the connection of the case, the output end of the reverse shaft and the drive end of the speedometer. There are three main types of seals on the transmission: seals for static parts (such as flat gaskets), seals for rotating parts (such as rotary shaft seals) and seals for reciprocating parts (such as grooved seals).

I. Seals for static parts

A seal for static parts (static seal) is a seal applied between the surfaces without clearance fit. Extremely small movements due to pressure and temperature changes, expansion and contraction, normal wear, and shock and vibration may not be considered for such seal. Figure 8.75 shows examples of seals for different purposes, including flange seal, plug seal and pipe seal. Figure 8.76 shows typical static seal forms, including O ring, cutting ring, four-blade ring and metal ring.

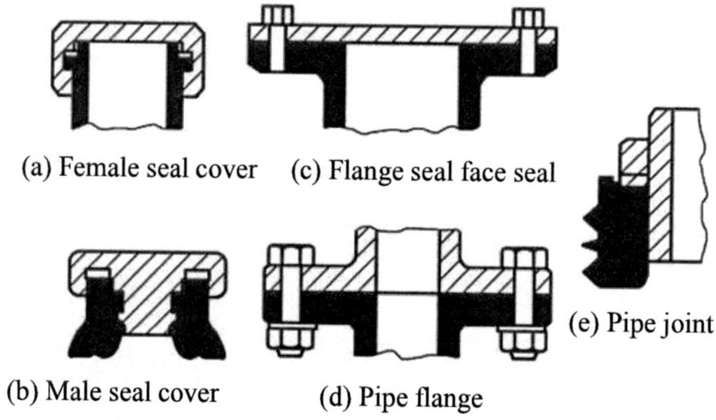

Fig. 8.75 Seals for different purposes

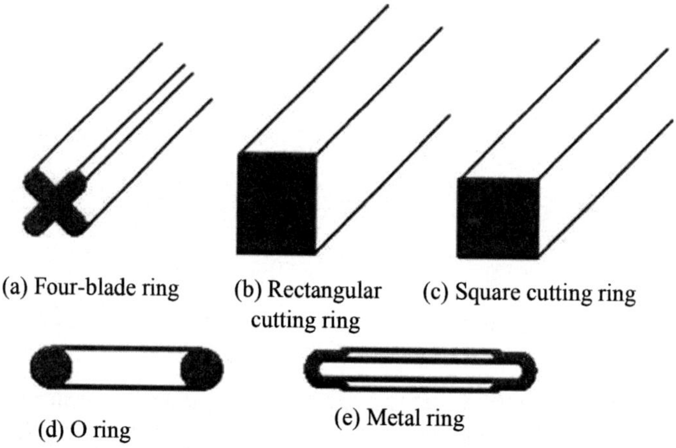

Fig. 8.76 Typical static seal forms

O rings are preferred when temperature, pressure, liquid and geometry conditions permit. The circular cross section of the O ring is easy to adapt to the geometry of hydraulic applications. In general, 15–70 lb (1 lb = 0.4536 kg) of load per inch (1 in = 0.0254 m) of linear length can quickly form a seal line or belt, effectively sealing 1500 lb/in^2 pressure without the use of a support ring. With the proper seal cartridge design and the use of a support ring, the pressure range can be extended beyond 3000 lb/in^2.

An O ring of a given inner diameter may have several different cross sectional sizes. In most static seal applications, any cross section suitable for assembly position can seal. In general: ① The maximum available cross section suitable for the installation site shall be selected; ② Installation damage, abrasion and rolling or twisting are rare in large cross sections; ③ Large cross sections are more tolerant to intermittent high temperatures; ④ Large cross sections are more tolerant to stretching and compressing. However, when the stretch rate exceeds 5%, leakage may occur. Material deterioration can occur if it is subject to high temperature or used in barely suitable liquids in addition to a high stretch rate. Excessive stretch rate will reduce the area of the cross section and lead to the ellipse of the cross section.

The surface oil seal is used to seal the parting face of the transmission case and cover. It can be a preformed seal (flat gasket, metal flanged gasket) or an unformed seal (sealant). After sizing on the sealing surface, the sealant can take on the desired shape; after being fitted to the sealing surface, the flat gasket will inevitably change shape, which ensures that the sealing surface matches the microstructure of the sealed part surface and closes the pores in the sealing material.

The flat gasket (soft gasket) usually adopts fabric oil seal (e.g. paper oil seal, sealing board) and fiber-reinforced oil seal (e.g. aramid fiber). The sealants (liquid oil seals) include chemically curing sealants (e.g. anaerobic sealant, silicone) and non-curing sealants (ethylene glycol sealant).

II. Seals for rotating parts

In the transmission, as long as there is no pressure oil in the through hole of the spindle, the synthetic rubber rotary shaft seal shall be adopted; otherwise, the through hole must be sealed to prevent to prevent splashing oil from escaping.

The clutch is in the state of rotation, the hydraulic oil needs to enter the clutch hydraulic cylinder through the relative rotation face, and the seal of the relative rotation face needs to use rotary seal. The sealing ring is generally used for the rotary seal and made of alloy cast iron and PTFE. Figure 8.77 shows an alloy cast iron rotary seal ring and Fig. 8.78 shows a PTFE rotary seal piston ring.

Teflon seal ring is made of PTFE and widely used in modern automotive automatic transmission. It is characterized by soft material, high abrasive resistance, heat resistance and friction performance and low cost. Because of soft material and easy

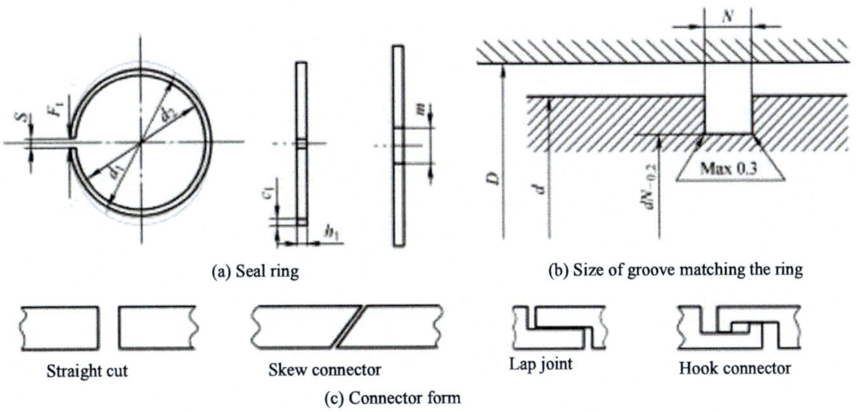

Fig. 8.77 Alloy cast iron rotary seal ring

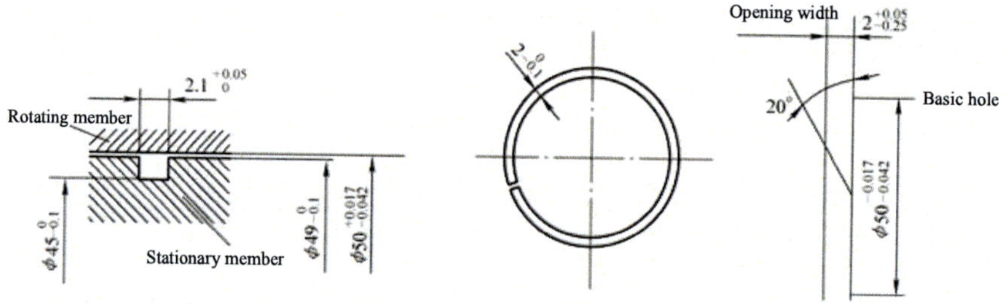

Fig. 8.78 PTFE rotary seal piston ring

deformation, it can still be well sealed in the groove of a circular shaft or hole even if the size of the shaft or hole is somewhat out of round. However, this material is prone to being scratched or punctured by metallics. The friction resistance of the PTFE piston ring is lower than that of the alloy cast iron piston ring. Figure 8.79 shows the comparison of the friction factors of these two piston rings with the change curve of relative angular velocity.

III. Seals for reciprocating parts

The seal for reciprocating parts mainly refer to the seal between the hydraulic cylinder and piston in the shift clutch and brake, with good sealing and small moving resistance. Such seal has three forms:

(1) Alloy cast iron piston ring, with small moving resistance and poor sealing.
(2) Rubber O ring, with good sealing, large moving resistance and difficult manufacturing control. If the compression amount is too small, the O ring is poor in sealing; if the compression amount is too large, the moving resistance is large.

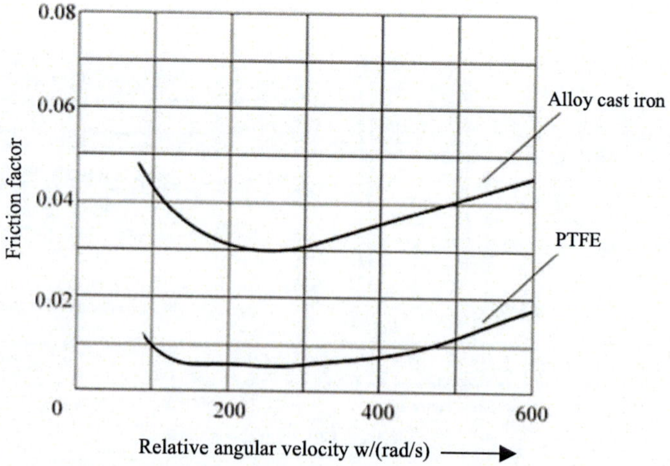

Fig. 8.79 Change curve of piston ring friction factor with relative angular velocity

(3) Lip seal, with small moving resistance, good sealing and troublesome manufacturing.

The lip seal is shown in Fig. 8.80. Figure 8.80a shows the sectional form of the lip seal; Fig. 8.80b shows the clearance between the piston and the hydraulic cylinder, the minimum clearance a_{min} shall ensure small moving resistance and the maximum clearance a_{max} shall ensure small leakage; Fig. 8.80c shows the outward lip seal mounted on the piston; Fig. 8.80d shows the inward lip seal mounted on the hydraulic cylinder.

IV. Transmission sealing detection

The actual transmission sealing detection methods are as follows:

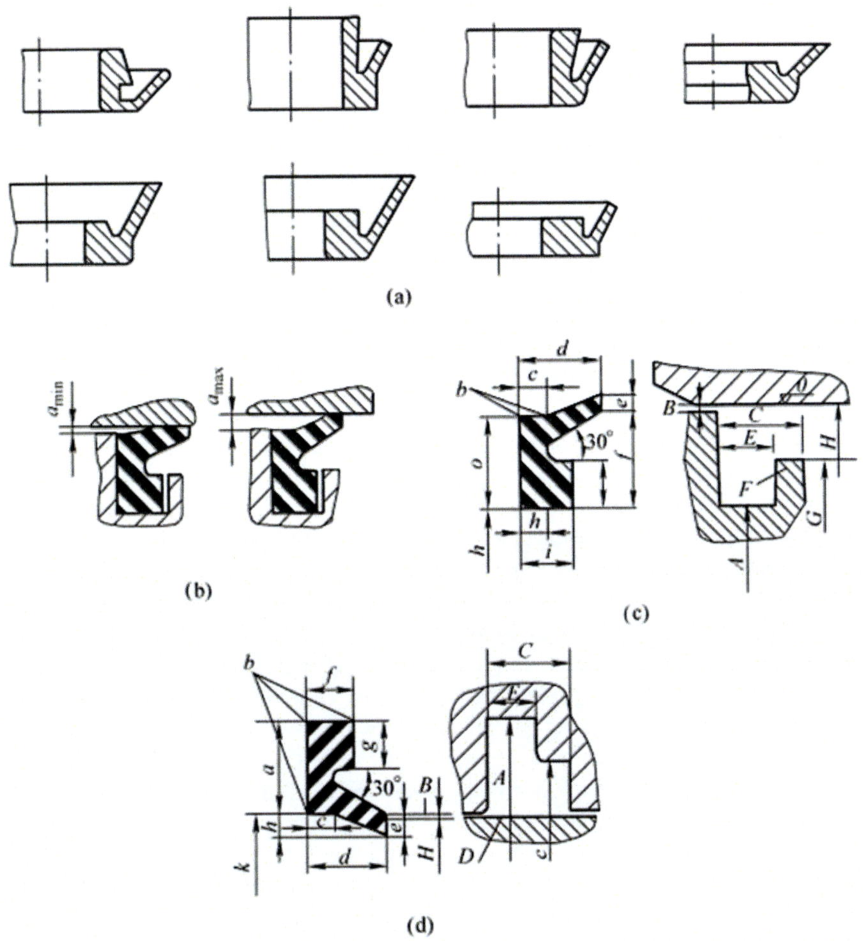

Fig. 8.80 Lip seal

(1) Differential pressure measurement. With the medium of air, the test is to test the pressure loss in the case.
(2) Water bath test. The immersion test is conducted with slight overpressure in the case.
(3) Helium leak test. The inspiration test is conducted using a leak detector with the medium of helium.
(4) Mass flow rate measurement to determine the leakage rate.

Common causes for leakage are as follows:

(1) Damage of sealing elements during assembly, such as distortion, shearing, etc.
(2) The sealing surface is damaged or does not meet the technical requirements.
(3) Mixed loading or installation error.
(4) Turnover of the seal lip of the rotary shaft seal during assembly, especially in the case of double oil seals.
(5) Misplacement or deflection of the rotary shaft seal during assembly.

8.8 Transmission Ventilation Design

With the increase of the engine speed, the internal oil temperature of the transmission will also increase, which will cause the expansion of the inner cavity gas and oil atomization, so that the pressure in the limited space of the inner cavity significantly increases. In this case, it is extremely important to use a reasonable breather to keep the transmission in normal working conditions, especially to ensure the sealing and lubricity at the rubber oil seal of the input and output shafts. The breather is to ensure the internal and external pressure balance of the transmission. A breather valve is installed on the transmission case. Considering the transmission working reliability and environmental pollution avoidance, it is not allowed to discharge the lubricating oil, oil vapor and oil mist from the breather valve, and the water, dirt and dust are not allowed to enter the transmission from the breather valve.

The breather valve assembly shall be preferred at the highest position of the box members. When this condition cannot be met due to structural constraints, the requirements may not be implemented, and the splashing lubricant shall be avoided. If not, baffles shall be added to block. The breather valve is usually classified into normally closed and normally open types. Figure 8.81 shows a normally closed breather valve assembly and its cover plate can be open under the pressure of 5–10 kPa. Figure 8.82 shows normally open breather valve assemblies. After the breather valve cover and metal breather pipe in type A normally open breather valve assembly are assembled, the breather valve cover shall be able to easily rotate and move axially around the shaft; under the action of 50 N pressure, the impacted plugs and valve bodies evenly spaced along the circumference in the type B normally open breather valve assembly and normally closed breather valve assembly shall not fall off and the plugs shall easily rotate and move axially. The opening pressure of type A normally open breather valve assembly is (4 ± 0.05) kPa.

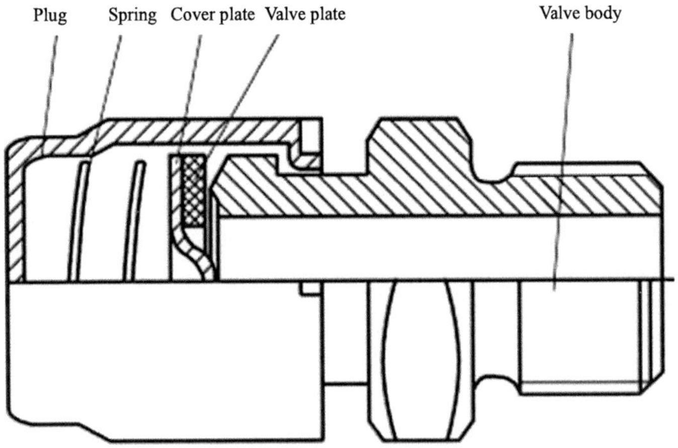

Fig. 8.81 Normally closed breather valve assembly

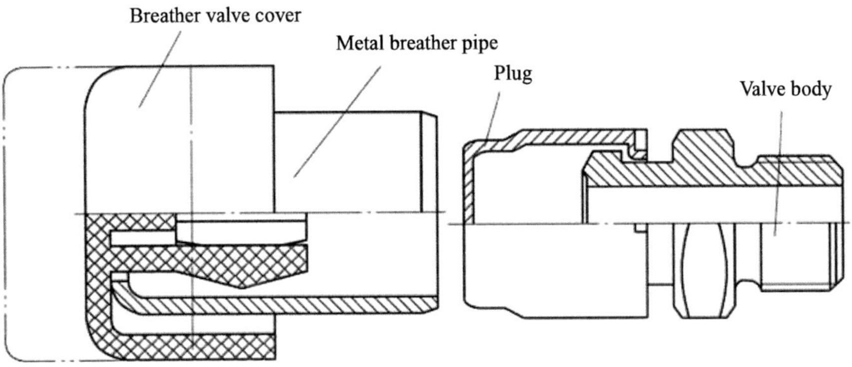

(a) Type A normally open breather valve (b) Type B normally open breather valve

Fig. 8.82 Normally open breather valve assembly

8.9 Transmission Tests

I. Purpose and type of transmission tests

The transmission is a key device of power system and its performance directly affects the vehicle power performance, economy, reliability and other performance. The transmission test is mainly to determine its performance and technical parameter and correctly reflect the performance of the transmission, such as power performance, efficiency, reliability and so on. The design performance, reliability and technological feasibility of the transmission are mainly verified in the new transmission design and development process; the quality stability and product consistency of the transmission are mainly verified in the transmission quality control process; the effectiveness of technological improvement of the transmission is mainly verified in the transmission quality improvement process; in addition, the material base database shall be established to provide the basic data for the forward design.

Depending on the test object, the transmission tests can be classified into gear test, powertrain test and related parts test. The gear tests mainly include bending fatigue strength test, contact fatigue strength test, gluing resistance test, lubrication test, efficiency test, transmission error test and vibration noise test; the powertrain tests mainly include the shift performance test, fatigue life test, contact spot test, case deformation test and transmission vibration noise test; related parts tests mainly include the related tests of the bearing, synchronizer and clutch. The typical tests are described below.

II. Typical transmission tests

1. Gear bending fatigue strength test

The main form of gear fatigue failure is bending fatigue fracture of gear teeth. Therefore, its bending fatigue characteristic is the key basic data to realize the anti-fatigue design and reliability design of the powertrain. At present, there is a lack of data on gear fatigue characteristics in China, and there are some differences between foreign gear materials and Chinese gear materials in smelting, machining, heat treatment and other aspects. The gear bending fatigue strength test is conducted on an electromagnetic resonance fatigue machine shown in Fig. 8.83, which is characterized by high loading frequency, high efficiency, accurate test load, accurate load point, evenly distributed load along the tooth width, simple and convenient clamping.

The gear bending fatigue curve can be divided into the inclined section of gear bending fatigue curve and the horizontal section of gear bending fatigue curve (gear bending endurance fatigue strength). Figure 8.84 shows the bending fatigue curve of a gear, which is the basis for checking the bending fatigue strength of the gear.

2. Gear contact fatigue strength test

The gear contact fatigue strength test is usually carried out on the back-to-back gear fatigue machine shown in Fig. 8.85. The torque loading mode is divided into hydraulic loading and torque lever loading. With the characteristics of high support stiffness, accurate loading, easy installation and energy-saving drive, the machine can verify the effect of gear design, gear material, processing technology and heat treatment parameter improvement on gear life; test the gear contact fatigue strength and verify the gear quality; test S–N curve of gear material to provide basic data for forward design.

A high-speed heavy-duty gear is generally required to have high strength and long life and designed according to the contact fatigue strength and bending strength. However, with the improvement and development of the gear material performance, heat treatment process level, processing method and surface treatment technology, the gear contact fatigue strength and bending strength have been able to meet the application requirements, while the gluing failure often occurs suddenly under the

Fig. 8.83 Electromagnetic resonance fatigue machine and fixture

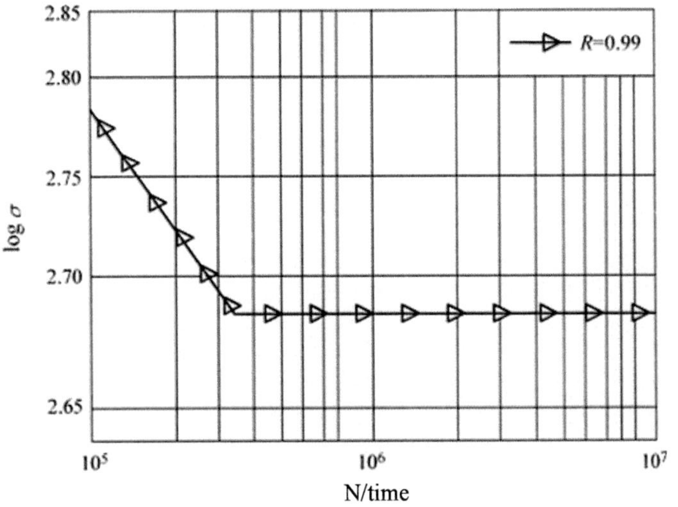

Fig. 8.84 Bending fatigue curve of a gear

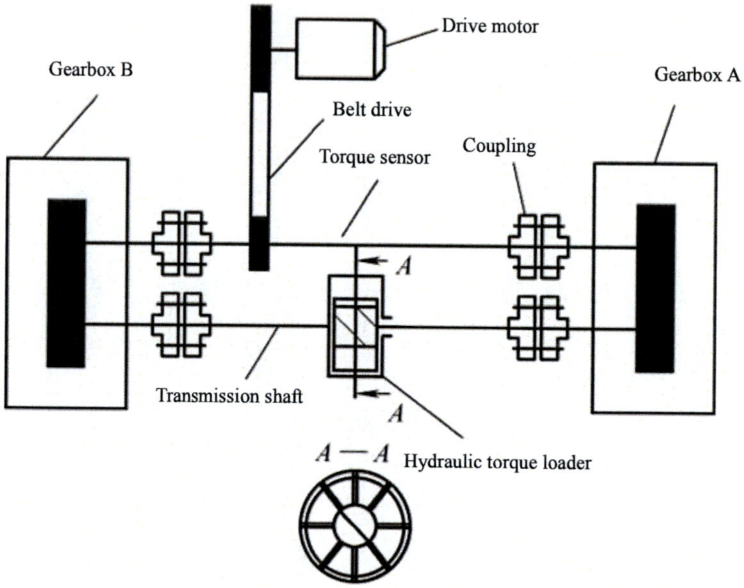

Fig. 8.85 Back-to-back gear fatigue machine

high-speed heavy-duty condition, thus limiting the bearing capacity and service life of the gear. The machine can also be used for ① gluing test to determine the gluing failure mechanism of various materials and lubricant combinations; ② selection of preventive measures against gluing failure to help designers select appropriate lubricants and additives and optimize the part geometry; ③ quantitative calculation criterion to develop and calculate the parts gluing failure criterion from the perspective of actual application.

3. **Transmission lubrication test**

Transform the transmission and set an inspection window shown in Fig. 8.86 in the bearings and key positions. Simulate the simulation of each gear of the transmission in the driving at horizontal condition, 15° uphill, 30° uphill, 15° downhill, 30° downhill, 15° left turn and 15° right turn and at the speed of 800 r/min, 1000 r/min, 1500 r/min, 2000 r/min, 3000 r/min, 4000 r/min, respectively.

4. **Transmission efficiency test**

The transmission efficiency test is mainly to test the transmission efficiency under different temperature, speed and torque. The test equipment shall include: ① a motor stand having a drive and a power absorption device; ② a drive and loading device, with the torque control accuracy of ±1%, measurement accuracy of ±0.5%, speed control accuracy of ±5 r/min and measurement accuracy of ±1 r/min; ③ a transmission fluid temperature control device, with the fluid temperature control accuracy of ±5 °C and

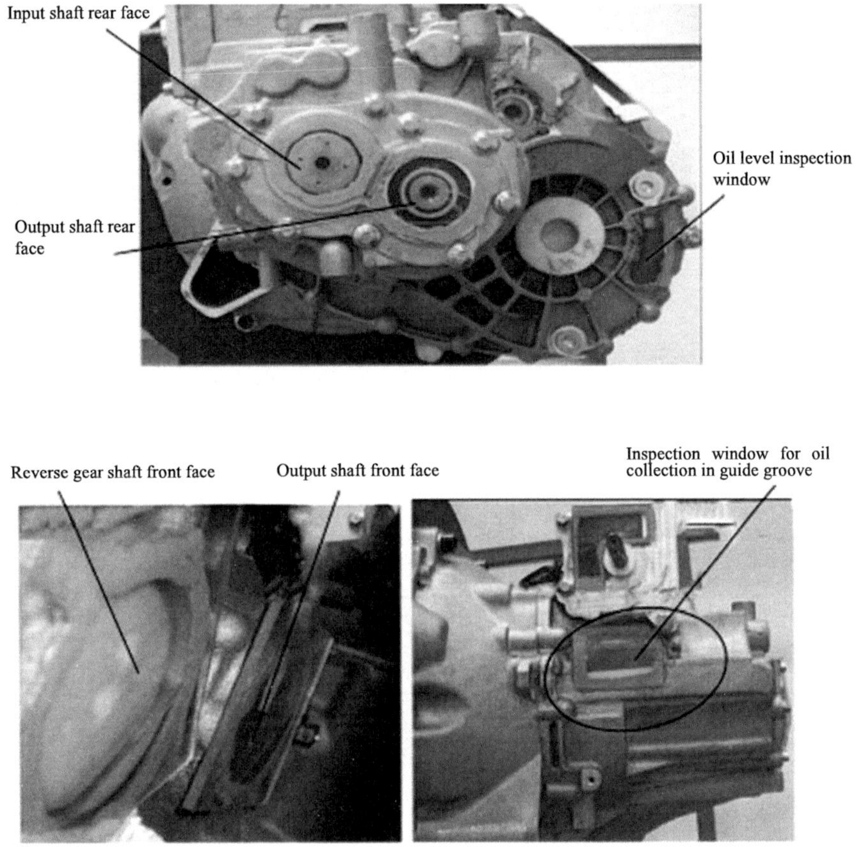

Fig. 8.86 Transmission set with an inspection window

measurement accuracy of ± 1 °C; ④ a data recording device to record the stand and oil temperature control system and transmission signals.

The efficiency test requirements are as follows: ① Install the transmission installed at the same angle as the vehicle; ② Control the oil temperature within a predetermined range; ③ Run in according to the regulations; ④ Change the test speed, input torque and test oil temperature in a certain way; ⑤ Record the transmission efficiency of each gear under different test oil temperature, test speed and input torque.

Bibliography

1. Chen Y, Yamamoto A, Omori K (2007) Improvement of contact fatigue strength of gears by tooth surface modification processing. In: Besançon: 12th IFToMM World Congress
2. Zhenrong Tao (2007) The search of calculating method of contact fatigue strength for gear. Mach Des Manuf 7:15–17
3. Wu L, Liu G, Wang B (1992) Probabilistic FEM for gear reliability calculation Part I: calculation of tooth root bending fatigue strength. Mech Drive 16(2):1–4
4. Wu L, Liu G, Wang B (1993) Probabilistic FEM for gear reliability calculation (Part II: calculation of tooth surface contact fatigue strength). Mech Drive 17(3):16–19
5. Gorokovsky VI, Bowman C, Gannon P E (2008) Deposition and characterization of hybrid filtered arc/magnetron multilayer nanocomposite cermet coatings for advanced tribological applications. Wear 265(5–6):741–755
6. Tian Y, Qu J, Qin L, et al (2011) Research status on gear surface strengthening technology. Hot Working Technol 40(24):211–215
7. Chen Z, Liu X, Liu Y, et al (2014) Ultrathin MoS_2 nanosheets with superior extreme pressure property as boundary lubricants. Sci Rep 5
8. Yoshita M, Ikeda A, Kuroda S (2004) Improve-ment of CVT pulley wear resistance by Micro-Shot Peening. JATCO Techn Rev 5:51–59
9. Wankai Shi, Hongwei Jiang, Datong Qin et al (2009) Friction and wear performance of superfine manganous phosphate conversion coating. Tribology 29(03):267–271
10. Wang CM, Liau HC, Tsai WT (2007) Effects of temperature and applied potential on the microstructure and electrochemical behavior of manganese phosphate coating. Surf Coat Technol 102(2–3):207–213
11. Huang J, Xu S, Xie S (2011) The design of automatic transmission control system of electric vehicle. J Jinggangshan Univ (Nat Sci) 32(1):100–103
12. Cheng N (2007) Design and selection manual of reducer and transmission. China Machine Press, Beijing
13. Ling L, Huang Y (2014) Topology optimization design of gearbox housing in electric bus. Appl Mech Mater 574:173–178
14. Huang Xiangshan, Xie Songjing (2013) Hydraulic-driven based the design of the transmission assemblies. Res Explor Lab 32(2):247–250
15. Ruan Zhongtang (1999) Design and selection guide of CVT. Chemical Industry Press, Beijing
16. Hu Z, Li B, Chen W (2011) Optimized design of gearbox cover of diesel engine based on modal analysis. Auto Eng (7):59–61
17. Chunxiang Luo (2004) Research into distribution of transmission ratio in design of gearbox. J Southwest Uni Natl (Nat Sci Ed) 30(3):377–380
18. Sun W, Zhang X, Su Q (2010) The Design of gearbox gelatinizing automated system. Auto Electron 26(5–2):165–172
19. Wang H, Li X, Niu S (2006) Design of gearbox in travel par wheel crane. Mech Eng (10)
20. Chen Y, Zang L, Ju D, et al (2017) Research status and development trend on strengthening technology of high strength automobile gear surface. China Surf Eng 30(1):1–15

Chapter 9
Transmission Fluid

With the development of automobile transmission technology, higher performance requirements have been put forward for the transmission fluid. In particular, with the continuous improvement of fuel economy, the transmission fluid tends to develop towards low viscosity. More and more OEMs require transmission fluid to meet lifetime requirements. At present, the automobile transmissions mainly include MT, AT, CVT, DCT and AMT. Due to different structure and working principles, the transmissions have different requirements for the transmission fluid. The MT and AMT use the same transmission fluid, so the transmission fluid mainly includes manual transmission fluid (MTF), automatic transmission fluid (ATF), continuously variable transmission fluid (CVTF) and dual clutch transmission fluid (DCTF), with the performance requirements shown in Table 9.1.

9.1 MTF

MT is the earliest, the most reliable and the most efficient transmission. With the increased requirement for vehicle lightweight and fuel economy, the transmission volume is decreased and the transmission fluid capacity is reduced, resulting in the increase of gear temperature, thus putting forward higher requirements for transmission fluid temperature. In addition, due to the extensive use of efficient synchronizers, synchronizer life, pitting resistance, abrasion resistance and other indexes also put forward higher requirements for the transmission fluid.

MTF falls into two broad categories: by SAE viscosity and by API performance. In addition, SAE and ASTM introduced MT-1 and PG-2 specifications, in which, MT-1 is a mechanical transmission fluid specification, higher than APIGL-4 vehicle gear oil in quality and improving the thermal stability, oxidation resistance, cleanliness, wear resistance, sealing material adaptability, and bronze fitting. The PG-2, higher than APIGL-5 vehicle gear oil in quality, can be used for the drive axle lubrication.

Y. Chen, *Automotive Transmissions*, Springer Tracts in Mechanical Engineering,
https://doi.org/10.1007/978-981-15-6703-2_9

Table 9.1 Performance requirements for various transmission fluids

Performance requirements		MTF	ATF	CVTF	DCTF
Synchronizer performance		O			O
Jitter resistance			O		O
Friction factor between metals				O	
Basic performance	Low-temperature fluidity	O	O	O	O
	Shear stability	O	O	O	O
	Thermo oxidative stability		O	O	O
	Clutch friction/stability		O	O	O
	Wear protection	O	O	O	O
	Foam resistance	O	O	O	O
	Material compatibility	O	O	O	O

Note ○ means that the item is required

9.2 ATF

ATF is the lubricating oil for AT, which plays a role in lubrication, cooling, sealing and vibration reduction. It is also a hydraulic fluid mainly to transfer the power. It is required to have high fluidity and fast response to the oil pressure control. In order to make ATF applicable to AT, the characteristics listed in Table 9.2 are proposed for ATF. Figure 9.1 shows the friction endurance requirements of the wet clutch.

9.3 CVTF

Since the CVT cone disk pressurization, transmission and speed regulation are implemented by high pressure oil and the CVTF plays the role of lubrication, power transmission and speed regulation simultaneously, the CVTF, in addition to high lubrication performance, must have the proper viscosity, viscosity index, shear stability, friction factor, traction coefficient, low temperature fluidity and viscosity temperature and pressure characteristics to meet the metal belt drive requirements and ensure proper slip rate. Foreign auto manufacturers and CVT manufacturers have developed the corresponding CVTF, using the form of OEM production or self-production. For example, Nissan and Honda commissioned SHELL and Rishi respectively to produce CVTF, while Toyota developed and produced CVTF by itself. CVT manufacturers own the intellectual property of CVTF.

The CVTF is used for the cone disk pressurization, transmission and speed regulation system. Through the action of hydraulic cylinder and piston on two pairs of drive and driven cone disks, the metal belt is clamped and friction is generated to transfer motion and torque. Therefore, the CVTF shall not only have good lubrication performance to reduce friction and wear, but also have appropriate adhesion and

Table 9.2 ATF characteristic requirements

Performance index	Practical application requirements	Additives used
Proper viscosity, high viscosity-temperature characteristic, shear stability and low temperature fluidity	Low temperature fluidity when used as a hydraulic fluid, and high viscosity-temperature characteristic with proper viscosity for the lubrication of the gear, shift and clutch	Viscosity-temperature characteristic improver, fluidity improver and purification dispersant
Oxidation stability	Inhibit the oxygenolysis of oil and prevent the formation of precipitation	Antioxidant
Wear resistance	Inhibit the wear of gear, shaft, hydraulic pump, etc.	Anti-wear agent
Foam resistance	Prevent forms from forming in the lubrication system	Antifoaming agent
Proper friction characteristics	Reduce shift impact and excessive sliding friction in the clutch engagement	Friction blender
Material compatibility	Prevent rubber seals from significant expansion, contraction and hardening	Sealant
Rust resistance	Prevent rust of iron parts	Antirust agent
Corrosion resistance	Make the metal surface inert to prevent corrosion	Metal inert agent and anticorrosive

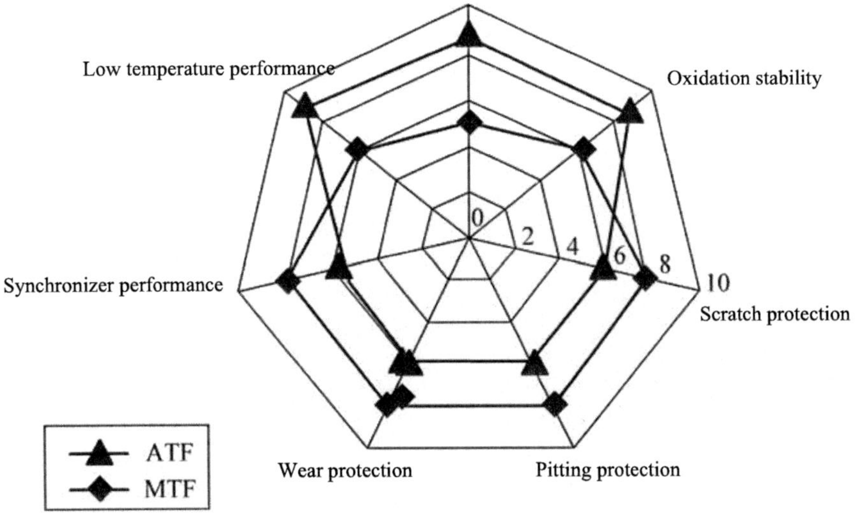

Fig. 9.1 Friction endurance requirements of wet clutch

Table 9.3 Requirements of different types of CVT for CVTF

CVT type	Metal belt CVT	Chain CVT	Ring-disk roller CVT
Powertrain	Friction drive, medium surface pressure	Friction drive, medium and high surface pressure	Traction drive, high surface pressure
Lubrication state	Boundary lubrication—mixture lubrication	Boundary lubrication—mixture lubrication	EHL (elasto-hydrodynamic lubrication)
Lubricating oil used	CVTF: base oil: mineral oil and synthetic oil	CVTF: base oil: mineral oil	Special traction oil Base oil: special synthetic oil
Performance requirements	Prevent wear and tear between metal belt and cone disk; high friction factor between metal belt and cone disk	High friction factor between metal belt and cone disk; good sintering resistance; noise suppression	Prevent the high pressure surface from falling off; high traction coefficient

internal friction to achieve accurate and variable power transmission. The traditional ATF cannot be used as CVTF. CVTF must be a combination of the base oil with excellent performance and additives, so as to achieve the optimal balance of friction factor and traction coefficient and maximize the working characteristics of CVT. The CVTF shall be used at a certain temperature and high temperature and its service life is usually required to be more than 100,000 km. It has high requirements for antioxidation stability, shear resistance, viscosity temperature and pressure characteristics, and lubrication and anti-wear performance, and strict requirements for low temperature performance and viscosity index. The lack of proper CVTF will result in deviation of the slip rate of the metal belt drive, failure of accurate transmission and speed regulation and failure to the CVT to present the best operating characteristics; meanwhile, poor lubrication will occur, resulting in abnormal wear of the metal belt, cone disk and other related parts, shortening the service life of CVT. In addition, different types of CVT have different requirements for CVTF, as shown in Table 9.3.

9.4 DCTF

DCT is divided into DDCT and WDCT with different requirements for the lubricating oil performance. The DDCT is mostly used in the low torque drive. At this time, DCT is similar to a conventional MT. MTF can meet the basic lubrication requirements only with focus on the gear and bearing lubrication instead of the lubrication of the clutch friction material. By contrast, the oil for the WDCT shall protect the gears and ensure normal working of the dual clutch. Therefore, the special lubricating oil formulation is required to combine the performance characteristics of MTF and ATF, highlight the friction characteristics and the protection of gear, bearing and synchronizer and other performance requirements and must meet the basic requirements of lubricating oil, such as heat stability, oxidation stability and rust protection. In addition, the oil change period required by different OEMs is also different, and the short oil change period is about 45000 km for the WDCT produced currently and the long period is lifetime.

It is a complex process for the lubricating oil manufacturers to complete the DCTF design. With the friction and wear balancing performance and strong bearing capacity, the lubricating oil can be able to control the corrosion, oxidation and sludge formation. It is also very important to choose the base oil (e.g. mineral oil) and synthetic oil and viscosity improver. Meanwhile, the dispersant, detergent, antioxidant and other components shall be added to reduce the formation of sediments. The content of additives may be adjusted according to different parts, friction material, viscosity and other requirements or the specific requirements of OEM. The DCTF is superior to the current transmission fluids in terms of friction characteristics, durability, environmental protection performance, wear resistance and bearing wear protection. The synthetic base oil is used completely or partially for the DCTF currently, but it is the future development trend to use class III base oil. With the extension of the oil change period, improving durability will be the main objective of designing the DCTF. The lubricating oil and additives suppliers must work closely with the OEMs to develop the next generation DCT oil specifications. Compared with ATF, DTF has more balanced performance. Figure 9.2 shows the friction endurance requirements of DCTF wet clutch.

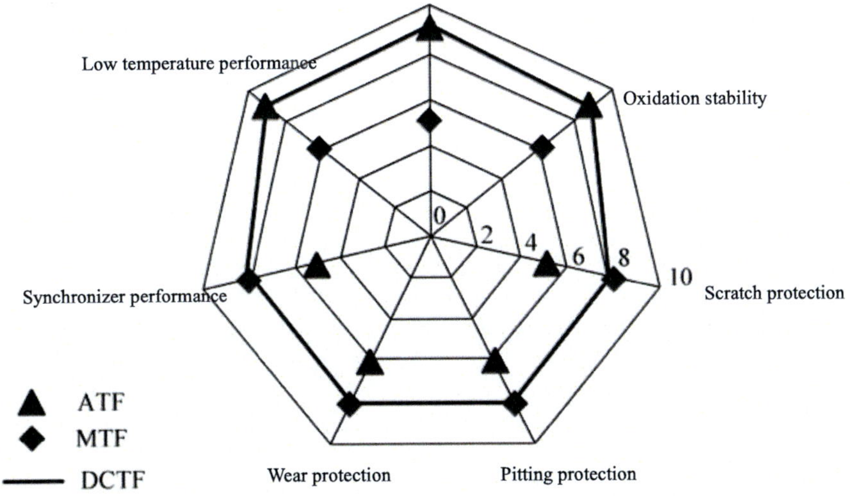

Fig. 9.2 Friction endurance requirements of DCTF wet clutch

9.5 Performance Requirements and Tests for Transmission Fluid

I. Performance requirements for transmission fluid

Different transmissions have different requirements for transmission fluid. Table 9.4 lists the performance parameters and evaluation indexes of a DCTF.

II. Tests for transmission fluid

Table 9.4 Performance parameters and evaluation indexes of a DCTF

No	Performance parameter		Evaluation index
1	Viscosity	Kinematic viscosity @100 °C/(mm^2/s)	6.1–6.5
		Kinematic viscosity @40°C/(mm^2/s)	29 (max)
		Dynamic viscosity/(mPa·s)	5000 (max)
		Viscosity index	175 (min)
2	Pour point/°C		−54 (max)
3	Component content	Calcium content ppm	100–150
		Phosphorus content ppm	210–300
		Sulfur content ppm	1200 (max)
		Moisture content ppm	500 (max)
4	Flash point (COC)/°C		190 (min)
5	Foaming performance (foam tendency/foam stability)/(mL/mL)	24 °C	≤10/0
		93.5 °C	≤50/0
		24 °C after cooling	≤10/0
6	Cleanliness		–/16/13 (max)
7	Evaporation loss (mass fraction, %)		≤6
8	TAN/(mgKOH/g)		≤1.7
9	Oxidizability DKA (mass fraction, %) (170 °C, 192H)		Viscosity change (40 °C and 100 °C) ≤ 10% TAN change 1.2 mgKOH/g (max)
10	Shear stability (20 h)/(mm^2/s)		≥5.8
11	Copper corrosion (130 °C, 3 h)		2A (min)
12	Rust protection		No rust
13	Density/(kg/m^3)		820–840

Note 1 ppm = 1 mg/kg

1. **Friction characteristic test**

The friction characteristic is the comprehensive balance performance of AT and DCT shift sensation, dynamic torque load and friction endurance. The ATF and DCTF with good performance are required to keep the friction characteristic unchanged within the whole operating temperature range. Figure 9.3 shows the friction endurance test. The dynamic friction torque, static friction torque, maximum friction torque and engagement time are measured through the meshing test of the actual friction plate, friction disk and hub, so as to evaluate the friction characteristics of ATF or DCTF. The friction characteristic of DCTF is generally evaluated according to the static friction factor and dynamic friction factor. The DCTF shall have the static friction factor and the dynamic friction factor matching the friction material. Too small dynamic friction factor will lead to the sliding increase in clutch engagement and large starting torque loss. In order to reduce the loss in torque transmission, the dynamic friction factor between the clutch plates shall be increased during the clutch engagement, but cannot be too large, so as not to increase the friction and wear between the clutches. Moreover, the static friction factor should not be too large; otherwise it will cause a fierce increase in torque and make the shift feel not smooth enough in the low speed phase of the clutch. Therefore, when adding friction modifier

Fig. 9.3 Friction endurance test

in DCTF, it is necessary to consider not only the smooth engagement of clutch and the reduction of torsional vibration, but also the torque loss during engagement, so as to balance the two.

2. **Wear Resistance Test**

Wear resistance is the basic performance requirement of the transmission for oil. In order to meet the needs of meshing gear lubrication, MTF, DCTF must have good lubrication, to prevent gear pitting and wear. Therefore, the resistance to pitting, abrasion and wear of DCTF is required to reach at least the same level of MTF. Figure 9.4 shows a pitting test.

3. **Bearing test**

The bearing test is carried out on the bearing test machine, as shown in Fig. 9.5. The rotating parts are driven by the adjustable speed motor and a certain load and speed are exerted on the bearing to check whether the lubricating oil can ensure the good condition of the bearing under such circumstance. Figure 9.6 shows the bearing damage picture.

Fig. 9.4 Pitting test

Fig. 9.5 Bearing test

4. **Synchronizer durability test**

Similar to MTF, the DCTF shall also have appropriate synchronizer durability. SSP180 synchronizer test is adopted to investigate the synchronizer durability of oil according to the friction factor and wear extent of the synchronous ring, as shown in Fig. 9.7. Currently, the synchronous ring is mainly made of brass, molybdenum, paper, sintered bronze and carbon composite material. The shift comfort and durability are different depending on the material. The carbon composite material is best, followed by paper, molybdenum and sintered bronze, and the brass is worst. Therefore, MTF and DCTF must match the material and structure of the synchronizer to achieve the durability of the synchronizer.

5. **High speed friction and shift test**

The high speed friction and shift test is carried out on the high speed friction and shift test bench shown in Fig. 9.8 to check the wear of the main parts of the DCT such as gear, friction plate, steel plate and shift actuator after the test under high speed and low torque. Figure 9.9 is the gear wear picture after the test and Fig. 9.10 is the friction plate and steel plate wear picture. As can be seen from the figures, the gear wear, friction plate wear and steel plate wear after the test are all within the acceptable range, proving that the DCTF meets the high speed friction and shift requirements.

Fig. 9.6 Bearing damage picture

In addition to the above test items, there are the following tests: DCT and AT anti-jitter durability test to prevent jitter of the clutch or brake in engagement or disengagement; shear stability test to reduce the oil change times; foam resistance test to reduce the transmission loss; aniline point test to prevent the rubber swelling; oil compatibility test to test whether the influence of the oil on the performance of the rubber parts, plastic parts and metal parts is acceptable.

9.6 Selection of Transmission Fluid

To select an appropriate transmission fluid, it is required to known the transmission type first; then make clear the operating environment, performance requirements and working conditions of the transmission fluid; carry out physical and chemical test, simulation test and bench test for the selected transmission fluid to judge whether it meets the expected performance indexes; finally carry out the real vehicle test to verify the rationality of the selected transmission fluid.

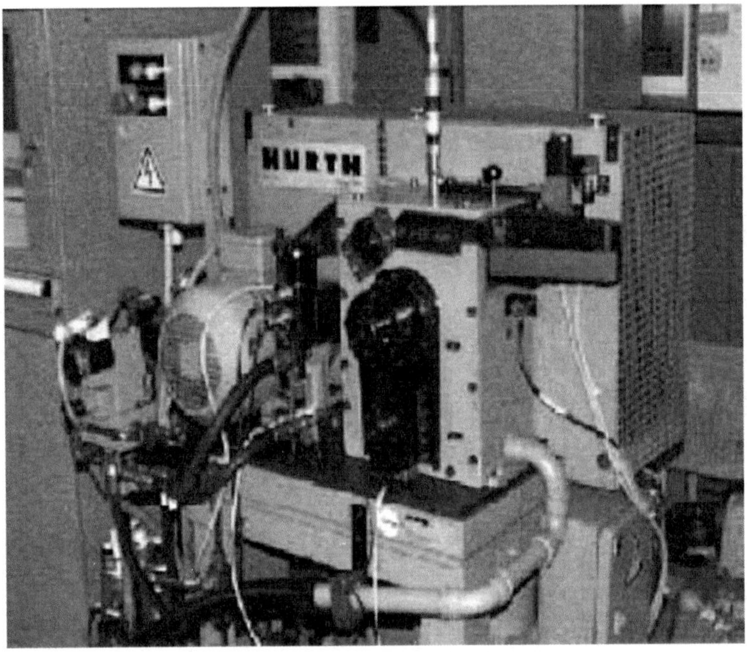

Fig. 9.7 Synchronizer test

Fig. 9.8 High speed friction and shift test

Fig. 9.9 Gear wear picture

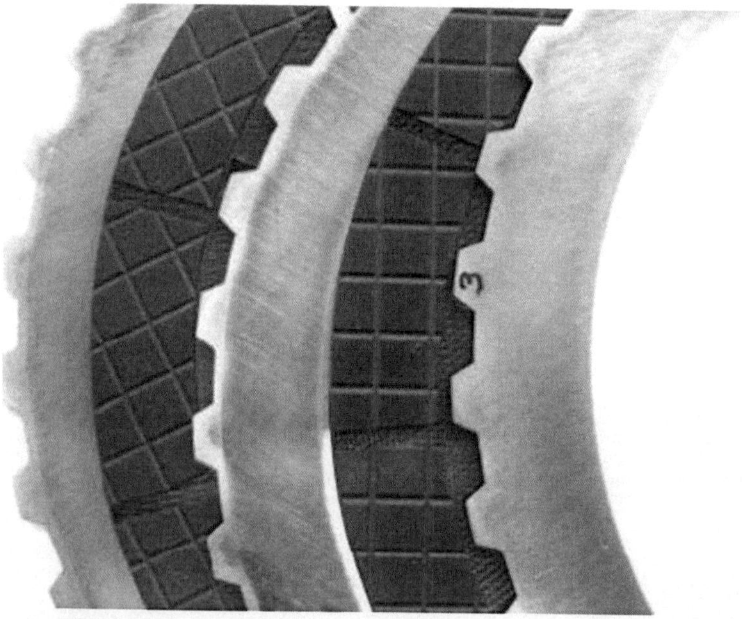

Fig. 9.10 Friction plate and steel plate wear picture

Choosing the right transmission fluid helps lubricate the transmission, reducing friction and saving energy. In general, low viscosity transmission fluid is more energy efficient than high viscosity transmission fluid and multi-speed transmission fluid is more energy efficient than single-speed transmission fluid, provided that lubrication requirements are met. With good low-temperature fluidity and high-temperature lubrication in a wide temperature range, the multi-speed transmission fluid can reach high viscosity transmission fluid level in terms of low-temperature fluidity and reach high viscosity transmission fluid level in terms of high-temperature lubrication. The transmission fluid in the mainstream models is mostly multi-speed transmission fluid.

Bibliography

1. Bigang Du, Yi Z, Huitian Z (2004) The Reasons of ATF Quality Abnormity and Prevention Measures. Lubr Eng 3:104–105
2. Xisheng Fu, Yuanqing P (2009) Development of automobile transmission and its performance requirements for transmission fluid. Pet Prod Appl Res 27(3):4–12
3. Guoru Z, Bin L, Lin S (2009) Lubricating oil for automobile powertrain—development trend of manual transmission fluid (MTF). Pet Prod Appl Res 27(4):4–9
4. Lin S, Wanying Li, Qunji X (2010) Tribological properties of lubricants for manual transmission gearbox. Tribology 30(2):179–183
5. Yaping H, Zhimin G (2000) ATF. Automob Transp 26(7):20–21
6. Yanning W (2011) Characteristics and use of automobile ATF. Silicon Valley 9:121
7. Gengbo T, Yong C, Daguo L et al (2011) Study on dual clutch transmission fluid. Lubr Eng 36(6):107–110
8. Wen W, Kufang W, Xinglin Y et al (2012) Quality requirements and test methods of dual clutch transmission fluid. Lubr Oil 1:50–55
9. Wu Lijun Xu, Mingxin ZM (2004) Analysis of ferrographical fault diagnosis methods for the transmission of TBM. Lubr Eng 5:60–62
10. Qiping Z, Yuliang S (2006) The performance and use for hydrodynamic oil and automatic transmission fluid. Chin Hydraul Pneum 2006(8):59–62
11. Kržan B, Vižintin J (2003) Tribological properties of an environmentally adopted universal tractor transmission oil based on vegetable oil. Tribol Int 36(11):827–833

Chapter 10
Design of Hydraulic Torque Converter

10.1 Working Principle and Characteristics of Hydraulic Torque Converter

I. Working principle of hydraulic torque converter

The hydraulic torque converter is located at the most significant end of the AT and is mounted on the engine flywheel, acting like a clutch with MT. It is a vane transmission with the liquid as the medium. When the hydraulic torque converter is working, the interaction between the active wheel blade and working liquid is used to realize the interconversion of mechanical energy and liquid energy (that is, the impeller continuously absorbs the power the internal combustion engine and transfers it to the turbine), and the transmitted torque is changed through the change of the liquid moment of momentum. With the ability to continuously change the speed and torque, it is a continuously variable transmission (CVT) most successfully used in the vehicles, fundamentally simplifies the manipulation; makes the vehicle start smoothly, accelerate quickly and gently with excellent vibration damping performance, thus prolonging the life of the powertrain and improving the ride comfort, average vehicle speed and driving safety; during stall, it has the maximum torque ratio, which prevents engine flame-out and more importantly, greatly improves the vehicle passability.

As shown in Fig. 10.1, the hydraulic torque converter mainly consists of the rotatable impeller 4, turbine 3 and fixed guide wheel 5. The shape of these parts is shown in Fig. 10.2. All active wheels are precisely cast adopted aluminum alloy, or made by stamping and welding with steel plates. The impeller 4 is integrated with the hydraulic torque converter housing 2 and bolted to the flange at the rear end of the engine crankshaft 1. The hydraulic torque converter housing 2 is made into two halves and welded into one after assembly (some are bolted). The starting gear ring 8 is provided outside the housing. The turbine 3 is connected with other parts of the powertrain through the driven shaft 7. The guide wheel 5 is fixed on the guide wheel fixing casing 6. After all the active wheels are assembled, a circular body with circular section is formed.

Y. Chen, *Automotive Transmissions*, Springer Tracts in Mechanical Engineering,
https://doi.org/10.1007/978-981-15-6703-2_10

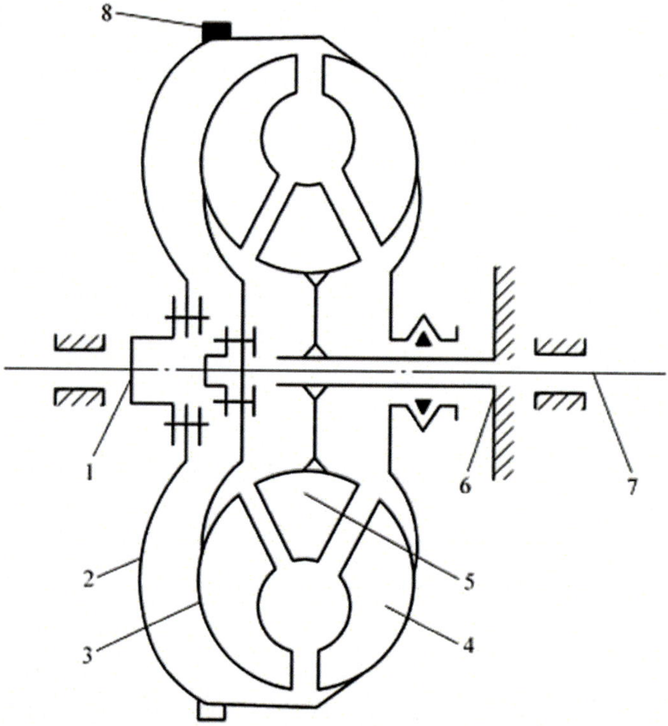

Fig. 10.1 Structure diagram of hydraulic torque converter. 1—engine crankshaft, 2—hydraulic torque converter housing, 3—turbine, 4—impeller, 5—guide wheel, 6—guide wheel fixing casing, 7—driven shaft, 8—starting gear ring

Like the hydraulic coupler, when the hydraulic converter works normally, the working fluid stored in the ring cavity not only has circular motion around the hydraulic torque converter shaft, but also has the circular flow in the circulation circle along the direction indicated by the arrow in Fig. 10.3, so that the torque can be transferred from the impeller to the turbine.

The hydraulic torque converter can transfer the torque, and change the torque value output by the turbine with the turbine speed (reflecting the driving speed of the vehicle) when the impeller torque is unchanged.

The reason why hydraulic torque converter can convert the torque is that it has an additional guide wheel mechanism compared with the coupler in structure. During the circular flow, the fixed guide wheel gives a moment of reaction to the turbine, so that the output torque of the turbine is different from the input torque of the impeller.

The working principle of the hydraulic torque converter is illustrated by the expanded view of the torque converter active wheel. As shown in Fig. 10.3, the center lines of flow paths of all circulation circles are expanded into a straight line on the same plane. In the expanded view, the impeller B, turbine W and guide wheel D form three annular planes. For the convenience of explanation, the engine speed and

1—engine crankshaft 2—hydraulic torque converter housing 3—turbine 4—impeller 5—guide wheel 6—guide wheel fixing casing 7—driven shaft 8—starting gear ring

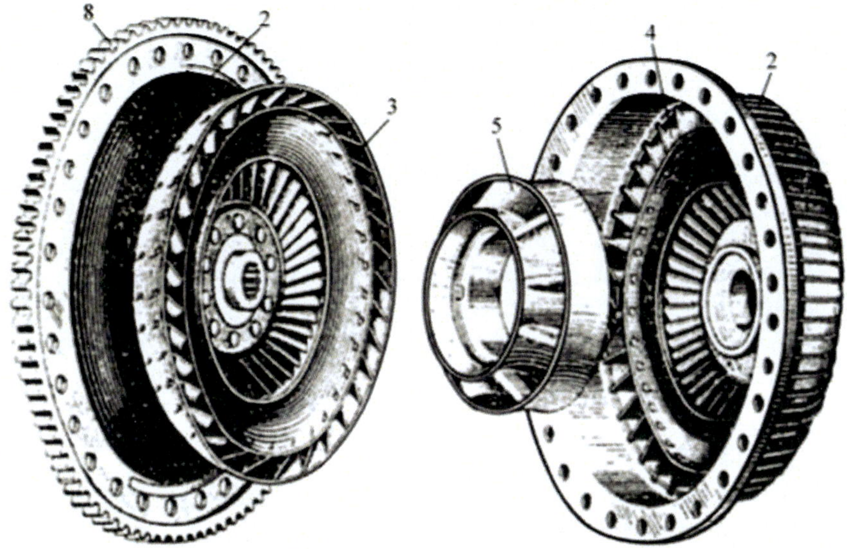

Fig. 10.2 Main parts of hydraulic torque converter (with the legends the same as Fig. 10.1)

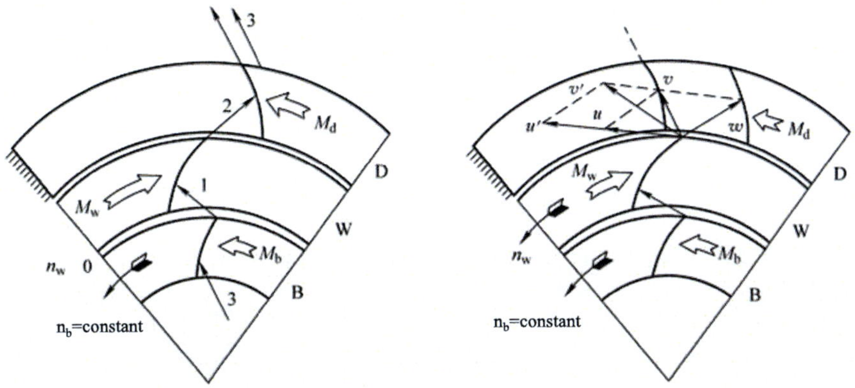

a) When n_b= constant and n_w=0

b) When n_b=constant and n_w gradually increases

Fig. 10.3 Working principle of hydraulic torque converter

load remain unchanged. That is, the speed n_b and torque M_b of the torque converter impeller are constant.

The vehicle starting condition is discussed first. The turbine speed is zero at the start, as shown in Fig. 10.3a and the working fluid rushes to the turbine blade at a certain absolute speed in the direction of arrow 1 in the figure under the action of the impeller. Since the turbine is stationary, the fluid flows out of the turbine along the blade and towards the guide wheel, as shown in arrow 2. Then the fluid flows from the fixed guide wheel blade to the impeller along the direction of arrow 3. When the fluid flows through the blade, it is forced by the blade and its direction changes. The torques of the impeller, turbine and guide wheel on the fluid are set to M_b, M'_w and M_d, respectively. According to the force balance condition of the fluid, $M'_w = M_b + M_d$.

Since the torque M_w of the fluid on the turbine is equal to M'_w in opposite direction, the turbine torque M_W is numerically equal to the sum of the impeller torque M_b and guide wheel torque M_d. Obviously, the turbine torque M_w is greater than the impeller torque M_b, that is, the hydraulic torque converter plays a role in increasing the torque.

When the driving force generated on the drive wheels by the output torque of the torque converter through the powertrain is sufficient to overcome the vehicle breakaway force, the vehicle starts and accelerates and the associated turbine speed n_w increases from zero. As shown in Fig. 10.3b, the fluid has the relative velocity ω along the blade at the outlet of the turbine and the convected velocity u along the circumference, so the absolute velocity v of the fluid flowing towards the guide wheel blade shall be the combined velocity of the two. Because the original impeller speed is constant and only the turbine speed changes, the relative velocity ω at the outlet of the turbine is unchanged and only the convected velocity u changes. The absolute velocity v of the fluid flowing towards the guide wheel blade will also gradually tilt to the left with the increase of the implicated velocity u (i.e., the increase of the turbine speed n_w), making the torque on the guide wheel gradually decrease. When the turbine speed increases to a certain value, the fluid from the turbine rushes to the guide wheel along the outlet of the guide wheel. The fluid does not change direction when flowing through the guide wheel, so the guide wheel torque M_d is zero and the turbine torque is equal to the impeller torque, i.e. $M_w = M_b$.

If the turbine speed n_w continues to increase, the absolute velocity v' of the fluid will continue to tilt to the left and the direction of the guide wheel torque is opposite to that of the impeller torque. The turbine torque is the difference of the two torques $M_w = M_b - M_d$. That is, the output torque of the torque converter is smaller than the input torque. When the turbine speed n_w increases to equal to the impeller speed n_b, the working fluid will stop flowing in the circulation circle and cannot transmit power.

II. Typical hydraulic torque converters

1. Three-element integrated hydraulic torque converter

The structure of the three-element integrated hydraulic torque converter is shown in Fig. 10.4. It is composed of impeller 8, turbine 5 and guide wheel 9, with the maximum torque ratio (that is, the torque ratio when the turbine speed is zero) of 1.9–2.5.

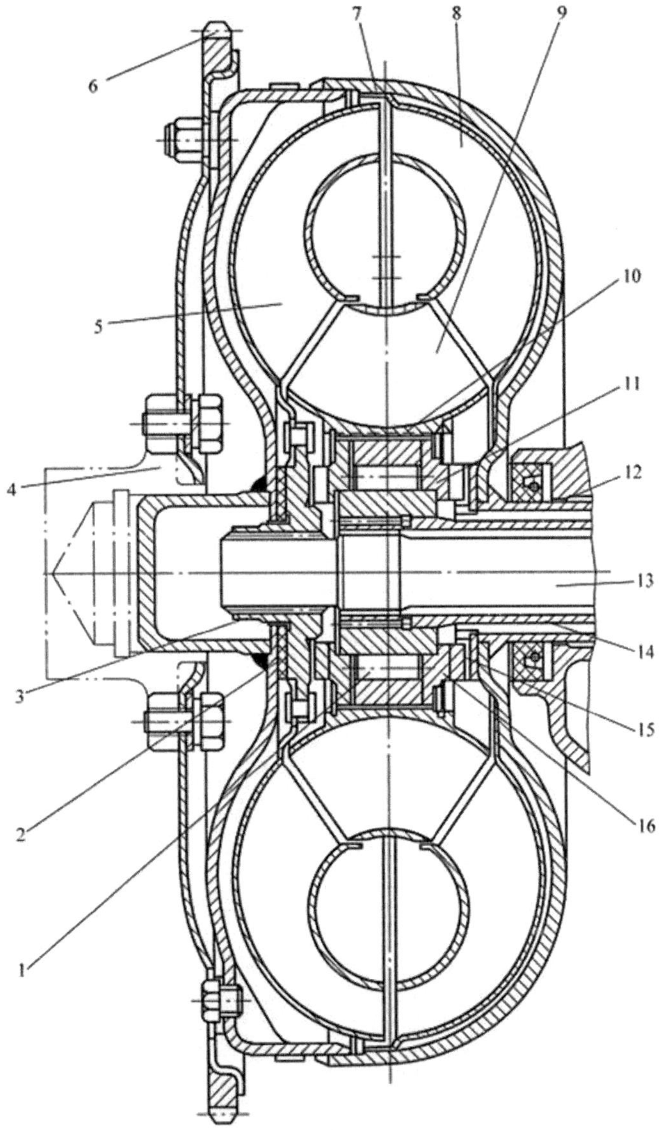

Fig. 10.4 Structure of three-element integrated hydraulic torque converter. 1—Roller, 2—pourable resin chock, 3—turbine hub, 4—flange, 5—turbine, 6—starting gear ring, 7—torque converter housing, 8—impeller, 9—guide wheel, 10—OWC outer race, 11—OWC inner race, 12—Impeller hub, 13—torque converter output shaft, 14—guide wheel fixing casing, 15—thrust washer, 16—OWC cover

The torque converter housing 7 is welded by the front and rear halves. The front end of the housing is connected to a tray fitted with a starting gear ring 6 and screwed to flange 4 at the rear end of the crankshaft. The screws are distributed unevenly on the circumference to maintain the original relative position of the torque converter and the crankshaft after repair and disassembly, so as not to destroy the dynamic balance.

The impeller 8 is equipped with radial flat blades. The impeller hub 12 welded on the impeller housing can rotate freely. The turbine 5 has sloping curved blades. The turbine hub 3 riveted with the turbine housing is splined with the output shaft 13 of the torque converter. The impeller and turbine blades and housing are steel plate stamping parts. The blade and inner ring are spot-welded and brazed to the housing. The guide wheel is cast in aluminum alloy and is fixedly connected to the OWC outer race 10.

The roller type one-way clutch of the hydraulic torque converter, as shown in Fig. 10.5, consists of the outer race 2, inner race 1, roller 5 and stainless steel laminated spring 6. The guide wheel 3 is riveted on the outer race 2 with rivet 4 (or with spline). The inner race 1 is splined with the fixing casing (14 in Fig. 10.4), so the inner race is fixed. The inner surface of the outer race 2 has several eccentric arc surfaces. The roller 5 is often pressed by the laminated spring 6 to the narrower end of the raceway between the inner and outer races, so as to wedge inner and outer races.

In case of low turbine speed and large difference from the impeller speed, the fluid from the turbine impacts on the guide wheel blade, trying to make the guide wheel 3 rotate clockwise. Since the roller 5 is wedged at the narrow end of the raceway, the guide wheel and the OWC outer race 2 are clamped and fixed together on the inner race 1. At this point, the hydraulic torque converter plays the role of increasing the torque. When the turbine speed rises to a certain degree, the impact

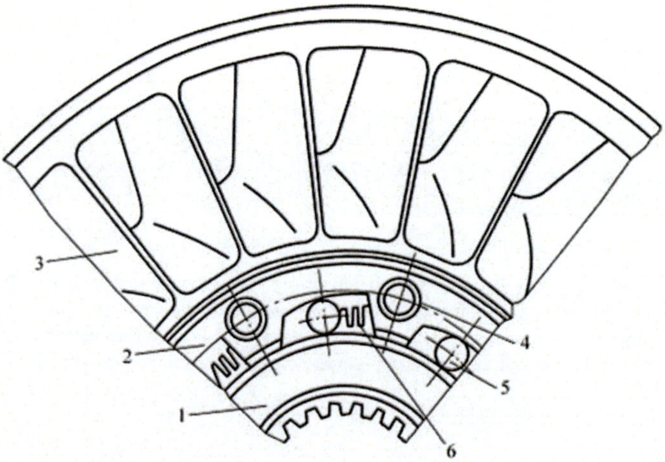

Fig. 10.5 Roller type one-way clutch of hydraulic torque converter. 1—inner race, 2—outer race, 3—guide wheel, 4—rivet, 5—roller, 6—laminated spring

force of the fluid on the guide wheel is reversed, so the guide wheel rotates freely in the same direction with respect to the inner race and the turbine. At this point, the converter switches to the coupler condition. Such torque converter which can switch to the coupler condition is called the integrated hydraulic torque converter. It is used since the coupler has higher efficiency than the torque converter in high gear ratio. Efficiency refers to the ratio of the output power to input power of the fluid drive. The law of the change of torque converter efficiency η_b and coupler efficiency η_0 with the gear ratio i is shown in Fig. 10.6, which also shows the curve of the change of the torque ratio K with the gear ratio i. It can be seen from the figure that, in the range of the gear ratio $i < i_K = 1$ (gear ratio when torque ratio $K = 1$), the torque converter efficiency is higher than the coupler efficiency; when $i > i_K = 1$, the torque converter efficiency η_b drops quickly, while the coupler efficiency η_0 continues to increase. The integrated hydraulic torque converter works according to the hydraulic torque characteristics at low speed. When the gear ratio reaches i_K, it works according to the coupler torque characteristics, thus increasing the range of high frequency.

The active wheels of the torque converter work in a closed chamber filled with hydrodynamic drive oil, which is not only the working medium but also the lubricant and coolant of the hydraulic elements. To prevent cavitation, a certain compensation pressure shall be maintained in the chamber, whose value varies according to the torque converter, usually within 0.25–0.7 MPa. Cavitation is the formation of bubbles in a working fluid when the pressure drops below the saturated vapor pressure of the fluid at that temperature. When the bubbles in the fluid move to the region with high pressure along with the flow, the bubbles quickly break under the impact of the surrounding oil, and then condense into liquid, resulting in a vacuum due to the sudden reduction of the volume. The surrounding fluid particles then fill up the space at an extremely high rate. At this moment, the fluid particles collide strongly

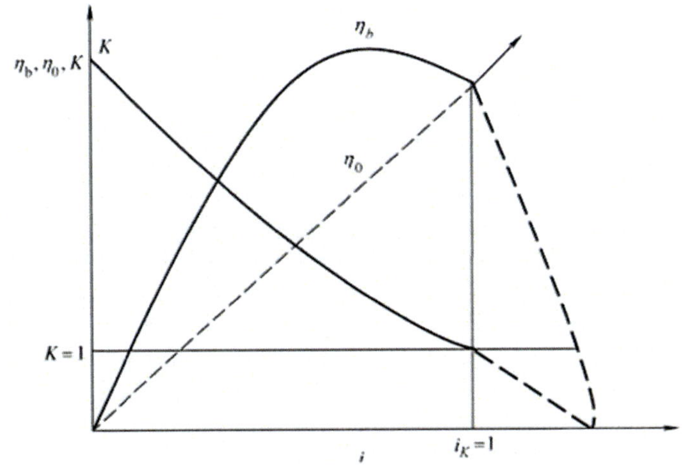

Fig. 10.6 Characteristics of three-element integrated hydraulic torque converter

with each other, producing obvious noise. Meanwhile, high local pressure is caused, causing the metal particles on the blade surface to be broken. It can be seen that cavitation will affect the normal operation of the torque converter and reduce its efficiency accompanied by noise. Therefore, adequate compensation pressure must be maintained in the working chamber.

The compensation oil with a certain pressure output from the hydraulic pump enters from the gap between the guide wheel and the impeller through the annular cavity between the guide wheel fixing casing 14 (Fig. 10.4) and the impeller hub 12, flows out from between the turbine and the guide wheel and leads to the cooler through the annular cavity between the fixing casing 14 and the torque converter output shaft 13 to cool the working fluid.

Because of the compensation pressure, the axial force on the active wheel is larger. For this purpose, a non-ferrous metal thrust washer 15 is installed at the end of the guide wheel, and a wear-resistant pourable resin chock 2 is installed between the turbine hub and the housing.

The three-element integrated hydraulic torque converter is of simple structure, reliable operation and stable performance, with the maximum efficiency up to 92%. When it switches to a coupler, the efficiency of the high gear ratio zone can reach 96%. Therefore, it has been widely used in limousines, and gradually used in large passenger cars, dump trucks and engineering vehicles.

2. **Four-element integrated hydraulic torque converter**

If some torque converters with large starting torque ratio use above three-element integrated hydraulic torque converter, the efficiency decreases significantly in the section between the maximum efficiency condition and the starting point of the coupler condition. To avoid this disadvantage, the guide wheel can be divided into two and mounted on separate OWC to form a four-element integrated hydraulic torque converter.

Figure 10.7 is the schematic diagram of four-element integrated hydraulic torque converter. When the turbine speed is low, the fluid flow at the outlet of the turbine impacts on the concave surfaces of the two guide wheel blades in the direction shown in v_1 in Fig. 10.7b. At this point, the OWCs of the two guide wheels are locked, the guide wheels are fixed and work under the torque converter condition. When the turbine speed increases to a certain extent and the fluid speed is v_2, the impact force of the fluid flow on the first guide wheel is reversed, and the first guide wheel rotates in the same direction with the turbine due to the release of the OWC. At this point, only the second guide wheel still plays the torque conversion role. When the turbine speed continues to approach the impeller speed, i.e. the fluid speed v_3, the second guide wheel is also subject to the reverse impact force of the fluid and rotates in the same direction as the turbine and the first guide wheel. Then the torque converter fully switches to the coupler condition.

The characteristics of a four-element integrated hydraulic torque converter are a combination of the characteristics of two torque converters and a coupler (Fig. 10.8). In the section with the gear ratio 0–i_1, two guide wheels are fixed, and the blades of

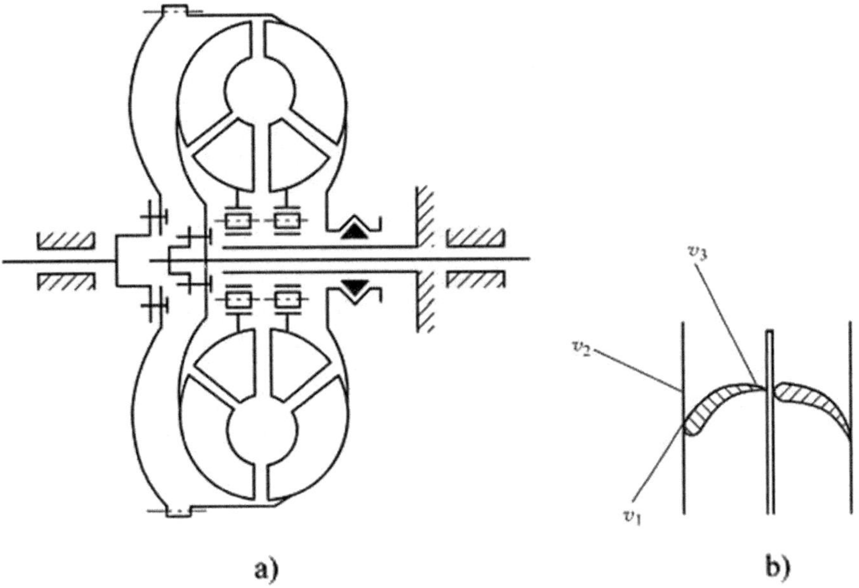

Fig. 10.7 Schematic diagram of four-element integrated hydraulic torque converter

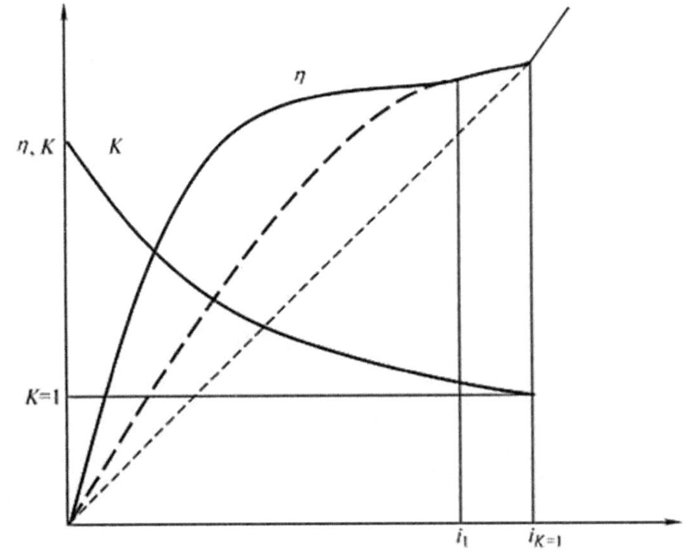

Fig. 10.8 Characteristics of four-element integrated hydraulic torque converter

the two guide wheels form a blade with a greater bending degree to ensure that a large torque ratio can be obtained under the condition of low gear ratio. In the section with the gear ratio i_1–$i_{K=1}$, the first guide wheel is released and the guide wheel with small blade bending degree in the torque converter works. At this time, a high efficiency can be obtained. At the gear ratio $i_{K=1}$, the torque converter is transferred to the coupler condition and its efficiency is improved linearly.

3. **Hydraulic torque converter with lockup clutch**

Because of the speed difference and hydraulic loss between the turbine and the impeller of the torque converter, the efficiency of the torque converter is not as high as that of the mechanical transmission and the fuel economy of the vehicle using the torque converter in normal driving is poor. The hydraulic torque converter with lockup clutch (Fig. 10.9) can be used to improve the efficiency of the torque converter at high gear ratio. The active part of the lockup clutch is composed of force transfer plate and hydraulic cylinder piston (i.e. pressure plate), which rotate together with the impeller; the driven part is the clutch driven plate mounted on the turbine hub spline. After entering via the oil channel, the hydraulic oil pushes the piston to the right and presses the driven plate and the lockup clutch is engaged. Then the impeller and the turbine are connected and rotate as a whole and the torque converter does not work. When the oil pressure is removed, the two are separated and the torque converter returns to normal operation.

When the vehicle starts or drives on a bad road, the lockup clutch can be released to make the torque converter work, in order to give full play to the advantage that the hydrodynamic drive can automatically adapt to the dramatic change in the driving resistance. When the vehicle is driving on a good road, the lockup clutch shall be engaged so that the input shaft of the torque converter is rigidly connected with the output shaft, i.e. direct mechanical drive. At this time, torque ratio $K = 1$ and torque converter efficiency $\eta = 1$, so as to improve the driving speed and fuel economy of the vehicle.

When the lockup clutch is engaged, the OWC is disengaged and the guide wheel rotates freely in the fluid flow. If the OWC is canceled, when the impeller and turbine rotate as a whole, the guide wheel will still be fixed, resulting in hydraulic loss increase and efficiency reduction.

III. Characteristics of hydraulic torque converter

The torque ratio K, efficiency η and capacity factor C_P corresponding to the speed ratio i are often used to represent the characteristics of the hydraulic torque converter, as shown in Fig. 10.10.

Figure 10.10 can accurately represent the basic performance of a series of hydraulic torque converters with different speeds, sizes and similar mechanical properties. The torque ratio curve $K = f(i)$ reflects the increase of the turbine torque T_T relative to the impeller torque T_P at different speed ratios and also reflects the adaptivity of the torque converter to the external load change. The K value under two conditions is used to evaluate the adaptability: stall torque ratio K_0 when $i = 0$; and speed ratio i_m when torque ratio $K = 1$ (speed ratio in coupler condition).

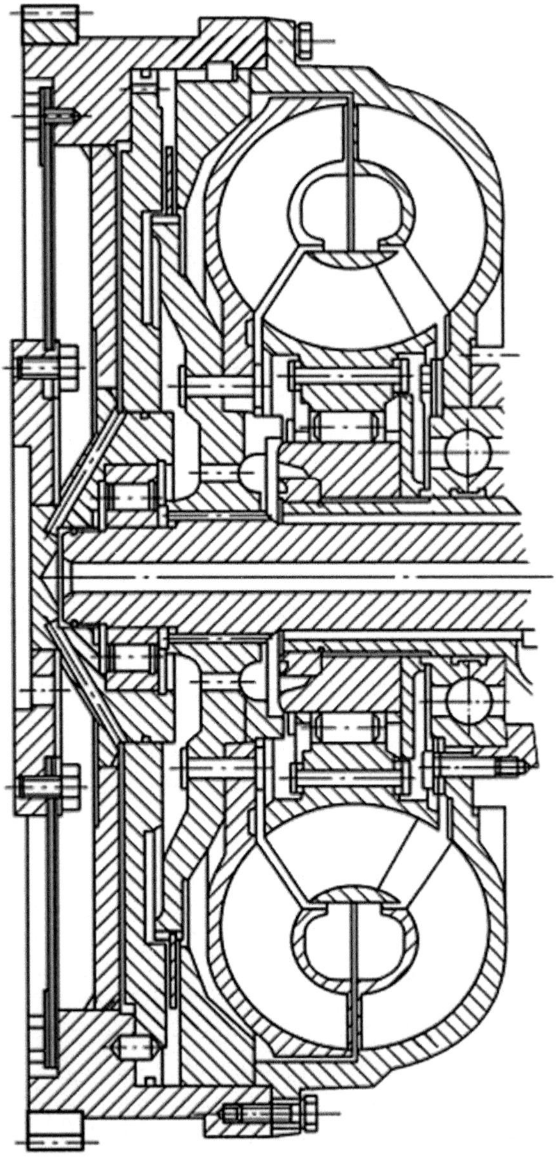

Fig. 10.9 Hydraulic torque converter with lockup clutch

$\eta = f(i)$ can represent the loss in power transmission and the concepts relating to the economy. It is the most important performance that decides the structure development and design theory improvement. There are two evaluation indexes: maximum efficiency $\eta_{\max}$, $\eta_{\max} = \eta^*$ for the single-phase torque converter and $\eta_{\max} = \eta_{\mathrm{maxm}}$ in the coupler condition for the integrated torque converter; and width of high-efficiency range, namely, speed ratio range $d_P = i_{P2}/i_{P1}$ when the efficiency is

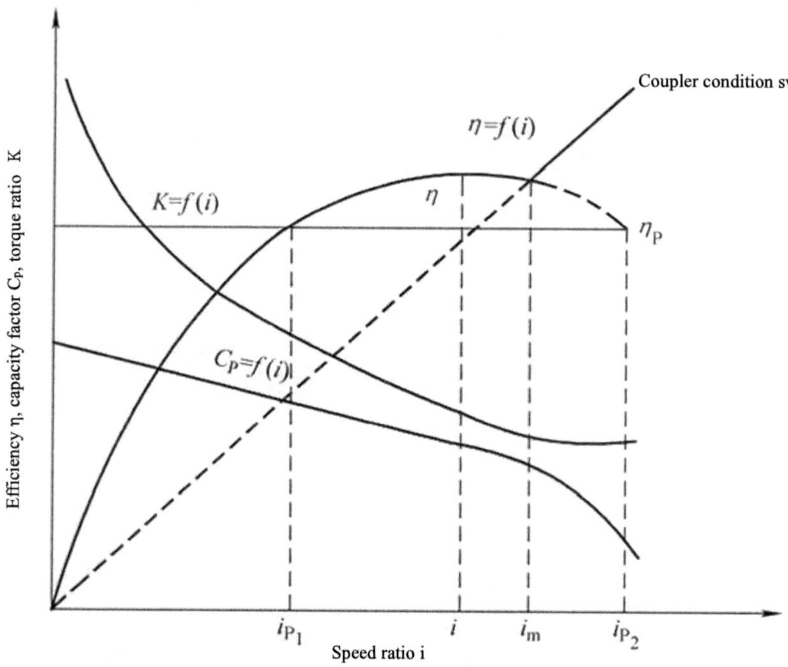

Fig. 10.10 Characteristics of hydraulic torque converter

not lower than 80%. The two indexes are often contradictory. The cars often work under high speed ratio, the greater $\eta_{\max}$, the better the economy, so the requirement for $\eta_{\max}$ is more important than d_P; off-road vehicles are on the contrary. In the quantification, the probability distribution of the torque converter conditions shall be studied first before an accurate conclusion is drawn. In recent years, the torque converter can only work in the starting, acceleration, or bad road and is often in slip condition or locking conditions (mechanical drive) in other cases, suggesting higher pursuit of $\eta_{\max}$.

$C_P = f(i)$ curve represents the transmitted torque at the corresponding input speed and reflects its load performance. The large C_P value indicates that the torque converter size and mass can be small when the same power is transmitted. The permeability property T can be expressed as

$$T = C_{PO}/C_{PM} \tag{10.1}$$

where,

C_{PO}, C_{PM}—capacity factor at zero speed and coupler operating point.

T reflects the influence of torque and speed changes of the turbine shaft on the torque and speed changes of the impeller shaft, namely the permeability.

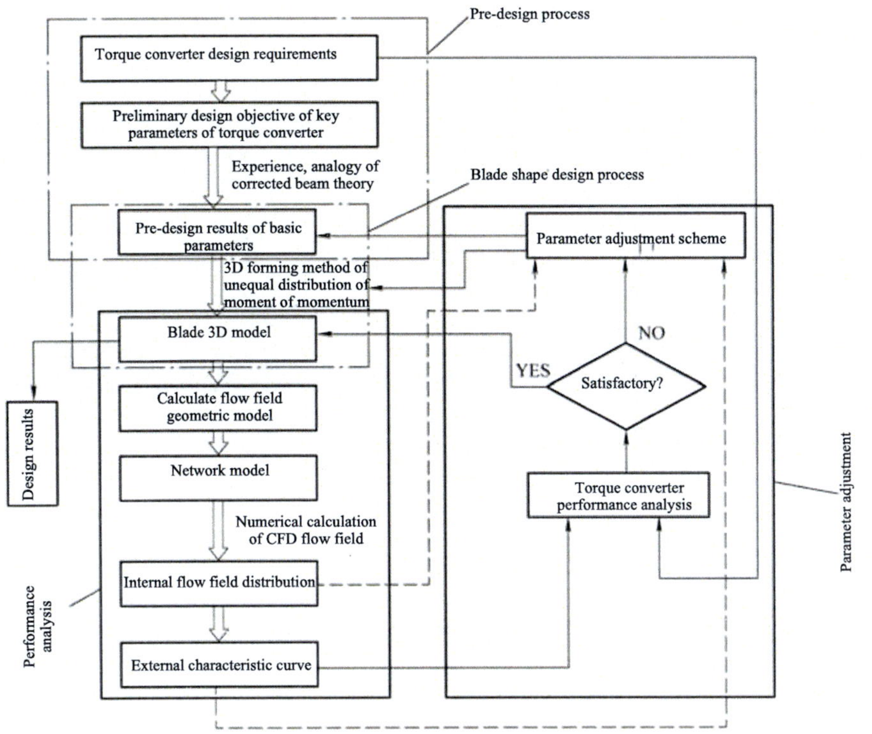

Fig. 10.11 Design flow of modern design method for torque converter

IV. **Design flow of hydraulic torque converter**

With the development of the CAD and CFD technology, a large number of trial production and test steps in the traditional design method can be realized by computer simulation, thus greatly saving the design time and cost. Meanwhile, the advanced blade design method makes the torque converter design and manufacture more accurate. Based on this, a modern design method for the torque converter is proposed, and its design flow is shown in Fig. 10.11, mainly including four steps: pre-design; blade shape design; numerical simulation and analysis of internal flow field of hydraulic torque converter, i.e. performance analysis; parameter adjustment.

The modern design theory of hydraulic torque converter is described based on the four steps below.

10.2 Pre-design of Hydraulic Torque Converter

The main purpose of the pre-design is to preliminarily determine the size of the circulation circle and the basic parameters such as the inlet and outlet angles of each blade so as to create conditions for the three-dimensional blade forming design. The pre-design process follows the experience, analogy and other methods of traditional design methods, and the parameters determined in the pre-design only serve as the initial conditions of blade forming design. In the subsequent design, these parameters will be adjusted according to the predicted performance. Therefore, the quality of theoretical pre-design will not affect the final design result, but accurate pre-design will greatly simplify the parameter adjustment. For this reason, the beam theory of energy head loss correction is proposed to improve the accuracy of the preliminary prediction of the torque converter performance.

I. Circulation circle

Figure 10.12 shows the circulation circle and blade angle of a three-element two-phase single-stage hydraulic torque converter. The shape of the circulation circle has a great influence on the structure size of the torque converter and the manufacture of active wheels. In order to meet the needs of the car FF layout scheme, the axial size of the circulation circle is strictly controlled, so it has mostly changed from round to flat. The circulation circle also develops from an arc to three arcs. An excellent prototype, if available for reference in the design, may be zoomed in or out according to the similarity principle to adapt to the optimum matching requirements of new models. If new design is required, the required capacity factor $C_P = T_P/n_P^2$ according to the engine characteristics, in which, T_P and n_P are the torque and speed of the impeller respectively. Then the approximate value of effective diameter D is obtained

$$D = \sqrt[5]{C_P/k_\lambda} \tag{10.2}$$

where,

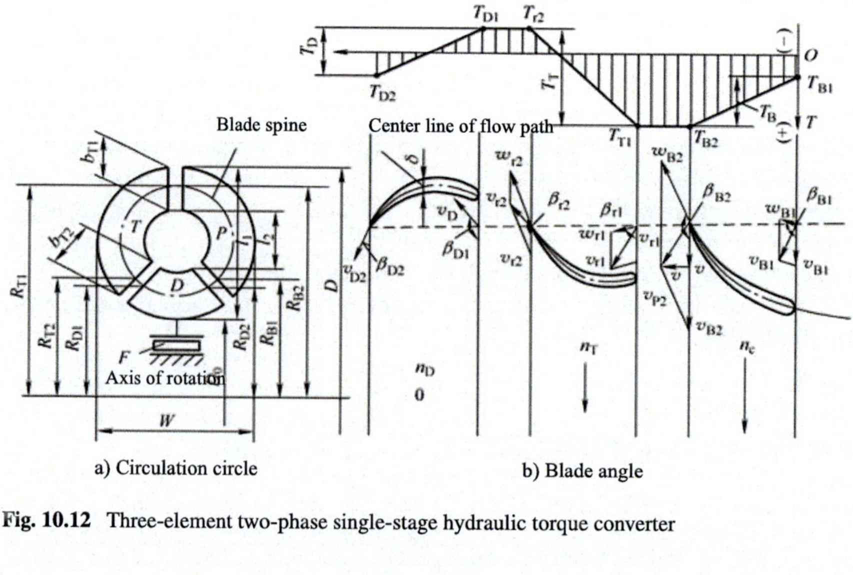

a) Circulation circle b) Blade angle

Fig. 10.12 Three-element two-phase single-stage hydraulic torque converter

k_λ—coefficient related to the torque converter shape, blade shape, liquid density and speed ratio;
D—effective diameter (m), which is an important dimension of torque converter and has important influence on the capacity factor, circulation circle and rubber size (mm).

The passage shape shall ensure that the passage area S of the cross section along the flow line is constant. If the impeller outlet radius R_{P2} is a characteristic radius, the relative area $s_n = S_n/R_{P2}^2$, selected in the range of 0.70–0.90; in addition, considering the installation requirements of the OWC, the relative characteristic radius $r_{D2} = R_{D2}/R_{P2}$, $r_{T2} = R_{T2}/R_{P2}$, with the value in the range of 0.5–0.65.

II. **Parameters of hydraulic torque converter**

For a high-performance hydraulic torque converter, the energy loss due to the interaction between the blade and fluid flow shall be kept to a minimum.

1. **No impact loss**

No impact loss shall be realized in the calculation condition. According to the unitary beam theory $\Sigma H_y = 0$, the inlet and outlet fluid flow angles of the active wheels at the non-impact inlet can be obtained. β_{D2} has narrow optimum range and is easy to select or determine according to formula (10.4), so the following formula is a function of the guide wheel outlet angle

$$\begin{aligned}
\cot\beta_{P_1} &= \left(-\frac{r_{P_1}}{q} + \frac{\cot\beta_{D_2}}{sD_2}\right)s_{P_1} \\
\cot\beta_{P_2} &= \left[\frac{1}{q^*}\left(\frac{k_\lambda}{q^*\rho} - 1\right) + \frac{r_{D_2}\cot\beta_{D_2}}{s_{D_2}}\right]s_{P_2} \\
\cot\beta_{T_1} &= \left[\frac{1}{q^*}\left(\frac{k_\lambda}{q^*\rho} - i^*\right) + \frac{r_{D_2}\cot\beta_{D_2}}{s_{D_2}}\right]s_{T_1} \\
\cot\beta_{T_2} &= \left\{\frac{1}{q^*}\left[\frac{k_\lambda}{r_{T_2}q^*\rho}(1 - K_y^*) - r_{T_2}i^*\right] + \frac{r_{D_2}\cot\beta_{D_2}}{r_{T_2}s_{D_2}}\right\}s_{T_2} \\
\cot\beta_{D_1} &= \left[\frac{k_\lambda}{r_{T_2}q^{*2}\rho}(1 - K_y^*) + \frac{r_{D_2}\cot\beta_{D_2}}{r_{T_2}s_{D_2}}\right]s_{D_1}
\end{aligned} \tag{10.3}$$

$r_{n1} = R_{n1}/R_{P2}$, $r_{n2} = R_{n2}/R_{P2}$, $s_{n1} = S_{n1}/R_{P2,}^2$ $S_{n2} = S_{n2}/R_{P2}^2$, $q = Q/R_{P2}^3 W_P$. The subscript n stands for impeller, P for turbine, D for guide wheel; q for relative flow and i^* for gear ratio. This dimensionless expression is convenient for study.

2. **Minimum ventilation loss**

Whenever there is fluid flow, there is ventilation loss. The optimum efficiency can only be achieved by controlling the ventilation loss to a minimum. The ventilation loss mainly depends on the blade angle, flow rate and flow area. For given i^* and k_λ*, to achieve the maximum η_y*, i.e. maximum torque ratio k_y*, the extreme value conditions are

$$\frac{\partial K_y^*}{\partial \cot \beta_{D_2}} = 0, \frac{\partial K_y^*}{\partial q^*} = 0, \frac{\partial K_y^*}{\partial s} = 0$$

Through simultaneous solution of the equation

$$\cot \beta_{D_2} = \frac{sr_{D_2}}{q^*}\sqrt{\frac{\xi_{mP}}{\xi_{mp} + \xi_{mD}}} = \frac{1}{m^*}\sqrt{\frac{\xi_{mp}}{\xi_{mp} + \xi_{mD}}} \tag{10.4}$$

where,

ξ_{mP}—turbine ventilation loss coefficient;
ξ_{mD}—guide wheel ventilation loss coefficient.

The test shows that when $m^* = q^*/s > 0.35$, the ventilation loss increases; when $m^* < 0.17$, vortexes will appear in the blade, easily causing additional losses. Therefore, the optimal range is $0.17 < m^* < 0.35$.

III. **Beam theory of energy head loss correction**

1. **Basic beam theory**

In the beam theory, when the hydraulic torque converter is constant, the action torque and energy head of each active wheel are required. The key is to obtain the variation of the circular flow under different working conditions of the hydraulic torque converter, which can be obtained according to the energy balance equation of fluid flow

$$H_P + H_T - \sum h = 0 \tag{10.5}$$

where,

H_P—theoretic energy head of impeller (m);
H_T—actual energy head of turbine (m);
$\sum h$—Total energy head loss (m).

From the Euler equation

$$H_P = \frac{U_{P2}V_{P2\theta} - U_{P1}V_{P1\theta}}{g}$$
$$H_T = \frac{U_{T2}V_{T2\theta} - U_{T1}V_{T1\theta}}{g} \tag{10.6}$$

The subscript θ indicates tangential.

In the beam theory, the energy head loss is divided into two basic losses to be calculated. The first kind of loss is related to the relative velocity of the fluid flow and is proportional to the square of the flow, which is called friction loss. The second kind of loss is related to the impact angle of the fluid flow and is proportional to the square of the loss velocity, which is called impact loss.

The calculation formula of friction loss h_{mc} is

$$h_{mc} = \lambda \frac{L}{4R_y} \frac{W^2}{2g} \tag{10.7}$$

where,

L—passage length (m);
R_y—hydraulic radius, ratio of flow area to wetted perimeter (m);
λ—friction factor;
W—relative speed (m/s).

The calculation formula of impact loss h_{cj} is

$$h_{cj} = \phi_y \frac{(\Delta V)^2}{2g} \tag{10.8}$$

where,

φ_y—impact loss coefficient;
Δv—impact loss velocity (m/s).

The energy head and energy head loss are substituted into the energy balance Eq. (10.5) to obtain the circular flow Q. Thus, the active wheel torque can be obtained according to the following formulas

$$\begin{aligned} T_P &= \rho Q(V_{P2\theta} r_{P2} - V_{P1\theta} r_{P1}) \\ T_T &= \rho Q(V_{T2\theta} r_{T2} - V_{T1\theta} r_{T1}) \\ T_S &= \rho Q(V_{S2\theta} r_{S2} - V_{S1\theta} r_{S1}) \end{aligned} \tag{10.9}$$

2. **Energy head loss correction**

(1) Friction loss: the shape of the circulation circle of the hydraulic torque converter has a certain effect on the flow friction loss of the torque converter. However, this factor cannot be considered since the friction loss formula in the straight pipe is adopted for the friction loss calculation in the traditional beam theory. Of course, the flow inside the torque converter is extremely complex three-dimensional flow, and it is difficult to grasp these losses correctly. In this paper, the friction loss equation of curved pipe with high reliability is used to simply reflect the change of friction loss caused by the change of passage curvature.

The impeller passages are approximate to curved lines, taking impeller for example, as shown in Fig. 10.13. The friction loss can be expressed as

Fig. 10.13 Curved pipe model parameters

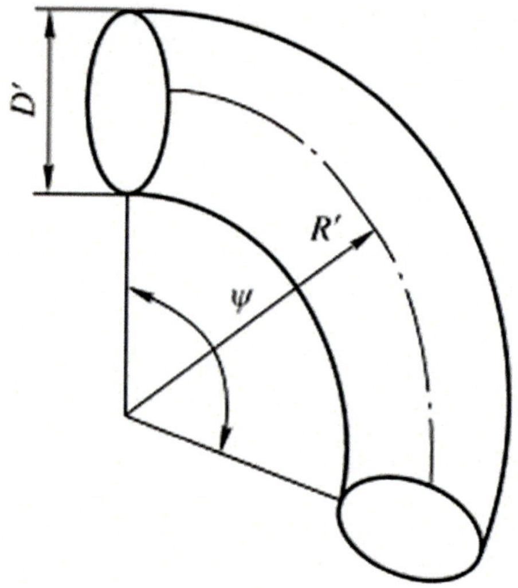

$$h_{Pmc} = \varsigma_P \frac{W^2}{2g} \tag{10.10}$$

where, the friction loss coefficient ζ_P is determined by the Reynolds number R_e, curved pipe curvature radius R', curved pipe central angle ψ and curved pipe diameter D'

$$\varsigma_P = A\psi \mathrm{Re}^{-0.2}\left(\frac{R'}{D'}\right)^{0.9} \tag{10.11}$$

where,

A—centrifugal passage correction factor.

The passage friction loss of the turbine and guide wheel is calculated in the same method as the impeller. When the impeller passages are approximate to curved lines, the curved line diameter may be determined by following the principle of equal flow area, and the curved line curvature radius shall be selected to ensure that the shape of the curved line is similar to the circulation circle.

(2) Impact loss at the impeller inlet: the large influence of the flow separation of the suction surface at the guide wheel outlet on the performance of the torque converter can be reflected by the impact loss at the impeller inlet, so the impact loss at the impeller inlet has been corrected in this paper. It is the key to calculate the impact loss velocity in the impact loss formula (10.8). Figure 10.14 is the calculation diagram of impact loss velocity at the impeller inlet. From the figure

$$(\Delta V)^2 = \left(\left(V_{P1m}^* - V_{PIm}\right)^2 \left(V_{P1m}^* - V_{P1\theta}\right)^2\right) \tag{10.12}$$

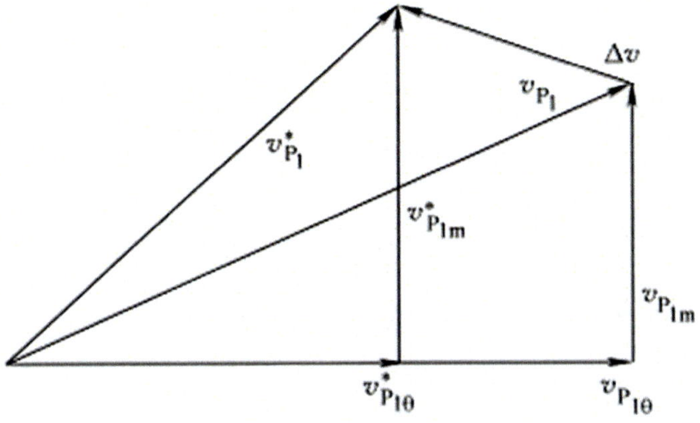

Fig. 10.14 Calculation diagram of impact loss velocity

where,

V^*_{P1m}, $V^*_{P1\theta}$—meridional velocity and tangential velocity of the fluid flow at the impeller inlet without impact.

In the traditional beam theory, the first item on the right side of the equal sign in the above equation—the difference of meridional velocity is ignored when calculating the impact loss velocity. In fact, a change in the circular flow will cause a big change in the item. To calculate the impact loss velocity, two component velocities without impact shall be calculated first. Figure 10.15 shows the velocity triangle at the impeller inlet and guide wheel outlet in case of no impact at the impeller inlet.

The condition for the entry of the fluid at the impeller inlet without impact is

$$\beta^*_{P1} = \beta_{P1b} \tag{10.13}$$

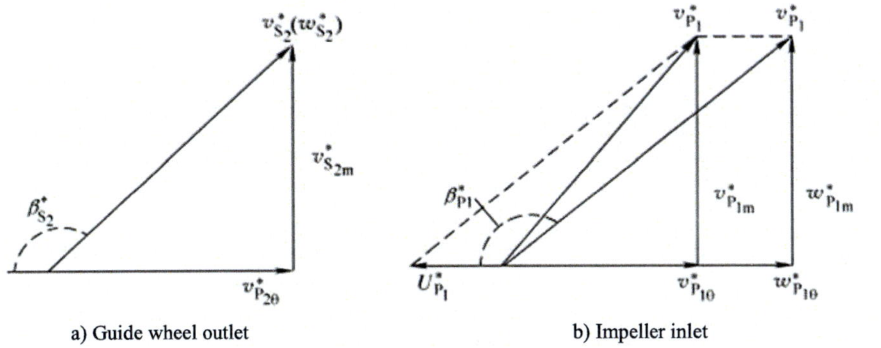

Fig. 10.15 Velocity triangle without impact

The fluid at the guide wheel outlet is considered to follow the blade angle, then

$$\beta_{s2}^* = \beta_{S2b} \tag{10.14}$$

According to the invariance principle of moment of momentum in the bladeless region

$$V_{S2\theta}^* r_{S2} = V_{P1\theta}^* r_{P1} \tag{10.15}$$

The following equation can be obtained from the velocity triangle

$$\frac{W_{P1m}^*}{W_{P1m}^*} = -\tan(\beta_{p1}^*) \tag{10.16}$$

$$\frac{V_{S2m}^*}{V_{S2m}^*} = -\tan(\beta_{S2}^*) \tag{10.17}$$

$$W_{P1m}^* = V_{P1m}^* = \frac{Q^*}{A_{P1}} \tag{10.18}$$

$$V_{S2m}^* = \frac{Q^*}{A_{S2}} \tag{10.19}$$

where,

A_{P1}, A_{S2}—flow area at the impeller inlet and guide wheel outlet.

The subscript θ indicates tangential and m indicates the projection onto the meridian plane.
Through simultaneous Eqs. (10.13)–(10.19), the component velocity of the fluid entering the impeller without impact is

$$V_{P1m}^* = \frac{\omega_P r_{P1} \tan(\beta_{P1m})}{\frac{A_{P1}}{A_{S2}} \frac{r_{P1}}{r_{S2}} \frac{\tan(\beta_{P1b})}{\tan(\beta_{S2b})} - 1} \tag{10.20}$$

$$V_{P1\theta}^* = \frac{\omega_P r_{P1}}{\frac{A_{S2}}{A_{P1}} \frac{r_{P1}}{r_{S2}} \frac{\tan(\beta_{S2b})}{\tan(\beta_{P1b})} - 1} \tag{10.21}$$

The formulas (10.20), (10.21) and (10.12) are substituted into the impact loss formula (10.8) to obtain the correction formula of the impact loss at the impeller inlet.

(3) Performance prediction comparison: for the W305 prototype torque converter, the traditional beam theory and the energy head loss correction beam theory are used for calculation respectively and compared with the test data. Figure 10.16 shows the performance prediction comparison.
Figure 10.16 shows that the energy head loss correction beam theory greatly improves the performance prediction accuracy of the efficiency and torque ratio,

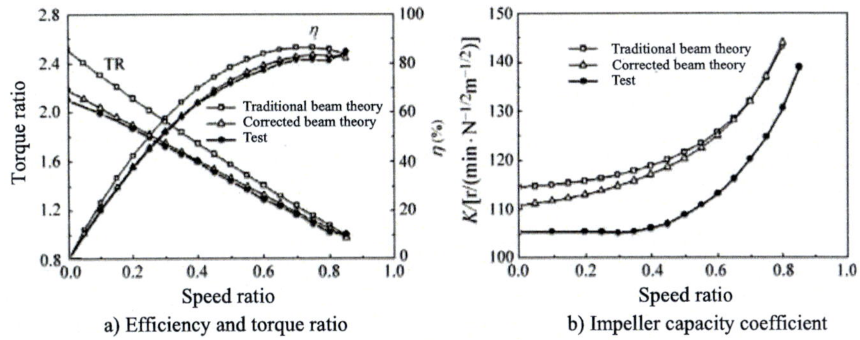

Fig. 10.16 Performance prediction comparison

the maximum relative error of the efficiency reduces from 9.35 to 2.44% and of the torque ratio reduces from 19.55 to 3.09%; with respect to the impeller capacity coefficient, the energy head loss correction beam theory does not change obviously.

Accurate one-dimensional performance prediction makes the pre-design process faster and more efficient. Although the numerical calculation of 3D fluid field can obtain enough accurate prediction performance, the one-dimensional method is still widely used in practical design because of its simple parameters and fast calculation. In practical applications, the one-dimensional method is mainly used to determine basic blade shape parameters such as inlet and outlet angles in the pre-design process. On this basis, more powerful tools and methods such as 3D flow field analysis and 3D blade forming are needed to obtain accurate and better design results.

10.3 Blade Shape Design

The blade shape design of hydraulic torque converter is carried out after the basic parameters such as circulation circle size and inlet and outlet angles of each blade are obtained. In the modern design method, these basic parameters are preliminarily determined by the pre-design process and further adjusted during the torque converter design process. The torque converter blade angle determined by the model change is shown in Table 10.1.

Table 10.1 Torque converter blade angle

Impeller blade		Turbine blade		Guide wheel blade	
Inlet angle	Outlet angle	Inlet angle	Outlet angle	Inlet angle	Outlet angle
133°	90°	46°	152°	93°	20°

The blade shape is a key factor that affects the performance of torque converter. Traditional design methods rely more on experience, and the design of blade shape by hand drawing is complicated and of low precision. With the development of modern processing technology, blade forming can be carried out directly according to the 3D model, thus accurately ensuring the spatial shape of the blade. The blade shape design method in this chapter may directly generate the 3D model of the blade finally.

I. Unequal distribution of moment of momentum

(I) Equal distribution of moment of momentum

The equal distribution of moment of momentum is the traditional circulation distribution method, whose theoretical basis is beam theory. It is considered that under the selected design speed ratio, for every increase of the same arc length in the center line of flow path on the circulation circle plane, the same moment of momentum shall be increased along the center line of flow path of the blade, so as to ensure a good flow condition in the passage. For each component of the torque converter, the torque formula can be written as

$$T = \rho \cdot Q \cdot (V_{2\theta} r_2 - V_{1\theta} r_1) \tag{10.22}$$

where,

Q—circular flow (L/s);
v_θ—tangential component of the absolute velocity (m/s), with the formula as

$$v_\theta = U + W_m \cot \beta \tag{10.23}$$

As shown in Fig. 10.17, equal distribution of moment of momentum is equal distribution of $(v_{2\theta} r_2 - v_{1\theta} r_1)$ along the center line of flow path to ensure that the increment of moment of momentum at each point is equal, so as to determine the spatial coordinates of each point. The specific steps are as follows:

(1) The center line of flow path is equally divided on the circulation circle plane and the element lines perpendicular to the center line of flow path are plotted through the equipartition points. The more equipartition points, the more accurate description of the blade shape.
(2) Calculate in the change in the moment of momentum at the inlet and outlet and calculate the corresponding moment of momentum at each equipartition point according to the incremental equality principle.
(3) Calculate the corresponding angle β at each equipartition point according to the formula (10.23).
(4) Determine the angles of the corresponding points on the inner and outer rings. According to the inverse potential flow theory, the velocity distribution law of the fluid flow on the cross section of passage can be described as

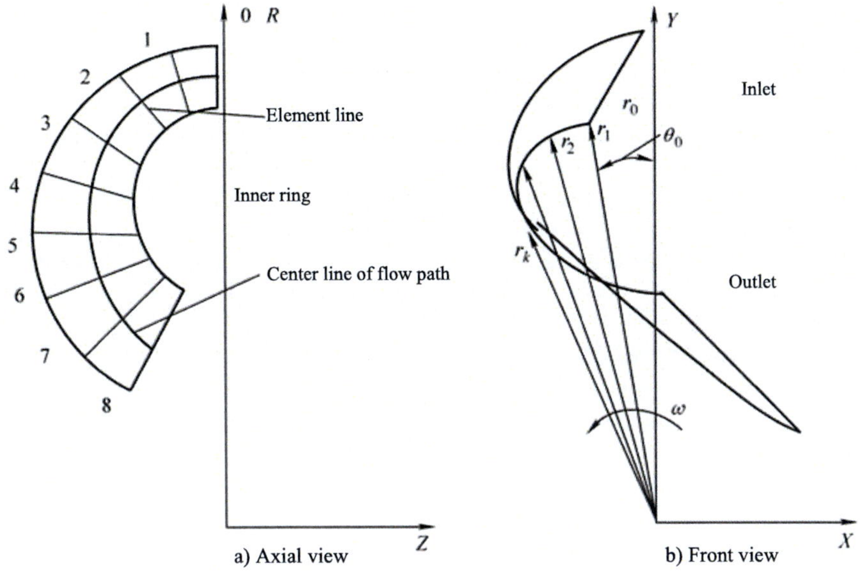

Fig. 10.17 Schematic diagram of equal distribution of moment of momentum

$$\frac{\cot\beta_C}{r_c} = \frac{\cot\beta}{r} = \frac{\cot\beta_s}{r_s} \tag{10.24}$$

where, the subscripts c and s represent the inner and outer rings respectively. The corresponding angle at each point on the inner and outer rings can be calculated according to the formula (10.24).

(5) Determine the blade shape according to the inner and outer ring radius and offset. The calculation formula of the offset x_k on any blade element line is as follows

$$x_k = r_k \sin\left(\theta_0 + \sum_{i=0}^{k} \frac{e\cot\beta_i}{r_i}\right) \tag{10.25}$$

where,

e—arc length between two adjacent points on the designed flow line;
θ_0—included angle between the axial plane of the starting point of the element line and the radial reference plane;
r_k—radius at which the element line intersects the designed flow line.

The subscript k is the serial number of the element line. Figure 10.17 illustrates this space geometry relation by taking the inner ring of the turbine as an example.

(6) Blade thickening. The basic principle of thickening is to make the blade surface smooth and make the flow area change gently.

(II) Unequal distribution of moment of momentum

The distribution law of moment of momentum actually determines the blade shape. Therefore, although the basic parameters such as the inlet and outlet angles of the blade are the same, different blade shapes will be generated by adopting different distribution methods, which will lead to changes in the overall performance of the torque converter. Based on this idea, the unequal distribution method proposed in this paper studies the distribution law of moment of momentum along the string direction. In order to study the influence of the distribution law of moment of momentum of each impeller on the torque converter performance, in this paper, the distribution law of moment of momentum of the impeller, turbine and guide wheel is changed respectively with the basic type of the torque converter designed by the equal distribution of moment of momentum, provided that other parameters are unchanged, so as to study its influence on the overall performance of the torque converter.

1. **Impeller**

The following dimensionless normalized moment of momentum $(v_\theta r)^*$ is defined for the impeller

$$(V_\theta r)^* = \frac{V_\theta r - V_{1\theta} r_1}{V_{2\theta} r_2 - V_{1\theta} r_1} \tag{10.26}$$

In order to analyze the influence of different distribution laws of moment of momentum on the torque converter performance, three special cases are selected in this paper for comparison. Figure 10.18a shows the comparison of three distribution laws of moment of momentum of the impeller. Among them, the impeller 2 is the traditional equal distribution scheme of moment of momentum; the moment of momentum of the first half of the impeller 1 changes greatly, indicating that this part of blade bears a large torque, so it is defined as the front loading scheme; similarly, the impeller 3 is defined as the rear loading scheme. To ensure the consistency of blade inlet and outlet angles, the variation law near the inlet and outlet is kept consistent when the distribution law of moment of momentum is changed (section AB and section CD in Fig. 10.18a). The same treatment method is adopted for the study of turbine and guide wheel subsequently.

The comparison of the torque converter efficiency, impeller capacity coefficient and torque ratio of three impellers is shown in Fig. 10.18b, c and d. As can be seen from the figures, impeller 3 has the highest maximum efficiency for the corresponding torque converter, impeller 1 is not far from impeller 3, and impeller 2 has the lowest maximum efficiency for the corresponding torque converter; the corresponding torque converter of impeller 1 has the maximum impeller capacity coefficient and starting torque ratio and of impeller 2 has the minimum impeller capacity coefficient and starting torque ratio.

It can be seen that the equal distribution of moment of momentum for the impeller is unfavorable to the efficiency and torque ratio performance of the torque converter, but it can obtain a small impeller capacity coefficient, which means that it can

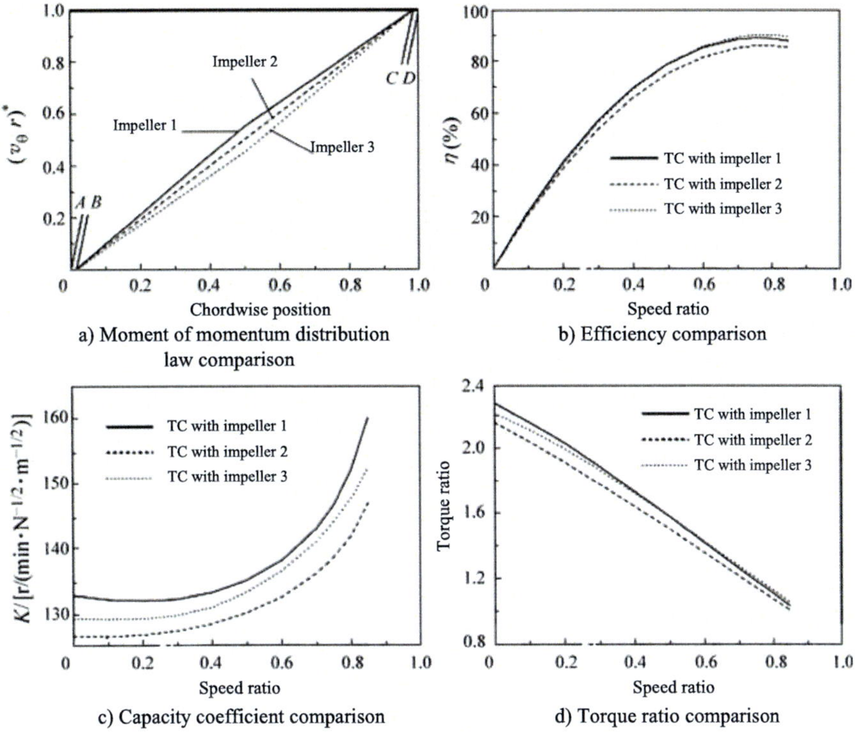

Fig. 10.18 Study on the distribution law of moment of momentum of impeller

match a motor with greater torque; the front loading scheme can be used to achieve higher maximum efficiency, starting torque ratio and large impeller capacity coefficient; while the rear loading scheme can be used to achieve maximum efficiency performance.

2. **Turbine**

The following dimensionless normalized moment of momentum $(v_\theta r)^*$ is defined for the turbine

$$(V_\theta r)^* = \frac{V_\theta r - V_{2\theta} r_2}{V_{1\theta} r_1 - V_{2\theta} r_2} \tag{10.27}$$

Figure 10.19a shows the comparison of distribution law of moment of momentum of three turbines. Turbine 2 is the traditional equal distribution scheme of moment of momentum; the moment of momentum of the first half of the turbine 3 changes greatly, which is defined as the front loading scheme; the turbine 1 is defined as the rear loading scheme.

Figure 10.19b, c and d show the comparison of the torque converter efficiency, capacity coefficient and torque ratio of three turbines respectively. As can be seen

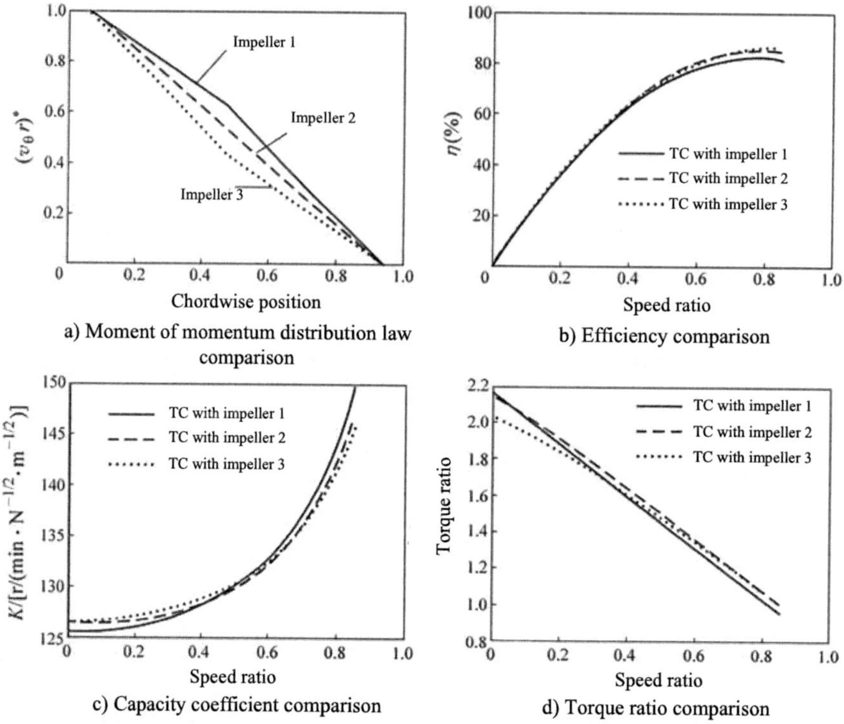

Fig. 10.19 Study on the distribution law of moment of momentum of turbine

from the figures, turbine 3 has the highest maximum efficiency for the corresponding torque converter, turbine 2 is not far from turbine 3 and turbine 1 has the lowest maximum efficiency for the corresponding torque converter; turbine 3 has the largest impeller capacity coefficient in the starting condition for the corresponding torque converter, turbine 2 is not far from turbine 3 and turbine 1 has the smallest impeller capacity coefficient, but after the high speed ratio is reached, the impeller capacity coefficient of the corresponding torque converter of turbine 1 is greater than that of turbine 2 and turbine 3; the corresponding torque converter of turbine 1 has the largest torque ratio in the starting condition and of turbine 3 has the smallest torque ratio.

It can be seen that the front loading scheme of turbine can be used to achieve the maximum efficiency performance, but the lowest starting torque ratio; the equal distribution scheme of moment of momentum can be used to achieve high efficiency performance and starting torque ratio; the rear loading scheme may be used to reduce the impeller capacity coefficient in the starting condition, but the efficiency performance also decreases obviously.

3. Guide wheel

The following dimensionless normalized moment of momentum $(v_\theta r)^*$ is defined for the guide wheel:

$$(V_\theta r)^* = \frac{V_\theta r - V_{1\theta} r_1}{V_{2\theta} r_2 - V_{1\theta} r_1} \tag{10.28}$$

Figure 10.20a is the shows the comparison of distribution law of moment of momentum of three guide wheels. Guide wheel 2 is the traditional equal distribution scheme of moment of momentum; the moment of momentum of the first half of the guide wheel 1 changes greatly, which is defined as the front loading scheme; the guide wheel 3 is defined as the rear loading scheme. Figure 10.19b, c and d show the comparison of the torque converter efficiency, impeller capacity coefficient and torque ratio of three guide wheels respectively. As can be seen from the figures, guide wheel 1 has the highest maximum efficiency for the corresponding torque converter, guide wheel 3 is not far from guide wheel 1 and guide wheel 2 has the lowest maximum efficiency for the corresponding torque converter; guide wheel 1 has the largest impeller capacity coefficient in the starting condition for the corresponding

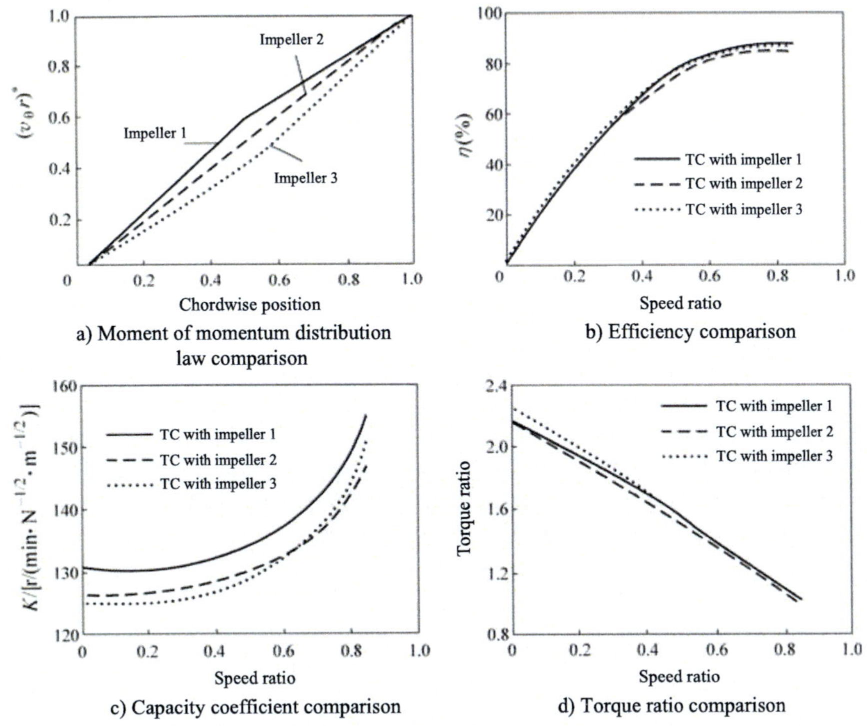

Fig. 10.20 Study on the distribution law of moment of momentum of guide wheel

torque converter, and guide wheel 3 has the smallest impeller capacity coefficient; the corresponding torque converter of guide wheel 3 has the largest starting torque ratio and of guide wheel 2 has the smallest starting torque ratio.

The front loading scheme can be used for the guide wheel to achieve maximum efficiency performance and largest impeller capacity coefficient; when the equal distribution scheme of moment of momentum is used, the efficiency performance and torque ratio are low; the rear loading scheme can be used to achieve high efficiency performance, largest starting torque ratio and low impeller capacity coefficient in starting condition.

4. Conclusion

The above comparative analysis shows that different distribution laws of moment of momentum have great influence on the torque converter performance. According to the three special distribution laws studied, the following conclusions can be drawn:

(1) To improve the efficiency performance, the rear loading scheme can be used for the impeller, front loading scheme for the turbine and guide wheel.
(2) To reduce the impeller capacity coefficient, the equal distribution scheme of moment of momentum can be used for the impeller, rear loading scheme for the turbine and guide wheel.
(3) To increase the starting torque ratio, the front loading scheme can be used for the impeller, rear loading scheme for the turbine and guide wheel.

It should be pointed out that, because the performance of the hydraulic torque converter is affected by the interaction between various components, it is not enough to compare different schemes of a single impeller to obtain the optimal torque converter performance in actual design and a variety of combining schemes must be comprehensively studied. It's a lot of work. With the further development of theoretical and experimental research, the research on unequal distribution of moment of momentum will be more perfect. The conclusions obtained in this section have important guiding significance for this blade shape design method.

Compared with the following 3D forming method, the advantage of the unequal distribution of moment of momentum is that the parameter change law adopted in the design is of some physical meaning, while the 3D forming method is only based on the pure geometric structure relation. However, in practical application, unequal distribution of moment of momentum is also faced with the problem of unreasonable distortion of blade when the blade shape is generated directly, while 3D forming method is more intuitive and reliable.

II. 3D forming method

(I) Geometrical principle

The blades of the hydraulic torque converter active wheels are generally spatially twisted, so in the blade shape design, the blade parameters such as real length, thickness and angle shall be correctly described first. The 3D forming method proposed in this paper mainly realizes the definition of blade shape through two transformations:

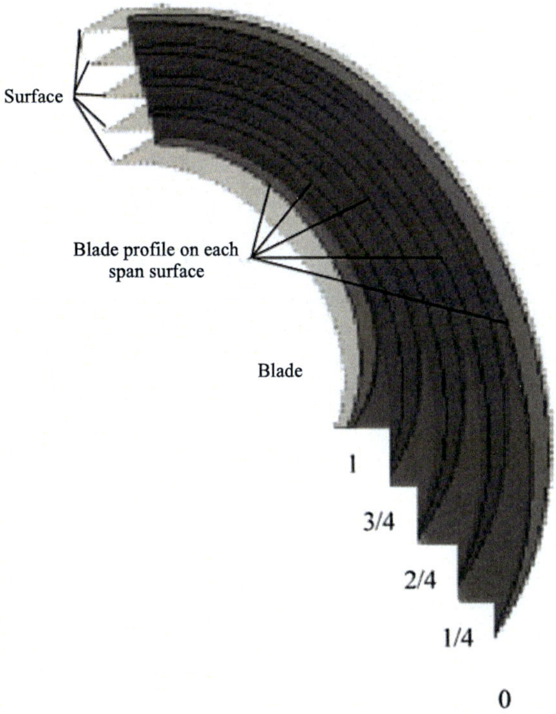

Fig. 10.21 Description of blade shape space

(1) The blade is segmented by using parallel span surfaces, so that the description of spatial blade shape can be realized by describing the blade section shape on each span surface, thus transforming the spatial curved surface into spatial curves. For example, in Fig. 10.21, the blade is segmented by five span surfaces (0, 1/4, 2/4, 3/4, 1), and the blade shape can be determined by the blade profile on each span surface.

(2) By using the method of isoclinic projection on multiple cylinders, the blade profile on the span surface is transformed into planar blade shape, thus transforming the spatial curve into a plane curve. The evolute obtained by isoclinic projection can keep the same length and inclination angle with the original curve Fig. 10.22 shows the isoclinic projection of the blade spine on the span surface 0. The isoclinic projection is also used in the traditional blade design method, but it mainly considers the inlet and outlet angles. The middle change process is manually plotted relying largely on the experience and the design process is tedious and of low precision. By defining the blade angle and thickness at each point on the blade spine, the three-dimensional coordinates of the blade surface points can be directly generated in the 3D forming method according to the space geometry relation, and then the 3D model of the blade is generated, not only ensuring the accuracy of the spatial blade shape and making parameterized definition and study on the blade shape.

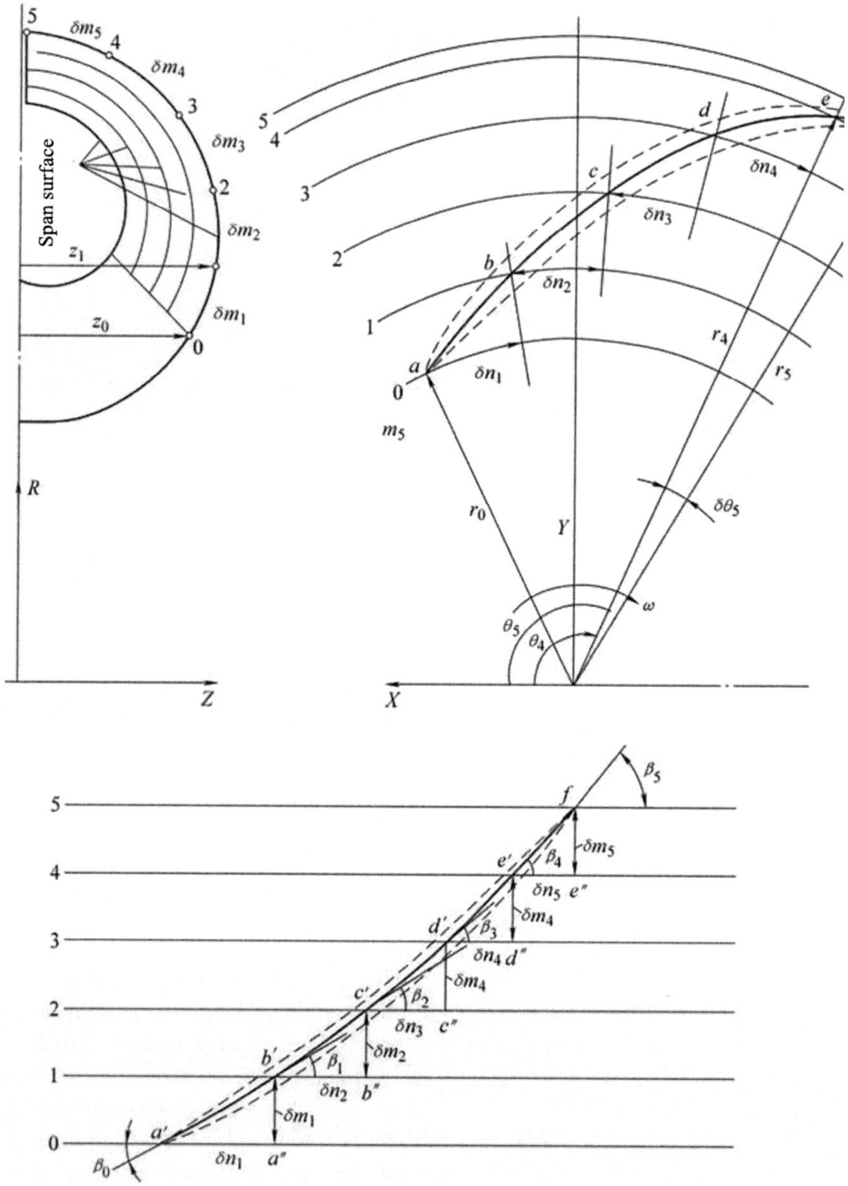

Fig. 10.22 Isoclinic projection on multiple cylinders

As shown in Fig. 10.22, the location parameters of the points on the blade spine are as follows:

x, *y*, *z*—coordinate values of *X*, *Y* and *Z* axes;
r—radius;
θ—rotation angle around *Z* axis from the *X* axis to the *Y* axis;
β—blade angle;
m—distance along the meridian curve;
s—percentage of distance along the curve ($0 \le s \le 1$, 0 at the inlet and 1 at the outlet);
c—real 3D length of curve.

The geometrical relationship among the parameters is as follows

$$\begin{aligned} \delta_m &= \sqrt{\delta_r \delta_r + \delta_z \delta_z} \\ \delta_c &= \sqrt{\delta_x \delta_x + \delta_y \delta_y + \delta_z \delta_z}, \quad c = \int_0^s \delta c ds \\ \beta &= 90^\circ - \operatorname{atan}\left(\frac{\delta\theta \cdot r}{\delta m}\right) \end{aligned} \tag{10.29}$$

Thus, after the division of the span surfaces, the shape of the blade spine can be determined by the angle parameter θ. When the blade is thickened, the corresponding point can be shifted by 1/2 of the thickness along the vertical direction of the spine. When processing the front and rear edges of the blade, it is also necessary to define the fillet radius to obtain reasonable blade shape. After obtaining the blade shape of each span surface, they are connected through the ruled surfaces to generate the 3D model of the blade shape.

(II) **Design process**

The specific design flow of 3D forming method is shown in Fig. 10.23. In general, it is sufficient to divide two or three span surfaces to determine blade shape, that is, to determine the distribution law of blade shape angle and thickness in the inner and outer rings or on the increased middle span surface.

In the design of the blade angle law, θ curve is the most basic definition of spatial position, but the blade angle β is a better indicator of the relationship with the performance in the blade design. Therefore, it is more intuitive and effective to describe the angle law of the blade shape using the blade angle. After the design of reasonable blade angle change law, the shape of the blade on the span surface is determined and the inclination angle of the spanwise blade can be adjusted by overall shift of the θ curve. The blade angle curve shall be kept gentle to ensure smooth blade shape.

The thickness distribution law mainly follows the principle of ensuring that the flow area is as consistent as possible. In the design, the statistical data of the torque converter blade thickness distribution can be used, or some mature blade thickness

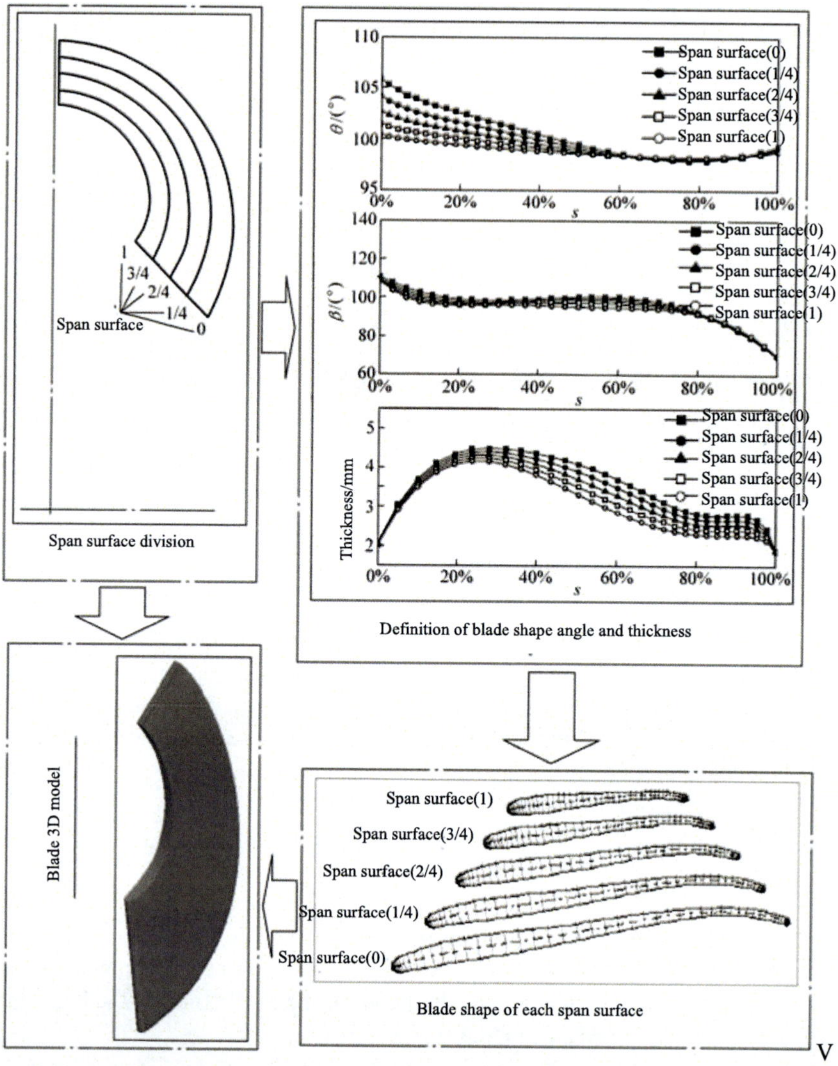

Fig. 10.23 Design flow of 3D forming method

distribution laws can be used, such as NACA series and Joukowski blade shape. The actual design shall be adjusted in detail according to the needs.

(III) Study on blade shape parameters

The basic angle parameters of the torque converter as the research object are shown in Table 10.1. In order to ensure that the basic angle parameters of the blade remain unchanged, three span surfaces are divided to define the blade shape, so the blade shape angle parameters on the middle span surface can be directly controlled. In

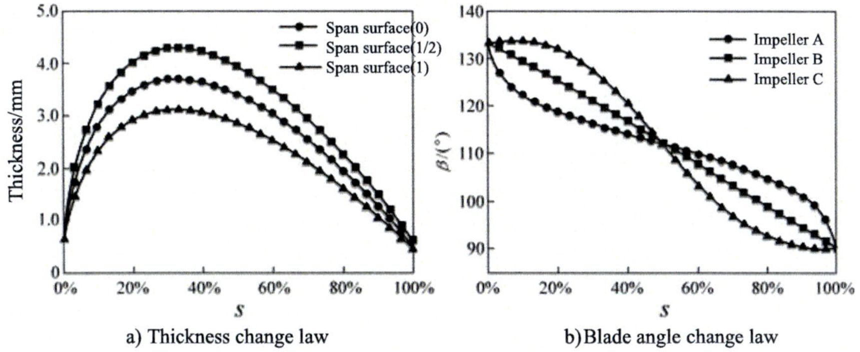

Fig. 10.24 Impeller blade parameters

order to prevent the distortion of blade shape, the change law of blade angle on the inner and outer rings is designed to be consistent with that on the middle span surface. When the change law of blade angle is described subsequently, only the change law of blade angle on the middle span surface is given.

1. **Impeller**

Three impellers used for comparative study adopt the same thickness change law. As shown in Fig. 10.24a, the chord length of the blade on the outer ring (i.e. span surface 1/2) is longer, and the maximum thickness is larger, about 4.3 mm; the blade on the inner ring (i.e. span surface 1) is relatively thin, with a maximum thickness of 3 mm, and the thickness distribution law of the whole blade remains consistent. For easy analysis, three special schemes are adopted for the blade angle change laws of three impellers, as shown in Fig. 10.25b, in which, the impeller B is linear and the basic form of comparative study. By integrating the blade angle from the inlet to the outlet along the chord direction, the angular displacement of the blade outlet edge relative to the inlet edge can be obtained. The three schemes in Fig. 10.25b can ensure that the relative angular displacement of blade inlet and outlet is the same, so that the performance comparison is only affected by blade shape.

The blade shapes of the three impellers are shown in Fig. 10.25a. The three impellers are respectively combined with the same turbines and guide wheels to form three torque converters. The blade angle change laws of the matched turbines and guide wheels are of linear type. The same method is also used for the comparative study on the turbine and guide wheel. The comparison of the efficiency, impeller capacity coefficient and torque ratio of three torque converters is shown in Fig. 10.25b, c and d.

As can be seen from the figures, in terms of the efficiency performance, the corresponding torque converter of the impeller C has the highest maximum efficiency and of the impeller A has the lowest maximum efficiency, but the change in maximum efficiency is small; the comparison of the impeller capacity coefficient K shows that the corresponding torque converter of the impeller C has the smallest impeller

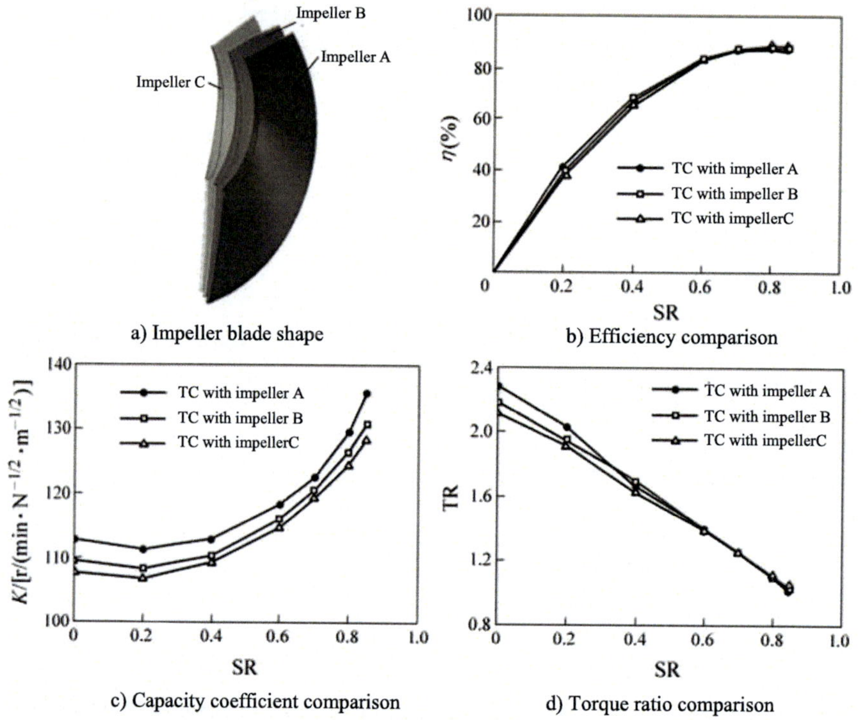

Fig. 10.25 Study on the impeller blade angle change law

capacity coefficient and of the impeller A has the largest; in terms of the torque ratio, the corresponding torque converter of the impeller A has the maximum starting torque ratio and of impeller C has the minimum starting torque ratio.

It can seen that the blade angle change law (e.g. impeller C) of gentle change at the front and rear edges and dramatic change in the middle may be used for the impeller to achieve high efficiency performance and small impeller capacity coefficient; the blade angle change law (e.g. impeller A) of dramatic change at the front and rear edges and gentle change in the middle may be used for the impeller to achieve large starting torque ratio and large impeller capacity coefficient; the blade angle change law is linear impeller, with the performance between the above two impellers.

2. **Turbine**

Three turbines adopt the same thickness change law, as shown in Fig. 10.26a. The maximum thickness of the blade shape on the outer ring (i.e. span surface 0) is about 8 mm and on the inner ring (i.e. span surface 1) is about 4 mm and the thickness distribution law of the whole blade remains consistent. Similar to the impeller, three special schemes are adopted for the blade angle change laws of three turbines, as shown in Fig. 10.26b, in which, the turbine B is linear and the basic form of comparative study.

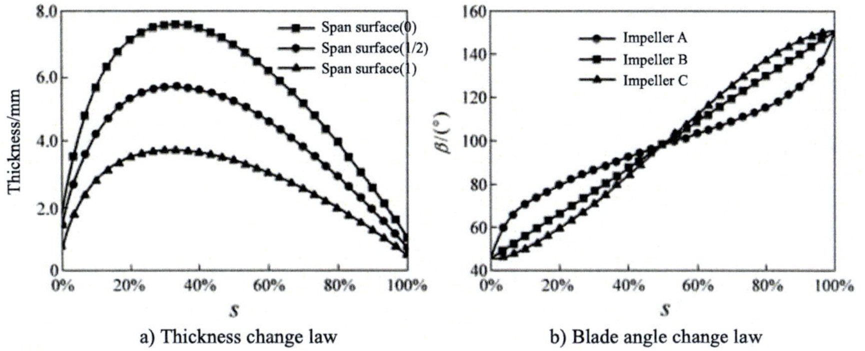

Fig. 10.26 Turbine blade parameters

The blade shapes of the three turbines are shown in Fig. 10.27a. The three turbines are respectively combined with the same impellers and guide wheels to form three torque converters, with the comparison of the efficiency, impeller capacity coefficient and torque ratio shown in Fig. 10.27b, c and d.

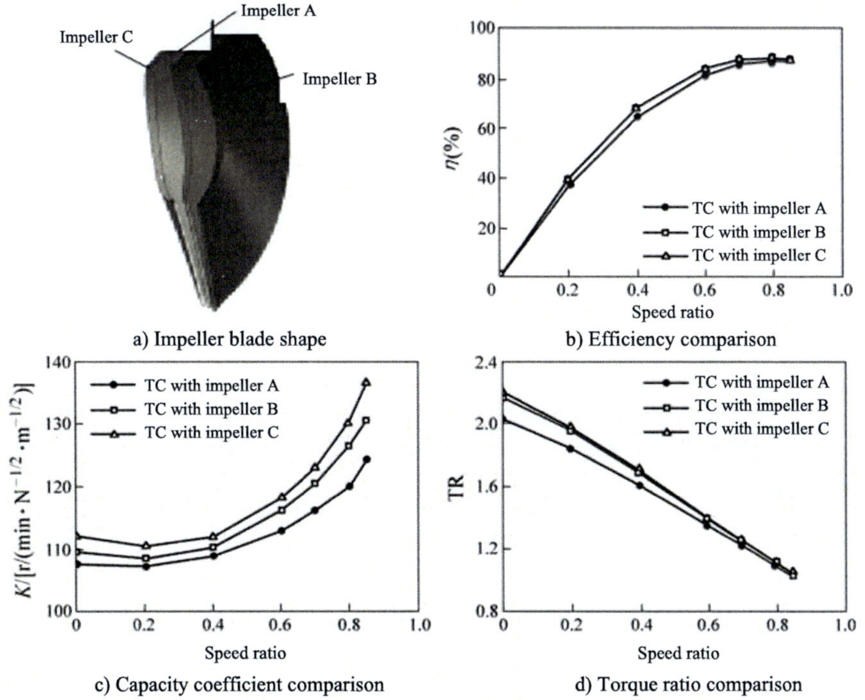

Fig. 10.27 Study on the turbine blade angle change law

In terms of the efficiency performance, the corresponding torque converter of the turbines C and B has the highest maximum efficiency, of the turbine B has higher efficiency than turbine C under the condition approaching the high efficiency point and of the turbine A has the lowest maximum efficiency; the comparison of the impeller capacity coefficient K shows that the corresponding torque converter of the turbine C has the largest impeller capacity coefficient and of the turbine A has the smallest; in terms of the torque ratio, the corresponding torque converter of the turbine C has the maximum starting torque ratio, of the turbine B has little difference in the starting torque ratio and of turbine A has the minimum starting torque ratio.

It can be seen from the above analysis that the blade angle change law (e.g. turbine C) of gentle change at the front and rear edges and dramatic change in the middle or the linear blade angle change law (e.g. turbine B) may be used for the turbine to achieve high efficiency performance and guarantee large starting torque ratio, the former of which has large impeller capacity coefficient; the blade angle change law (e.g. Turbine A) of dramatic change at the front and rear edges and gentle change in the middle may be used to achieve small impeller capacity coefficient and small starting torque ratio.

3. **Guide wheel**

Three guide wheels adopt the same thickness change law, as shown in Fig. 10.28a. The maximum thickness of the blade shape on the outer ring is about 6 mm and on the inner ring is about 5.4 mm and the thickness distribution law of the whole blade remains consistent. Similar to the impeller, three special schemes are adopted for the blade angle change laws of three guide wheels, as shown in Fig. 10.28b, in which, the guide wheel B is linear and the basic form of comparative study.

The blade shapes of the three guide wheels are shown in Fig. 10.29a. The three guide wheels are respectively combined with the same impellers and turbines to form three torque converters, with the comparison of the efficiency, impeller capacity coefficient and torque ratio shown in Fig. 10.29b, c and d.

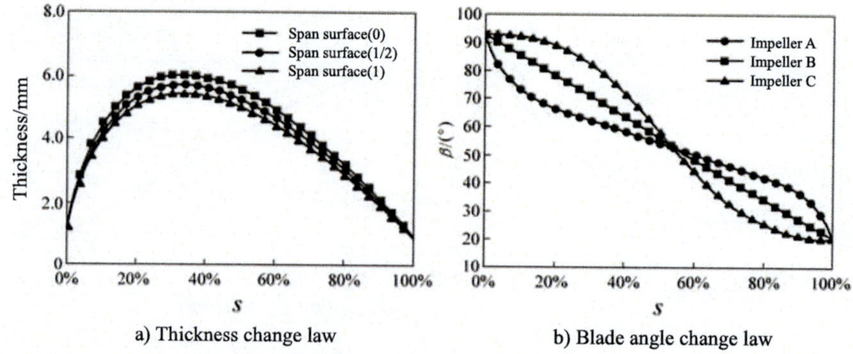

Fig. 10.28 Guide wheel blade parameters

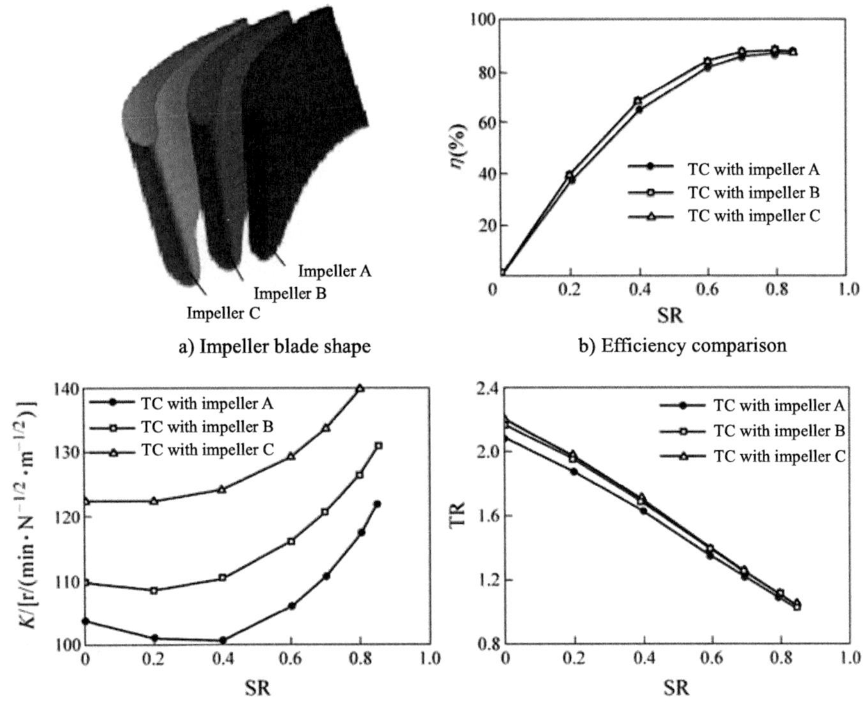

a) Impeller blade shape

b) Efficiency comparison

c) Capacity coefficient comparison

d) Torque ratio comparison

Fig. 10.29 Study on the guide wheel blade angle change law. SR—speed ratio

In terms of the efficiency performance, the corresponding torque converter of the guide wheel B has the highest maximum efficiency and of the guide wheel C has the lowest; the comparison of the impeller capacity coefficient K shows that the corresponding torque converter of the guide wheel C has the largest impeller capacity coefficient and of the guide wheel A has the smallest; in terms of the torque ratio, the corresponding torque converter of the guide wheel C has the maximum starting torque ratio, of the guide wheel B has little difference in the starting torque ratio and of guide wheel A has the minimum starting torque ratio.

It can be seen from the above analysis that the linear blade angle change law (e.g. guide wheel B) may be used for the guide wheel to achieve high efficiency performance; the blade angle change law (e.g. guide wheel A) of dramatic change at the front and rear edges and gentle change in the middle may be used to achieve small impeller capacity coefficient and small starting torque ratio; the blade angle change law (e.g. guide wheel C) of gentle change at the front and rear edges and dramatic change in the middle may be used to achieve large starting torque ratio.

4. **Conclusion**

(1) To improve the maximum efficiency of the torque converter, the options include: blade angle change law of gentle change at the front and rear edges and dramatic change in the middle for the impeller; blade angle change law of gentle change at the front and rear edges and dramatic change in the middle for the turbine; linear blade angle change law for the guide wheel. However, in general, it is not obvious to adjust the torque converter efficiency by changing the blade angle change law, and the more effective method is still to change the basic parameters such as blade inlet and outlet angles.
(2) To reduce the impeller capacity coefficient, the options include: blade angle change law of gentle change at the front and rear edges and dramatic change in the middle for the impeller; blade angle change law; blade angle change law of dramatic change at the front and rear edges and gentle change in the middle for the turbine; blade angle change law of dramatic change at the front and rear edges and gentle change in the middle for the guide wheel. By comparison, the guide wheel blade angle change law has the largest influence on the impeller capacity coefficient. Under the starting condition, the maximum change of the impeller capacity coefficient K_0 may be close to 20 r/min $N^{-1/2}$ $m^{-1/2}$, while the impeller and turbine have relatively small influence. Therefore, the impeller capacity coefficient can be adjusted in a large range by changing the guide wheel blade angle change law. This method is easier to ensure the stability of other performance parameters than the more used impeller outlet tip bending method.
(3) To increase the starting torque ratio, the options include: blade angle change law of dramatic change at the front and rear edges and gentle change in the middle for the impeller; blade angle change law; blade angle change law of gentle change at the front and rear edges and dramatic change in the middle for the turbine; blade angle change law of gentle change at the front and rear edges and dramatic change in the middle for the guide wheel.

Based on the above conclusions and taking into account the requirements of the torque converter design specification comprehensively, the W30 prototype is modified in this paper, and the blade shape parameter change law discussed above is partially adjusted to finally obtain the torque converter performance that meets the requirements.

The blade angle change laws subject to comparative study in this section are only several special cases and the thickness change law only follows the empirical data. In the actual design, the optimal change law shall be sought according to specific design objectives. Since the change law of blade parameters may be of any shape and the impellers interact with each other, it will be a lot of long-term work to study the influence of blade parameters on the torque converter performance in detail.

10.4 Numerical Simulation and Analysis of Internal Flow Field of Hydraulic Torque Converter

The hydraulic torque converter is a multi-stage steam turbine with closed flow passage, and its internal flow is complex three-dimensional viscous flow. Its flow field characteristics cannot be obtained accurately by traditional research methods. With the development of computational fluid mechanics (CFD) and computer software and hardware technology, the research on the internal flow field of hydraulic torque converter is becoming deeper and deeper. In this paper, the general fluid analysis software STAR-CD is used to conduct numerical simulation of the internal flow field of the hydraulic torque converter, which lays a foundation for the detailed analysis of the internal flow field characteristics of the hydraulic torque converter.

I. Internal flow field model of hydraulic torque converter

(I) Basic assumptions

The flow of working media in the hydraulic torque converter is very complex. In this paper, the internal flow field of hydraulic torque converter is analyzed with the following assumptions:

(1) According to the requirements for the hydrodynamic drive oil in the SAE standard, its density and viscosity change very little during the working process, which can be assumed to be constant, as $\rho = 890$ kg/m^3 and $\mu = 0.00425$ N s/m^2 respectively, meaning that the working medium in the hydraulic torque converter is typical incompressible viscous fluid.
(2) During normal operation, the temperature of the working medium in the hydraulic torque converter does not change much. In this paper, the influence of temperature is ignored, and only the velocity and pressure distribution of the flow field is studied.
(3) The torque converter is cast. All components can be considered as absolute rigid bodies. That is, the impeller, turbine and guide wheel have no relative displacement in the working process, and the inner, outer rings and blades are not deformed.
(4) In general, the flow of cooling oil is less than 0.2% of the circular flow, and the fluid leakage between the impellers is also very small compared with the circular flow, so it can be ignored.
(5) Under the same working condition, the flow field characteristics of each passage in the same impeller are the same. This assumption has two implications: periodic symmetry in space, which makes it possible to analyze only one passage when studying the flow field; steady state in time, that is, under the same working condition, regardless of the relative position of the impellers, the flow field characteristics of the same impeller are unchanged.

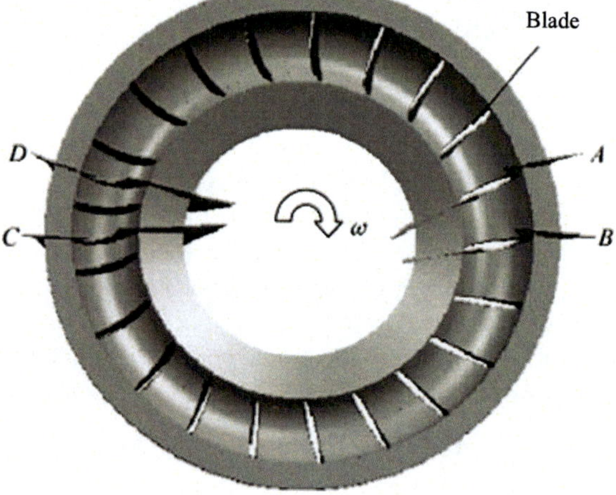

Fig. 10.30 Selection of calculation region

(II) Geometric model

The space between the inner ring, outer ring and blade of the three impellers, together with the bladeless region between the impellers, constitutes the working passage of the hydraulic torque converter. According to above assumption (5), only one passage space shall be selected for each impeller as the calculation region for analysis. The calculation region includes the passage part within the blade, and a short section of bladeless region before the blade inlet edge and after the blade outlet edge.

The selection scheme of the calculation region is illustrated with the impeller as an example below: the passage is divided at the middle position near the adjacent blades. The passage space, which is cut by two periodic symmetric surfaces, completely contains the blade area, as shown in the part between surfaces C and D in Fig. 10.30. Here, the difference of angular displacement between surfaces C and D shall be equal to the difference of angular displacement between adjacent blades, namely $2\pi/N$, where N is the number of blades of the impeller. For different impellers, there is no restriction on the relative position of surfaces C and D.

In this paper, the passage model is established for calculation by using the above scheme.

(III) Mesh model

In this paper, hexahedral elements are used to divide the mesh, which can ensure good convergence of numerical simulation. In order to make full use of computer memory resources, the mesh distribution is often uneven in the numerical calculation of practical engineering problems. In the regions where the expected change in the solved variables is more dramatic, the mesh distribution should be dense, while in the regions with gentle changes, the mesh distribution should be relatively sparse At this point, in order to ensure the calculation convergence and the accuracy of the calculation results, two issues should be noted: first, the width of each element in

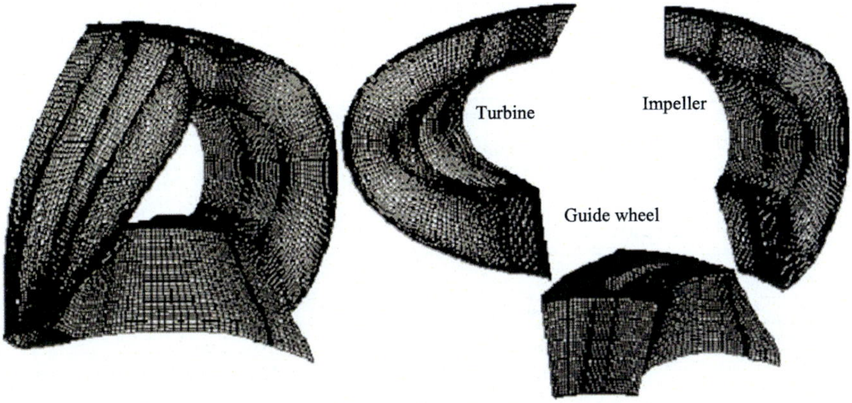

Fig. 10.31 Mesh model of calculation region

different directions should be kept in proper proportion; second, the change in the width of adjacent elements in the same coordinate direction should be kept within the appropriate range. Figure 10.31 shows the mesh model of calculation region.

(IV) Boundary conditions

There are three types of different boundary surfaces in the calculation region: blade surface, inner and outer ring surfaces; inlet and outlet surfaces; passage partition surfaces.

1. Wall boundary conditions

The blade surface, inner and outer ring surfaces are non-deformable solid walls. Therefore, these surfaces are set as non-slip wall boundaries relative to the interior meshes. Meanwhile, the influence of the surface roughness is ignored, and these solid walls are considered to be hydraulically smooth, and their surface roughness will not affect the flow conditions inside the flow field. STAR-CD provides two methods for the treatment of the near-wall flow field:

(1) In the two-layer model and low Reynolds number model, the conditions are directly added to the boundary layer to obtain the boundary layer distribution by solving the mass, momentum and turbulence equations, with the mesh model as shown in Fig. 10.32a. In this method, the near-wall region is treated in much the same way as the internal flow region except that non-slip conditions are applied to the surface of the boundary element. In addition, at a certain distance from the wall, there is a transformation from a high Reynolds number model to a low Reynolds number model ("the transformation position of two models" in the figure). Generally speaking, the transformation position varies along the wall. In Fig. 10.32, the mesh grid in the near-wall layer (NWL) is required to be high and its thickness must be sufficient to contain the actual near-wall flow region. However, the actual near-wall flow thickness is unknown, the number

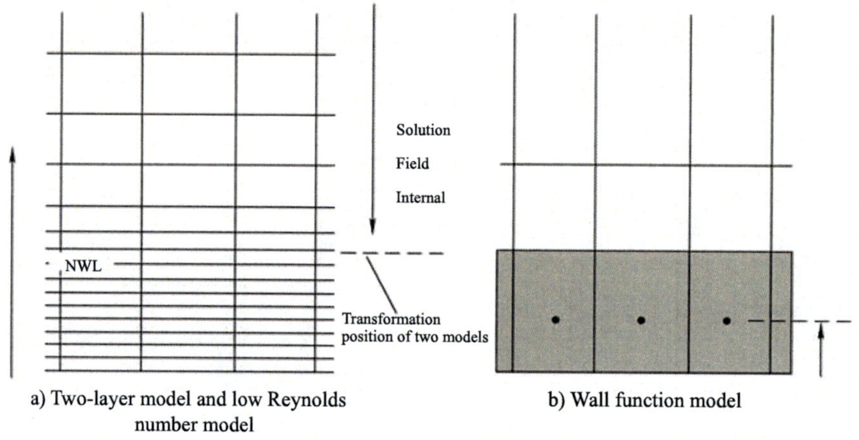

Fig. 10.32 Near-wall flow field treatment

of element layers in NWL needs to be adjusted several times in order to solve the reasonable distribution of velocity and other parameters.

(2) In the wall function method, special algebraic equations are used to describe the distribution of velocity, temperature and turbulence parameters in the boundary layer. These algebraic equations are called wall functions, with the mesh model shown in Fig. 10.32b. The main assumptions on which the wall functions are based are as follows: the changes in the velocity and other parameters mainly exist in the direction perpendicular to the wall; the influence of the pressure gradient and mass force is considered to be very small; the direction of shear stress on the layer is parallel to the velocity vector; there is a balance between the generation and dissipation of turbulent energy. The wall function method can save memory and calculation time and is widely used in engineering turbulence calculation. The wall function method is used in the standard $k - \varepsilon$ model.

2. **Mixing plane theory**

The mixing plane theory is used in this paper for unified calculation of different rotation speeds of the components in the hydraulic torque converter. The mixing plane theory is a treatment method proposed by Denton to calculate the multistage turbomachinery. In practical problems, the sliding interface between components changes with time. When the mixing plane is used instead of the sliding interface, the steady state calculation can be carried out by eliminating the time-variation by means of circumferential averaging.

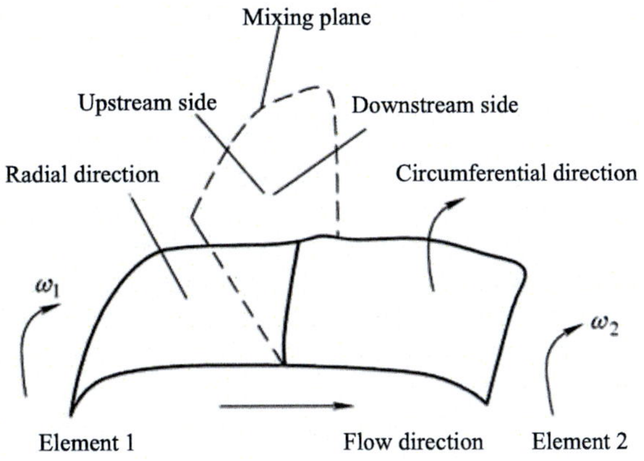

Fig. 10.33 Schematic diagram of mixing plane

Figure 10.33 is the schematic diagram of typical mixing plane, with the upstream and downstream belonging to different components. The upstream outlet surface has pressure boundary conditions, while the downstream inlet surface has inlet boundary conditions. The velocity calculated on the upstream outlet surface is distributed in the circumferential direction of the same radius for area averaging, and the value obtained is taken as the downstream inlet boundary condition. Similarly, the pressure calculated on the downstream inlet surface is distributed in the circumferential direction of the same radius for area averaging, and the value obtained is taken as the upstream pressure boundary condition. It is iterated in this way in the calculation process until convergence.

According to the mixing plane theory, the boundary settings for the inlet and outlet surfaces of each component are shown in Fig. 10.34, in which, Inlet represents the inlet boundary condition, and Pressure represents the pressure boundary condition. A lot of applications show that the numerical simulation of the hydraulic torque

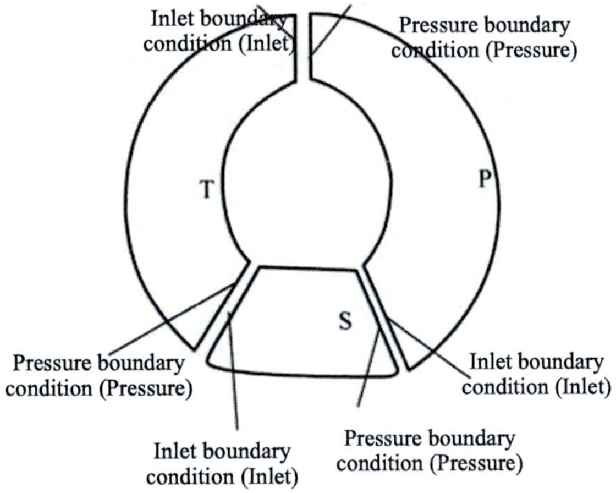

Fig. 10.34 Mixing plane boundary settings

converter using the mixing plane theory has sufficient precision for both external characteristic parameters and internal flow field characteristics.

3. **Periodic boundary condition**

The passage partition surfaces are periodic symmetric and the included angle between two adjacent passage partition surfaces is equal to that between two adjacent blades. According to the assumptions made in the flow field, the flow field characteristics of the two surfaces are completely consistent, so the passage partition surface is set as the periodic boundary condition. The scalar values (such as pressure and temperature) are equal at points corresponding to periodic boundary, and vector values (such as velocity) are equal but shall be rotated to ensure that they are in the same direction relative to their respective boundaries. The specific implementation of periodic boundary condition in numerical calculation is to add an equation that limits periodic symmetry to the discrete equation set for solving the whole flow field space.

II. Numerical calculation of internal flow field of hydraulic torque converter

(I) Calculation steps

The numerical calculation of flow field is generally divided into three parts: preprocessing, solver calculation and post-processing, as shown in Fig. 10.35. The solver calculation belongs to the internal processing of the fluid analysis software, while the preprocessing and post-processing are the main work in the numerical calculation of flow field. In the numerical calculation of internal flow field of hydraulic torque converter, the preprocessing mainly includes the geometric modeling, the division of mesh model, the setting of boundary conditions and the calculation of initial flow field. The preprocessing has an important influence on the convergence and accuracy of numerical calculation. The post-processing refers to the further analysis of the numerical calculation results to obtain an in-depth understanding of the flow field and torque converter performance. In this paper, such wok is mainly completed in the preprocessor and postprocessor PROSTAR of STAR-CD.

A reasonable initial flow field can improve the convergence rate and accuracy of numerical calculation. In the flow field calculation of torque converter, the main

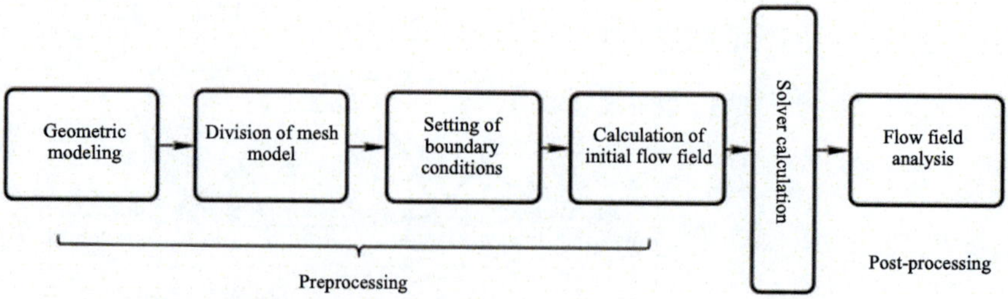

Fig. 10.35 Numerical calculation process

methods to obtain the initial flow field are as follows: the flow field calculated with at an assumed circulation velocity is taken as the initial flow field; the flow field after calculation of certain steps with the laminar flow model is taken as the initial flow field; the flow field calculated with adjacent speed ratio conditions is taken as the initial flow field. The above method is applied to the calculation in this paper, with the steps as follows: the medium speed ratio condition (speed ratio 0.4) is taken as the first calculation condition. When the condition is calculated, the initial flow field is obtained by assuming the circulation velocity and then subject to numerical solution. If it is convergent, other conditions may be calculated; if it is divergent, the initial flow field may be unreasonable and the flow field after calculation of certain steps (e.g. 50 steps) with the laminar flow model may be taken as the initial flow field, and then a turbulence model is set for solving again until convergence; after the medium speed ratio condition is calculated, the result is taken as the initial flow field to calculate the adjacent speed ratio conditions (speed ratio 0.3 and 0.5). By analogy, all operating points can be calculated.

(II) **Convergence criteria**

The numerical calculation of flow field is the process of solving a system of linear equations iteratively. When certain convergence criteria are met, it can be considered that the numerical solution is close enough to the exact solution to stop the iterative process. In steady-state iterative calculation, when the number of iterations is k, the residual r_φ^k on a specific element is the imbalance caused by incomplete calculation in the discrete control equation, i.e.

$$r_\varphi^k = \alpha\varphi_P^k - \sum \alpha_i \varphi_l - b \tag{10.30}$$

α, α_i and b are coefficients associated with flux, φ represents flux and superscript k represents the number of iterations. The standardized absolute residual R_φ^k is used to determine convergence in the actual calculation and is defined as follows

$$R_\varphi^k = \frac{\sum \left| r_\varphi^k \right|}{M_\varphi} \tag{10.31}$$

M_φ is the standardized coefficient. In STAR-CD, the default definition for M_φ is

$$M_\varphi = \sum_P \alpha_P \varphi_P^K \tag{10.32}$$

For a momentum equation, the velocity gradient at the element P is taken as φ_P to calculate the standardized coefficient; for a continuity equation, the maximum r_φ^k in the first ten iterations is selected as the standardized coefficient; for other scalar parameters, the scalar value at element P is taken as φ_P to calculate the standardized coefficient.

The residual is used to determine when to end the calculation and is judged on the basis that all residuals shall be less than the user-defined standard λ, i.e.

$$\max\left(R_{\varphi}^{k}\right) < \lambda \tag{10.33}$$

The default value of λ is 10^3. In addition to satisfying the requirements of residuals, the parameters in the flow field are also required to be close to constants, which together constitute sufficient conditions for convergence.

(III) **Result**

1. **Mesh-independent solution**

In the numerical calculation of practical problems, the workload of mesh generation is very large. Often after repeated debugging and comparison can the mesh suitable for the specific problems be obtained. The meshes used to obtain the numerical solutions should be dense enough that further refined meshes have little effect on the numerical results. Such numerical solutions are called mesh-independent solutions.

2. **Preliminary determination of accuracy**

The convergent numerical results describe the distribution of the parameters in the flow field. To judge whether the calculation results are accurate, it is necessary to conduct a detailed study on the flow field distribution based on the test results. In the next part, the flow field characteristics and overall performance of each component will be analyzed in detail.

The preliminary determination is only a simple check of the results according to the flow field characteristics of the hydraulic torque converter, but it is a necessary condition for the accuracy of the results. The preliminary determination generally includes two aspects:

(1) Circular flow balance check. For each component, the relative error of mass flow rate between the inlet and outlet surfaces must be less than 0.01%; for all three components, the relative error between the respective mass flow rates must be less than 0.1%.
(2) Torque balance check. The hydraulic torque converter has the characteristics of the torque balance, namely the turbine torque value shall be equal to the sum of the impeller torque value and the guide wheel torque value. For an accurate calculation result, the torque unbalance value shall be less than 0.2–0.3% of the impeller torque.

III. **Internal flow field analysis of hydraulic torque converter**

In order to deeply understand the working principle and flow field distribution of the torque converter to guide the model change, the typical distribution characteristics of the internal flow field and its mechanism of production will be studied with the impeller of W305 torque converter as an example and the overall performance of the impeller will be studied below. When it is necessary to make a comparative analysis of the flow field under different speed ratios, speed ratios of 0.00, 0.40 and 0.80 are selected as three typical conditions, representing starting condition, medium speed ratio condition and high-efficiency condition respectively.

(I) Internal flow field analysis of impeller

1. Geometric definition of passage

It is assumed that the direction of the passage from the outer ring to the inner ring is composed of innumerable non-intersecting layers of circular surfaces, which have similar geometric shapes with the inner and outer rings of the passage; in the direction of rotation, the passage is composed of rotating sweeps of the blade surface; from the inlet to the outlet, the passage is made up of numerous chord surfaces. Thus, three spatial coordinate directions of the passage are defined, as shown in Fig. 10.36: in the chordwise direction, the circular flow direction of the fluid is positive, called chord coordinate, which is 0 at the inlet and 1 at the outlet; in the pitchwise direction, the rotation direction is positive, called pitch coordinate, which is 0 on the pressure surface and 1 on the suction surface; in the spanwise direction, the direction from outer ring to inner ring is positive, called span coordinate, which is 0 on the outer ring and 1 on the inner ring. Accordingly, the surfaces perpendicular to the chord coordinate, pitch coordinate, and span coordinate are called the chord surface, pitch surface, and span surface, respectively. Only the chord coordinate of the impeller passage is selected below for analysis.

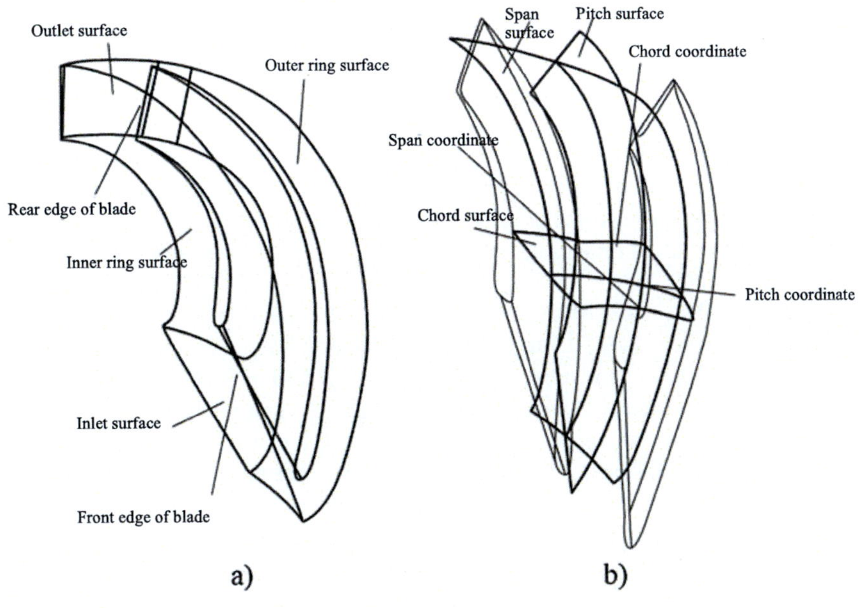

Fig. 10.36 Geometric definition of passage space

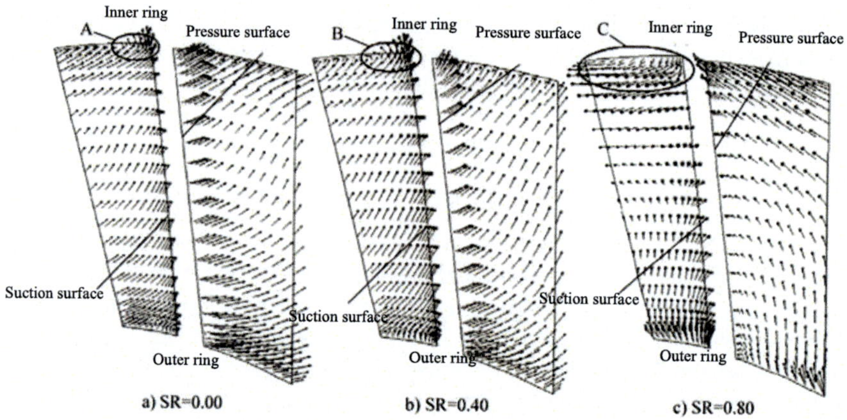

Fig. 10.37 Velocity distribution on the inlet surface of impeller blade

2. Flow field characteristics on chord surface of impeller passage

(1) Inlet surface of impeller passage blade: the inlet surface of impeller passage blade refers to the chord surface at the front edge of the blade and the relative velocity distribution on this surface at each speed ratio (SR) is shown in Fig. 10.37.

It can be seen from Fig. 10.37 that, at each speed ratio, the reverse flow regions (region A in Fig. 10.37a, region B in Fig. 10.37b and region C in Fig. 10.37c) appear at the included angle between the inner ring and the suction surface. There are two main reasons for this phenomenon: first, after the fluid flows out from the guide wheel outlet, the flow area increases. Therefore, the flow in this region can be considered as jet dispersion, and the low velocity zone and even reverse flow will appear at the outer boundary of the jet dispersion; second, the incurve of the inner ring wall creates a wake flow in this region. In order to improve the flow condition in this region, in addition to changing the circulation circle shape, it is also possible to adjust the guide wheel blade shape to change the distribution of incoming flow, so as to eliminate or weaken the occurrence of reverse flow.

Figure 10.38 shows the distribution of the static pressure on the inlet surface of the impeller blade. It can be seen that with the increase of speed ratio, the absolute value of static pressure increases gradually, but the pressure difference in the whole surface decreases gradually and the pressure distribution tends to be uniform. At the speed ratio, the pressure difference in the whole surface is relatively small and the maximum pressure difference is 2.046×10^5 Pa; at low speed ratio, the maximum pressure difference is 2.413×10^5 Pa. The existence of pressure difference will cause the fluid flow to deviate from the circular flow direction, thus forming secondary flow, which will be specially analyzed subsequently.

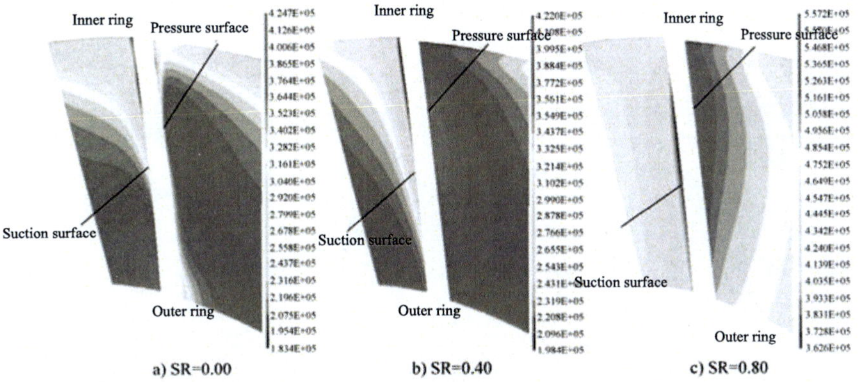

Fig. 10.38 Distribution of static pressure on inlet surface of impeller blade

(2) Chord surface in the middle of impeller passage: the relative velocity distribution on the chord surface in the middle of the impeller is shown in Fig. 10.39. It can be seen from the figure that, at the low speed ratio, the maximum velocity is on the outer ring side, while the velocity is low on the inner ring side, indicating that the flow in the inlet section is of jet structure, namely, near the inner ring is the wake zone and near the outer ring is the jet zone; at the high speed ratio, the flow velocity decreases due to the reduction of the circular flow, but this distribution is more pronounced and a large low velocity zone (A in Fig. 10.39c) appears near the included angle between the inner ring and the suction surface, which is mainly caused by the change in the passage curvature on the meridian surface.

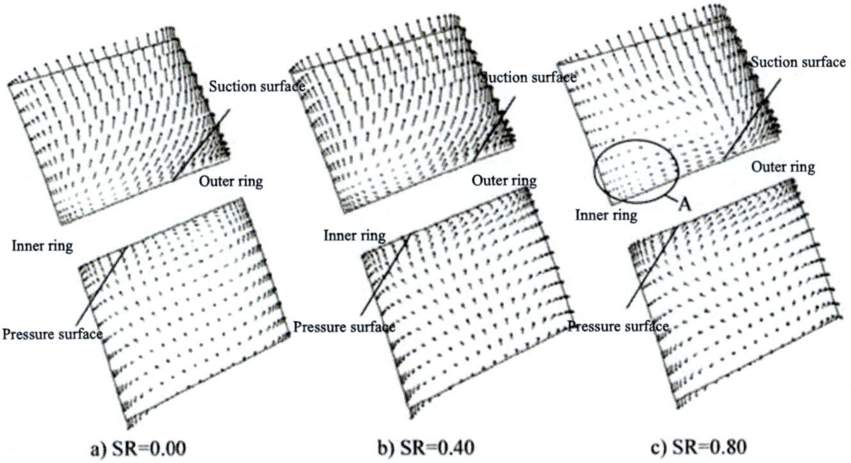

Fig. 10.39 Relative velocity distribution on the chord surface in the middle of impeller

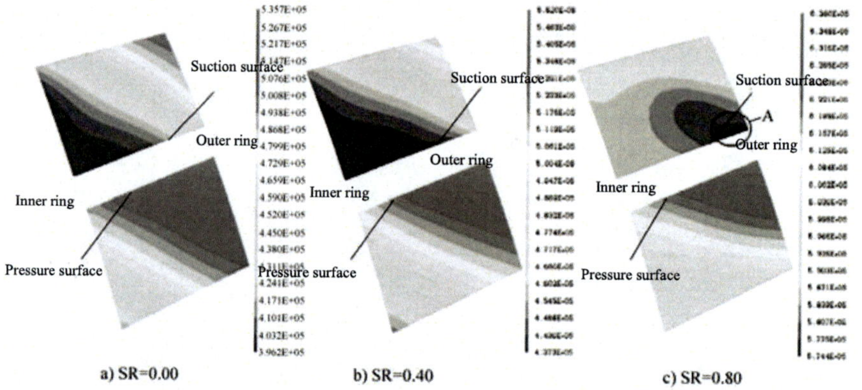

Fig. 10.40 Distribution of static pressure on chord surface in the middle of impeller

Figure 10.40 shows the distribution of the static pressure on the chord surface in the middle of impeller. As can be seen from the figure, at the low and medium speed ratio, the increase of speed ratio has little influence on the distribution structure of static pressure, the high pressure zone is located at the included angle between the pressure surface and the outer ring, and the low pressure zone is located at the included angle between the suction surface and the inner ring; at the high speed ratio, the high pressure zone is also located at the included angle between the outer ring and the pressure surface, but the local low pressure zone (A in Fig. 10.40c) appears on the suction surface near the outer ring. This abnormal local low pressure will cause large energy loss.

(3) Outlet surface of impeller passage blade: the outlet surface of impeller passage blade refers to the chord surface at the rear edge of the blade and the relative velocity distribution on this surface is shown in Fig. 10.41. The velocity at the angle between the pressure surface and the outer ring is higher, while the velocity at the angle between the suction surface and the inner ring is lower. The velocity distribution presents typical jet/wake structure, especially under the condition of high speed ratio. A fully developed jet zone appears on one side of the pressure surface, with large and evenly distributed velocity; a wake zone appears on one side of the suction surface, and the velocity is low in a large range close to the suction surface.

Due to the rotation of the passage, the flowing fluid in the rotating coordinate system is affected by the coriolis force. The coriolis force applied to the fluid per unit volume is expressed as follows

$$F_C = 2\rho\omega V \tag{10.34}$$

where,

F_C—coriolis force (N), in the opposite direction to the rotation direction;

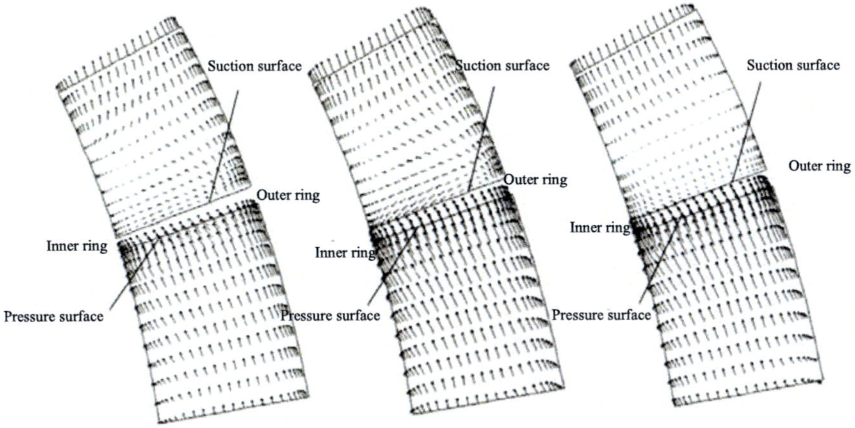

Fig. 10.41 Distribution of relative velocity on outlet surface of impeller blade

v—flow velocity (m/s);
ω—impeller speed (rad/s).

Formula (10.34) indicates that the coriolis force is a function of the flow velocity. When the flow velocity is not distributed evenly, namely $\partial V/\partial \theta \neq 0$, the effect of coriolis force will make the velocity distribution more uneven; on the contrary, if the velocity is distributed evenly, the coriolis force has less influence on it.

When the fluid flows into the impeller passage, the boundary layer on the suction surface and pressure surface start to develop at the same time and the effect of the coriolis force is shown in Fig. 10.42. The coriolis force on the pressure surface will inhibit the development of the boundary layer, while the coriolis force on the suction surface will promote the growth of the boundary layer. Therefore, the effect of coriolis force on the fluid in the passage results in a low-speed wake zone on the suction surface and a high-speed jet zone on the pressure surface. With respect to the impeller, the passage mainly bends in the meridian plane. The fluid will be affected by the centrifugal force when flowing through the curved passage and the centrifugal force on the fluid per unit volume is $F_S = \rho v^2/r$, where, the centrifugal force F_S is outward perpendicular to the flow velocity, v is flow velocity and r is radius of curvature.

When the flow velocity is very large or the radius of curvature is very small, the effect of centrifugal force will be great, as shown in Fig. 10.43. The centrifugal force causes the boundary layer on the inner ring surface to grow, and the adverse pressure gradient generated causes the boundary layer to separate, resulting in a wake zone with lower pressure and velocity. Therefore, the effect of centrifugal force on the fluid in the passage results in a low-speed wake zone near the inner ring and a high-speed jet zone near the outer ring. Based on the above analysis results, the distribution position of the jet/wake structure is mainly determined by the relative

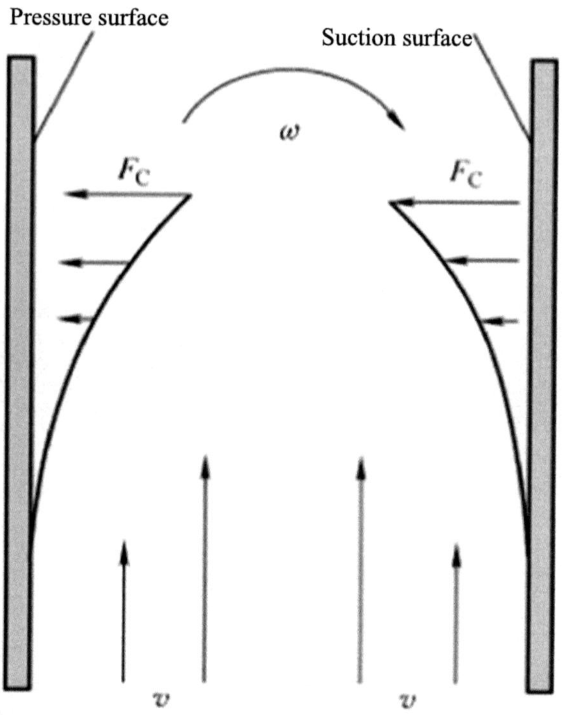

Fig. 10.42 Effect of coriolis force

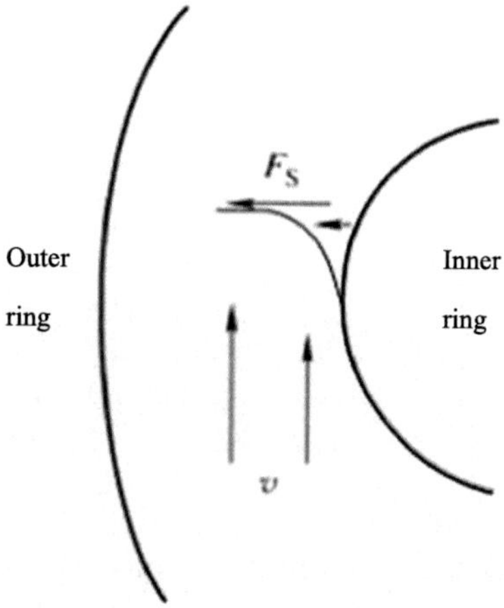

Fig. 10.43 Effect of centrifugal force

influence degree of rotation and bending of the passage. Generally, Rossby number is used to evaluate the relative influence of centrifugal force and coriolis force, namely

$$Ro = \frac{V}{\omega r} \propto \frac{V}{2\omega r} = \frac{\rho V^2/r}{2\rho\omega V} = \frac{F_S}{F_C} \tag{10.35}$$

For low Rossby number of impeller flow ($R_0 \leq 1$), the coriolis force dominates. Therefore, the influence of the boundary layer growth on the suction surface on the wake flow is greater than that on the flow separation in the inner ring, and the jet zone and wake zone will be located near the pressure surface and the suction surface respectively. For high Rossby number of low ($R_0 \geq 1$), the centrifugal force dominates. The flow separation in the inner ring caused by the passage bending has a greater influence on the flow, and the jet zone and wake zone will be located near the outer ring and the inner ring respectively. When the Rossby number is close to 1 ($R_0 \approx 1$), the effects of coriolis force and centrifugal force are similar. The jet zone will be located at the interface between the pressure surface and the outer ring, while the wake zone will be located at the interface between the suction surface and the inner ring.

With respect to the W305 prototype torque converter, at the speed ratio 0.00 or 0.40, $R_0 > 1$, the flow velocity is high near the outer ring and low near the inner ring; at the speed ratio 0.80, $R_0 \leq 1$, a jet zone appears on one side of the pressure surface, while a wake zone appears near the suction surface, which is consistent with the above analysis. The distribution of the static pressure on the outlet surface of the impeller is shown in Fig. 10.44. It can be seen that on the impeller outlet surface, the distribution of static pressure increases in proportion along the radial direction and the maximum pressure is located on the outer ring near the pressure side. With the increase of speed ratio, this distribution trend becomes more and more obvious. It can also be seen that with the increase of speed ratio, the static pressure on the impeller outlet surface rises gradually, but the maximum pressure gradient decreases and the maximum pressure difference decreases from 1.422×10^5 Pa at the speed ratio 0.00 to 1.096×10^5 Pa.

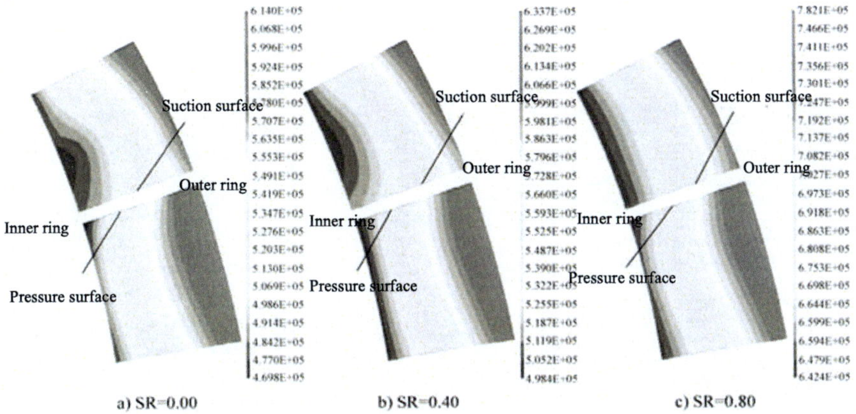

Fig. 10.44 Distribution of static pressure on outlet surface of impeller blade

(II) Overall analysis of impeller flow field

1. Secondary flow

When the fluid in the passage is affected by the transverse differential pressure, it will deviate perpendicular to the boundary direction, which can be regarded as the combination of the primary flow and the secondary flow perpendicular to it. Because of continuity, the secondary flow is not limited to the boundary layer, but also affects the primary flow zone. Figures 10.45 and 10.46 show the distribution of secondary flow on each chord surface under starting condition and high-efficiency condition respectively. The secondary flow on the blade inlet surface does not form a circular flow. It mainly flows from the outer ring to inner ring at low speed ratio and mainly flows from the suction surface to pressure surface at high speed ratio. The form of secondary flow is greatly affected by speed ratio, which is mainly caused by the change of incidence angle. In both starting condition and high-efficiency condition, the secondary flow on the middle chord surface has an obvious vortex

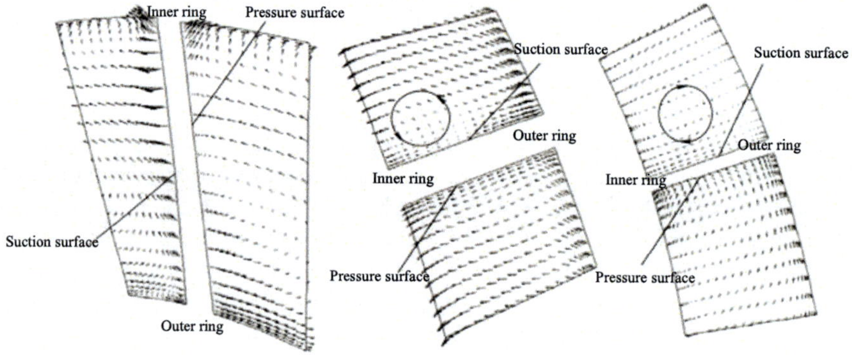

Fig. 10.45 Distribution of secondary flow under starting condition

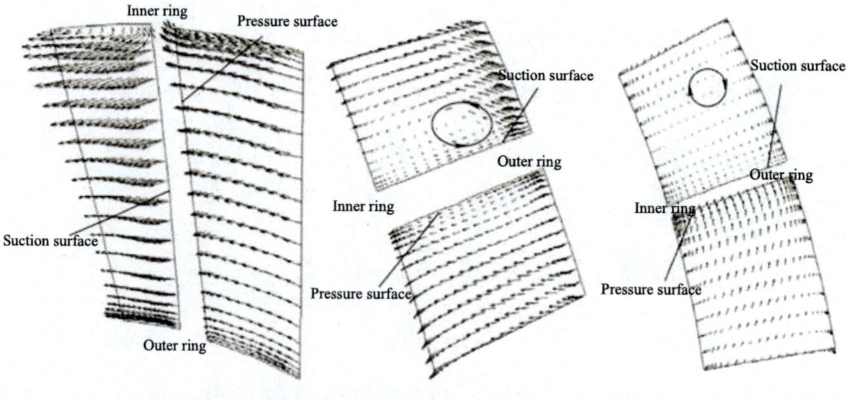

Fig. 10.46 Distribution of secondary flow under high-efficiency condition

rotating counterclockwise (seen from the downstream side), and a vortex rotating clockwise (seen from the downstream side) when it reaches the outlet surface of the blade. Gruver et al. found similar trends when they studied different torque converters. The secondary flow in the impeller passage is mainly caused by the vortex in viscous shear layer, the flow deflection, passage meridian plane bending, coriolis force caused by rotation, centrifugal force at the blade boundary layer and other comprehensive factors, which make the flow characteristics very complex.

2. **Loss analysis**

The above qualitative analysis of flow field characteristics gives us a deep and detailed understanding of flow field, which has the most direct guiding significance for improving the design performance. In order to study the flow condition more intuitively and evaluate it quantitatively, the paper introduces the method of loss analysis. Bernoulli equation points out that in the steady motion of the fluid without viscosity and heat conduction, the total energy per unit mass of the fluid remains unchanged along the same flow line, i.e.

$$H = P/\rho + 1/2W^2 + \psi = \text{Constant} \tag{10.36}$$

where,

P—static pressure;
ω—relative velocity;
ψ—body force potential energy per unit mass.

In the loss analysis of hydraulic torque converter, both experimental analysis and numerical research in recent years have used Bernoulli constant to construct the Rothalpy constant as the basis for quantitative evaluation of loss and calculation of impeller hydraulic efficiency. In rotating impeller, the body force of the fluid particle is mainly centrifugal force and the body force potential energy ψ is expressed as

$$\psi = -\frac{1}{2} f_{\text{centrifugal force}} r = -\frac{1}{2}\omega^2 rr = -\frac{1}{2}u^2 \tag{10.37}$$

where,

ω—rotation speed;
r—radius of a particle in a fluid from the rotary shaft;
u—peripheral velocity.

Thus, the Rothalpy constant is defined as

$$\text{Rothalpy} = \frac{p}{\rho} + \frac{1}{2}\omega^2 - \frac{1}{2}u^2 \tag{10.38}$$

For incompressible non-viscous flow, the Rothalpy value is a constant along the flow line; while for viscous flow, the difference of the parameter between two points

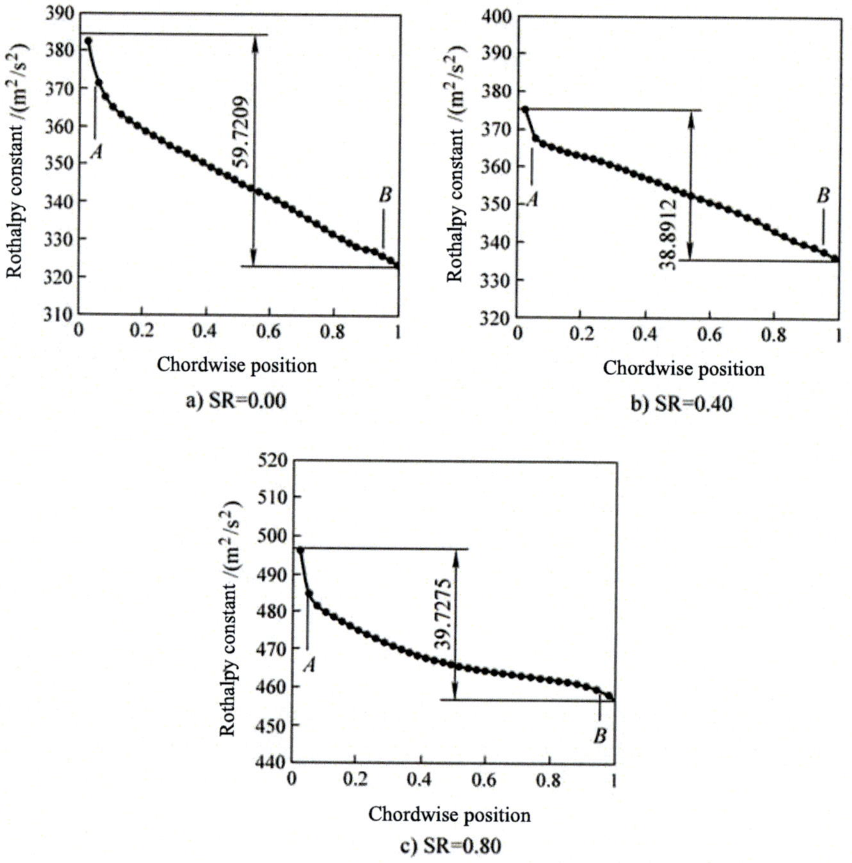

Fig. 10.47 Change of Rothalpy constant in chordwise direction of impeller

on the same flow line represents the hydraulic loss. Figure 10.47 shows the changes of Rothalpy constant from inlet to outlet at different speed ratios, where, A and B represent the front and rear edges of the blade respectively. It can be seen that, the Rothalpy constant changes dramatically before the fluid enters the blade passage, the section before point A. At this time, the fluid flows to the front edge of the blade in the bladeless region and the hydraulic loss is large due to the retardation at the front edge of the blade; as the fluid enters the blade passage, the Rothalpy constant changes gently, and the loss is small; when the fluid leaves the blade, that is, the region after point B, the decrease in the Rothalpy constant is due to the diffusion loss caused by the fluid from the bladed region to the bladeless region. Meanwhile, as can be seen from the Rothalpy constant difference at the inlet and outlet in the figure, when the speed ratio changes from 0.00 to 0.40, the hydraulic loss decreases, but when the speed ratio reaches 0.80, the hydraulic loss is slightly greater than that in

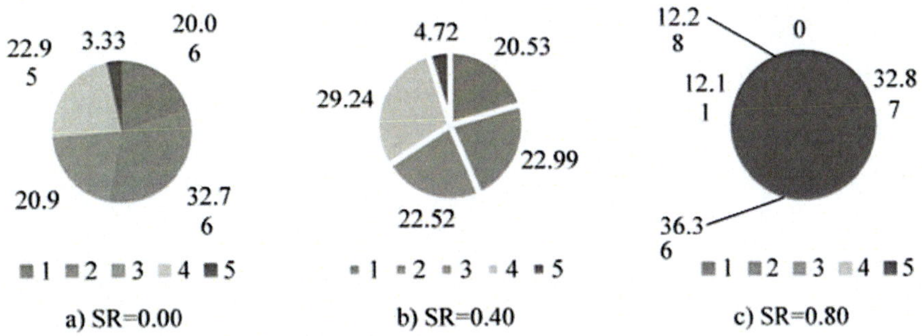

Fig. 10.48 Impeller passage loss distribution

the medium speed ratio condition, which is one of the reasons for the low maximum efficiency of W305 prototype torque converter.

In order to further describe the loss distribution of the whole passage, the passage is divided into the following five sections:

(1) Bladeless region in inlet section: from the passage inlet to the inlet surface of the blade passage, i.e. front edge of blade.
(2) Front section of blade passage: from the front edge of blade to 1/3 chord length of blade passage.
(3) Middle section of blade passage: from 1/3 chord length of blade passage to 2/3 chord length of blade passage.
(4) Rear section of blade passage: from 2/3 chord length of blade passage to rear edge of blade.
(5) Bladeless region in outlet section: from outlet surface of the blade passage, i.e. rear edge of the blade to the passage outlet.

The reduction of the Rothalpy constant in each section from the start position to the end position represents the hydraulic loss in each range. Figure 10.48 shows the distribution of the impeller passage loss in each range at different speed ratios. It can be seen from the figure that under various working conditions, the loss (range 1) caused by the fluid retardation at the front edge of the blade is large, which is mainly caused by unreasonable design of the front edge of the blade. For this problem, the front edge fillet of the blade has been improved in the model change. The loss in the front section of blade passage (range 2) accounts for the largest proportion, and the loss in this range is mainly understood as impact loss. With the increase of speed ratio, losses in the middle and back end of blade passage (ranges 3 and 4) account for a smaller and smaller proportion of the total loss. The loss in the bladeless region in outlet section (range 5) is relatively small and can be interpreted as diffusion loss.

Mass averaging is performed for the Rothalpy constant at the same span to obtain the spanwise Rothalpy constant distribution on each chord surface of the impeller, as shown in Fig. 10.49. The difference between the abscissa values of the inlet and outlet surface curves represents the hydraulic loss from the inlet to the outlet at the span. In various working conditions, the loss near the inner ring (vertical coordinate

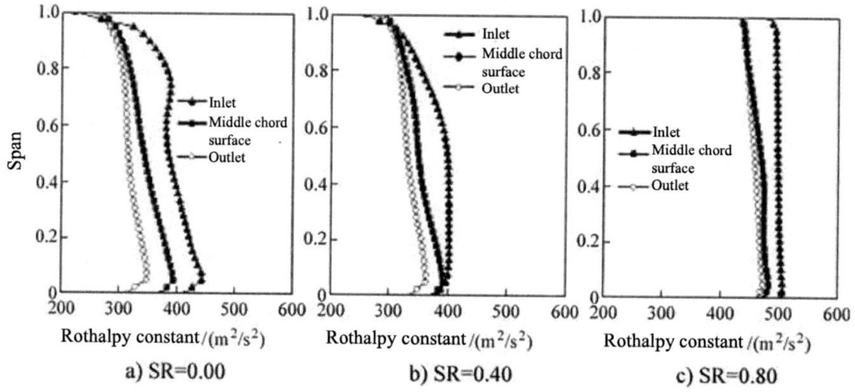

Fig. 10.49 Rothalpy constant distribution in spanwise impeller

close to 1.0) mainly occurs in the front half of the passage, that is, between the inlet and the middle chord surface; while the loss distribution near the outer ring changes greatly under different speed ratios.

10.5 Parameter Adjustment of Hydraulic Torque Converter

The predicted performance is compared with the design requirements, and the torque converter parameters are adjusted according to their influence law on the performance. These parameters include blade inlet and outlet angles, distribution law of moment of momentum in unequal distribution of moment of momentum, blade angle change law and thickness distribution law in 3D forming method. After the parameter adjustment, the blade forming step is entered again and the performance analysis and parameter adjustment are repeated until the torque converter performance meets the design requirements.

The performance indexes of the hydraulic torque converter include torque converter efficiency η, torque ratio TR and impeller capacity coefficient K. In the practical application, the torque converter performance is evaluated mainly through several typical conditions: starting conditions TR_0 and K_0, high-efficiency condition η^* and corresponding i^*. The penetrability coefficient $T = (K_M/K_0)^2$, where, the subscript M represents the coupler operating point.

I. Model change for parameter adjustment of W305 prototype hydraulic torque converter

W305 prototype torque converter is subject to model change using the modern design method below.

In order to analyze the working mechanism and performance of the torque converter, a CMM is firstly used to measure the points on each surface of the W305 prototype torque converter to obtain its geometric model, and then the method of numerical flow field calculation is used to analyze the internal flow field, and the direction of model change is sought in combination with the design objective. After the basic angle parameters are determined in the pre-design stage, a comparative study is made on the blade shape change law in order to obtain the optimal torque converter performance. Through multiple parameter adjustments and numerical simulation analysis, the final design result is called W305-1 hydraulic torque converter. The basic blade angle parameters before and after the model change are listed in Table 10.2.

The blade shape of W305-1 hydraulic torque converter is generated by 3D forming method. The thickness and angle change laws of each impeller are shown in Figs. 10.50, 10.51 and 10.52. The blade angle change law of dramatic change at the front and rear edges and gentle change in the middle is adopted for the impeller, the linear blade angle change law is adopted for the turbine and the blade angle change law of the guide wheel is partially adjusted based on the linear type.

The resulting blade shape is compared to the blade shape of W305 prototype hydraulic torque converter as shown in Fig. 10.53.

II. **Flow field analysis of impeller of W305-1 hydraulic torque converter**

Table 10.2 Comparison of basic blade angle parameters before and after the model change of torque converter

Angle parameter	Impeller inlet angle	Impeller outlet angle	Turbine inlet angle	Turbine outlet angle	Guide wheel inlet angle	Guide wheel outlet angle
W305 prototype	108°	88°	33°	142°	103°	25°
W305-1	133°	90°	46°	152°	93°	20°

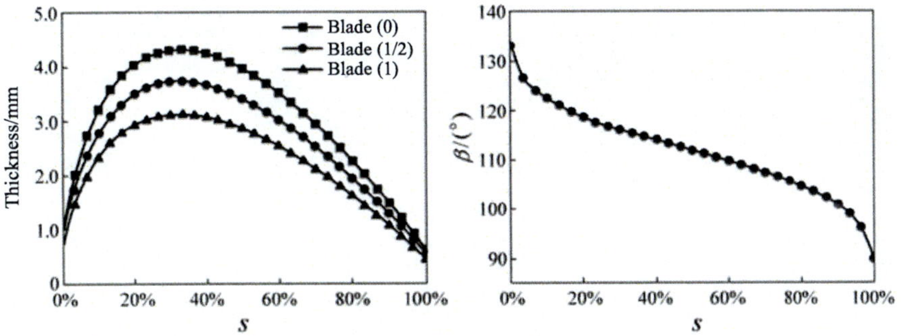

a) Thickness change law b) Blade angle change law

Fig. 10.50 Impeller blade parameters of W305-1 hydraulic torque converter

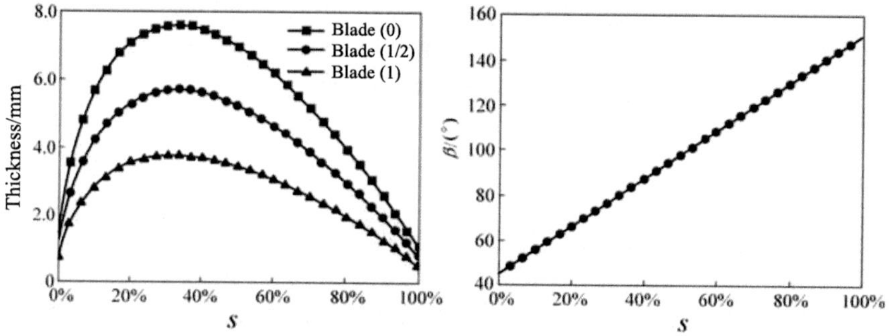

a) Thickness change law b) Blade angle change law

Fig. 10.51 Turbine blade parameters of W305-1 hydraulic torque converter

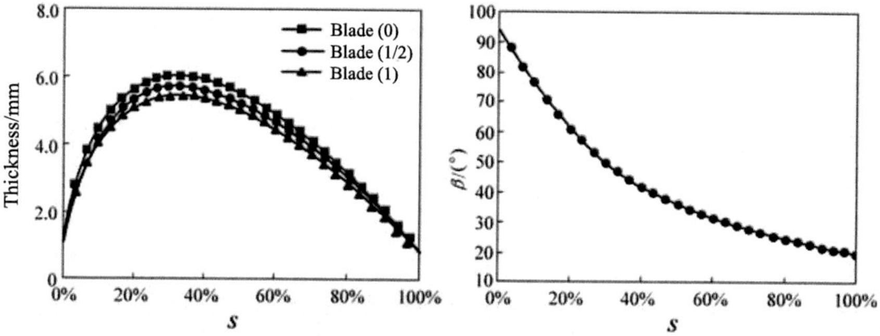

a) Thickness change law b) Blade angle change law

Fig. 10.52 Guide wheel blade parameters of W305-1 hydraulic torque converter

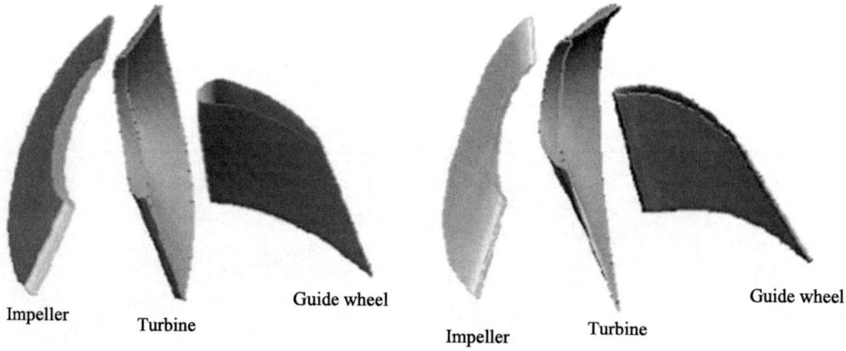

Fig. 10.53 Blade shape comparison

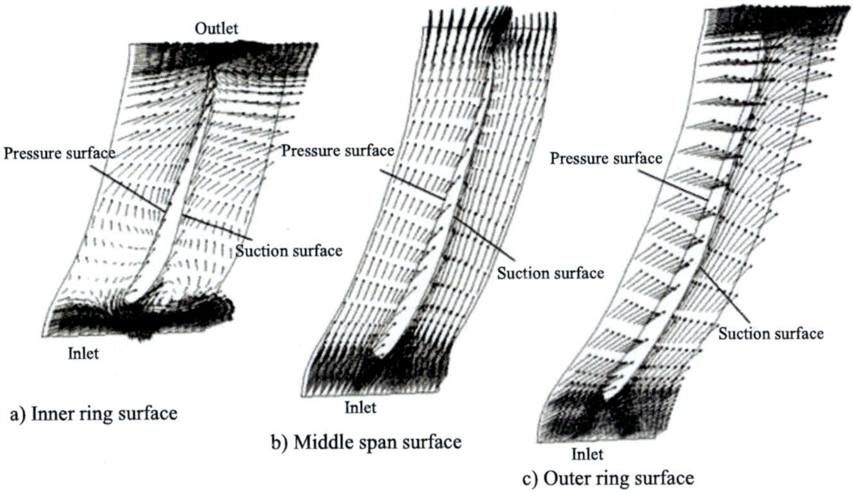

Fig. 10.54 Span surface velocity distribution of impeller under high-efficiency condition

To analyze the performance improvement mechanism, the internal flow field of impeller before and after the model change is analyzed with the hydraulic torque converter impeller as example.

The velocity distribution on the span surface of the impeller of W305-1 hydraulic torque converter under high-efficiency condition is shown in Fig. 10.54. Although the flow condition near the front edge of the blade on the inner ring surface is still poor, compared with the W305 prototype hydraulic torque converter, the existing various reverse flows and vortices have been greatly weakened or even eliminated. Some flow condition (e.g. low velocity zone at the front edge of the blade on the inner ring surface) in the impeller flow field is difficult to eliminate due to its own movement characteristics. Sometimes, the design objective of the impeller capacity coefficient is satisfied at the expense of the efficiency (for example, the wake zone will appear on the suction surface at the rear edge of the blade due to the blade tip bending). Therefore, the poor flow condition still exists in some regions in the model-changed impeller.

The loss analysis method mentioned above is used for the change of Rothalpy constant in chordwise direction of impeller and the Rothalpy constant is also used to evaluate the hydraulic loss in the passage. Figure 10.55 shows the change of Rothalpy constant of the impeller from input to outlet at each speed ratio. Compared with the prototype, the hydraulic loss at each speed ratio is greatly reduced and the abnormal phenomenon that the hydraulic loss of the prototype torque converter under high efficiency condition is greater than that at medium speed ratio is also eliminated.

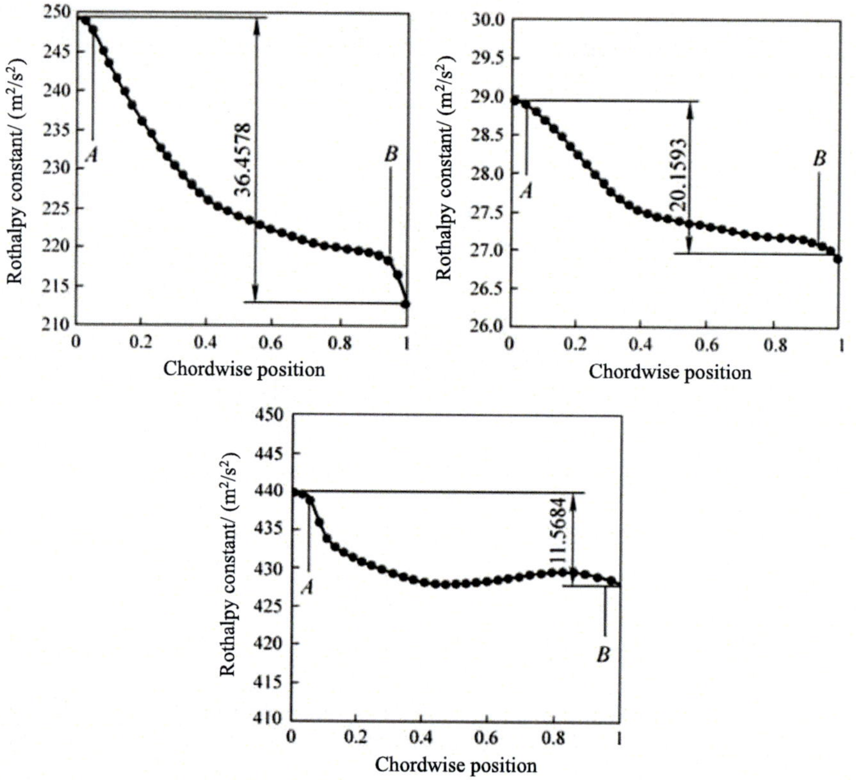

c) SR=0.80

Fig. 10.55 Change of Rothalpy constant in chordwise direction of impeller

10.6 Hydraulic Torque Converter Matching with Engine

The hydraulic torque converter has a great influence on the load distribution characteristics of the engine and the main performance of the vehicle. The power performance, fuel economy, exhaust pollution and other performance of the vehicle largely depend on whether the engine and torque converter work together well. Therefore, it is necessary to study their input characteristics, co-working range and stability, as well as their output characteristics. The co-working of the hydraulic torque converter and engine will be studied below to achieve the best match between them, so as to improve the performance of hydraulic traditional vehicles.

I. **Input characteristics of co-working**

According to the capacity factor curve $C_p = f(i)$ of the hydraulic torque converter shown in Fig. 10.10, the speed ratio i is given, C_P is found in the figure and then the impeller torque $T_p = C_p n_p^2$ is used to obtain the T_P value corresponding to different

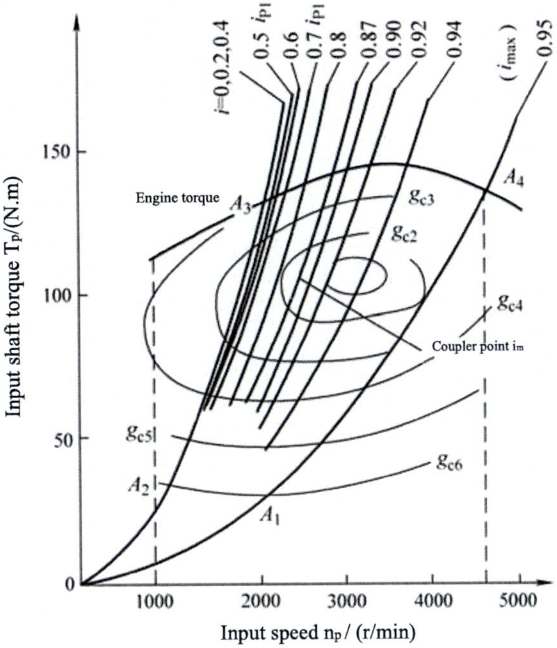

Fig. 10.56 Input characteristics of co-working of hydraulic torque converter and engine

input speed n_P. It is a parabola that goes through the origin. Different i has different C_p, which forms a set of load parabolas, including several special points: $i = 0$, i_{P1} and i_{P2} (corresponding to speed ratio at $\eta = 80\%$), i^* and $i_{\max}$. For the integrated torque converter, the coupler operating point i_m shall also be added. Then the engine load characteristics are plotted on the same chart in the same coordinate proportion. The intersection of two curves is the stable point of co-working, and the chart shows their input characteristics (Fig. 10.56). The area surrounded by $A_1A_2A_3A_4$ is the co-working range *a*.

For the integrated torque converter, the load lines of $\eta_{\max}$ and $i_{\max}$ shall be made intersect at the maximum engine power point A_4 and the $A_1A_2A_3A_4$ working range shall surround a good fuel economy zone. It is also necessary to check whether A_3 is in coincidence with or close to the maximum engine torque point T_{em} when $i = 0$ to improve the vehicle starting torque, or whether A_3 is in the left low n_e zone to reduce the noise. The height of point A_2 reflects the torque requirements for starting motor. These requirements are often contradictory. If the focus is on fuel economy zone, the acceleration performance is poor; on the contrary, if the focus is on the high power zone, the fuel economy will become worse, so we need to distinguish the priority according to the vehicle type. For a torque converter with lock, the matching relationship is different. Generally, the torque converter is locked and enters mechanical drive at i_m operating point, and a high velocity area shall be set aside at the intersection point between C_{PM} parabola and engine load characteristics in the coupler condition for working during locking. Allison also adopted the SCAAN

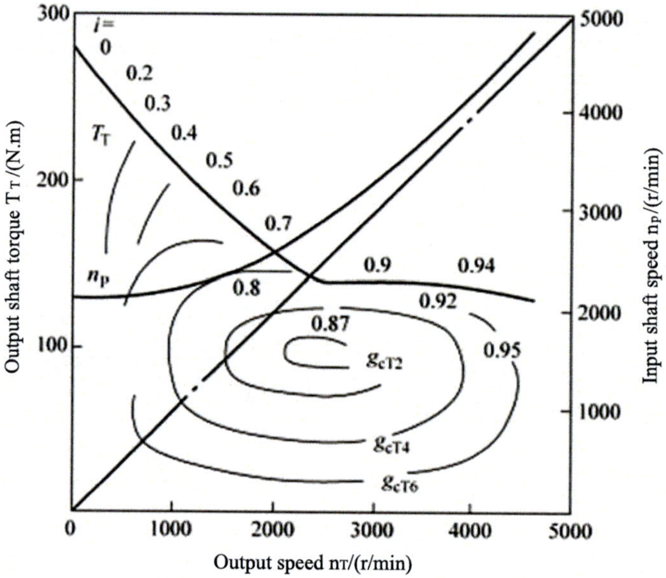

Fig. 10.57 Output characteristics of co-working of hydraulic torque converter and engine

program to evaluate the matching performance, which is based on the reliability index and gradeability. The practice has proved that this method is also successful.

Output characteristics of co-working: on the basis of input characteristics, $T_T = -KT_P$, $n_T = in_P$ and output specific fuel consumption $g_{eT} = g_e/\eta$ are used to obtain the output characteristics of co-working of the hydraulic torque converter and engine shown in Fig. 10.57, which are the basis for calculation of the vehicle traction characteristics. The quality of the output characteristics depends on the engine type, the torque converter characteristics, dimension D, and the input characteristics of co-working. Optimizing the matching of hydraulic torque converter and the engine enables it to work more in the high efficiency area, thus improving the economy and power performance of the co-working.

II. Adjustment of matching between hydraulic torque converter and engine

The co-working area can be adjusted when it is not ideal. The essence is to change the growth slope of the impeller torque T_P. As can be seen from $T_p = k_\lambda n_p^2 D^5$, there are the following adjustment ways:

(1) Change the effective diameter D, which is often used since D^5 has large and obvious influence on T_P.
(2) Change k_λ, i.e. change the blade angle β on the same premise to change the corresponding capacity, so as to meet the matching requirements. By changing D (coarse tuning) or β (fine tuning), the engine with different characteristics can be matched to realize the serialization and generalization of the hydraulic torque converter.

(3) Change the impeller speed n_P. Set an intermediate drive $i_{eP} = n_e/n_p$ between the engine and the torque converter, so that the original $n_p = n_e$ may be changed to new $n_p = n_e/i_{eP}$. n_p also has a large influence on T_P, which is also very effective for some models difficult in installation and layout.

Bibliography

1. Hua T (2005) Research on the modern design theory of torque converter. Jilin University, Changchun
2. Zhisheng Yu (2000) Automobile theory. China Machine Press, Beijing
3. Mitschke M, Wallentowitz H (2009) Automobile dynamics, vol 4. (Translated by Yinsan C, Qiang Y). Tsinghua University Press, Beijing
4. Hongxin Z (1996) Auto design. Version 2. China Machine Press, Beijing
5. Wangyu W (2017) Auto design. Version 2. China Machine Press, Beijing
6. Anlin G (2001) Automatic transmission (III)—locking and slip control of hydraulic torque converter. Autom Technol 7:1–4

Chapter 11
Planetary Gear Drive

With the development of automatic transmission technology, noise reduction and efficiency improvement have become important issues in the field of automotive technology. In AT, the planetary gear train, which can obtain a large gear ratio and meet the requirements of miniaturization and lightweight, has been widely applied, which puts forward higher requirements for the noise and efficiency of the planetary gear train. This chapter will introduce the theoretical calculation and tests of the transmission efficiency and vibration noise of the planetary gear train. The planetary gear train consists of an input shaft, an output shaft and a fixed shaft, all of which are aligned on the same axis. Figure 11.1 shows the power transmission route of the compound planetary gear train in an AT.

Three gear ratios can be obtained by controlling the engagement and disengagement of the clutch and brake in the double-row planetary gear train in Fig. 11.1:

(1) Fig. 11.1a shows the gear ratio 1:0.33 of gear 1.
(2) Fig. 11.1b shows the gear ratio 1:0.617 of gear 2.
(3) Fig. 11.1c shows the overdrive, with the gear ratio of 1:1.44.

See Table 11.1 for the main parameters of gears in the planetary gear train shown in Fig. 11.1.

The gears are made of SCr420H subject to carburizing and quenching heat treatment, with the surface hardness of about 58HRC.

It can be seen from the design parameters of the planetary gear in the AT that the planetary gear and center gear with high tooth and standard tooth are used respectively to change the contact ratio. In order to obtain different accuracy and surface roughness values of tooth surface, two sets of planetary gears and center gears are processed by two different processes (shaving and grinding), in which, one set is only subject to gear shaving, carburizing and quenching and the other set is processed with the same processing technology as the previous set before gear grinding. The former has the accuracy of JIS 3–4 and the tooth surface roughness of about R_aR_{max} 6 μm (R_a

Y. Chen, *Automotive Transmissions*, Springer Tracts in Mechanical Engineering,
https://doi.org/10.1007/978-981-15-6703-2_11

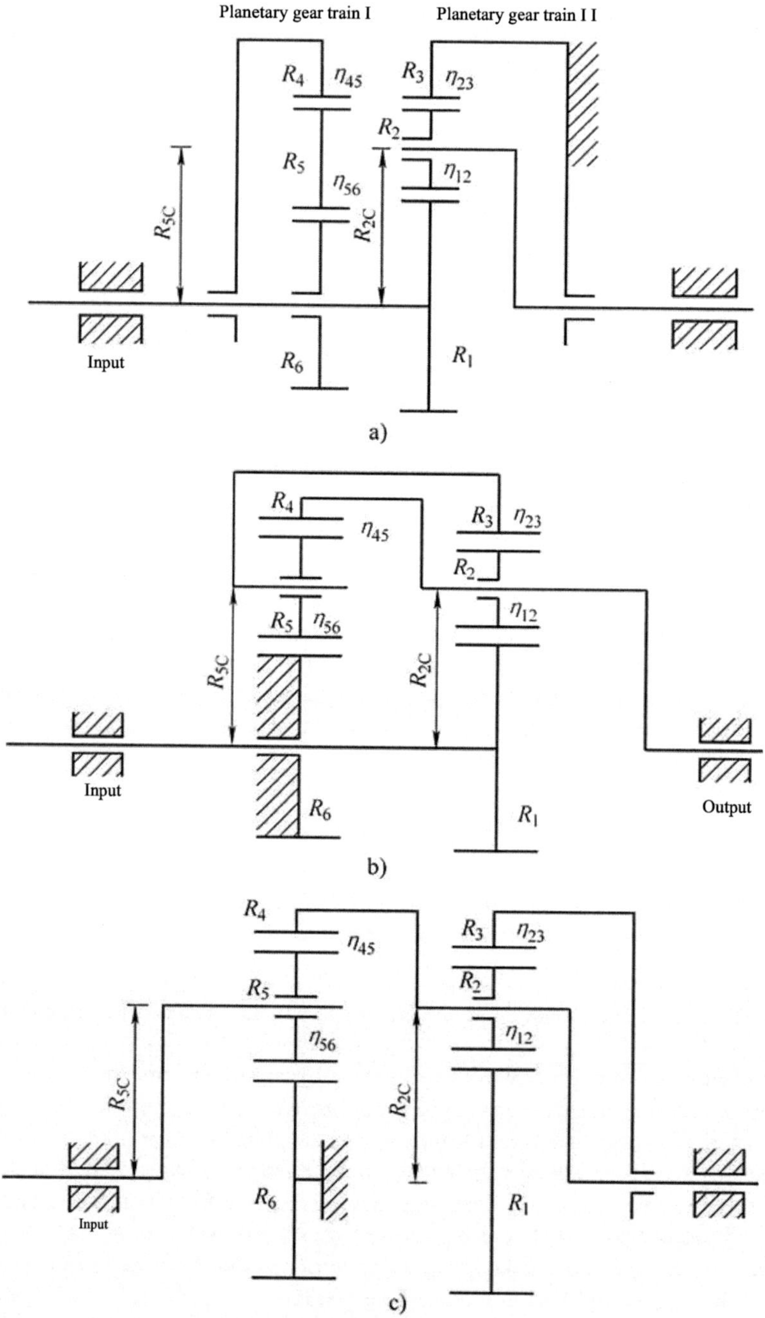

Fig. 11.1 Power transmission route of compound planetary gear train

Table 11.1 Main parameters of gears in planetary gear train

	Planetary gear train I			Planetary gear train II		
Gear type	Planetary gear	Internal gear	Center gear	Planet gear	Internal gear	Center gear
Number of gears	33	21	75	37	19	75
Normal module/mm	1.23			1.23		
Normal pressure angle	20°			20°		
Helical angle	23.3°			23.3°		
Pitch radius/mm	22.09	14.06	50.2	24.76	12.72	50.2
Effective face width/mm	17.0			17.0		
Profile contact ratio A	1.65	–	1.84	1.65	–	1.81
Profile contact ratio B	1.31	–	1.69	1.30	–	1.66
Longitudinal contact ratio	1.74	–	1.74	1.99	–	1.99

1.5 μm). The latter has the accuracy of JIS 0–1 and the tooth surface roughness of about R_aR_{max} 2 μm (R_a 0.5 μm).

11.1 Theoretical Calculation of Transmission Efficiency of Planetary Gear Train

The power transmission efficiency of the planetary gear train depends on the gear mesh efficiency (also known as the benchmark efficiency) when the planetary carrier is fixed. The higher the benchmark efficiency, the higher the efficiency of the planetary gear train. When meshing, the tooth surfaces in contact with each other will slip relative to other contact points other than the meshing point, so there is also tangential friction on tooth surface in addition to the normal force on tooth surface. The magnitude of tangential friction varies with the position of the meshing point on the meshing line. The existence of tangential friction on the tooth surface will reduce the power transmission efficiency and also affect the wear, vibration, noise and meshing effect of the gear teeth.

The benchmark efficiency η_{12} of the external meshed helical gear pair (center gear/planetary gear in the planetary gear train) is

$$\eta_{12} = 1 - \mu\pi\left(\frac{1}{z_A} + \frac{1}{z_B}\right)\left(\varepsilon_1^2 + \varepsilon_2^2 + 1 - \varepsilon_1 - \varepsilon_2\right)/\cos\beta \tag{11.1}$$

where,

μ—average friction factor of tooth surface;
z_A—number of teeth of center gear;
z_B—number of teeth of planetary gear;
ε_1—engaging-in contact ratio;
ε_2—engaging-out contact ratio;
β—helical angle.

The benchmark efficiency η_{23} of the internal meshed helical gear pair (planetary gear/gear ring in planetary gear train) is

$$\eta_{23} = 1 - \mu\pi\left(\frac{1}{z_B} - \frac{1}{z_C}\right)(\varepsilon_1^2 + \varepsilon_2^2 + 1 - \varepsilon_1 - \varepsilon_2)/\cos\beta \tag{11.2}$$

where,

z_C—number of teeth of gear ring.

Suppose the average friction factor $\mu = 0.08$, calculate the benchmark efficiency of two planetary gear sets in the planetary gear train for test using the formulas (11.1) and (11.2), substitute the data in Tables 11.2 and 11.3 and calculate to obtain:

(1) Benchmark efficiency of planetary gear/center gear pair in the first planetary gear set: $\eta_{56} = 0.984806$, benchmark efficiency of gear ring/planetary gear pair: $\eta_{45} = 0.992021$. The benchmark efficiency of the first planetary gear set is calculated as $\eta_{01} = \eta_{45}\eta_{56} = 0.976948$.
(2) Benchmark efficiency of planetary gear/center gear pair in the second planetary gear set: $\eta_{12} = 0.984245$, benchmark efficiency of gear ring/planetary gear pair: $\eta_{23} = 0.991144$. The benchmark efficiency of the second planetary gear set is calculated as $\eta_{02} = \eta_{12}\eta_{23} = 0.975527$.

Table 11.2 Gear parameters of first planetary gear set

Normal pressure angle	$\alpha = 20°$
Normal module	$m_n = 1.23$
Reference helical angle	$\beta = 23.262°$
Tooth addendum	$h_a = 1.17m_n$
Tooth dedendum	$h_f = 1.47m_n$
Number of teeth of sun gear	$z_A = 33$
Number of teeth of planetary gear	$z_B = 21$
Number of teeth of gear ring	$z_C = 75$
Normal modification coefficient of sun gear	$x_{n1} = -0.1539$
Normal modification coefficient of planetary gear	$x_{n2} = 0.1550$
Normal modification coefficient of gear ring	$x_{n2} = 0.1560$

Table 11.3 Gear parameters of second planetary gear set

Normal pressure angle	$\alpha = 20°$
Normal module	$m_n = 1.23$
Reference helical angle	$\beta = 23.2624°$
Tooth addendum	$h_a = 1.17m_n$
Tooth dedendum	$h_f = 1.47m_n$
Number of teeth of sun gear	$z_A = 37$
Number of teeth of planetary gear	$z_B = 19$
Number of teeth of gear ring	$z_C = 75$
Normal modification coefficient of sun gear	$x_{n1} = -0.2652$
Normal modification coefficient of planetary gear	$x_{n2} = 0.2160$
Normal modification coefficient of gear ring	$x_{n2} = 0.1560$

I. Calculation of gear efficiency in gear 1

In Fig. 11.1a, the gear ratio in gear 1 is

$$i_0 = \frac{z'_A}{z'_A + z'_C} \tag{11.3}$$

It is calculated that $i_0 = 0.3304$.

In Fig. 11.1a, the planetary gear R_2 is a drive gear and the center gear R_1 is a driven gear.

Taking into account the benchmark efficiency in Fig. 11.2, the friction power loss ΔL_1 between the center gear R_1 and planetary gear R_2 and the friction power loss ΔL_3 between the planetary gear R_2 and gear ring R_3 are calculated by means of the circumferential force.

When calculating the power loss, the power loss will not occur because the revolution motion will not cause the sliding between the tooth surfaces. On the center gear R_1, the input torque is T_i and the angular velocity is ω_1. The power transmission efficiency without the center gear revolution is

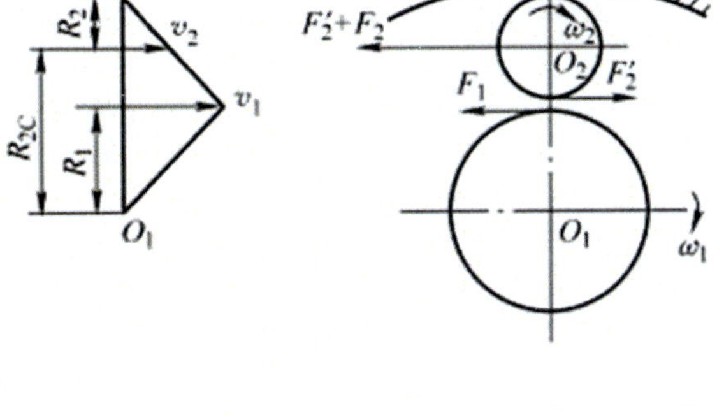

Fig. 11.2 Calculation of gear power loss in gear 1

$$\eta = 1 - \frac{\Delta L_1 + \Delta L_3}{T_i \omega_1} \tag{11.4}$$

where, ΔL_1 and ΔL_3 are calculated according to the angular velocity of each gear relative to the planetary carrier at the meshing point, namely

$$\Delta L_1 = F_1 R_1 \omega_1' (1 - \eta_{12}) \tag{11.5}$$

$$\Delta L_3 = F_2' R_2' \omega_2' (1 - \eta_{23}) \tag{11.6}$$

where,

ω'_1 and ω'_2—angular velocity of center gear and planetary gear relative to the planetary carrier.

The following relations

$$F_2' = \eta_{12} F_1;\ \omega_2' = \frac{\omega_1' R_1}{R_2};\ \omega_1 = \frac{\omega_1'(R_2 + R_3)}{R_3}$$

are substituted into the power transmission efficiency formula to obtain $\eta = 0.98361$.

II. **Theoretical efficiency calculation of planetary gear in gear 4**

The circumferential force on the gears in gear 4 is shown in Fig. 11.3.

In the figure, ω_{5C} is the angular velocity of planetary carrier. The relative angular velocities ω'_1 and ω'_5 are determined and the friction power loss ΔL_1 between the center gear and planetary gear and the friction power loss ΔL_3 between the planetary gear and gear ring are calculated by

$$\Delta L_1 = F_5 (1 - \eta_{56}) R_6 \omega_1' \tag{11.7}$$

$$\Delta L_3 = F_5' (1 - \eta_{45}) R_4 \omega_5' \tag{11.8}$$

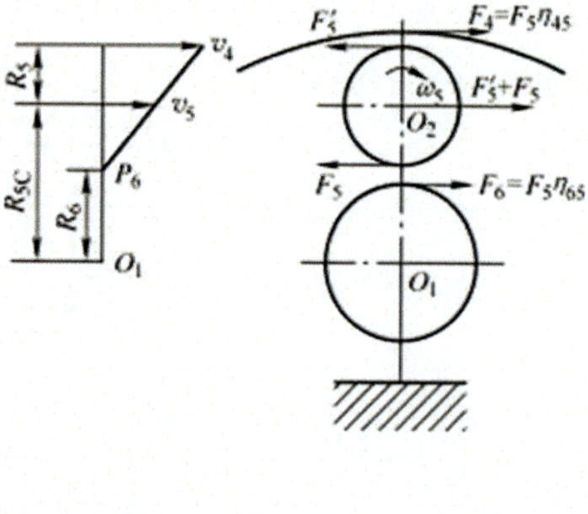

Fig. 11.3 Circumferential force acting on the gears in gear 4

where

$$\omega_1' = \omega_{5C};\ \omega_5' = \omega_4 - \omega_{5C}$$

The above and the following relations

$$\omega_4' = \frac{2R_{5C}}{R_4}\omega_{5C};\ F_5 = F_5';\ R_6 + 2R_5 = R_4$$

are substituted into the efficiency formula $\eta = 1 - (\Delta L_1 + \Delta L_1)/(T_5\omega_5)$ to obtain $\eta = 0.992921$.

III. **Derivation of theoretical efficiency formulas of gears in gear 2**

1. **Derivation of approximate formulas for theoretical efficiency not using mechanism coincidence**

The circumferential force acting on the gears in gear 2 is shown in Fig. 11.4. The relationship between the circumferential forces of each gear is

$$\begin{cases} F_{22} = F_{21} = F_{2C}/2 = F_1 = F_3 \\ F_4 = F_{52} = F_5 = F_6 = F_{5C}/2 \\ F_{5C}R_{5C} = F_3R_3 = F_1R_3 \\ F_4R_4 = F_{2C}R_{2C} \end{cases} \tag{11.9}$$

The circulation power is

$$L_C = R_{2C}\frac{R_3R_4}{2R_{2C}R_{5C}}F_1\omega_{2C} = \omega_3R_3F_3 \tag{11.10}$$

According to Fig. 11.2b, there is the following angular velocity relation

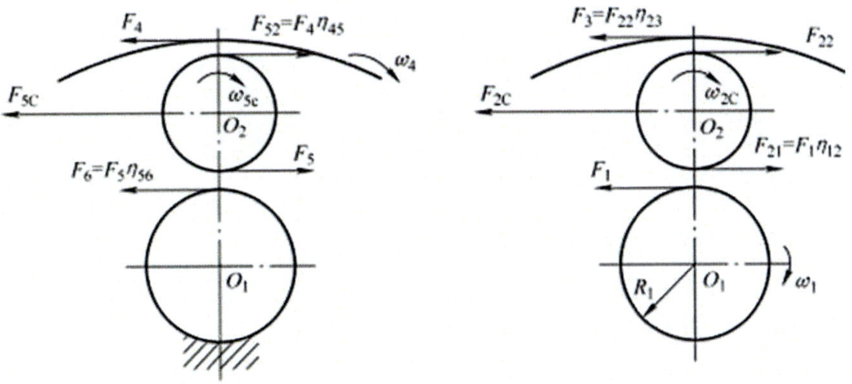

Fig. 11.4 Circumferential force acting on the gears in gear 2

$$\omega_6 = 0; \omega_{5C} = \omega_3; \omega_4 = \omega_{2C}; \omega_4 R_4 = 2\omega_{5C} R_{5C}$$

Taking into account the benchmark efficiency of the mechanism in Fig. 11.4, the power loss ΔL_{12} between the center gear R_1 and planetary gear R_2 in the second planetary gear set and the power loss ΔL_{23} between the planetary gear R_2 and gear ring R_3; the power loss ΔL_{45} between the gear ring R_4 and planetary gear R_5 in the first planetary gear set, and the power loss (friction heat loss) ΔL_{56} between the planetary gear R_5 and fixed center gear R_6 are calculated by means of the circumferential force. The revolution motion of the planetary gear is independent of the slip between the tooth surfaces. Here, the revolution rate C is introduced, so the proportion of the rotation component of the input angular velocity is $1 - C$.

The calculation process of the revolution rate is as follows: the angular velocity ω_{2C} of planetary carrier (R_{2C}) and the angular velocity ω_3 of gear ring R_3 are positive. If $\omega_{2C} < \omega_3$, the revolution rate is

$$C_3 = (2\omega_3 - \omega_{2C})/\omega_3 \tag{11.11}$$

If $\omega_{2C} < \omega_1$, the revolution rate is

$$C_1 = \omega_{2C}/\omega_1 \tag{11.12}$$

If $\omega_{5C} < \omega_4$, the revolution rate is

$$C_4 = \omega_{5C}/\omega_4 \tag{11.13}$$

The approximate formula for the efficiency of the dual-planetary gear set in Fig. 11.2b is

$$\eta = 1 - \frac{\Delta L_{12} + \Delta L_{23} + \Delta L_{45} + \Delta L_{56}}{F_{2C} R_{2C} \omega_{2C} - F_{22} R_3 \omega_3} \tag{11.14}$$

where

$$\begin{cases} \Delta L_{12} = F_1(1 - \eta_{12})(1 - C_1)\omega_1 R_1 \\ \Delta L_{23} = F_{22}(1 - \eta_{23})(1 - C_3)\omega_3 R_3 \\ \Delta L_{45} = F_4(1 - \eta_{45})(1 - C_4)\omega_4 R_4 \\ \Delta L_{56} = F_6(1 - \eta_{56}) C_4 \omega_4 R_6 \end{cases} \tag{11.15}$$

Finally, it is obtained that $\eta = 0.977355$.

2. **Derivation of approximate formulas for theoretical efficiency using mechanism coincidence**

In the dual-planetary gear set shown in Fig. 11.2b, the power will be shunted only when the center gear of the planetary gear set is fixed. A part of power flows back to R_{2C} from the gear ring R_4 and generates power cycle between the first and second planetary gear sets. The power passing through the planetary carrier R_{2C} is greater than the input/output power. The gear ring R_3 and center gear R_1 are the drive part and R_{2C} is the driven part. The second planetary gear set makes up a differential mechanism.

In the following calculation process, the dual-planetary gear set is divided into two to calculate its efficiency.

(1) Component 1: as shown in Fig. 11.5, the sun gear R_1 is fixed, the theoretical circulation power is input from R_3 and output from R_{2C}. In this process, the efficiency is η_{2C3}
The circulation power is

$$L_C = L_{C3} = F_3 R_3 \omega_3 = \omega_3 R_3 F_3 \tag{11.16}$$

Output power

$$L_{C3} = L_{C3} \eta_{2C3} \tag{11.17}$$

(2) Component 2: as shown in Fig. 11.6, the gear ring R_3 is fixed and the input drive power of the center gear R_1 is

$$L_1 = F_1 R_1 \omega_1 \tag{11.18}$$

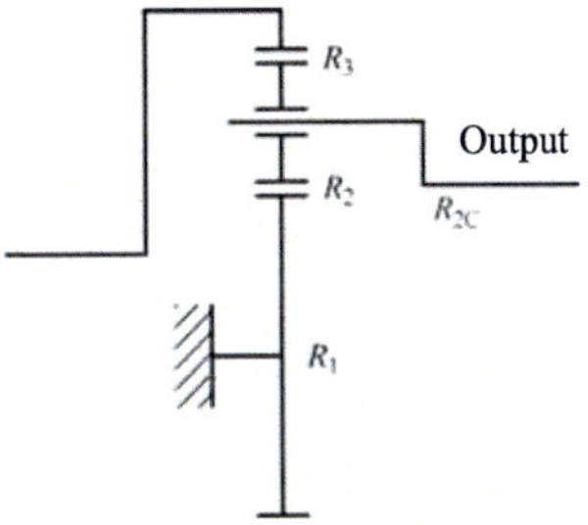

Fig. 11.5 Analysis of component 1

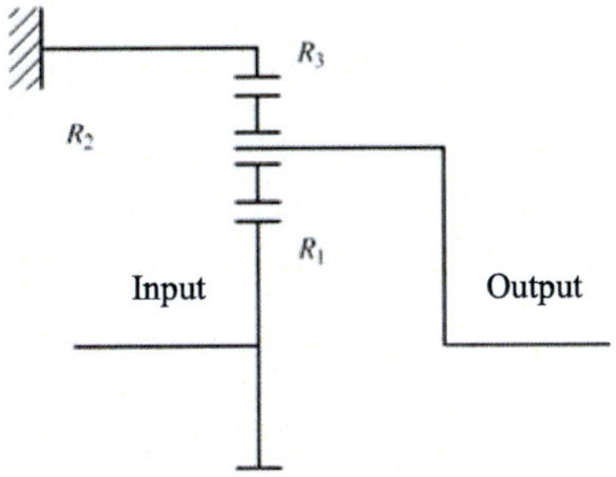

Fig. 11.6 Analysis of component 2

The power is output from the planetary carrier R_{2C} and the efficiency in the process is η_{2C1}. Then the output power is

$$L_{C1} = F_1\omega_1 R_1\eta_{2C1} \tag{11.19}$$

(3) Theoretical transmission efficiency (Figs. 11.7, 11.8 and 11.9)

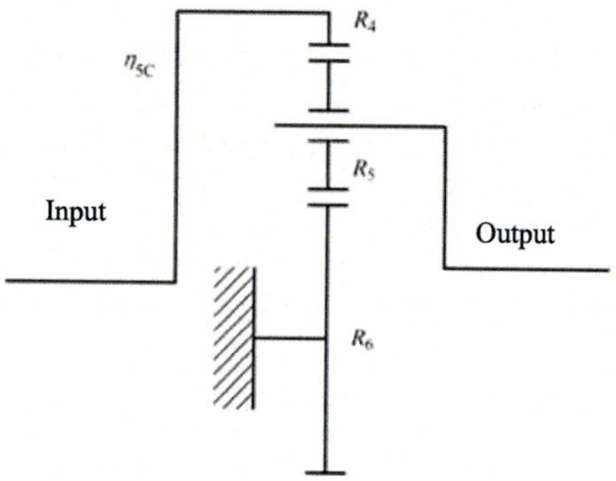

Fig. 11.7 R_6 fixed

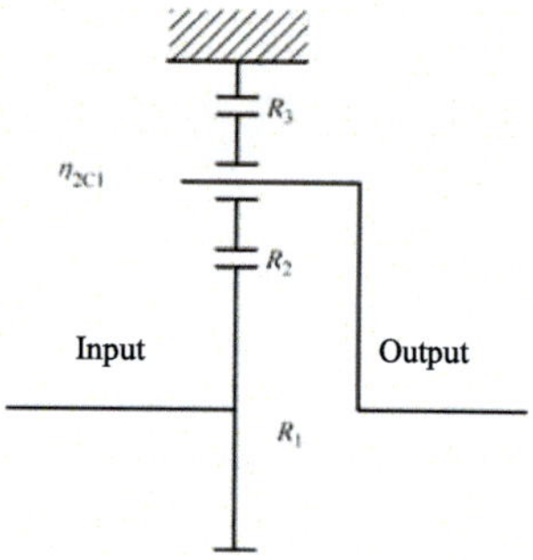

Fig. 11.8 R_3 fixed

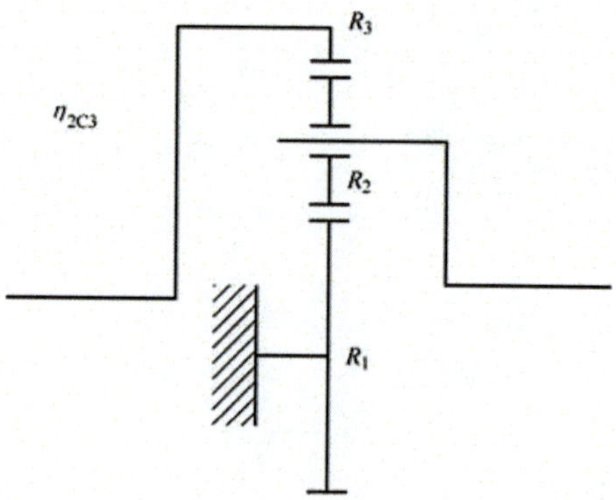

Fig. 11.9 R_1 fixed

$$\eta = \frac{L_{C1} + L_{C3} - L_{C3}/\eta_{5C}}{F_1 R_1 \omega_1} \tag{11.20}$$

where,
η_{5C}—efficiency when the center gear R_6 in the first planetary gear set is fixed, driven by the gear ring R_4 and output by the planetary carrier R_{5C}.

Then

$$\begin{cases} L_{C1} = F_1 R_1 \omega_1 \eta_{2C1} \\ L_{C3} = F_3 R_3 \omega_3 \eta_{2C3} \\ L_{C3} = L_C = F_3 R_3 \omega_3 \end{cases} \tag{11.21}$$

$$F_{3th} = F_{1th} \tag{11.22}$$

$$\omega_3 = \frac{R_4 \omega_{2C}}{2R_{5C}} = \frac{R_1 R_4 \omega_1}{2R_2 C_2 R_{5C} - R_3 R_4} \tag{11.23}$$

$$\eta_{5C} = \frac{\eta_{O1} + i_O}{1 + i_O} \tag{11.24}$$

$$i_O = R_4/R_6 \tag{11.25}$$

Figure 11.9 R_1 fixed It is calculated that $\eta_{5C} = 0.99296$.

$$\eta_{2C1} = \frac{1 + \eta_{O2} i_O}{1 + i_O} \tag{11.26}$$

$$i_O = R_3/R_1 \tag{11.27}$$

It is calculated that $\eta_{2C1} = 0.98361$.

$$\eta_{2C3} = \frac{\eta_{O2} z'_A + z'_C}{z'_A + z'_C} \tag{11.28}$$

It is calculated that $\eta_{2C3} = 0.99192$.
Through the above formulas, $\eta = 0.97042$.

IV. Theoretical efficiency calculation considering bearing loss

The main reasons that affect the efficiency of planetary gear include tooth surface friction loss, bearing friction loss, lubricant churning loss and air drag loss. To analyze the transmission efficiency of tooth surface of planetary gear, it is necessary to find out the bearing friction loss, lubricant churning loss and air drag loss, among which, the lubricant churning loss and air drag loss can be measured by no-load test. The bearing loss in the transmission is theoretically analyzed below.

(I) **Analysis of bearing friction**

The bearing friction moment can be measured experimentally and there are four types of friction:

(1) Pivoting friction.
(2) Friction caused by minute slip of rotary parts.
(3) Slip contact friction.
(4) Lubricant viscosity.

The friction loss varies with the bearing form, working speed and lubrication conditions.

The bearing friction can be divided into dynamic friction and static friction, and the dynamic friction is related to working load and speed. Items (1), (2) and (3) above, namely, the contact between bearing metal parts are determined by the load; item (4) is determined by the speed.

(II) **Transmission efficiency of thrust bearing**

The needle roller of thrust bearing under axial load will produce corresponding friction force. In the case of good lubrication, the sliding friction factor is about 0.006; in case of poor lubrication, the friction factor is about 0.015.

1. **Transmission efficiency of thrust bearing in gear 1**

Figure 11.10 shows the installation positions of the bearings in the planetary gear train for test. The arrows in Fig. 11.11 indicate the direction of the thrust on the center gear and gear ring of the gears in gear 1 (single planetary gear set).

Through the analysis of the thrust on the planetary gear train in Fig. 11.11, the center gear R_1 and gear ring R_3 in Fig. 11.1a have thrust effect on the thrust bearings 5 and 6 (Fig. 11.10) and the transmission efficiency is

$$\eta_{b1} = 1 - \frac{F_1 \tan\beta R_{h5}\mu_{r1}(\omega_1 - \omega_{2C})}{F_1 R_1 \omega_1} \tag{11.29}$$

$$\eta_{h2} = 1 - \frac{F_1 \tan\beta R_{b5}\mu_{r1}\omega_{2C}}{F_1 R_1 \omega_1} \tag{11.30}$$

where,

μ_{r1}—friction factor of thrust bearing;
R_{b5}—rotation radius of the roller center of bearing 5;
R_{b6}———rotation radius of the roller center of bearing 6.

2. **Transmission efficiency of thrust bearing in gear 2**

Figure 11.12 shows the direction of the thrust of the 2 speed planetary gear train on the center gear and gear ring. The thrust of the planetary gear train in Fig. 11.12 is analyzed. The thrust bearings 5 and 6 are subject to the thrust of the center gear R_1 and gear ring R_3 in Fig. 11.1b and the transmission efficiency is

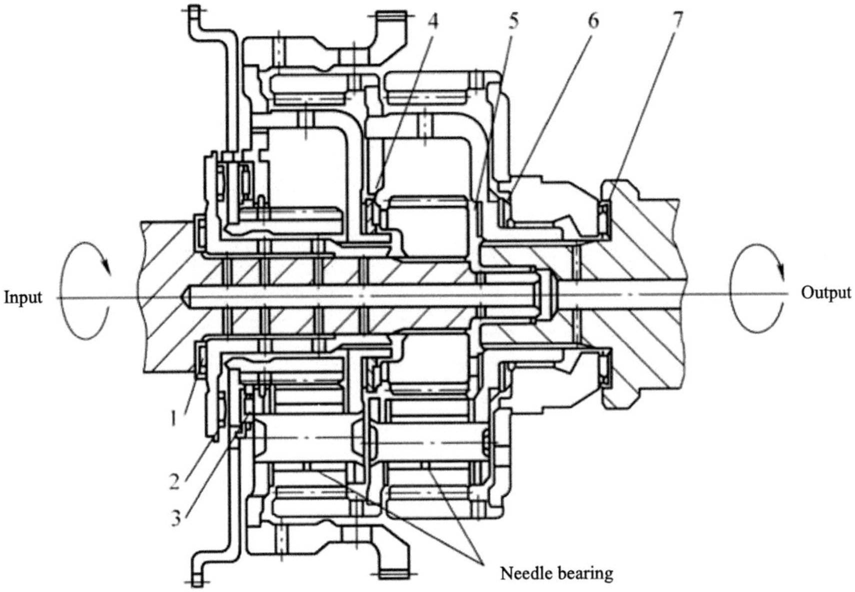

Fig. 11.10 Installation positions of the bearings in the planetary gear train for test. 1–7—bearing

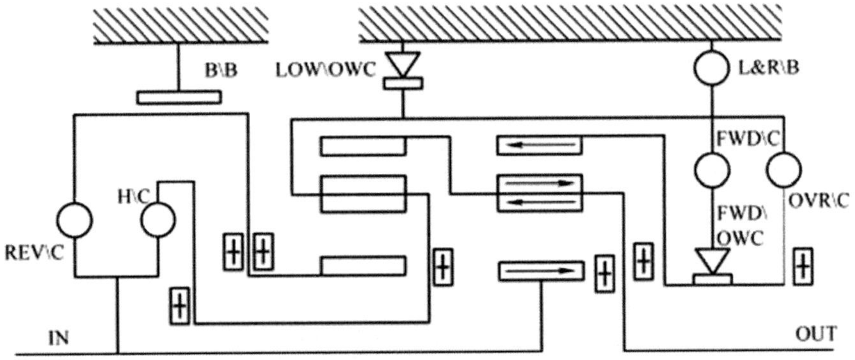

Fig. 11.11 Force on the center gear and gear ring in gear 1

$$\eta_{h3} = 1 - \frac{F_{3th} \tan \beta R_{b5} \mu_{r1} \omega_3}{F_1 R_1 \omega_1} \tag{11.31}$$

The thrust bearing 4 is subject to the thrust of the gear ring R_4 and the transmission efficiency is

$$\eta_{b4} = 1 - \frac{F_4 \tan \beta R_{b4} \mu_{r1} (\omega_1 - \omega_{2C})}{F_1 R_1 \omega_1} \tag{11.32}$$

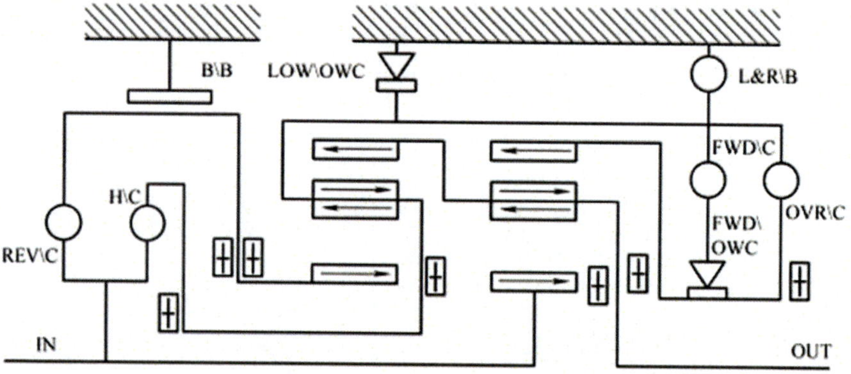

Fig. 11.12 Force on the center gear and gear ring in gear 2

where,

R_{b4}—rotation radius of the roller center of bearing 4.

(III) **Transmission efficiency of needle bearing**

The crowded needle bearing installed on the planetary gear is free of retainer, inner ring and outer ring, maximizing the miniaturization, increasing the effective radius and greatly increasing the rated load of the crowded needle bearing. The planetary gear and crowded needle bearing are shown in Fig. 11.13. Figure 11.14 is the installation diagram of crowded needle bearing.

The crowded needle bearing has no retainer, so the roller will tilt, which will produce axial thrust acting on the inner ring and the outer ring in opposite direction. Because the length of the needle roller is larger than its diameter, the outer ring flange

Fig. 11.13 Planetary gear and crowded needle bearing

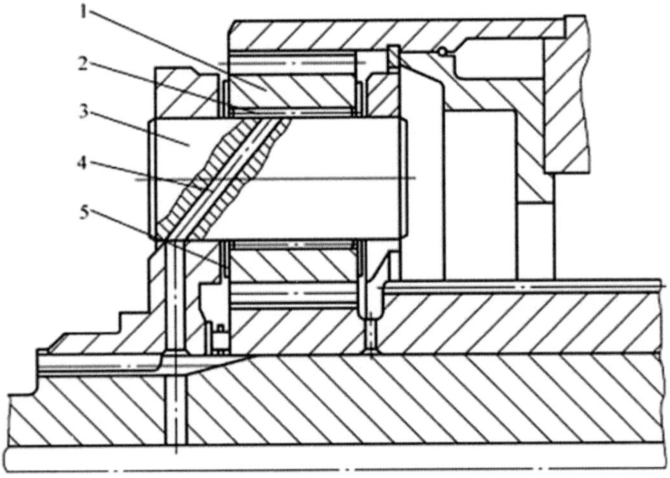

Fig. 11.14 Installation diagram of crowded needle bearing. 1—milled planetary gear (carburizing and quenching), 2—needle roller, 3—carburized and quenched shaft, 4, 5—oil passage

is unable to limit the rolling direction of the needle roller, which may easily cause the tilt of the needle roller and will have a large influence on the friction.

The gears used in the planetary gear train of AT are helical gears, and the axial force generated by the meshing of the gear ring and planetary gear is opposite to that generated by the meshing of the center gear and planetary gear, which will aggravate the tilt trend of roller. The bearing load caused by the tilt of the needle roller accounts for a portion of the bearing axial load. The friction resistance caused by the tilt of the needle roller will increase the friction force.

1. **Transmission efficiency of needle bearing in gear 1**

The approximate formula for the transmission efficiency η_{B6} of the crowded needle bearing used by the planetary gear in the second planetary gear set is

$$\eta_{b6} = 1 - \frac{F_{P1} R_{b1} \mu_{r1}}{F_1 R_1 \omega_1} \frac{R_1 (1 - C_1) \omega_1}{R_2} \tag{11.33}$$

$$F_{P1} = F_1 \sqrt{1 + [(4R_2/B_1) \tan \beta]^2} \tag{11.34}$$

where,

F_{P1}—compound radial load of planetary gear R_2 acting on the needle bearing;
μ_{r1}—friction factor of thrust bearing;
ω_1—angular velocity of input shaft;
C_1—revolution rate, $C_1 = \omega_{2C}/\omega_1$, in which, ω_{2C} is the angular velocity of output shaft.

$$C_1 = \omega_{2c} \div \omega_1. \tag{11.35}$$

F_1—circumferential force acting on the sun gear R_1;
B_1—needle roller length of the needle bearing on the planetary gear R_2;
R_{b1}—bearing bore diameter;
β—helical angle of planetary gear;
R_1—pitch radius of sun gear R_1;
R_2—pitch radius of planetary gear R_2.

2. Transmission efficiency of needle bearing in gear 2

In the 2 speed dual-planetary gear set shown in Figs. 11.1b and 11.15, 10 bearings are subject to the radial and axial loads by the gears and support the components in the planetary gear train, in which, 7 bearings are crowded needle bearings.

When the radial and axial loads act on the crowded needle bearings in the second planetary gear set, the theoretical transmission efficiency is

$$\eta_{b7} = 1 - \frac{F_{P1} R_{b1} \mu_{r1}}{F_1 R_1 \omega_1} \frac{R_1(1 - C_1)\omega_1}{R_2} \tag{11.36}$$

The theoretical transmission efficiency of the crowded needle bearings in the first planetary gear set is

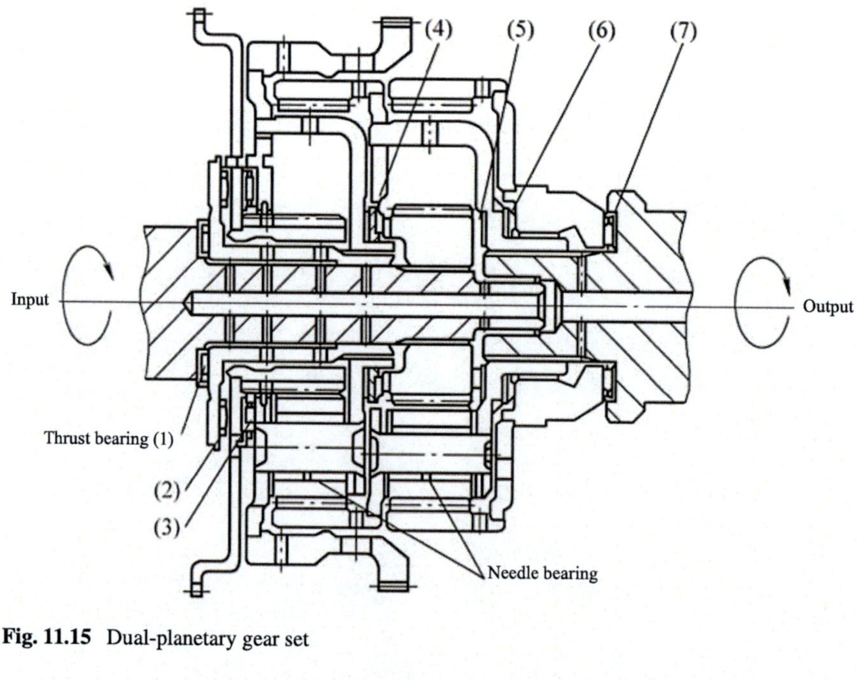

Fig. 11.15 Dual-planetary gear set

$$\eta_{b8} = 1 - \frac{F_{P2} R_{b2} \mu_{r2}}{F_1 R_1 \omega_1} \frac{R_4 (1 - C_4) \omega_4}{R_5} \tag{11.37}$$

$$F_{P2} = F_1 \sqrt{1 + [(4R_2/B_2) \tan \beta]^2} \tag{11.38}$$

where,

F_{P2}—compound radial load of planetary gear R_5 acting on the needle bearing;
μ_{r2}—friction factor of thrust bearing;
C_4—revolution rate, $C_4 = \omega_{5C}/\omega_4$, in which, ω_4 is the angular velocity of gear ring R_4 in the first planetary gear set;
ω_{5C}—angular velocity of revolution of planetary gear R_5 in the first planetary gear set;
B_2—needle roller length of the needle bearing on the planetary gear R_5;
R_{b2}—bearing bore diameter;
R_4—pitch radius of gear ring R_4;
R_5—pitch radius of planetary gear R_5.

V. **Theoretical calculation of overall efficiency**

1. **Overall efficiency of 1 speed gears**

The theoretical overall efficiency η_{rth} of the 1 speed gears (single planetary gear set) is

$$\eta_{rth} = \eta_{th} \eta_{b1} \eta_{b2} \eta_{b6} \eta_0 \tag{11.39}$$

where,

η_{th}—tooth surface benchmark efficiency;
η_{b1}, η_{b2}—transmission efficiency of thrust bearing;
η_{b6}—efficiency of needle bearing;
η_0—idle efficiency of 1 speed single planetary gear set test.

Affected by the lubricant churning loss and the air drag loss from the gear drive, it is difficult to calculate the idle efficiency η_0, which is mainly affected by the speed and is unrelated to the load, so its value can be tested through a blank experiment.

2. **Theoretical overall efficiency of 2 speed gears**

The theoretical overall efficiency of the 2 speed gears (dual-planetary gear set) is

$$\eta_{rth} = \eta_{th} \eta_{b3} \eta_{b4} \eta_{b5} \eta_{b7} \eta_{b8} \eta_0 \tag{11.40}$$

where,

η_{b3}, η_{b4} and η_{b5}—transmission efficiency of thrust bearing;
η_{b7} and η_{b8}—efficiency of needle bearing;
η_0—idle efficiency of 2 speed dual-planetary gear set test.

11.2 Transmission Efficiency Test of Planetary Gear Train

The transmission used in the test is of dual-planetary gear set type, as shown in Fig. 11.15. The transmission has four gears and its power transmission route is shown in Fig. 11.16.

In order to reduce the impact of vibration and elastic deformation, the transmission case is required to be stiff enough; a coupling is arranged on the fixed part or the transmission shaft to ensure that the gears mesh evenly when working. Each part shall have high manufacture and assembly accuracy, and the position error at the center of rotation of the input and output shafts shall be controlled below 0.025 mm.

Three lubricants are used in the test. A is mineral oil commercially available for ordinary lubrication, B is the automobile ATF and C is the steam turbo-compressor lubricant (mineral oil). The ATF is mainly used in this test.

The relationship between the kinematic viscosity and temperature of the three lubricants is shown in Fig. 11.17.

The composition and principle of the power absorption type gear efficiency test bench for the test are shown in Fig. 11.18. The drive motor drives the test planetary gear train to work through the CVT, V-belt, coupling and torque estimator. The load is loaded on the output shaft of the planetary gear train by the drum brake via the output torque estimator and coupling.

Before the test, the bench shall run smoothly for 20 min, and then gradually load from the no-load state for the test. The speed of the input shaft is controlled within the range of 500–3000 r/min, and the torque is controlled within the range of 0–85 N m. In order to make the lubrication mode consistent with AT, the oil pumping shall be used for forced lubrication, and the three injectors near the center gear and one injector in the upper part of the center gear are used to provide lubricant for the planetary gear set. The lubricant temperature at the injectors is determined by a temperature sensor.

In fact, it is difficult to determine the gear transmission efficiency. Although it is considered to calculate the transmission efficiency (overall efficiency) through the theoretical efficiency calculation and test method of the meshing loss, the influence factors and mechanical relations are not easy to study. The structure of the planetary gear train is more complex and it is more difficult to accurately determine its transmission efficiency.

Under the forced lubrication conditions, the power loss types affecting the overall efficiency η of the planetary gear train are as follows:

(1) Meshing tooth surface friction loss ΔL_1.
(2) Bearing friction loss ΔL_2.
(3) Churning loss ΔL_3 caused by the rotating parts such as gear churning the lubricant.
(4) Air drag loss ΔL_4 generated in the rotation of parts such as gear.

In the above items, (1) and (2) are load influence, (3) and (4) are speed influence. The overall efficiency is

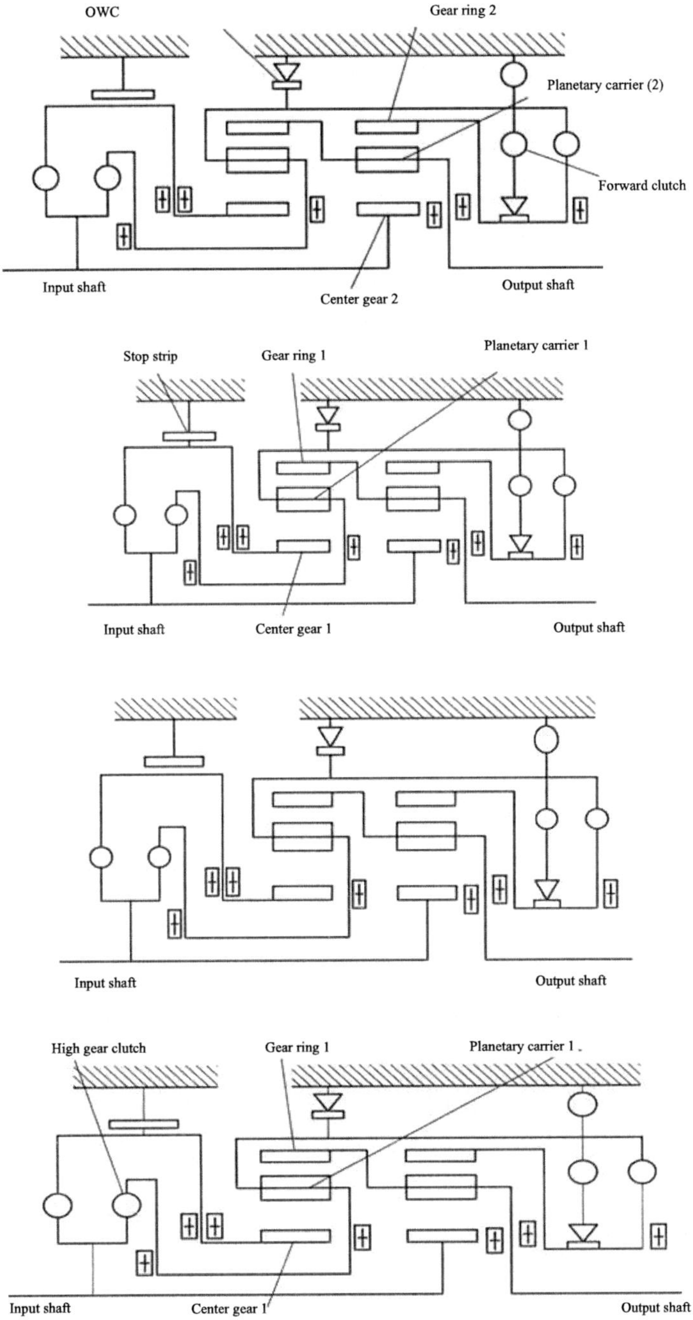

Fig. 11.16 Power transmission route of 4 gears

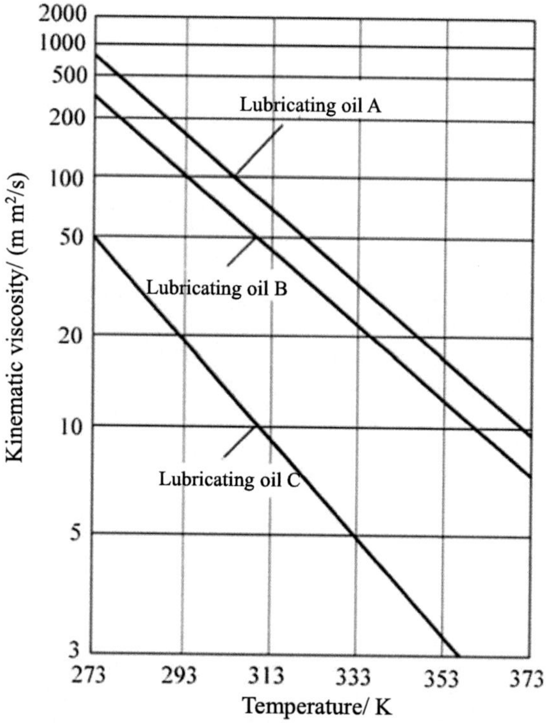

Fig. 11.17 Relationship between the kinematic viscosity and temperature of three lubricants

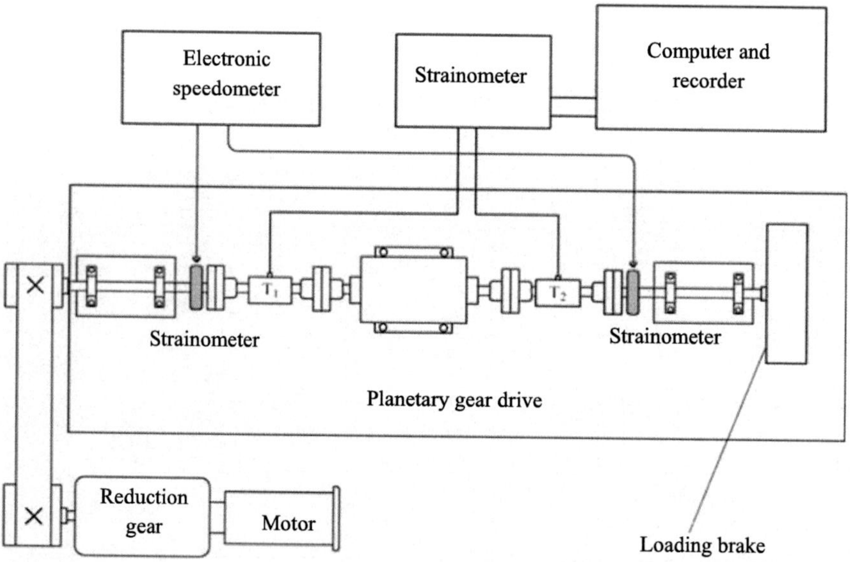

Fig. 11.18 Composition and principle of the power absorption type gear efficiency test bench

$$\eta = 1 - \frac{\Delta L_1 + \Delta L_2 + \Delta L_3 + \Delta L_4}{T_i \omega_1} \tag{11.41}$$

where,

T_i—input torque;
ω_1—drive angular velocity.
$\Delta L_3 + \Delta L_4$ may be measured in no-load state and ΔL_2 may be calculated from the bearing loss formula.

After the power loss $\Delta L_3 + \Delta L_4$ in the idling is measured, the theoretical calculation efficiency mentioned above is compared with the measured efficiency to check the influence of the power cycle and other factors on the overall efficiency.

The single and dual-planetary gear sets are tested respectively, and the influence of the variables controlled and adjusted, including speed, load, gear accuracy, gear contact ratio, lubricant type, temperature and flow, on the transmission efficiency is analyzed finally.

I. **Influence of churning loss on overall efficiency**

At the same temperature, the higher the viscosity of the lubricant, the greater the churning loss, as shown in Fig. 11.19. The lower the temperature of the lubricant, the greater the churning loss, as shown in Fig. 11.20. As can be seen from Fig. 11.21, at the same temperature, the higher the lubricant flow and gear speed, the greater the churning loss.

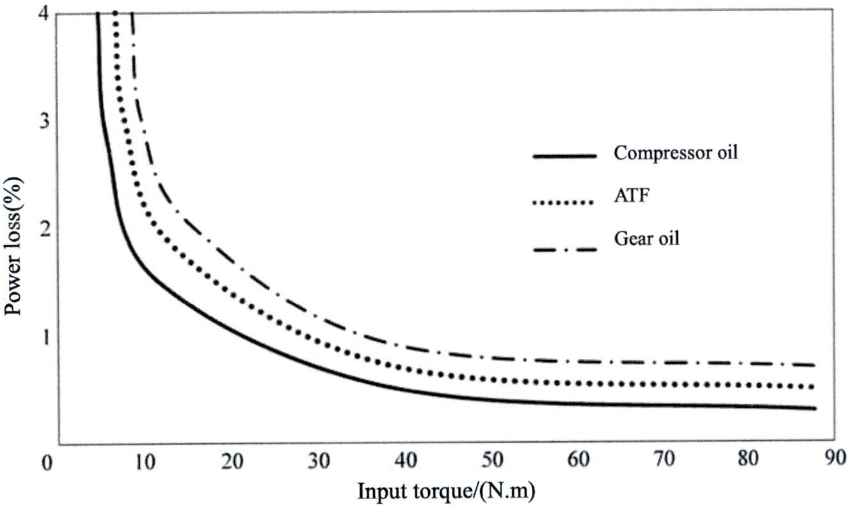

Fig. 11.19 Relationship between lubricant viscosity and churning loss

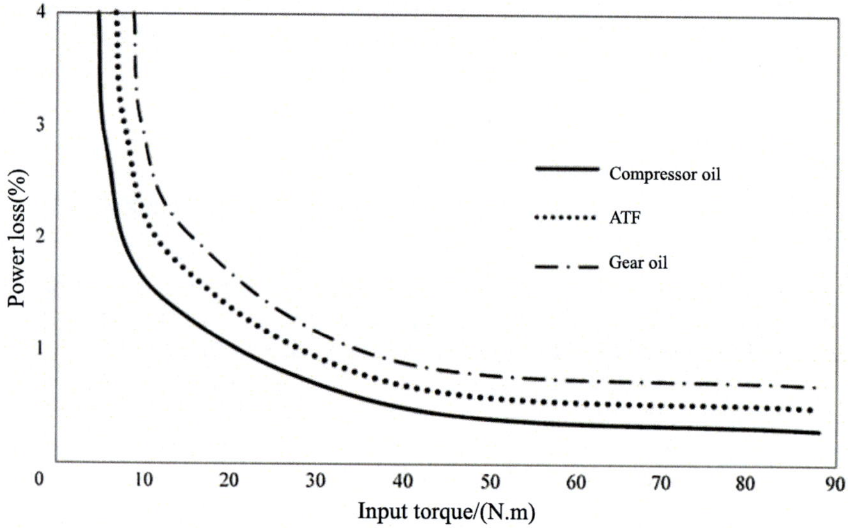

Fig. 11.20 Relationship between lubricant temperature and churning loss

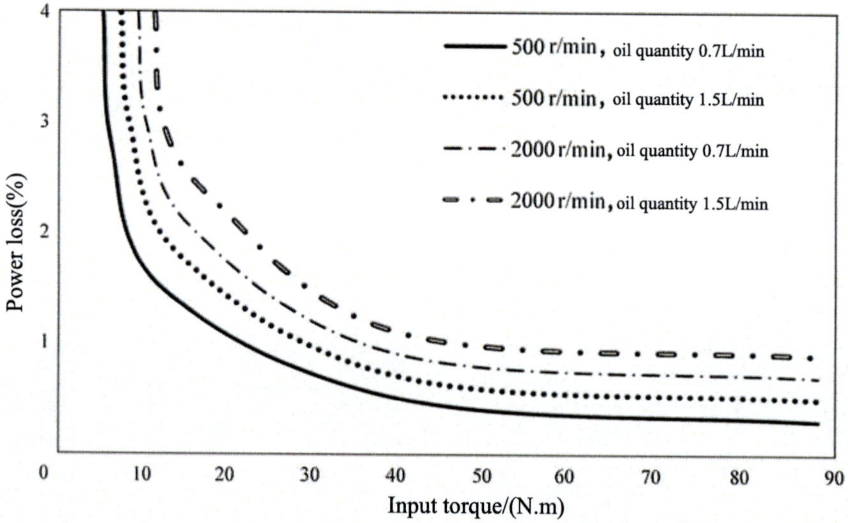

Fig. 11.21 Relationship between lubricant flow and gear speed and churning loss

II. **Influence of processing technology on overall efficiency**

The set of center gear and planetary gear of 1 speed gear for the test is shaved and the other set is ground. The test results are shown in Fig. 11.22 and the transmission efficiency of gears manufactured by the two processing technologies is slightly different. The transmission efficiency of the ground gears is about 0.5% higher than

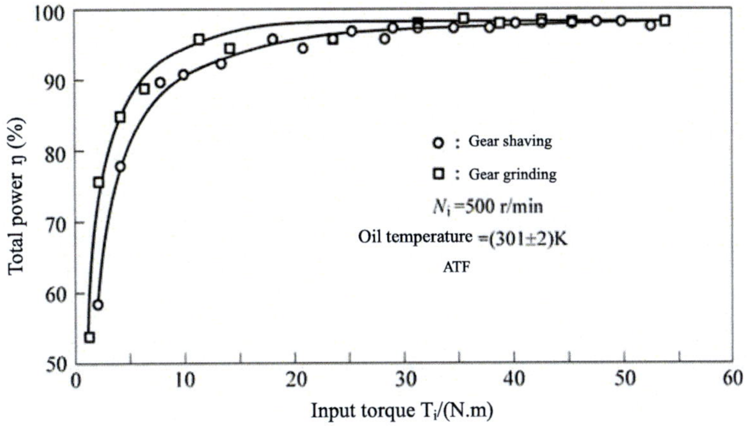

Fig. 11.22 Influence of processing technology on overall efficiency

that of the shaved gears. This is because the tooth surface of the shaved gears is rough and less accurate with large tooth surface friction factor.

III. **Influence of speed on overall efficiency**

The 1 speed gear is selected for the test, with the gear ratio of 1:0.33, the processing technology of gear shaving and the input speed of 500, 1500 and 2000 r/min. The lubricant with high viscosity is selected.

The test results are shown in Fig. 11.23. The corresponding transmission efficiency has little difference at the speed of 500 and 1000 r/min; but it decreases obviously at the speed 2000 r/min. It is speculated that the lubricant churning loss and the air drag loss are related to the square of the speed.

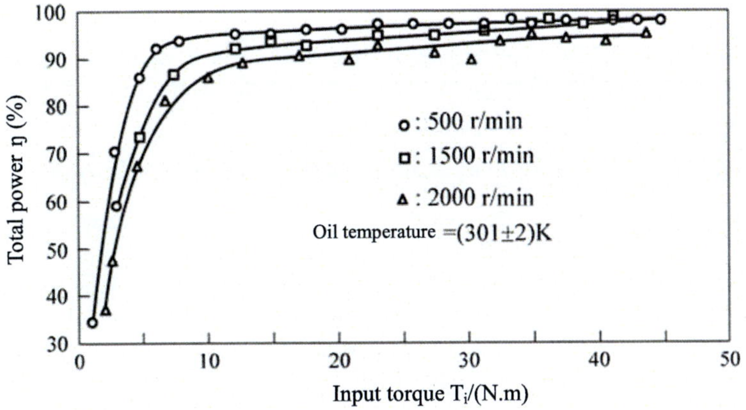

Fig. 11.23 Influence of speed on overall efficiency

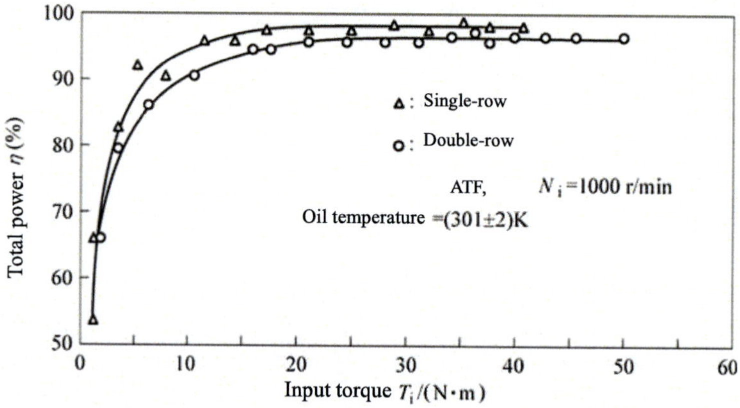

Fig. 11.24 Influence of single and dual-planetary gear sets on overall efficiency

IV. Influence of single and dual-planetary gear sets on overall efficiency

Figure 11.24 shows the comparison of the transmission efficiency test results of single and dual-planetary gear sets. The processing technology is gear grinding. The center gear of the first planetary gear set in the dual-planetary gear set is fixed and the second planetary gear set is of differential form. The 2 speed power cycle transmission route of a power ($F_{22}R_3\omega_3$) is: gear ring R_3 → planetary gear R_2 → planetary carrier R_{2C} → gear ring R_4 → planetary carrier R_{5C} → gear ring R_3. The efficiency reduction caused by the power cycle is also analyzed in the theoretical calculation of the previous section.

V. Influence of contact ratio on overall efficiency

The test only changes the gear contact ratio of the planetary gear train and the influence of the contact ratio on the overall efficiency is shown in Fig. 11.25. The contact ratio 1.30/1.66 is represented by "□", while the contact ratio 1.65/1.82 is represented by "Δ". The efficiency decreases in the large contact ratio because the increase in the relative slippage of two meshing teeth results in the friction loss increase.

VI. Influence of lubricant temperature and flow on overall efficiency

The influence of the lubricant temperature on the transmission efficiency is shown in Fig. 11.26. The lubricant temperature is high and the total efficiency is high at low load because of low lubricant viscosity and small churning loss at high temperature.

The influence of the lubricant flow on the transmission efficiency is shown in Fig. 11.27. When the lubricant flow is increased, the overall efficiency will decrease significantly; but after the load is increased, the influence of the lubricant flow on the efficiency will decrease. It is because that the increase in the lubricant flow will make the viscous resistance increase and the churning loss is not affected by the load on the tooth surface.

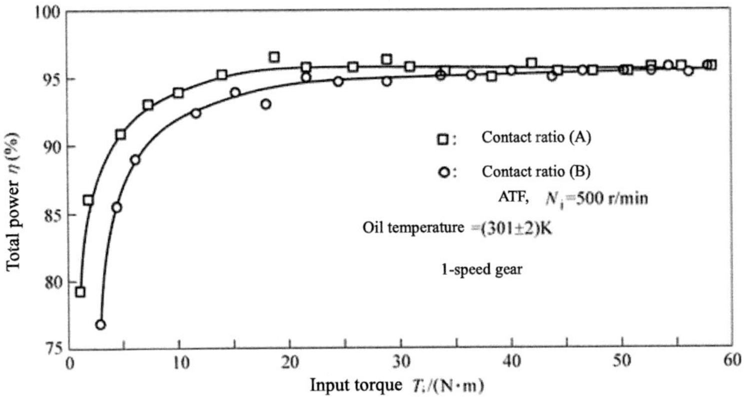

Fig. 11.25 Influence of contact ratio on overall efficiency

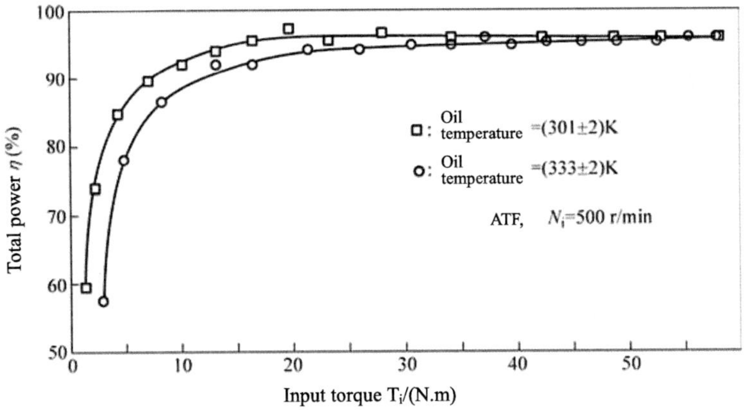

Fig. 11.26 Influence of lubricant temperature on overall efficiency

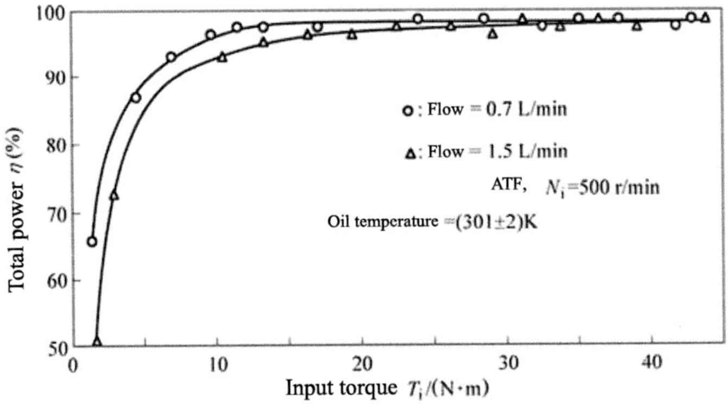

Fig. 11.27 Influence of lubricant flow on overall efficiency

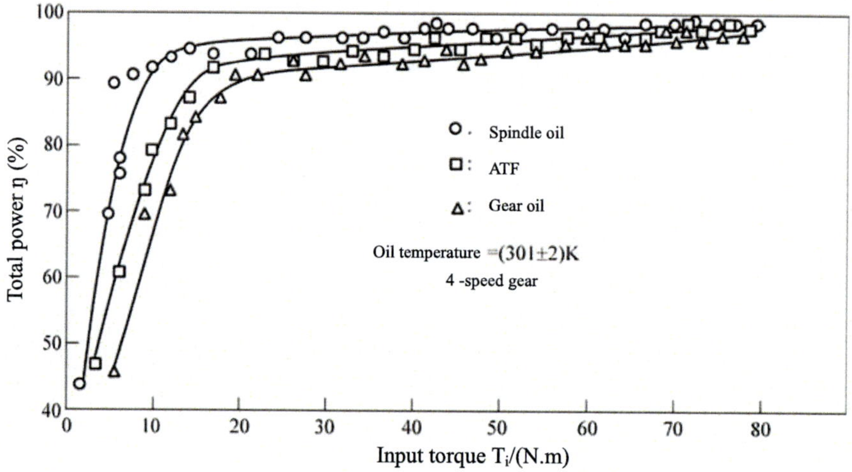

Fig. 11.28 Influence of lubricant type on overall efficiency

VII. Influence of lubricant type on overall efficiency

The 4 speed gear is selected for the test, with the gear ratio of 1:1.44, the processing technology of gear grinding and the input speed of 500 r/min. The test results are shown in Fig. 11.28. Among the three lubricants applied, that with low viscosity has high overall efficiency, but the influence of viscosity on transmission efficiency is decreased with the increase of load.

Even though the difference in the kinematic viscosity of the lubricant is up to 6 times, the influence of the lubricant type on the transmission efficiency is not significant when the medium load or above is applied.

VIII. Theoretical calculation and test results analysis of overall efficiency

Figure 11.29 shows the theoretical efficiency and test efficiency of the input torque and tooth surface meshing of the 1 speed gear at the input speed 500 r/min, as well as the relationship between the efficiencies with consideration to the churning loss and bearing loss. The input torque is 0–50 N m. The tooth surface benchmark efficiency η_{th} of the gear is calculated with the average friction factor between tooth surfaces $\mu = 0.08$ and marked with straight line in Fig. 11.29; the dash-dotted line represents the efficiency curve including churning loss and air loss; the dotted line represents the efficiency calculated after considering the needle bearing friction factor $\mu_{r1} = 0.0025$ and thrust bearing friction factor $\mu_{r2} = 0.008$ ($\eta_{rth} = \eta_{th}\ \eta_{b1}\ \eta_{b2}\ \eta_{b6}\ \eta_0$).

μ_{r1} is inversed by the test value of η_{rth} to obtain $\mu_{r1} = 0.011$. That is, the measured overall efficiency under this friction factor is at the same level with the calculated value.

$\mu_{r2} = 0.0025$ is the friction factor of a needle bearing with a retainer at low speed operating without tilt of the needle roller. In fact, the crowded needle bearing without retainer used for the AT will make the needle roller tilt due to the radial and

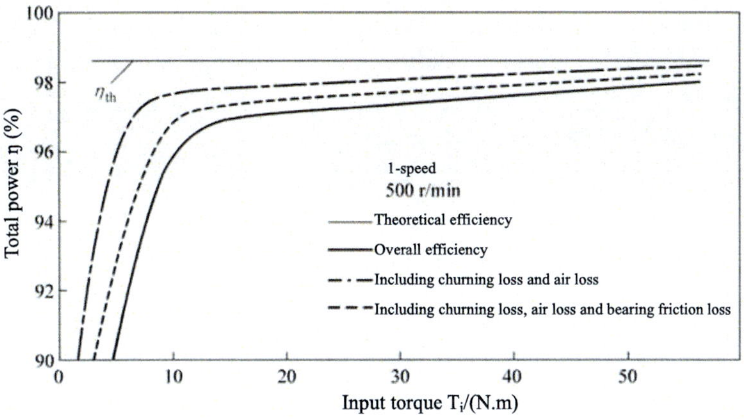

Fig. 11.29 Analysis of 1 speed gear test results

axial loads in the planetary gear train. That is, the roller rotates in the tilt condition, which will lead to an increase in the rotational resistance under unchanged load, i.e. increase in the friction factor. Moreover, when the speed reaches above a few hundred revolutions per minute, the influence of the needle roller tilt on friction factor will be more serious. Therefore, the needle bearing friction factor $\mu_R = 0.010$ inversed from the test is reliable. The analysis of the 2 speed gear test results is shown in Fig. 11.30.

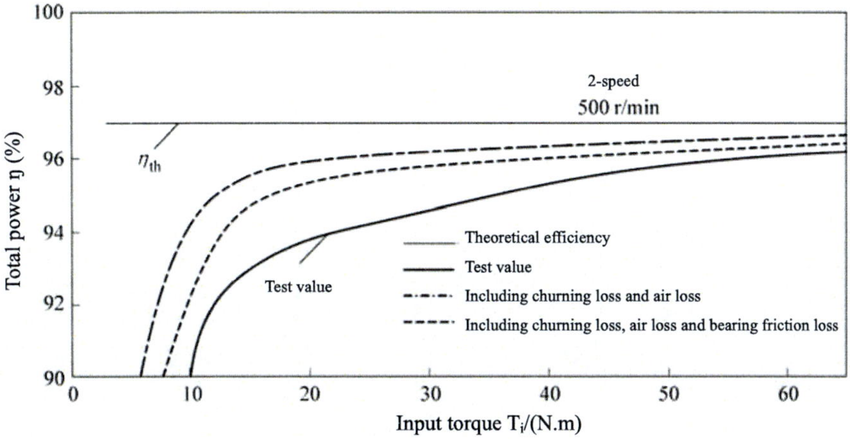

Fig. 11.30 Analysis of 2 speed gear test results

11.3 Theoretical Calculation of Vibration and Noise of Planetary Gear Train

Gear noise is the main problem of AT noise. From low speed to high speed, noise is generated in gear engagement under various operating conditions. The planetary gear train is a gear mechanism in which several planetary gears are meshed with the center gear and gear ring simultaneously and its characteristics are different from the external meshed gear.

Due to the pitch error, tooth contour error and tooth profile error in the planetary gear train, even if the input speed is constant, the output speed will fluctuate periodically, which will affect the fatigue life of the mechanism and generate noise.

The noise is generated in gear engagement when the vehicle is driving under various conditions. How to reduce noise and vibration is an important subject in transmission design.

I. **Planetary gear layout**

In the planetary gear train, the number of planetary gears shall satisfy a certain relation: the angle between the lines of the center of two adjacent planetary gears and the center of the planetary gear train shall be an integral multiple of $\frac{360°}{z_A+z_C}$; the planetary gears shall be spaced at the same intervals and the number of planetary gears shall be an integral multiple of $\frac{360°}{z_A+z_C}$, as shown in Fig. 11.31. Here, z_A is the number of teeth of center gear and z_C is the number of teeth of gear ring.

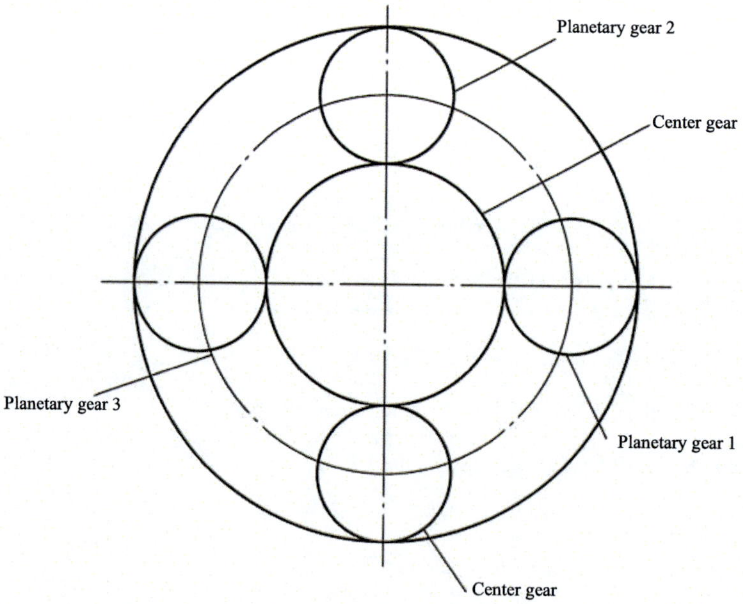

Fig. 11.31 Planetary gear layout requirements

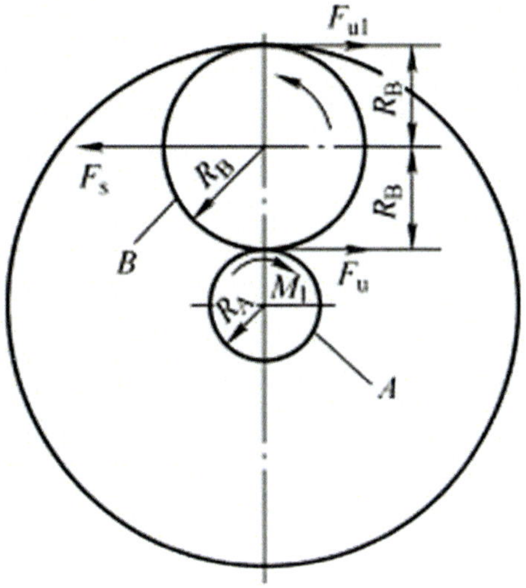

Fig. 11.32 Force on single planetary gear

The number of teeth of planetary gear z_B is determined by

$$\begin{cases} z_C - z_B = z_A + z_C \\ z_B = (z_C - z_A)/2 \end{cases} \tag{11.42}$$

II. Planetary gear mesh cycle

There is a certain error in the gear. Even if the uniform angular velocity is input, the output angular velocity will also fluctuate periodically, which may affect the fatigue life of the planetary gear train and cause noise problems.

As shown in Fig. 11.32, the circumferential force of the center gear acting on the planetary gear is $F_u = M_1/R_A$ and of the gear ring acting on the planetary gear is F_{u1}. According to the torque balance, F_u and F_{u1} have the same magnitude and direction; according to the force balance, the force between the planetary carrier and planetary gear is $F_s = 2F_u$.

As shown in Fig. 11.33, in the meshing process, a single planetary gear is subject to two dynamic loads from the gear ring and sun gear and these two dynamic loads often have different cycles. According to the phase relation, the resultant forces acting on different planetary gears at the same time would be different.

III. Analysis of meshing phase difference and torsional vibration of planetary gear

When the number of teeth of center gear and the number of teeth of gear ring can be exactly divided by the number of planetary gears, different planetary gears have the same meshing phase and have the same load at the same time. It can be seen from the force analysis in Fig. 11.33 that the resultant force of the force acting on the gear ring and center gear respectively is 0 and the resultant moment on the center gear is

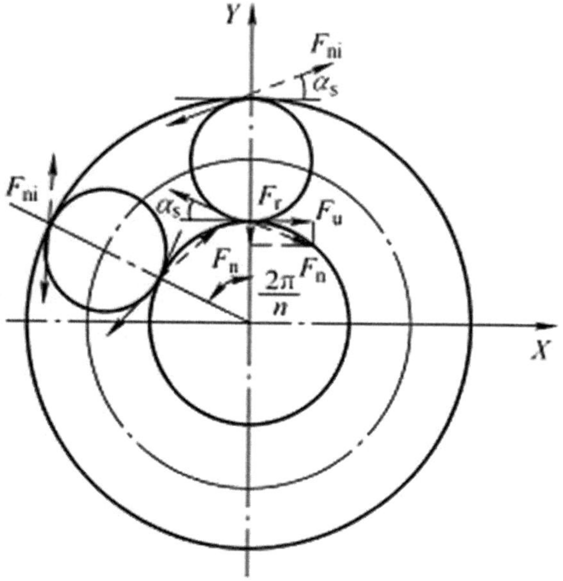

Fig. 11.33 Force on two planetary gears

$$\sum M = nR_aF_n\cos\alpha_n \tag{11.43}$$

where,

R_a—pitch radius of center gear;
F_n—normal force on tooth surface;
α_n—pressure angle.

The relationship between normal load change and torsional vibration of center gear can be obtained by the above equation.

When the number of teeth of center gear and the number of teeth of gear ring cannot be exactly divided by the number of planetary gears, planetary gears bear different load at the same time due to different meshing phases, which will affect the resonance of the gear ring, and the effect of manufacturing error will be smaller.

The meshing phase difference $\Delta\theta$ can be changed by changing the relationship among the number of teeth of center gear, the number of teeth of gear ring and the layout angle of the planetary gear. The gear ring of the planetary gear train in the AT for the passenger vehicle is of thin-walled structure and the deformation under the load may make the gear ring meshed evenly with the planetary gears, so the vibromotive force of the planetary gear train is mainly affected by the meshing of the center gear and planetary gear.

To make the phase difference exist, the number of teeth shall meet the conditions under the premise of evenly distributed planetary gears

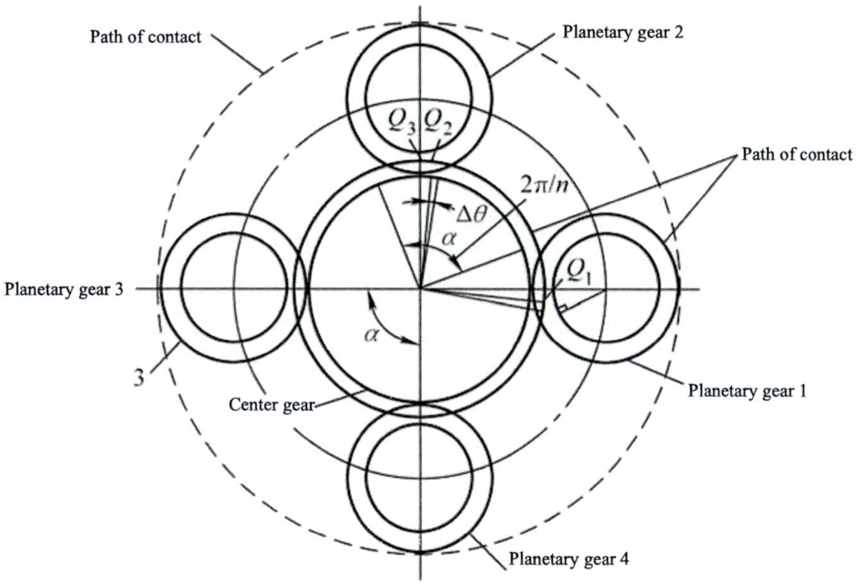

Fig. 11.34 Meshing phase difference between center gear and planetary gear

$$\begin{cases} (z_A + z_C)/n = \text{integer} \\ z_A/n \neq \text{integer} \end{cases} \tag{11.44}$$

Figure 11.34 is the schematic diagram of the meshing phase difference between the center gear and planetary gear. It is assumed that the point Q_2 is located on the planetary gear 2 and has no phase difference with the starting meshing point (point Q_1) at the dedendum of planetary gear 1 and the addendum of center gear. If the actual meshing point Q_3 of the planetary gear 2 is made rotate to Q_2 around the gear train center, the angle $\Delta\theta$ of rotation is the meshing phase difference.

The noise of the planetary gear train is mainly from the torsional vibration in the system. If the meshing phase of all planetary gears is different, the periodic vibration of planetary gear meshing with the center gear and gear ring will be significantly reduced.

The phase difference between the jth planetary gear and the center gear at the beginning of meshing of a tooth is

$$\Delta\theta = kj \times 2\pi Z'/n \tag{11.45}$$

where,

Z'—remainder of division of the number of teeth of sun gear by the number of planetary gears;
j—planetary gear code;
k—harmonic order.

When the harmonic order is k, the vibromotive force F_{kj} of the meshing between the jth planetary gear and the center gear is

$$F_{kj} = F_{nkj} \sin\left(k\omega t + kj \times 2\pi Z'/n\right) \quad (11.46)$$

where,

F_{nkj}—amplitude of the force acting on the gear (maximum value).

The relationship between the total torque of the force on the planetary gears and the pitch conforms to the sine law. If the amplitude of the loads is equal, the total torque is

$$\begin{aligned} M_k &= \sum_{j=1}^{n} F_{nk} R_B \cos\alpha_n \\ &= F_{nk} R_B \cos\alpha_n \sum_{j=1}^{n} \sin\left(k\omega t + kj \times 2\pi Z'/n\right) \end{aligned} \quad (11.47)$$

where,

R_B—pitch radius of planetary gear;
α_n—pressure angle.

If M_k is 0 and the calculated value of the meshing transfer error caused by the rigidness is 0, the torsional vibration will not appear.

Figure 11.35 shows 1 speed planetary gear train with phase difference. Because of the phase difference, the sinusoidal forces generated in meshing cancel each other, and the vibromotive force of meshing is 0.

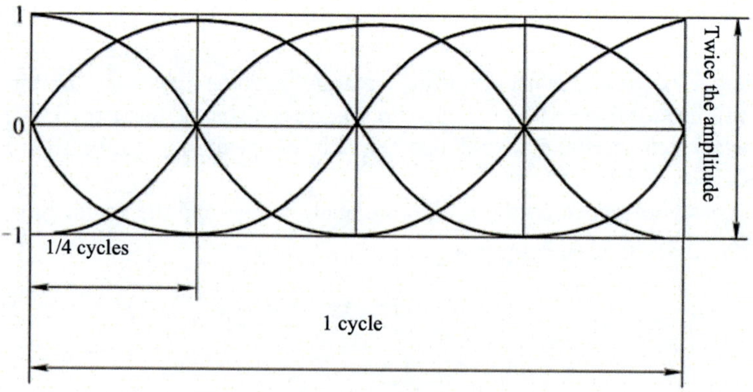

Fig. 11.35 1 speed planetary gear train with phase difference

IV. Calculation of phase difference of test planetary gear train

The 1 speed planetary gears are evenly distributed and have proper meshing phase difference, so that the vibromotive force at the meshing positions cancels each other. In case of no manufacturing error in the gear, the vibromotive force of meshing is 0 theoretically.

The meshing phase difference of the planetary gear train is

$$\Delta\theta = \frac{2\pi}{z_A}H - \frac{2\pi}{n} \tag{11.48}$$

where,

H—an integer;
n—number of planetary gears;
z_A—number of teeth of center gear, with the formula as

$$z_A = nH \pm 1 \tag{11.49}$$

It is calculated from formula (11.48) that the meshing phase difference of the 1 speed gear is $\Delta\theta = 2.432°$.

The meshing frequency is one of the reasons that affect the gear noise and vibration. Through spectral analysis of the gear noise and vibration frequency, the peak occurs at the doubling of the meshing frequency. The noise and vibration are also affected by gear dynamic load.

The meshing frequency f_z of the 1 speed planetary gear is the number of meshing teeth of the planetary carrier (output) per revolution. The number of meshing teeth can be calculated from the relative speed between the center gear and the planetary carrier

$$(n_A - n_S)z'_A = \left(n_A - \frac{n_A z'_A}{z'_A + z'_C}\right)z'_A = \frac{n_A z'_C z'_A}{z'_A + z'_C} \tag{11.50}$$

where,

z'_A—number of teeth of sun gear in second planetary gear set;
z'_C—number of teeth of gear ring in second planetary gear set;
n_A—sun gear speed (r/min);
n_S—planetary carrier speed (r/min).

Input shaft speed in gear 1 (r/min) $n_i = n_A$ and meshing frequency

$$f_z = \frac{n_i z'_C z'_A}{60(z'_A + z'_C)} \tag{11.51}$$

Through calculation

$$f_z = 0.413n_i$$

The number of meshing teeth of the 4 speed planetary gear is calculated from the relative speed between the gear ring and the planetary carrier

$$(n_C - n_S)z_C = \left(\frac{n_S(z_A + z_C)}{z_C} - n_S\right)z_C = n_S z_A \quad (11.52)$$

where,

z_A—number of teeth of center gear in second planetary gear set;
z_C—number of teeth of gear ring in second planetary gear set;
n_S—planetary carrier speed (r/min);
n_C—gear ring speed (r/min).

Input shaft speed in gear 4 (r/min) $n_i = n_S$ and meshing frequency

$$f_z = \frac{n_i z_A}{60} \quad (11.53)$$

Through calculation

$$f_z = 0.55n_i$$

Because there are several pairs of gears meshed in a gear assembly, there will be more than one vibration frequency. To analyze the vibration frequency, it is necessary to calculate the meshing times of each gear pair.

Meshing times of each gear pair in gear 1: gear ring $1.321n_i$; center gear $2.678n_i$; planetary gear $1.304n_i$.

Meshing times of each gear pair in gear 4: center gear $3n_i$; gear ring $1.32n_i$; planetary gear $1.571n_i$.

11.4 Vibration and Noise Test of Planetary Gear Train

A power absorption type gear test bench is used to test the planetary gear trains with phase difference (1 speed) and without phase difference (4 speed) respectively. The vibration and noise of the planetary gear train can be measured by changing the meshing rate, accuracy, load and speed.

The power absorption type gear test bench is shown in Fig. 11.36. In order to accurately measure the noise generated by the planetary gear train, the motor, the reducer and the planetary gear train in the dashed box are placed in the sound insulation box for sound insulation, so as to reduce the dark noise to less than 6 dB.

The lubricant is ATF, which is supplied to the planetary gear train from three injectors near the center gear and one injector in the upper part of the center gear,

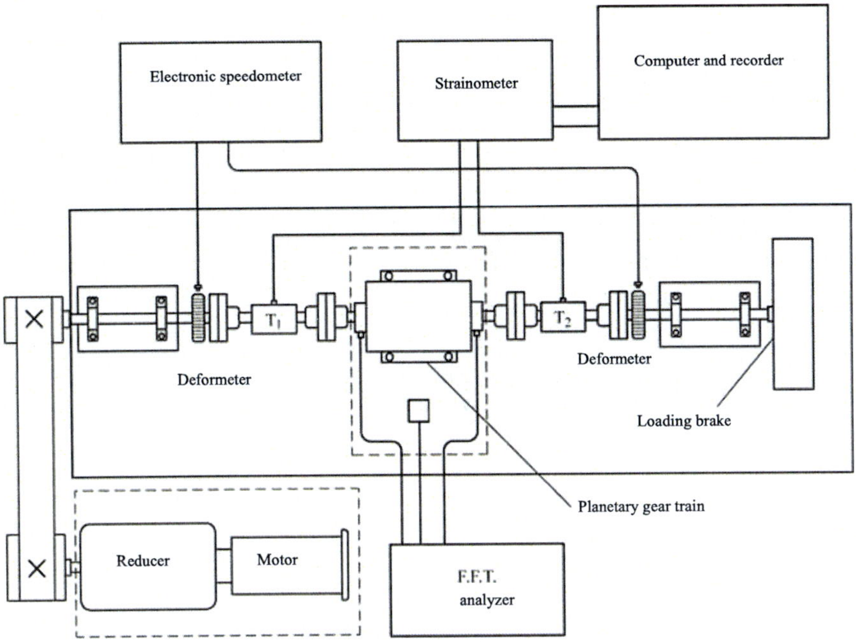

Fig. 11.36 Power absorption type gear test bench

at the flow rate of 1.5 L/min and the temperature of (32 ± 2) °C. The input torque range is 0–80 N m, and the speed of high speed shaft (input shaft) is 500–2400 r/min. The speed changes 100 r/min each time, and the torque changes 5 N m each time. An acceleration vibration sensor is installed near the supports of the input shaft and the output shaft to measure the vibration acceleration and noise of the planetary gear train.

I. Small load test for shaved gears

Figures 11.37 and 11.38 show the results of the small load test for shaved gears. The test gear parameters are the high-tooth gear pair A in Table 11.1. "□" means 1 speed gear (with phase difference) and "Δ" means 4 speed gear (without phase difference). The input speed is 500–2400 r/min, the input torque in gear 1 is 20 N m, and in gear 4 is 43 N m.

As can be seen from the figure, the noise level of the planetary gear train with phase difference is 3–8 dB lower than that without phase difference, and the vibration level is 2–4 dB lower. The test results show that the existence of phase difference can significantly reduce the torsional vibration, and the design of phase difference is an effective way to reduce the vibration and noise of the planetary gear train.

II. Large load test for shaved gears

Figures 11.39 and 11.40 show the results of the large load test for shaved gears. The input torque in gear 1 is 40 N m, and in gear 4 is 80 N m. It can be seen from the

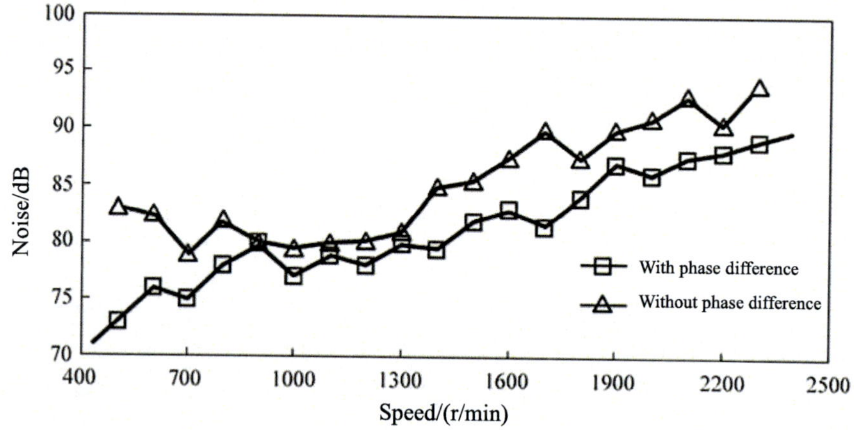

Fig. 11.37 Influence of shaved gear (small load) phase difference on noise

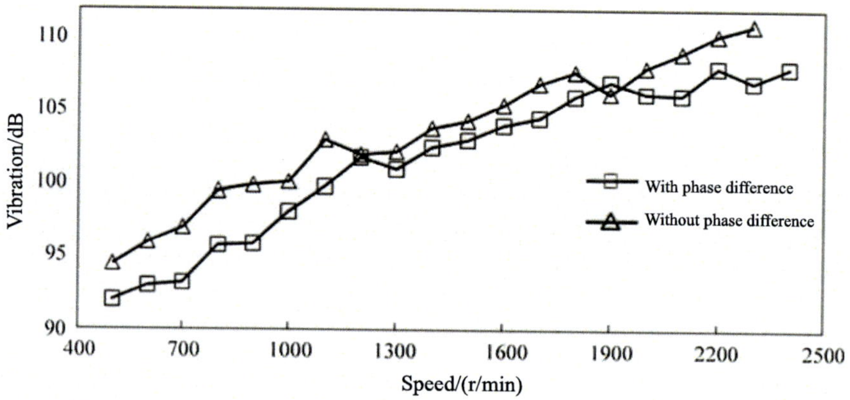

Fig. 11.38 Influence of shaved gear (small load) phase difference on vibration

figure that under the same operating conditions, compared with the gear grinding process, the change law of the two curves is almost the same. It can be considered that the elastic deformation of gear teeth is the main cause for vibration and noise under large load conditions.

III. **Small load test for ground gears**

The planetary gear and center gear for test are ground by the MAAG gear grinding machine to the accuracy up to JIS 0–1. The test results are shown in Figs. 11.41 and 11.42. As can be seen from the figure, the noise level of the planetary gear train with phase difference is 6–10 dB lower than that without phase difference within the test speed range and the accuracy of ground gear has relatively small influence on the test results; in the case of phase difference and no phase difference, the vibration level is

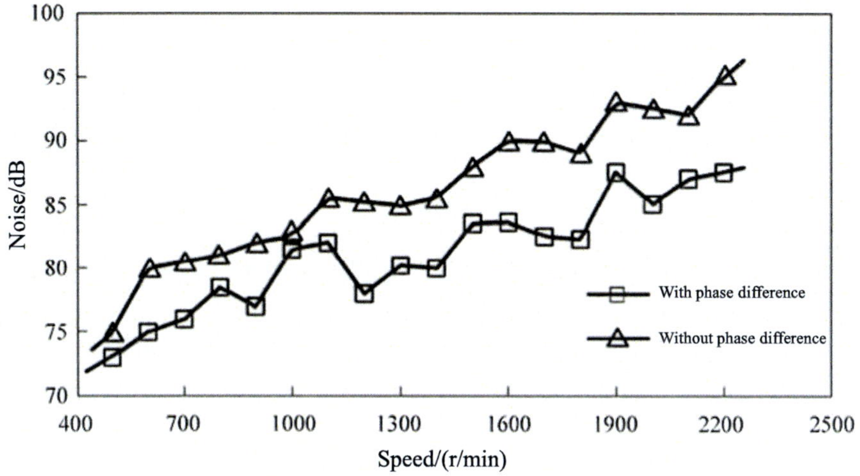

Fig. 11.39 Influence of shaved gear (large load) phase difference on noise

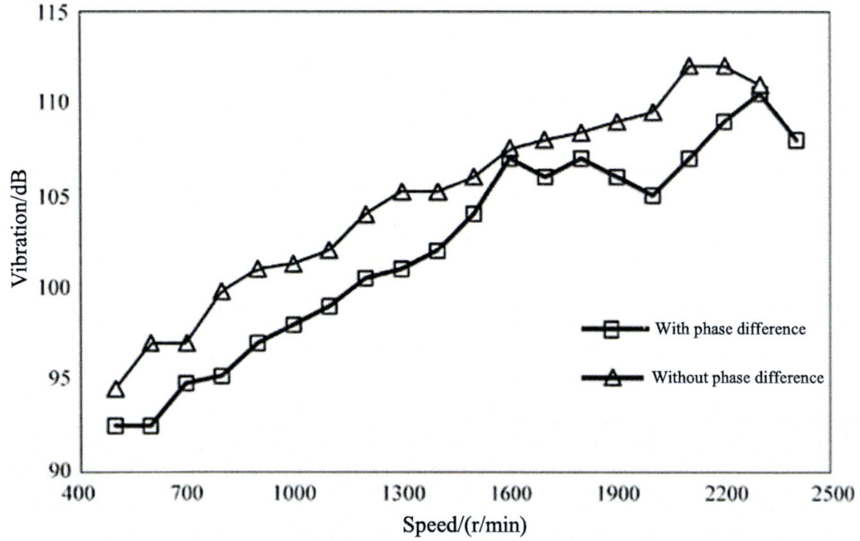

Fig. 11.40 Influence of shaved gear (large load) phase difference on vibration

basically the same with the increase of the speed. The vibration level of gear with 1 gear (with phase difference) is 2–4 dB lower than that with 4 gears (without phase difference).

Tables 11.4 and 11.5 list the vibration and noise level data of the 1 speed (with phase difference) and 4 speed (without phase difference) in the frequency range of 0–10 kHz and 0–5 kHz respectively.

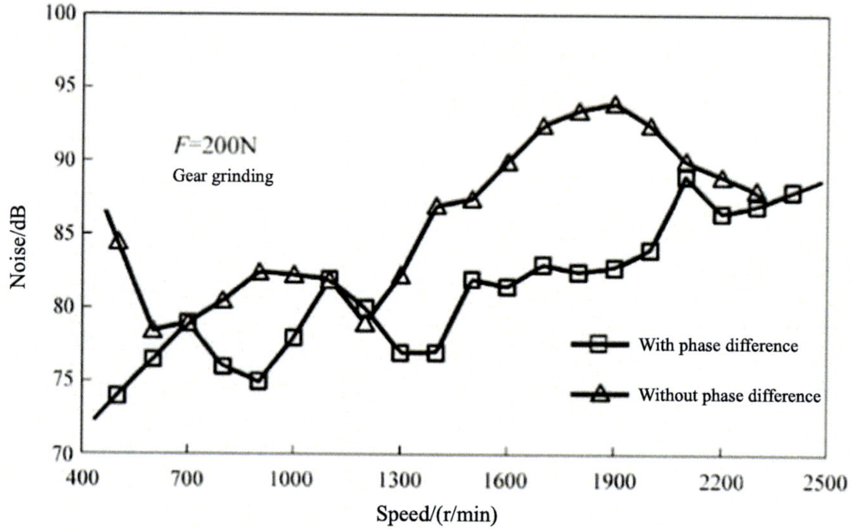

Fig. 11.41 Influence of ground gear (small load) phase difference on noise

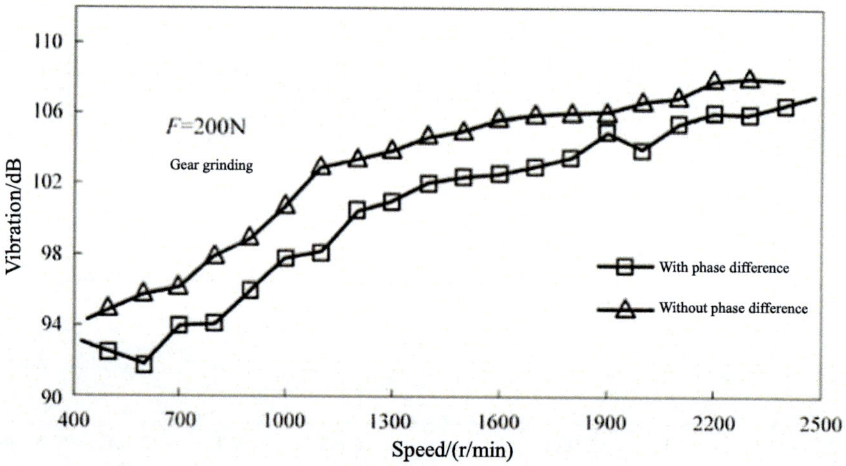

Fig. 11.42 Influence of ground gear (small load) phase difference on vibration

IV. Large load test for ground gears

It is set in the test that the input speed is 500–2400 r/min, the input torque in gear 1 is 40 N m, and in gear 4 is 80 N m (tangential load 370 N). The noise level curve from the test is shown in Fig. 11.43. As can be seen from the figure, the noise level of the planetary gear train with phase difference is 3–9 dB lower than that without phase difference. Compared with the test results when the tangential load of the tooth

Table 11.4 Noise level data

Speed/(r/min)	Noise/dB (0–10 kHz)		Speed/(r/min)	Noise/dB (0–5 kHz)	
	1 speed gear	4 speed gear		1 speed gear	4 speed gear
500	74.21	84.68	500	71.30	83.28
600	74.21	78.76	600	76.48	83.68
700	79.19	79.21	700	78.27	82.72
800	72.17	81.00	800	75.66	82.76
900	75.19	82.73	900	81.27	83.15
1000	78.03	82.50	1000	77.69	77.94
1100	82.22	82.00	1100	78.30	79.52
1200	79.97	78.83	1200	76.68	80.82
1300	76.95	82.68	1300	80.66	82.34
1400	76.93	87.45	1400	81.35	86.02
1500	81.58	87.59	1500	82.80	86.79
1600	81.28	90.55	1600	83.95	91.70
1700	83.72	82.92	1700	84.63	91.57
1800	82.53	93.71	1800	81.88	93.71
1900	83.51	94.00	1900	86.19	95.52
2000	84.26	92.55	2000	88.11	92.64
2100	89.06	90.29	2100	86.26	91.96
2200	86.40	89.14	2200	87.44	87.98
2300	87.09	87.98	2300	90.04	90.66
2400	88.37		2400	89.18	

surface is 200 N, the influence of phase difference on the noise level is more obvious after the increase of input torque.

Figure 11.44 shows the influence of phase difference on vibration. As can be seen from the figure, within the range of speed change, the amplitude of vibration level change measured in the two tests is basically the same, with a difference of 3–5 dB. However, the corresponding speed of the peak value at vibration level is different from that in Fig. 11.43.

V. Influence of torque on vibration and noise of planetary gear train with phase difference

The influence of the torque on the vibration and noise of the planetary gear train at the speed 1200 r/min is shown in Figs. 11.45 and 11.46 respectively. It can be seen from the figure that the torque has little influence on noise. When the load torque is increased, the difference of vibration level tends to increase.

Table 11.5 Vibration level data

Speed/(r/min)	Noise/dB (0–10 kHz)		Speed/(r/min)	Noise/dB (0–10 kHz)	
	1 speed gear	4 speed gear		1 speed gear	4 speed gear
500	92.27	94.69	500	90.90	93.49
600	91.62	95.41	600	92.79	94.78
700	93.87	96.45	700	93.57	95.38
800	94.27	98.30	800	94.29	96.91
900	95.68	99.53	900	95.83	97.94
1000	97.47	101.00	1000	96.67	98.71
1100	98.53	102.96	1100	97.11	103.91
1200	100.37	103.05	1200	97.64	101.90
1300	100.94	104.10	1300	99.22	101.91
1400	102.10	104.68	1400	99.93	102.15
1500	102.46	105.37	1500	99.92	103.44
1600	102.82	105.68	1600	100.19	103.22
1700	102.98	106.07	1700	101.20	102.58
1800	103.12	106.10	1800	101.70	103.29
1900	105.44	106.32	1900	103.74	103.90
2000	104.33	107.10	2000	104.94	105.52
2100	105.84	107.74	2100	104.90	105.45
2200	106.76	108.56	2200	103.60	108.21
2300	106.17	108.77	2300	103.39	106.97
2400	107.09		2400	107.96	

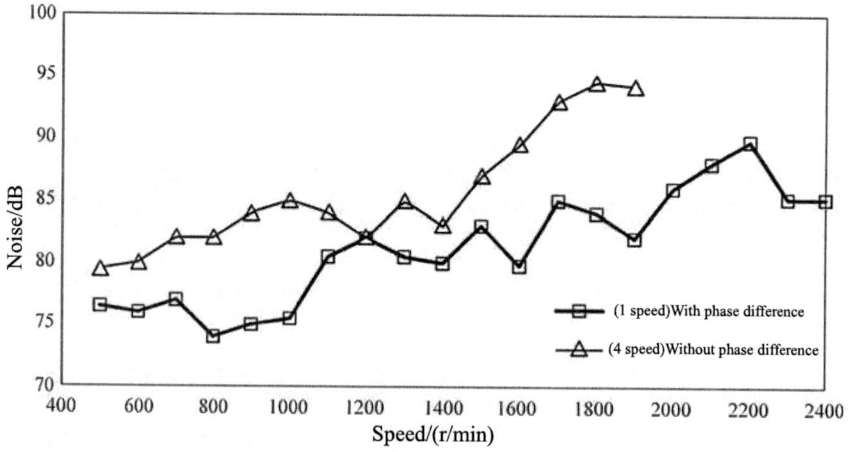

Fig. 11.43 Influence of ground gear (large load) phase difference on noise

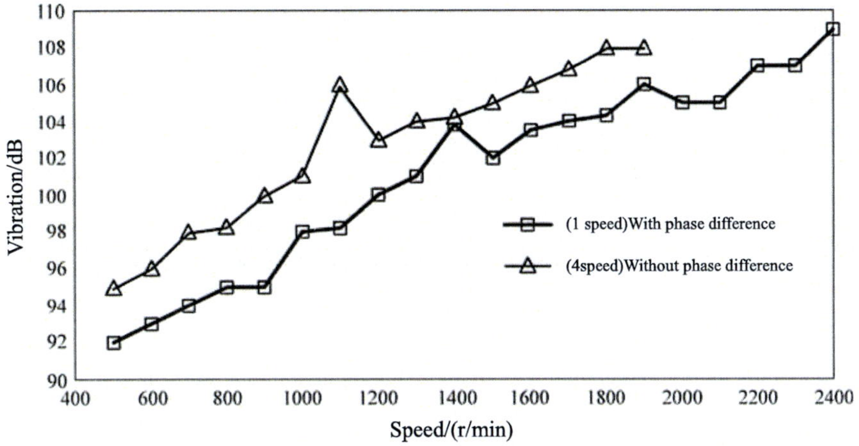

Fig. 11.44 Influence of ground gear (large load) phase difference on vibration

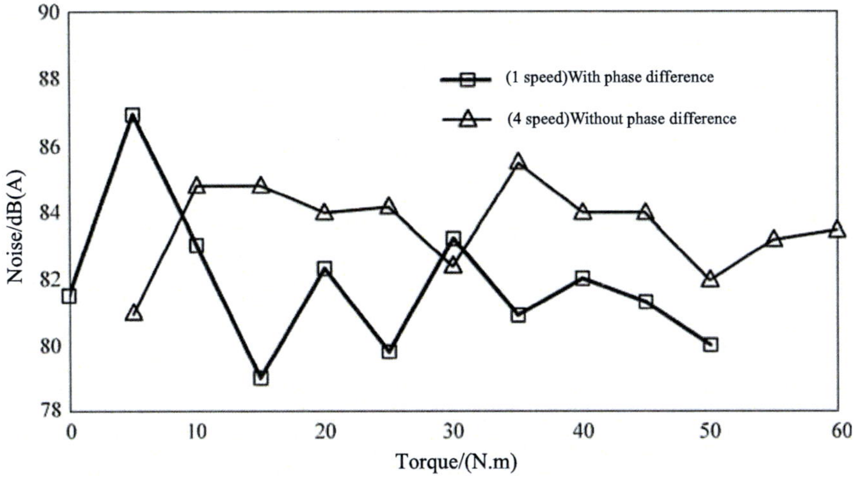

Fig. 11.45 Influence of torque on noise

VI. **Influence of contact ratio on vibration and noise of planetary gear train with phase difference**

See Table 11.1 for the parameters of 1 speed in the test. High-tooth gear pair A (1.65/1.81) and standard gear pair B (1.30/1.66) are selected, and the input shaft speed is 500–2400 r/min. It can be seen from Figs. 11.47 and 11.48 that two gear pairs are tested respectively with 200 N tangential load on tooth surface, and there is no significant difference in noise and vibration levels. The reason is that the contact ratio of the 1 speed planetary gear train is large, and the gear offsets the vibromotive force caused by the stiffness and flexibility.

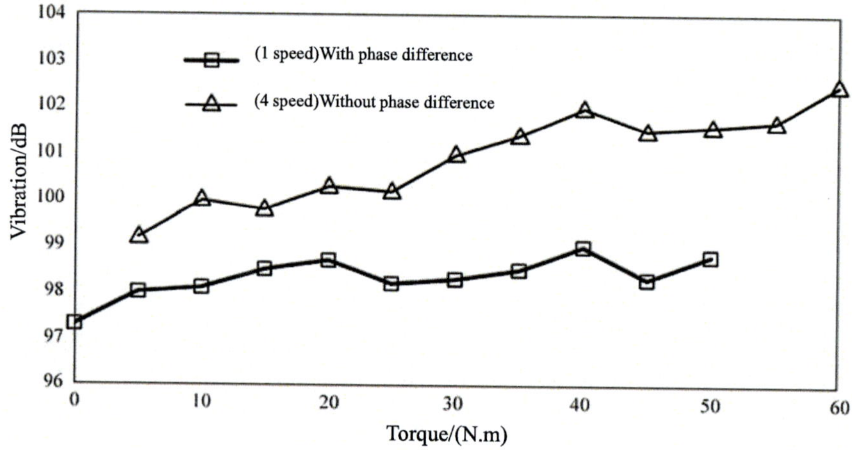

Fig. 11.47 Influence of contact ratio on vibration

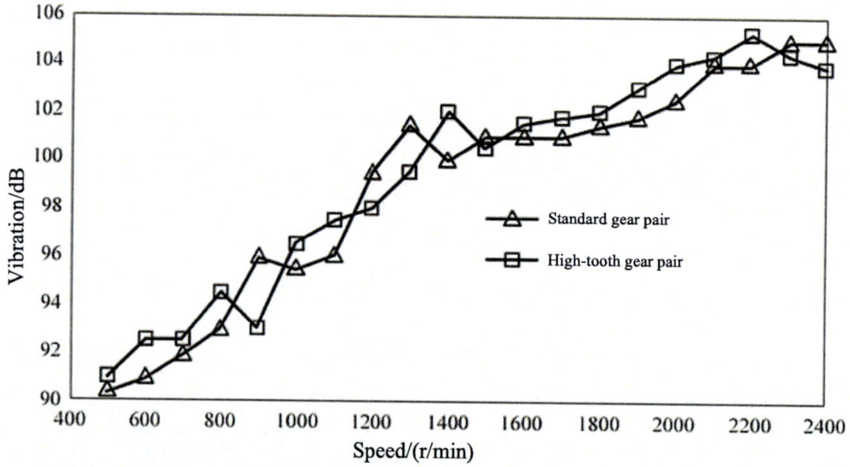

Fig. 11.46 Influence of torque on vibration

VII. Influence of accuracy on vibration and noise of planetary gear train with phase difference

In the two gear pairs used in the test, one dual-planetary gear set of planetary gear and center gear is subject to gear shaving, carburizing and quenching and the other set is ground by MAAG gear grinding machine after gear shaving, carburizing and quenching. The accuracy of the former is JIS 3–4 and the tooth surface roughness is about R_{amax} 6 μm (R_a 1.5 μm); the accuracy of the latter is JIS 0–1 and the tooth surface roughness is R_{amax} 2 μm (R_a 0.5 μm). The test results are shown in Fig. 11.49 and 11.50. As can be seen from the figures, the vibration and noise level of the ground

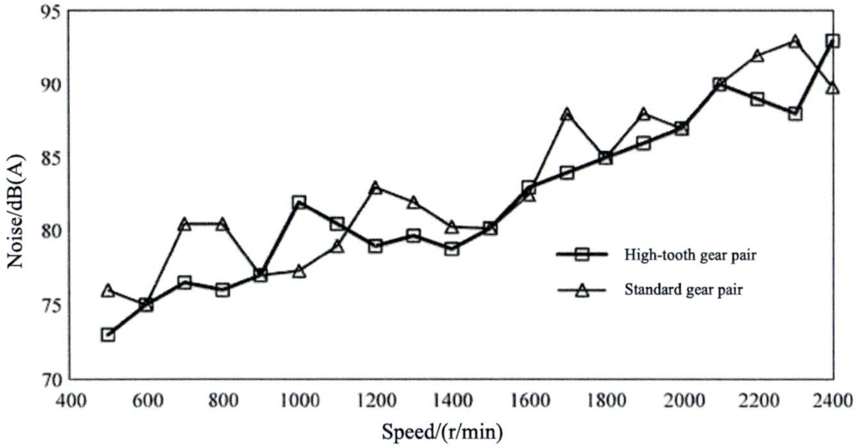

Fig. 11.48 Influence of contact ratio on noise

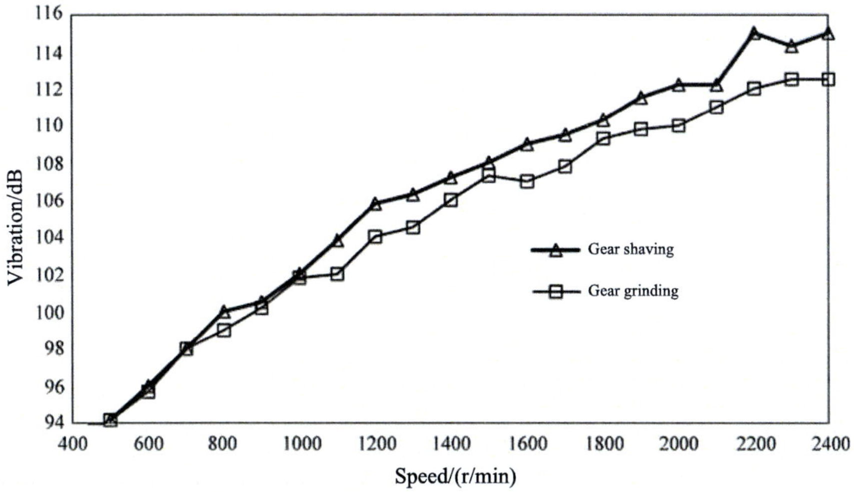

Fig. 11.49 Influence of accuracy on vibration

high-accuracy gear is 2–3 dB lower than that of the gear without grinding. Especially when the speed reaches above 1000 r/min, the difference of vibration level is more obvious.

VIII. **Influence of torque on vibration and noise of planetary gear train of different accuracy with phase difference**

The gear used in the test is 1 speed planetary gear train with phase difference, and there are two kinds of processing technology, namely gear shaving and grinding. As shown in Figs. 11.51 and 11.52, the input shaft speed is 1200r/min, and the input

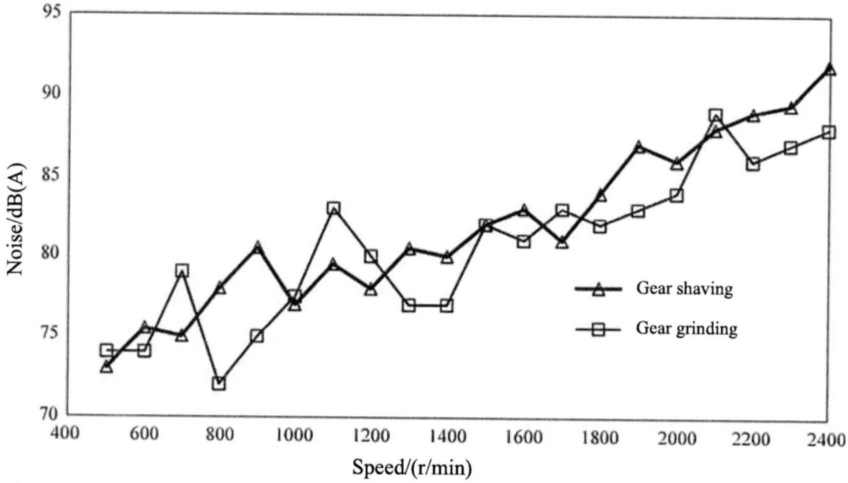

Fig. 11.50 Influence of accuracy on noise

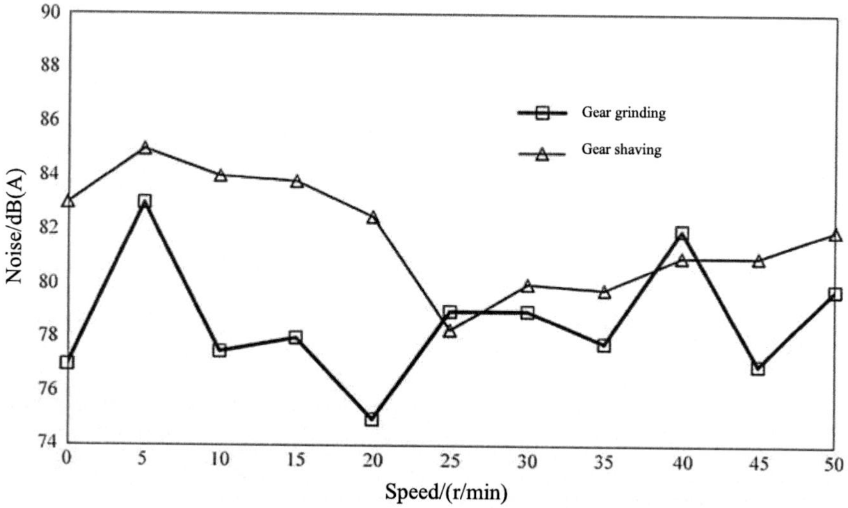

Fig. 11.51 Influence of torque on the noise of planetary gear train of different accuracy

torque change range is 0–50 N m. As can be seen from the figures, when the input torque exceeds 10 N m, the vibration and noise levels tend to increase with the increase of torque. In the test torque range, the vibration noise of the ground gear is lower.

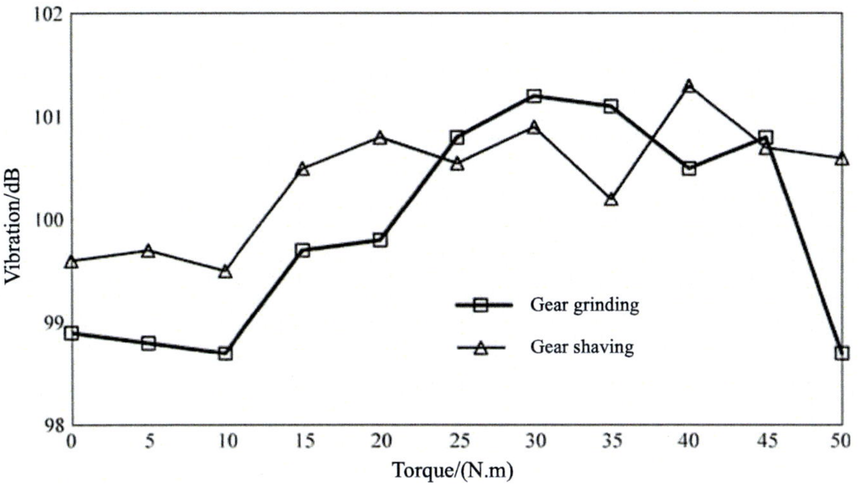

Fig. 11.52 Influence of torque on the vibration of planetary gear train of different accuracy

IX. Analysis of vibration and noise test results

Figures 11.53 and 11.54 show the spectral analysis results of the vibration and noise of the 1 speed planetary gear with phase difference. Test conditions: gear grinding, tangential load 200 N and input shaft (high speed shaft) speed 1600 r/min.

The meshing frequency of the planetary gear is 660 Hz, and the peak of vibration and noise acceleration level corresponding to the frequency doubling could not be clearly identified. The reason is that there is phase difference in the meshing of planetary gear and center gear, and the vibration caused by phase difference is exactly offset with the torsional vibration caused by the planetary gear.

Figures 11.55 and 11.56 show the spectral analysis results of the vibration and noise of the 4 speed planetary gear without phase difference and the test conditions are the same as those for spectral analysis of 1 speed. Compared with that with phase

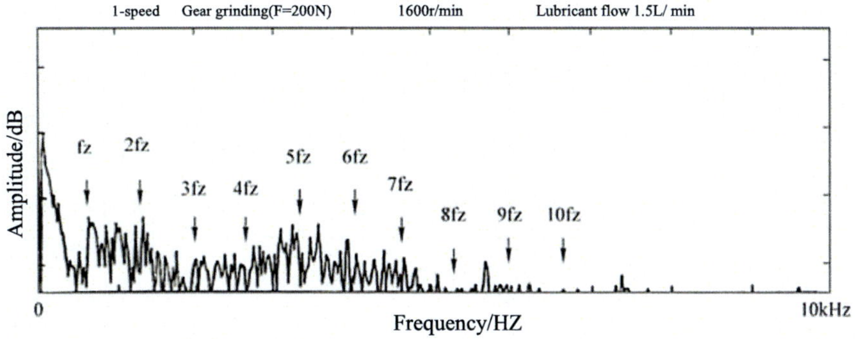

Fig. 11.53 Spectral analysis results of vibration of 1 speed planetary gear

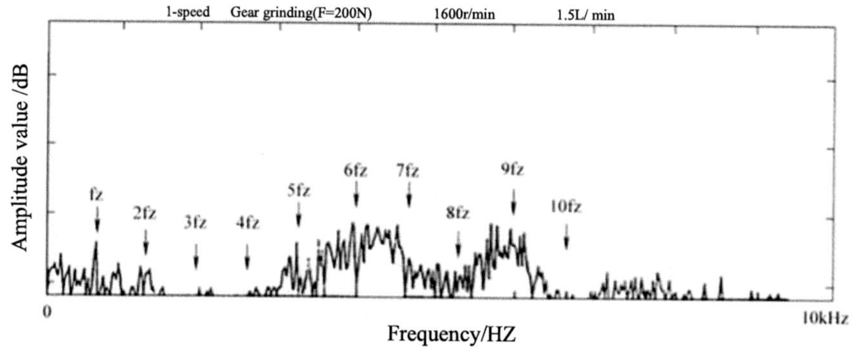

Fig. 11.54 Spectral analysis results of noise of 1 speed planetary gear

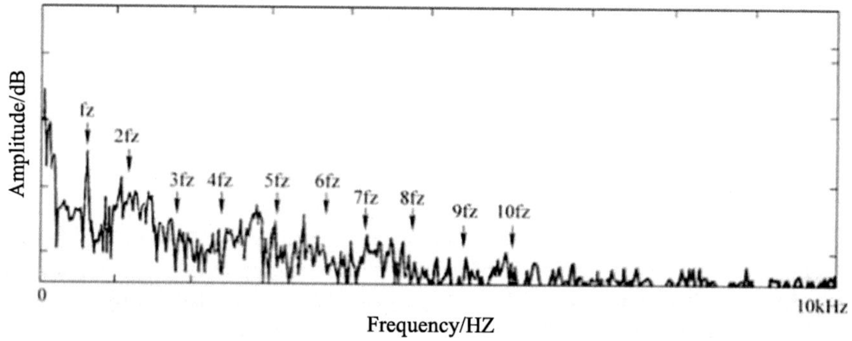

Fig. 11.55 Spectral analysis results of vibration of 4 speed planetary gear

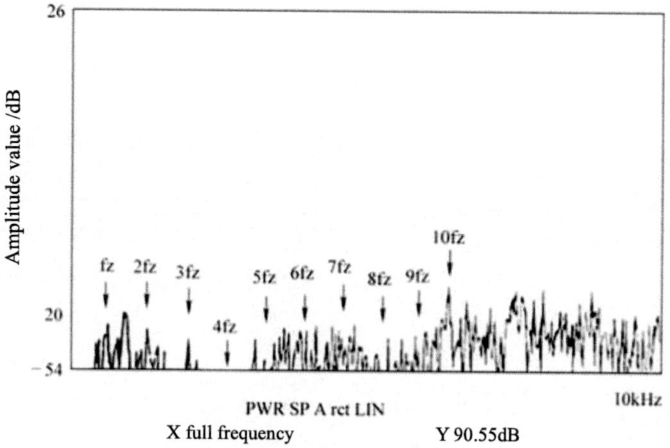

Fig. 11.56 Spectral analysis results of noise of 4 speed planetary gear

difference, the noise level without phase difference is about 9 dB higher, and the vibration acceleration level is about 3 dB higher. The meshing frequency of planetary gear is 611 Hz, and the corresponding vibration and noise levels are much higher when the harmonic order is 1–3 times than that with phase difference. It can be seen from the figure that the circumferential vibration acceleration of the planetary gear train is formed by the superposition of vibration of the integer multiple frequencies in the meshing frequency. The total stiffness of the planetary gears without phase difference is a multiple of the number of planetary gears and the vibromotive force will increase in the periodically varying meshing.

Bibliography

1. Mitschke M, Wallentowitz H (2009) Automobile dynamics (Translated by Yinsan C, Qiang Y). Tsinghua University Press, Beijing
2. Hongxin Z (1996) Auto design. Version 2. China Machine Press, Beijing
3. Wangyu W (2017) Auto design. Version 2. China Machine Press, Beijing
4. Li C (2005) Study on efficiency of planetary gear train of automatic transmission. Chongqing University, Chongqing

Chapter 12
Electronic Control System of Automatic Transmission

12.1 Introduction to AT Electronic Control System

The AT electronic control system, as the brain of the AT, is mainly to communicate with other VOBCs through various sensors to obtain real-time transmission, engine and vehicle operating state, achieve accurate shift control and clutch control according to the vehicle speed, accelerator pedal and other signals, free the driver from the heavy manual shift and achieve good driving comfort and better fuel economy.

The AT electronic control system, with the block diagram shown in Fig. 12.1, usually consists of a sensor, a control unit and an actuator. Firstly, the sensor detects the information of the transmission condition and driving condition and converts it into an electrical signal. Then, the control system analyzes and processes the electrical signal according to the shift law, makes a judgment and issues a shift instruction for shift control. The coordination control mode of "PID + self-learning" is used in the shift process. On the one hand, parameter adjustment is implemented through feedback control to ensure the shift quality of in the shift process. On the other hand, according to the shift control effect, the parameters are further self-learned and optimized to prepare for the next shift. Repeat this way and it can continuously approach the ideal shift process and constitute a complete closed loop. It may also provide interfaces for other peripheral control units, such as the engine control unit (ECU) and electronic stability control (ESP) unit to synergistically improve vehicle driving comfort, safety and other functions.

The AT electronic control system provides all the functions that meet the OBD diagnosis regulations, and can diagnose the electrical rationality of the components and subsystems, and output the diagnosis results according to the regulations or customer requirements.

Y. Chen, *Automotive Transmissions*, Springer Tracts in Mechanical Engineering,
https://doi.org/10.1007/978-981-15-6703-2_12

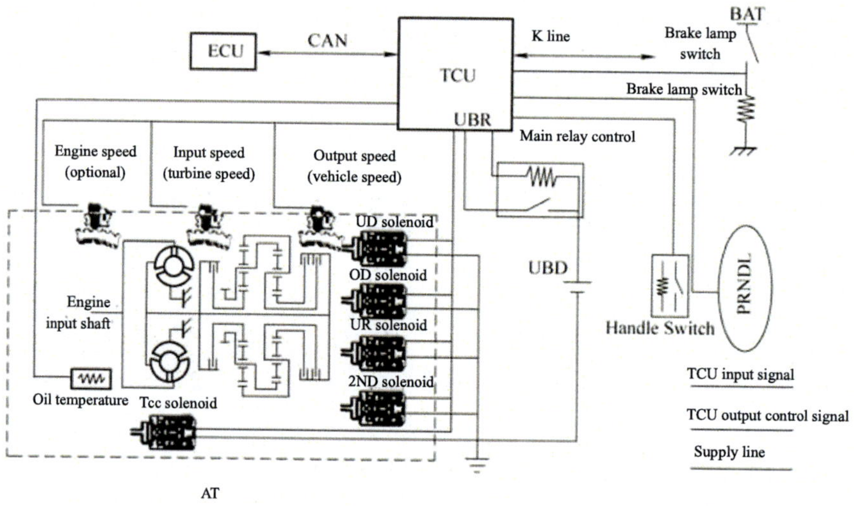

Fig. 12.1 Block diagram of AT electronic control system

With the complexity of vehicle control technology in recent years, the capacity of software is increasing. To facilitate reuse of software, AT electronic control system adopts a hierarchical structure that covers sensor, actuator characteristics, and microcomputer functions (changes in characteristics and functions have no impact on the software).

1. **Software structure**

Figure 12.2 shows the hierarchical structure example of the vehicle control software. The current software structure is based on the different concepts of vehicle manufacturers and software suppliers, but in order to develop in the unified direction, standardization activities are increasingly active at home and abroad.

2. **Operating system (OS)**

The OS is a module that manages the allocation of information processing resources for the applications to reduce the software development and management cost. The OS can be used to easily use the external development software, construct a multiprocessing system with time limit, cover different processors and improve the recycling, portability and reliability to reduce the cost.

The main features of the OS include: task scheduling in the form of priority (execution priority), management of different time limits for applications, management of intertask shared resources (exclusion of access to the same resource between the seemingly parallel-running applications), intertask communication and event communication (support intertask coordination). Functions previously implemented in application processing are handled by an OS independent from the application software.

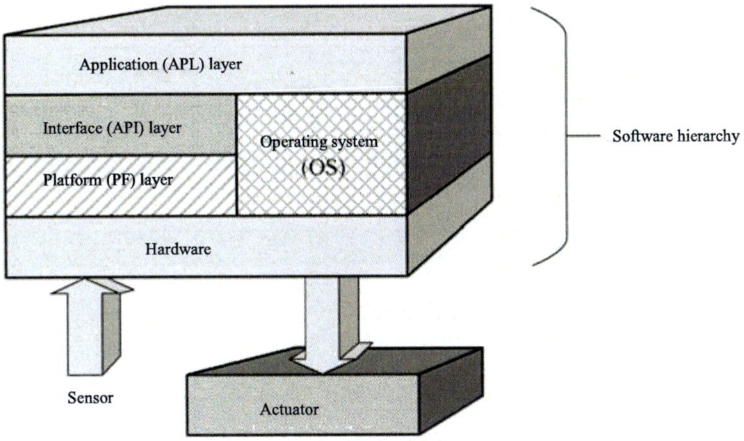

Fig. 12.2 Hierarchical structure example of software

In addition, the standardized real-time OS in the auto industry is characterized by highly real-time implementation (a simplified style dedicated to automobiles), the emphasis on quantification (corresponding to stock savings), and the on-board communication networking. As the vehicle control software, OS is constantly carrying out the standardization work aiming at the optimal parts.

3. **Platform (PF) layer**

The PF layer is the part of the software that uses hardware functions such as microcomputer peripherals to manipulate the input of the sensor signals and the output of the actuators. The computer peripherals vary widely depending on the type of microcomputer, but are designed to perform best in a particular system.

For example, when a drive solenoid valve is driven by Pulse Width Modulation (PWM), the best method is selected based on the drive characteristics (dynamic range, precision, etc.) of the existing computer peripherals and actuators before design.

According to the specified time, the calculated count value is compared with the cycle set value and duty cycle (task) set value, and the PWM output time chart shown in Fig. 12.3 is used to make the output reverse to realize the PWM output action.

4. **Interface (API) layer**

The API layer is the part that operates the exchange of data between APL and PF. When the APL layer and the PF layer process different data attributes (data form, unit system, etc.), data conversion at the API layer can be used to realize data exchange between the existing APL software and the existing PF software.

For example, when a solenoid valve duty cycle or a stepping motor can be used to control the bypass air flow, the control value of air flow (in m^3/min) can be converted into control signal data by driving the signal solenoid valve based on the respective flow and actuators.

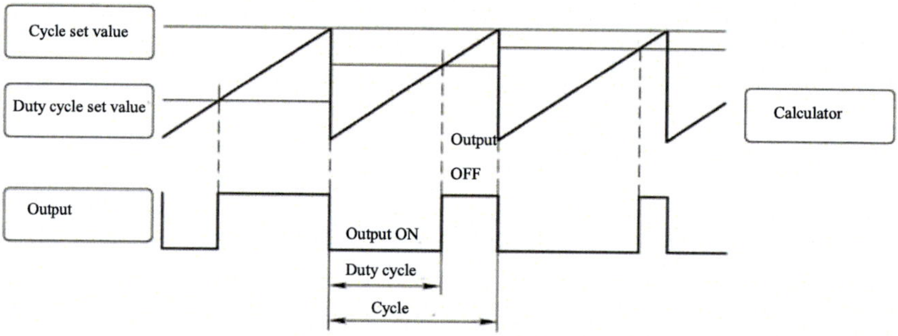

Fig. 12.3 PWM output time chart

5. **Application (APL) layer**

The APL layer is installed in the actual control system through the methods described in the preceding paragraph and the software developed. By standardizing the data interface at the API layer in advance, the reuse of APL software can be realized, thus giving full play to the advantages of high development efficiency.

For example, in the bypass air flow control, the calculation processes are standardized in the form of air flow control value (m^3/min), so that they are not affected by the air flow control method, thus ensuring the independence of APL components.

6. **Hierarchical software architecture**

According to the development characteristics of the TCU system, hierarchical software architecture is proposed to provide a friendly interface for the upper programs, thus ensuring the realization of different AT control requirements on the same platform, as shown in Fig. 12.4. This hierarchical architecture can enhance the functions of the underlying software and weaken the coupling between the underlying software and the application layer software; this architecture is also in line with the development trend of software architecture. For different applications, the architecture can tailor and upgrade the underlying software, which reduces the increase in the cost of upper software modification and improves the efficiency of AT software development.

12.2 AT Control System Development

With the improvement of control requirements, the application of the control system becomes more and more extensive and its position is more and more important. As a result, the control system is becoming more and more sophisticated and complex, especially with larger software system. The resources needed for the control system (development cycle, cost and number of developers) increase year by year, and

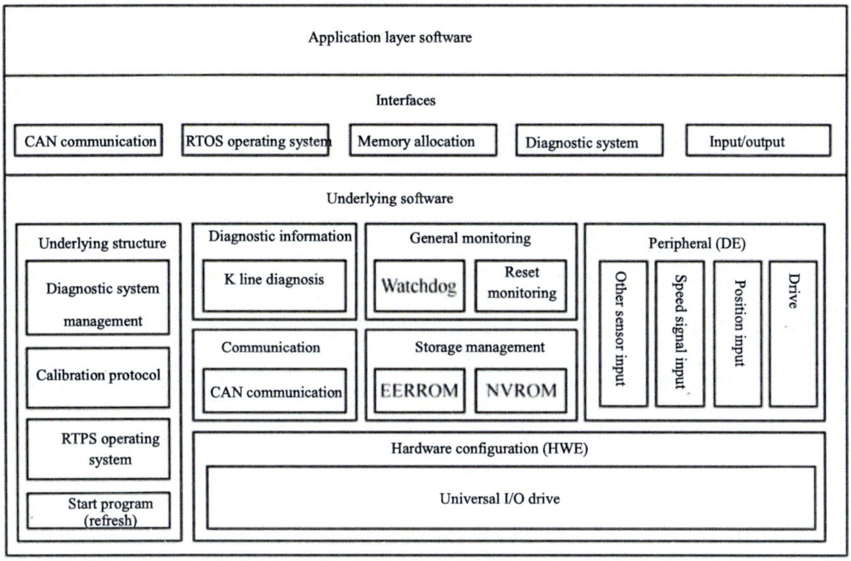

Fig. 12.4 Underlying software architecture

improving the software development efficiency has become an important issue. Making full use of the "model" is one of the means to solve the above problem. This process is called model foundational development. Model is the product of abstraction of physical devices. The objects of model include not only control objects but also control devices. The model foundational development has the following advantages:

(1) Easily implement high performance control.
(2) Validate the control system at the initial stage of development.
(3) Make full use of automation.

The following is a description of the transmission control development process and development approaches.

I. Development process

Figure 12.5 is the development process of the transmission control system. Previously, the control system was designed after the basic design of control objects such as engine transmission was completed. However, in the development mode shown in Fig. 12.5, there are many cases that the system as a whole cannot meet the design requirements due to the limitation of control objects. Therefore, it is better to develop control objects and control strategies or systems gradually after the overall design of the system.

Figure 12.6 shows a V-shaped diagram of the development process from a broad software development perspective. The following is a description of the contents of each process in the figure.

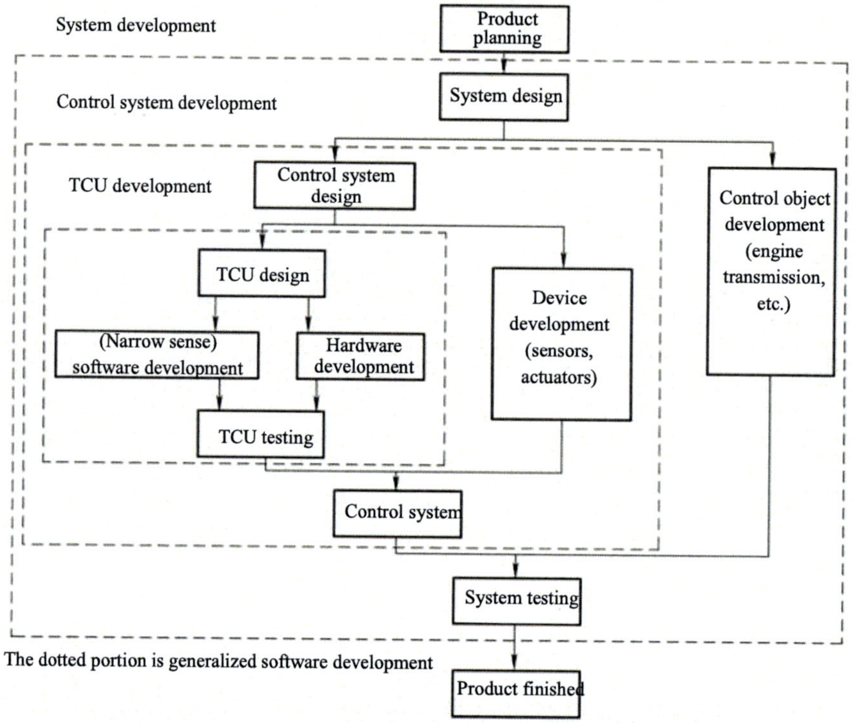

Fig. 12.5 Development process of transmission control system

Fig. 12.6 V-shaped diagram of development process

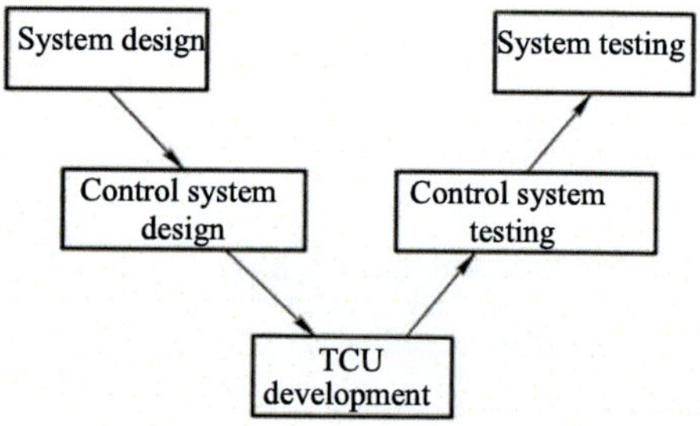

II. System design

The system design is equivalent to the initial stage of system development, which is the stage of designing the overall system functions according to the information such as commodity planning and regulations.

The system design subprocess is shown in Fig. 12.7, and each process is summarized as follows.

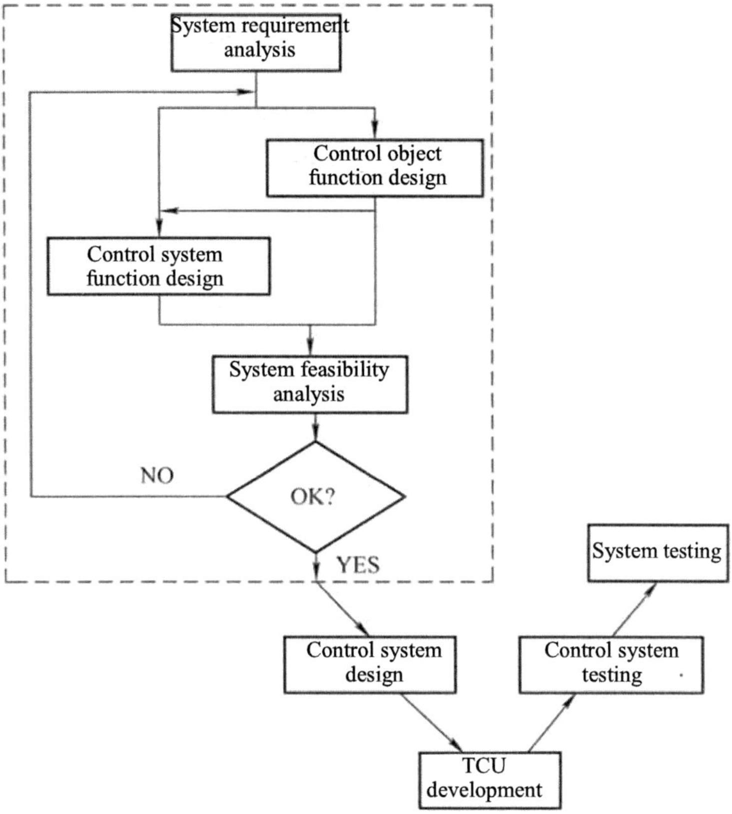

Fig. 12.7 System design subprocess

(1) System requirement analysis: the functions required for the whole system shall be set first. The ideal state is that the functions can be set quantitatively, but they can only be set qualitatively in most practical cases. However, even when a decision cannot be made, it is important to set a feature as a temporary target and promote development. In addition, in addition to the target function, it is also important to grasp the constraints.
(2) Control object function design: when the control system is designed, the control objects such as engine and transmission should be grasped functionally. When the control objects and the control system are developed in parallel, the specific design of the control objects is started after completing the function setting of the control objects. On the other hand, there are many control system designs for existing control objects. In this case, it is necessary to fully understand the control objects and master their functions.

(3) Control system function design: according to the set system function requirements and the set or mastered control object functions, the functions required for the control system shall be designed.
(4) System feasibility analysis: through the combination of control objects and control system, it is necessary to fully study whether it can achieve the functions required by the whole system. It would be a huge waste of time and resources to build on a system that cannot be implemented. In addition, due to the unconscious focus on the realization of major functions in the design process, it is easy to neglect the research on the harm of side effects, which needs full attention.

III. Control system design

The function requirements of the control system are set in the system design stage, where the specific design that satisfies relevant performance is made.

Figure 12.8 shows the control system design stage subprocess. Here mainly introduces the software development, and no longer explains the equipment development and TCU hardware development.

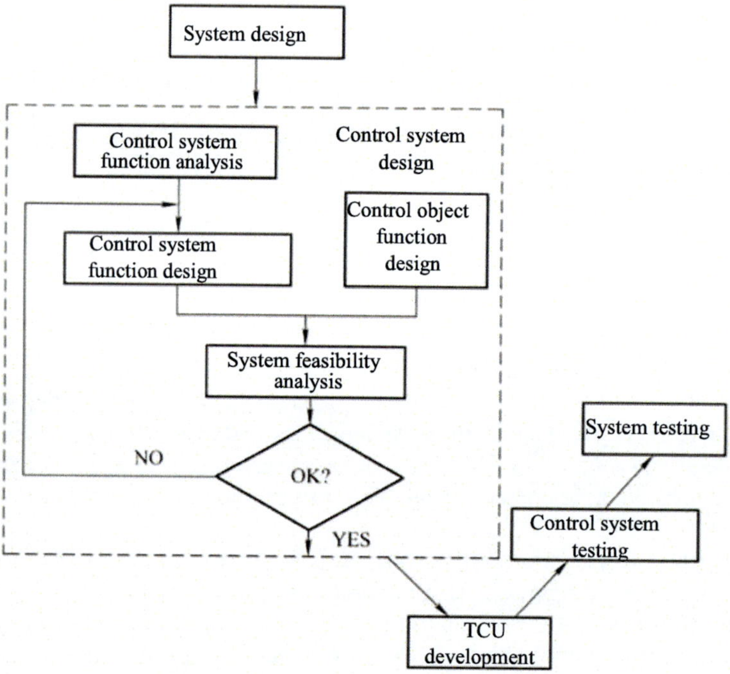

Fig. 12.8 Control system design subprocess

1. **Control system function analysis**

The AT electronic control system usually consists of the sensor, control unit and actuator. The sensor transmits the detected vehicle state, road conditions and other information to the control unit, and the control unit issues a target instruction based on such information, which is finally completed by the actuator. The control unit can adopt rule-based or optimized control algorithm. In addition, the AT electronic control system may also provide interfaces for other peripheral control units, such as the engine control unit (ECU) and electronic stability control (ESP) unit to synergistically improve vehicle driving comfort, safety and other requirements.

At the initial stage of development, developers only care about whether the control system is working properly, but tend to ignore the security issues in case of failure. However, failure protection is a very important part in the development of control system, which needs to be fully discussed from the initial stage.

First of all, it is necessary to determine the level or degree of danger the user falls into when the system fails, and what level of failure protection should be carried out. In the power transmission control system, the general design idea is to adopt a failure mode that will not cause danger even if there is a failure or misoperation in the control system. In advance, the action state of important components of the system in failure is studied, and the parts that produce the dangerous failure mode are double-configured, or the failure detection is carried out through other information, so as to prevent the occurrence of dangerous actions.

In addition, FMEA (Failure Mode Effective Analysis) is generally used for failure detection. The principle of this method is to study the working state of each part of the system when a failure occurs.

2. **Control system function design**

Functional strategy is the core of the control system. According to the function requirements of the control system completed in the system design process, the control functions required to enter the TCU are designed. Function requirements are "what should be done". In addition, since the control system is actually loaded into the TCU, the content regarding "how should do" must be documented. By repeating the different levels of "what should be done" to "how should do", we gradually reach the actual possible level, as shown in Fig. 12.9.

The process of reducing to a functional unit is called modularization, and it is important to determine control specifications that are easy to understand and maintain. In the design of modules, modules should be arranged from the functional point of view, while the minimum amount of module interfaces shall be fully considered to achieve the overall functions.

The main description methods of the control functions are described below:

(1) Data flow: data flow diagram.
(2) Control flow: state transition diagram and flow chart.

The function of the data flow diagram is to describe the control functions along the flow direction of a control variable, which is the basic form of the control specification

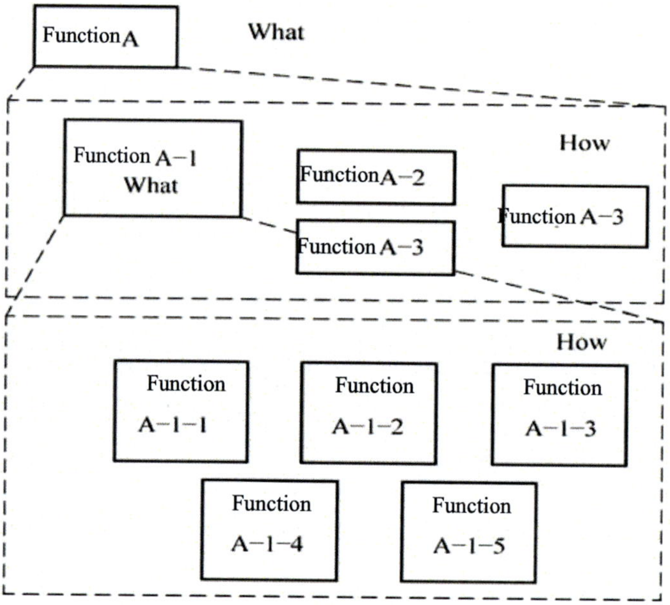

Fig. 12.9 Hierarchical structure of control specifications

when the control object is a successive event system. At this point, the input/output of the variable and its relationship are relatively clear, and it has a good affinity with the method of structured analysis. Its disadvantage is that it is not clear when and under what conditions to proceed. Figure 12.10 shows the data flow diagram.

The state transition diagram is suitable for describing the control functions of discrete event systems, such as human operation or failure state. This description makes it easy to show clearly what state can exist and how it changes under what conditions. However, in contrast to the data flow diagram, it has the disadvantage of not being able to clearly reflect the flow of variable. Figure 12.11 shows the state transition diagram.

The flow chart describes the control functions along the processing flow, which has the best affinity with the actual writing code. This description is also difficult to clearly reflect the flow of variable. Figure 12.12 is a flow chart.

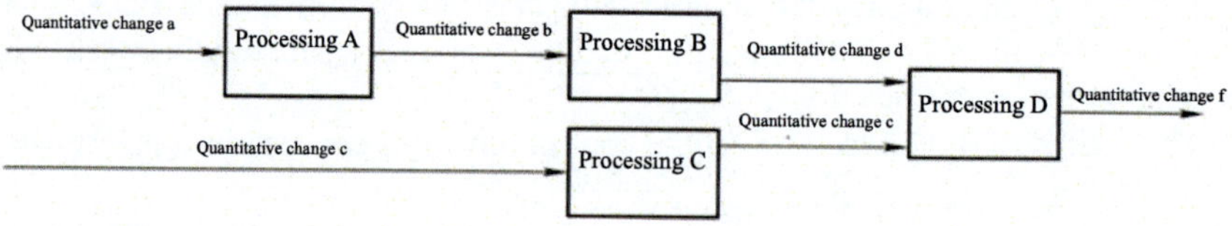

Fig. 12.10 Data flow diagram

Fig. 12.11 State transition diagram

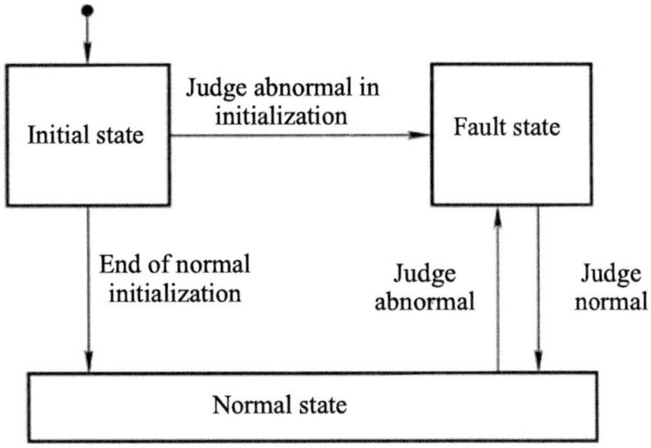

Fig. 12.12 Flow chart

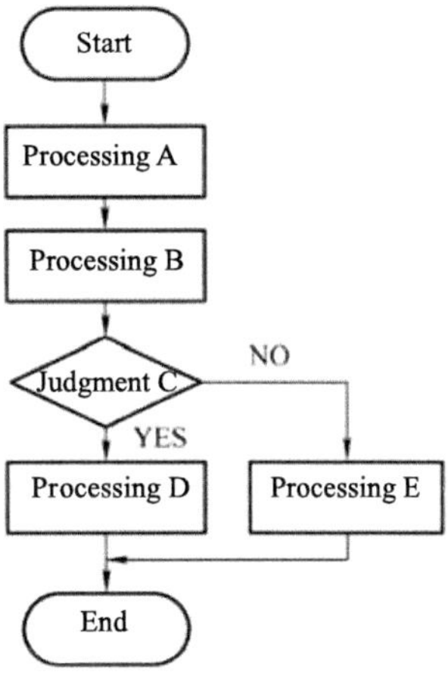

It is necessary to consider the characteristics of the control before deciding which description method to choose. In actual control, there are both successive event systems and discrete event systems. In this case, it is necessary to consider the actual computer assembly problem again. Therefore, the combination of the above descriptions is often used.

In addition, the control functions require data such as variables and constants, as well as information such as association attributes, in addition to the description described so far. Such information is called a data dictionary. The required attribute information, including the units and addresses of variables and constants, and conversion formulas to physical quantities when using fixed decimal points, will be used in the code generation and validation process. Therefore, it is very important to manage data dictionary uniformly in the development of control system. It should be noted that the data dictionary here differs from the definition of the data dictionary used in the data flow diagram that accompanies the software design related to information processing.

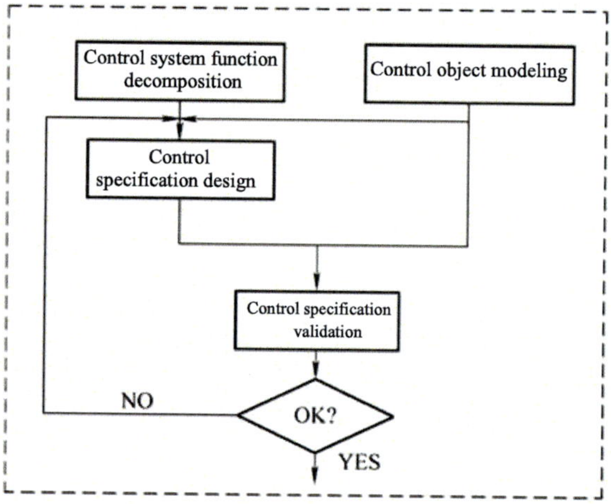

Fig. 12.13 Control object function design subprocess

In the model foundational development, the design results can be directly used as the control device model that can be simulated after the above description is perfected with tools. In addition, efficient use of automatic code generation is one of its advantages.

3. **Control object function design**

The control object function design subprocess is shown in Fig. 12.13. The control object modeling has two purposes: one is for the design of control specifications; the other is for the validation of control specifications. The control object functions are designed or grasped in the system design process; in this process, the model being simulated needs to be materialized. However, instead of pursuing absolute precision, the convenience of modeling and emphasis on computing time are more needed in the process of control object function design.

4. **System feasibility analysis**

The closed-loop simulation can be used to analyze the feasibility of the joint simulation of the control device model and the control object model. At this time, the validation script should be completed from two aspects: validation of analysis process and validation of control functions. The former is used for feasibility analysis with unknown design content, called black-box testing, which is the main method for system feasibility analysis; the latter is used for feasibility analysis with known design content, called white-box testing. In addition to the feasibility analysis of control functions, the system feasibility analysis also includes the software feasibility analysis. The index used to measure feasibility is called coverage, including decision coverage and condition coverage.

Decision coverage (branch coverage): with a view to branches of control functions, showing the percentage of branches of an actual calculation at least once.

Condition coverage: with a view to the theoretical calculation that includes state transition, showing the portfolio ratio of actual testing across the portfolio.

IV. TCU development (TCU writing)

From the point of view of software development, TCU development refers to the process of preparing the actual writing code into TCU according to the control functions. Loading software is generally classified into platform and application program. The former is equivalent to the operating system, mainly dependent on the TCU hardware; the latter runs on the basis of the platform and records the control content. The TCU writing subprocess is shown in Fig. 12.14.

(1) Code generation: C language is mainly used in the transmission control now. The codes were written manually before. However, with the recent development of system tools, the use of system tools for automatic programming has become popular. Only when the control functions are described as actual feasible models in the system tools can automatic coding be used.
(2) Code testing: the generated code may be tested due to its possible flaws. For example, for small units of software, testing for consistency between control functions and codes is called unit testing. In the model foundation development, the consistency can be tested through simulation because of the control device model with dynamic control functions.
(3) Compilation and link: the validated coding group shall be combined and loaded into the TCU. Because this process is the same as a traditional and general software development process, it is not specified here.
(4) TCU testing: even though the testing is conducted in small units, the overall testing is also required after combination. This testing can be done on software only, or in the combined loading in TCU.

V. Control system testing

The previous process has tested whether the TCU works according to the control functions. The next step is to combine the sensor, actuators and the TCU loaded with the code, and then test whether they can act as a control system according to the instructions. The specific testing methods include HILS (Hardware in the Loop Simulation) and real vehicle testing. The control system testing subprocess is shown in Fig. 12.15.

1. **Hils**

HILS means to test with physical devices in one part of the simulated closed loop. Figure 12.16 shows the composition of the HILS using actual ECU. In this case, the engine, transmission and vehicle control object models are combined into a real-time simulator to carry out the testing similar to that using real vehicles.

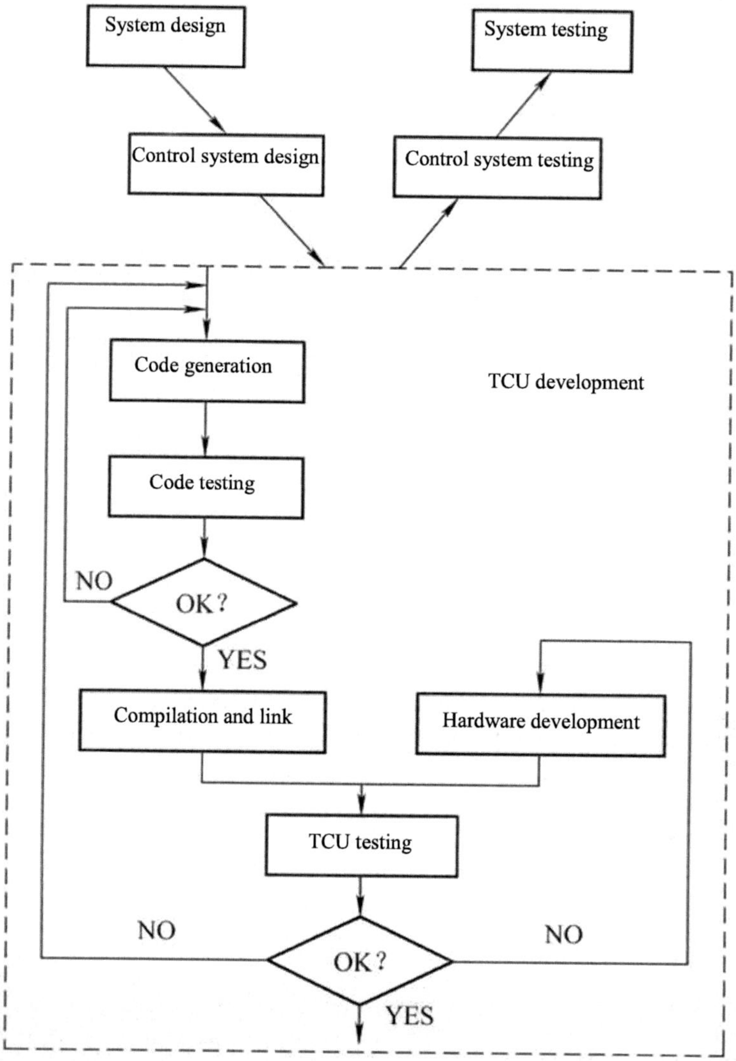

Fig. 12.14 TCU writing subprocess

The advantage of HILS is that it can improve testing safety, reproducibility and inclusion, but it also has the disadvantage of limiting the testing range due to the accuracy of control object model.

In HILS, ECU may be used as a real machine, and the actual sensors and actuators may also be used as required. A portion of the control objects can also be used as a real machine, depending on the situation.

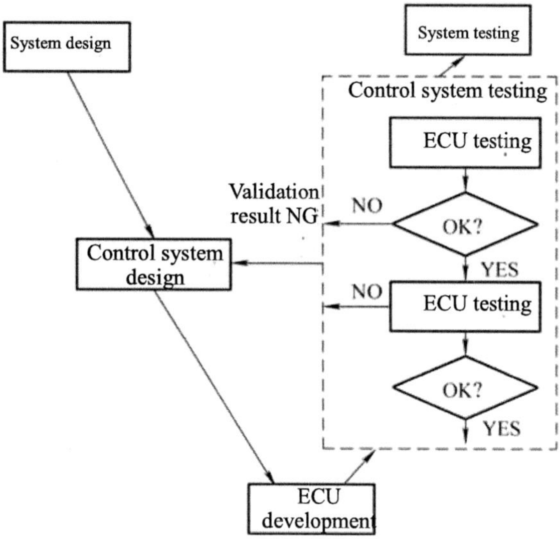

Fig. 12.15 Control system testing subprocess

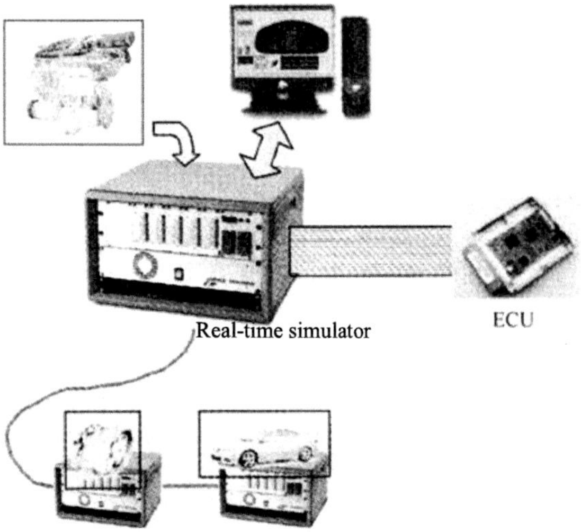

Fig. 12.16 Composition of HILS

2. **Real-time simulation testing**

Due to the limitations of HILS, the actual system needs to be tested. The specific method is to test the real vehicle after using the test bench. With the improvement of the accuracy of the control object model and the reduction of modeling time, the importance of actual system testing has decreased, but it has not reached the level that the actual system testing is no longer needed at this stage.

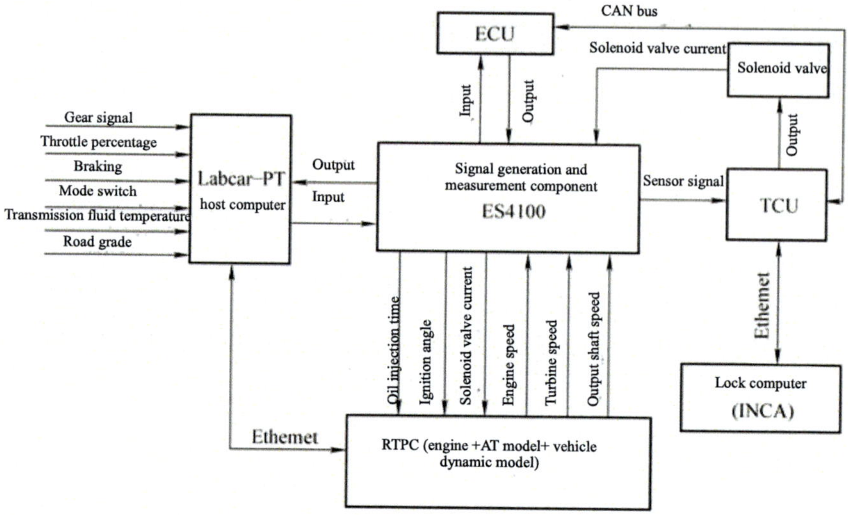

Fig. 12.17 AT Real-time simulation system

In addition, due to the testing environment, there are more cases in which HILS is used as a link of ECU testing, while real vehicle testing is used as a link of system testing.

A full vehicle model (prototype integrating a controlled AT) is set up through Simulink and the real-time HILS system consists of Labcar system and transmission control unit (TCU) by means of ETAS, as shown in Fig. 12.17. The principle of the real-time simulation system is as follows: a real-time simulation code is generated by the simulation model on the Host PC through compilation and downloaded to the RTPC via networking. The RTPC makes real-time operation of the model and sends the model operation variables, such as the engine speed, transmission input and output shaft speed, and signals required by the TCU and engine control unit (TCU) to the signal generation and measurement component ES4100. ES4100 converts the received signal variable into the corresponding actual physical signal and sends it to TCU through an adapter. TCU receives the relevant input signal and controls the solenoid valve to perform the corresponding action according to the software program therein. The solenoid valve current is detected by ES4100 current detection board and the measured value is resent to the RTPC model. The AT model calculates the pressure and torque of the relevant clutch/brake based on the solenoid valve current, and then recalculates the engine speed, transmission input and output shaft speed and other signals, thus forming an AT closed-loop system with the TCU.

As shown in Fig. 12.18, the hydraulic torque converter lockup operation is carried out in the real-time simulation system. It can be seen from the data that the control process has the same trend with the real vehicle testing results and similar data. Therefore, in the development process, the real-time simulation system can be used for logical validation and pre-matching of function modules, as well as simulation

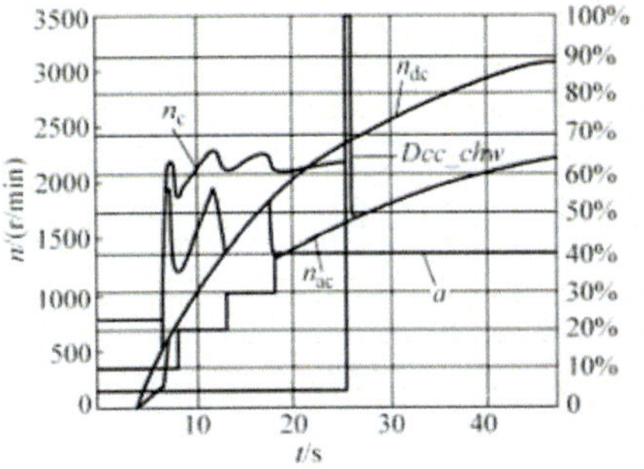

a) Simulation test bench test results

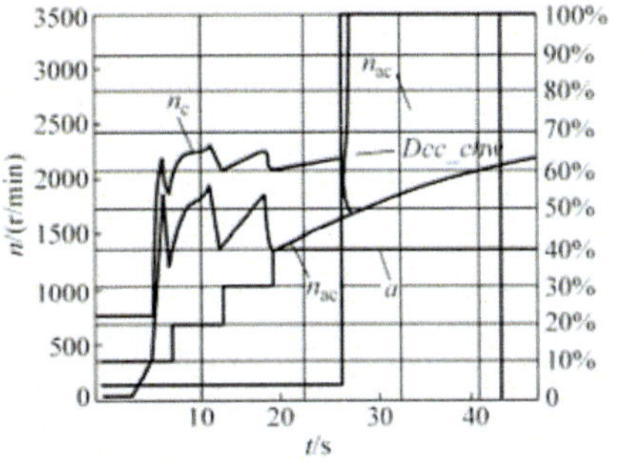

b) Real vehicle test results

Fig. 12.18 Comparison of testing results between simulation test bench and real vehicle

and testing of special conditions, without any operational risk; in the software batch production stage, automatic test program is developed on this basis to replace repeated manual testing.

VI. System testing

The system testing is in the final stage of system development. It is to combine the control systems and control objects as a whole system to test whether it meets the set system requirements. The system testing subprocess is shown in Fig. 12.19.

1. Calibration of control constants

The action of the ECU software is determined by the control algorithm and the accompanying control constants. Based on the characteristics of different control objects of each vehicle, the control constants to achieve the best control performance are obtained step by step. This process is called the calibration of control constants. Calibration has a broader meaning such as the calibration of different vehicles or of each individual, but the calibration described in this section refers to the calibration of control constants. There are many cases in which the control constants cannot be represented in the numerical form of scalar. In this case, they can be represented in the one-dimensional to three-dimensional graphs as shown in Fig. 12.20. A graph above two-dimensional level is called a MAP.

The system testing has a significant impact on system performance and requires a lot of development resources, especially for engine system development.

First, design is made on the computer. The values obtained from the design parameters, the monomer characteristics of the sensor and the actuators are set as control constants. Then the initialization and the maximum and minimum values of the constants are set. A constant will also change constantly in value during development, so it is important to note that it has the same concept of initialization and maximum and minimum values as a variable. In the development of model foundation, the full use of simulation method can improve the control constant precision at the initial stage of development and the efficiency of subsequent development.

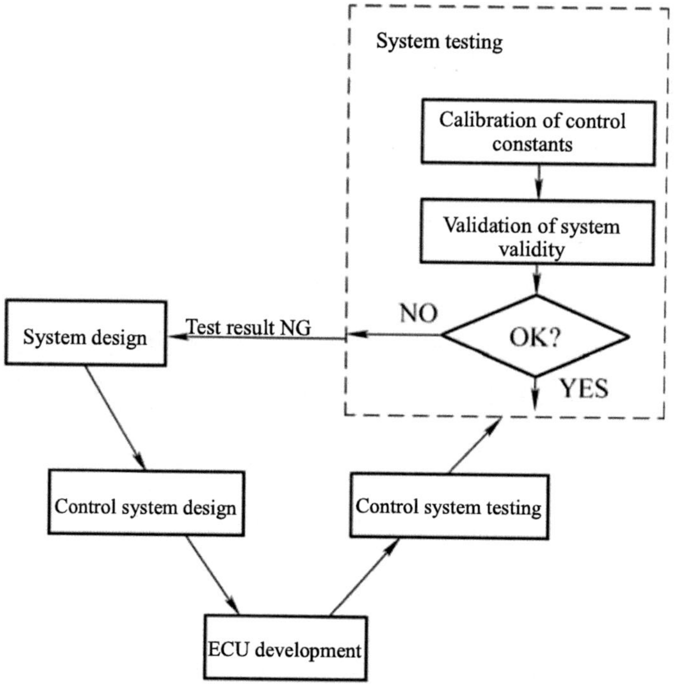

Fig. 12.19 System testing subprocess

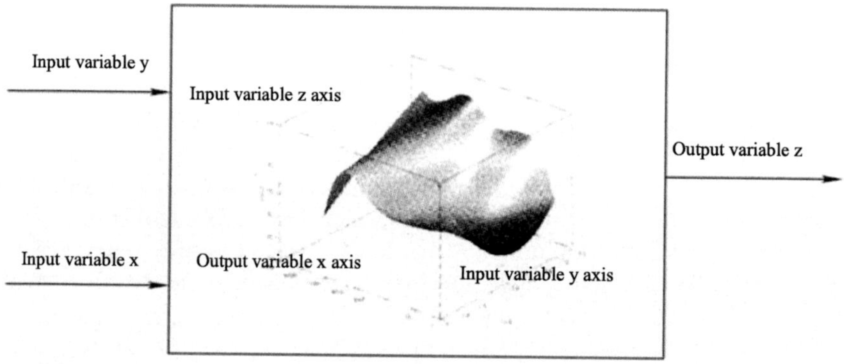

Fig. 12.20 3D graph

Secondly, calibration is carried out on the actual equipment, including steady-state calibration and transient calibration.

1. Steady-state calibration: there are two ways of steady-state calibration: one is to make the control object produce the actual action, and to search for the best control constant manually; the other is to take the initial steady-state characteristics of

the control object as the model, and to set the optimal control constant according to the model, namely model base calibration. The former is effective in the case of few parameters, while the advantages of the latter gradually emerge with the increase of control parameters.

The model base calibration steps are shown in Fig. 12.21.

(1) Set the measuring points using the testing planning. When more parameters affect the control constants, the total number of combinations becomes very large. For example, when there are 5 parameters and each need to reach 10 levels, the total number of combinations is 105, which is difficult to complete in reality. The testing planning is used to control the characteristics of the control object through the practical number of combinations of testing.
(2) Measure data by real machine testing. Measure the characteristics of the control object on the measuring points obtained in 1).
(3) Function approximation (modeling) of real vehicle characteristics. The types and times of the functions are set according to the characteristics of the control object and then the functions are subject to approximate calculation according to the data from the testing and the statistical approach.
(4) Model-based optimization results. Based on the modeled control object, the optimal point is calculated and searched. At this time, it should be noted that we should avoid the mistake of pursuing local best effect while thinking about constraints.

2. Transient calibration: transient calibration is to optimize the transient characteristics of a system. Since it is very troublesome to use real machines, especially real vehicles, for transient calibration, we should try to perfect the previous process to minimize the workload at this time. However, in areas where model accuracy is insufficient and where human perception is processed, the real machine must be used for transient calibration.

12.3 Validation of System Validity

The whole system is subject to final validation after the transient calibration. When requirements are not met, the system design process needs to be returned for development. But this will cause a huge waste of development resources and must be avoided. The solid and reliable testing by full use of the model at the early stage of development is to prevent iterative development.

VII. Model based modern transmission control software development

The running condition of the vehicle is complicated and changeable, and new power sources and actuators are introduced constantly. In order to achieve high-performance transient quality and energy saving performance, it is necessary to design a complex control system and carry out burdensome control parameter calibration. At present, the main control methods of vehicle control are open-loop control, feed-forward table

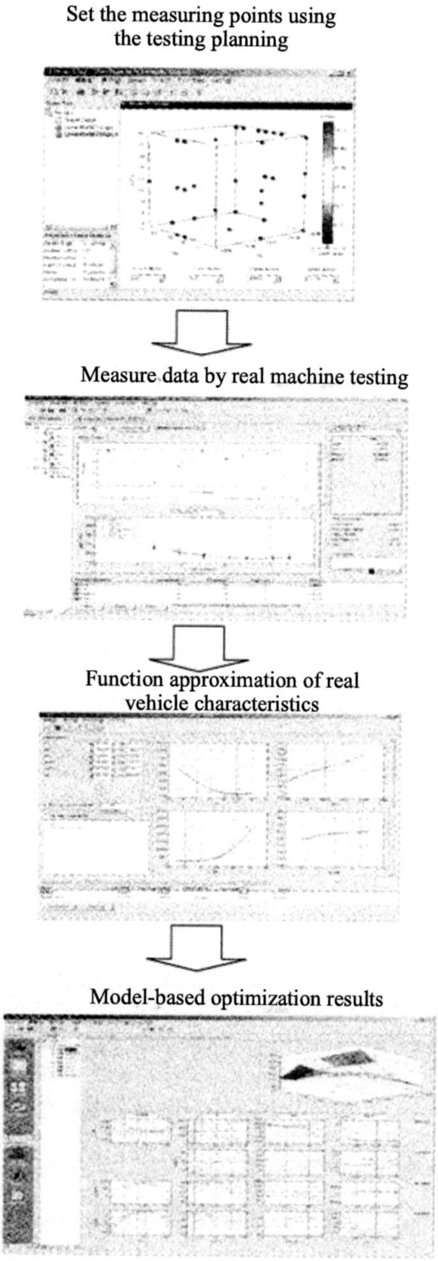

Fig. 12.21 Model base calibration steps

lookup, feedback PID, etc., and the calibration work of control software is huge. At present, the length of vehicle control system code has exceeded 10 million lines, and tens of thousands of parameters need to be calibrated. Parameter calibration requires a lot of time-consuming and laborious bench test and real vehicle test. Even so, the real vehicle calibration based on the static or quasi-static bench is still difficult to ensure the control performance of actual transient conditions.

Model Based Design (MBD) is an effective way to solve the above problems and has attracted the attention of the industry.

The simplified dynamic model for controller contains the main dynamic components of the system and can predict the dynamic behavior of the system. Therefore, the control system designed based on this model can improve the transient quality of the system. In addition, this design method can explicitly design the controller, which can greatly reduce the workload of control parameter calibration. Moreover, if the model error is corrected online, the performance can also be adaptive to ensure the performance quality in the whole life cycle. This kind of model has not been widely used in the actual development. The precedents reported include 4-Step-Approach of FEV, as shown in Fig. 12.22, in which the simplified model and the Modern Control Theory Approach based on the simplified model are used.

The 4-step-approach can improve the effectiveness of the controller design, which is divided into four steps:

(1) Extract a simplified model of all relevant components. This step is to simplify the control loop, including the dynamic parameters of the primary system, ignoring the influence of secondary factors. The model shall be processed linearly near the operating point so that the engineer can analyze the system parameters.
(2) A simplified model is used to analyze the system, including power performance and stability, with consideration to operation instructions and robustness. Typical methods of this kind of analysis include root locus method and

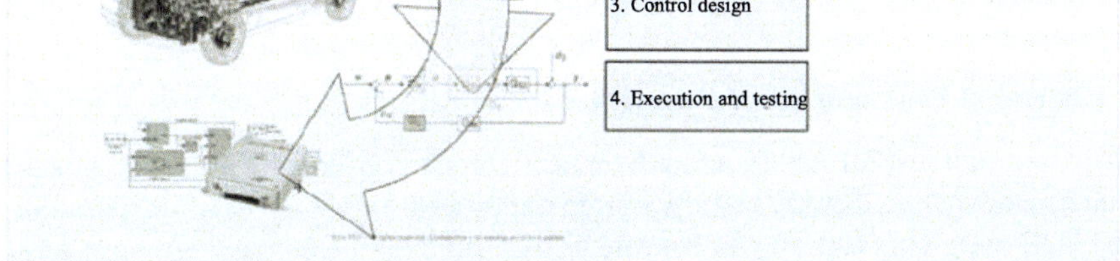

Fig. 12.22 4-step-approach controller design method

stability analysis based on Nyquist criterion. Different from common methods, the simplified model can also be used for frequency domain analysis.

(3) The ideal result of professional controller system structure and gain may be obtained through control design of the above system analysis.
(4) The designed and optimized controller is performed and tested on a vehicle or test bench.

In a structural approach to system identification and simplified modeling, the controller design and testing can shorten the functional development cycle and reduce costs, while reducing CPU and memory usage load.

VIII. Communication between ECUs

The high performance and complexity of the control increases the capacity of the electronic control unit and causes significant wiring harness installation and quality problems. To solve the above problems, the application of vehicle Local Area Network (LAN) is expanding rapidly. In the engine control system that needs real-time performance and extreme high reliability, Controller Area Network (CAN) gradually becomes the mainstream because of its high-speed communication performance and good error detection and disposal performance.

In addition, since 2003, relevant regulations in the United States have stipulated the external diagnostic (diagnostic system) devices of vehicles, requiring the diagnostic method to realize 500 KB/s high-speed communication, and it is expected that the communication speed will continue to be improved in the future.

As for other communication modes, the Flex Ray on-board network standard with higher communication speed (10 MB/s) and reliability is gradually adopted. In addition, the communication between ECU and intelligent actuators has begun to use cheaper Local Interconnect Network (LIN) bus.

12.4 Control Strategies of AT

Any AT (AT, AMT, DCT, CVT) requires a control system for automatic control. The control system is to automatically select the gear pursuant to the shift law during the vehicle driving according to the control lever position and the driving state of the vehicle, and to change the gear ratio of the transmission through the actuator, so as to realize the speed change or gear shift. The composition of the AT control system is shown in Fig. 12.23.

I. Control of hydraulic torque converter lockup clutch

At present, almost all AMTs adopt hydraulic torque converter with lockup clutch. The hydrodynamic drive is mainly used in the process of vehicle starting, climbing and shifting, giving full play to the role of hydraulic torque converter in increasing torque and realizing flexible transmission. Meanwhile, the power transmission by liquid may further reduce peak load and torsional vibration, extend the service life

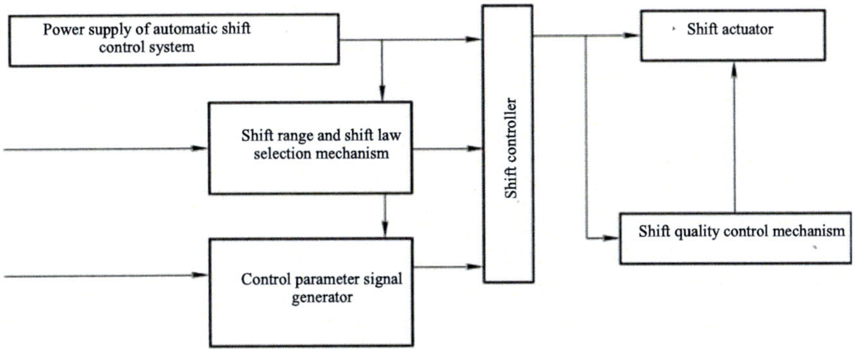

Fig. 12.23 Composition of AT control system

of the powertrain, and improve ride comfort, average vehicle speed, safety and passability. The lockup control of the lockup clutch realizes the transformation from hydrodynamic drive to mechanical drive, which is also a shift control in essence. That is, the control quality of the lockup process directly affects the smoothness of the vehicle shift. Therefore, it is of great significance to study the lockup process control of the lockup clutch.

12.5 Modeling of Hydraulic Torque Converter with Lockup Clutch

The lockup process quality of the lockup clutch mainly depends on the control over its engagement process. The engagement process of the lockup clutch is a sliding friction process in which the drive and driven parts experience varying speeds to the same speeds. Figure 12.24 shows the dynamic model of clutch engagement. From the figure

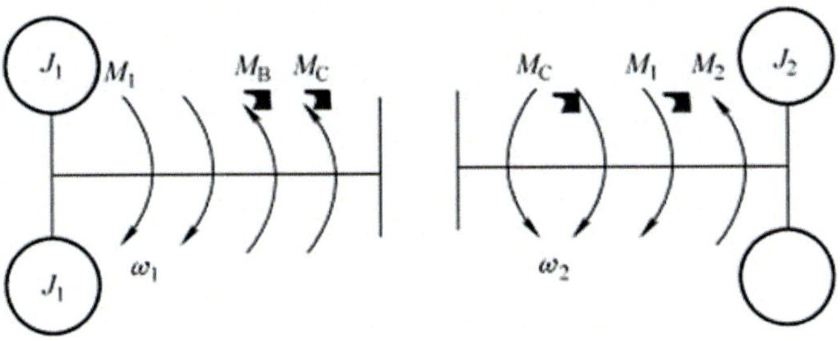

Fig. 12.24 Dynamic model of lockup clutch engagement

$$M_1 = M_B + M_c - J_1 \frac{d\omega_1}{dt} \tag{12.1}$$

$$M_2 = M_c + M_T - J_2 \frac{d\omega_2}{dt} \tag{12.1}$$

where, M_1—input driving moment of hydraulic torque converter;
J_1—rotational inertia of drive part of hydraulic torque converter;
ω_1—impeller angular velocity;
M_C—friction moment of lockup clutch;
M_B—moment on impeller;
M_T—moment on turbine;
M_2—moment of reaction on turbine shaft;
J_2—rotational inertia of driven part of hydraulic torque converter;
ω_2—turbine angular velocity.

When the lockup clutch is in the sliding friction state, its transmitted torque is

$$M_c = \mu_2 p_2 r_c \tag{12.3}$$

where, μ_C—sliding friction factor;
P_C—lockup clutch pressure;
r_C—equivalent action radius.

When the lockup clutch is engaged, its transmitted torque is

$$M_c = \mu_{c0} p_c r_c \tag{12.4}$$

where, μ_{C0}—static friction factor of lockup clutch.

According to the above analysis, the friction moment transmitted by the lockup clutch mainly depends on the friction factor, the lockup clutch pressure and the equivalent action radius.

12.6 Lockup Clutch Oil Circuit

The lockup clutch control oil circuit of TC521hydraulic torque converter is shown in Fig. 12.25. When the torque converter meets the lockup conditions, the electronic control unit TCU of the torque converter issues a PWM electrical signal to the lockup solenoid valve to control it to open and controls the hydraulic oil in the main oil circuit through the slide valve to enter the pressure cavity of the lockup clutch pressure plate. Meanwhile, the main oil pressure pushes the flow control valve of the torque converter upward, closes the oil outlet of the torque converter and connects the oil inlet with the lubricating oil circuit. In this way, the lockup clutch is engaged under the differential pressure and the power of the engine is directly transmitted to the driven plate (turbine) through friction. With the engagement of

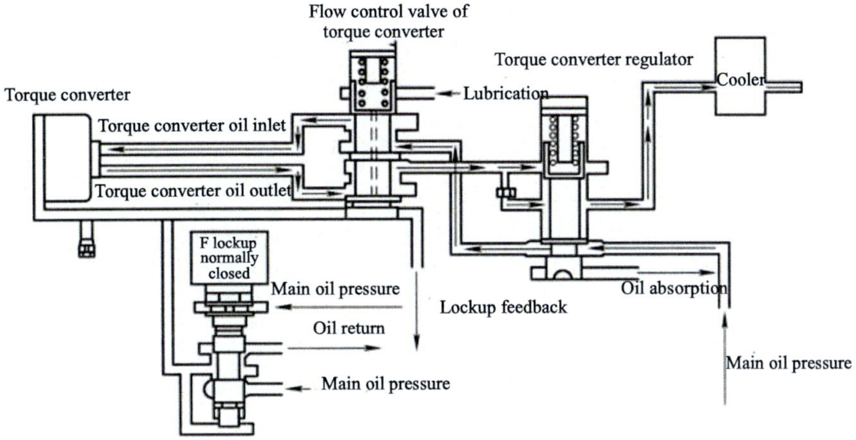

Fig. 12.25 Lockup clutch control oil circuit of TC521hydraulic torque converter

the lockup clutch, the torque converter impeller, turbine and turbine shaft rotate at the same speed with the engine and the mechanical drive replaces the hydrodynamic drive, thus improving the transmission efficiency. When the torque converter meets the unlocking conditions, the lockup solenoid valve powers off, the liquid chamber drains oil, the torque converter is unlocked and the hydrodynamic drive restores. Figure 12.25 shows that when the torque converter is locked, the lockup hydraulic oil is simultaneously diverted to the pressure regulator valve of the main oil circuit, and the oil pressure of the main oil circuit is lowered.

The system makes buffering control of the lockup clutch engagement process through the PWM valve and bilateral throttle valve. The structure of the digital pressure regulator valve is shown in Fig. 12.26, where, O is the drainage port, P_c is the oil supply port of lockup clutch and P_m is the main oil circuit port. It can be

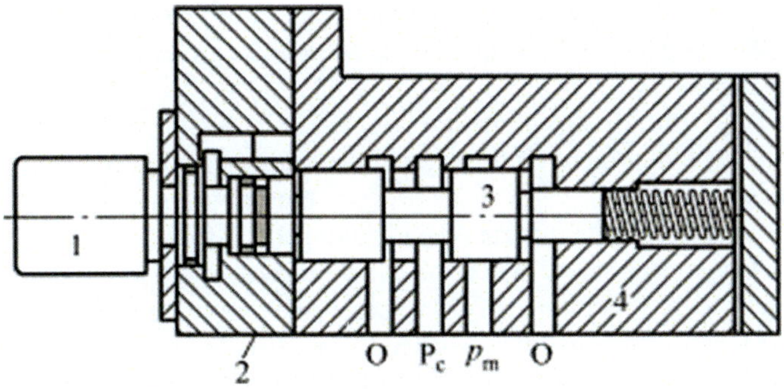

Fig. 12.26 Structure of digital pressure regulator valve. 1—Electromagnetic coil, 2—PWM valve, 3—valve element, 4—bilateral throttle valve. O—drainage port, PC—oil supply port of lockup clutch, Pm—main oil circuit port

seen from the figure that if the control oil pressure of the PWM valve rises, the slide valve will move to the right, the bilateral throttle valve opening is increased, the opening at the metering edge of the drainage port is reduced and the oil pressure in the intermediate valve chamber rises. Meanwhile, the oil pressure back fed to the right end of the slide valve also increases. If the resultant force of the oil pressure at the right end and the spring force exceeds the thrust at the left end, the slide valve moves to the right and the oil pressure in the intermediate valve chamber falls until a balance is reached. The slide valve will stabilize in a position to obtain a stable output oil pressure.

12.7 Study on Control Strategies for Lockup Process of Lockup Clutch

The slip-free lockup makes the hydrodynamic drive into mechanical drive. To absorb the vibration caused by engine or road impact, a torsional damper is installed in the clutch, and the compression plate is connected to the housing by elastic element to ensure that the compression plate is elastic along tangential and axial directions. Even so, the low-speed vibration of the engine is still difficult to attenuate. The lockup point is generally taken from the operating point of the coupler, and the efficiency loss is still high. However, too low speed ratio will not only cause the lockup to produce impact, but also cause the engine to stall when the vehicle is fast braking. Therefore, the initial lockup is limited to the narrow range of the high gear, high speed, and small throttle percentage. To extend the lockup range, the slip control is used to transition between the torque converter and full lockup conditions (Fig. 12.27), which is achieved by controlling the oil pressure on the lockup clutch friction plate (Fig. 12.28).

The staged control strategies for the lockup process of the lockup clutch are shown in Fig. 12.29:

(1) Rapid oil filling stage. At this stage, the solenoid valve is in a fully open stable state, the oil pressure rises quickly and the lockup clutch reaches the initial engagement state.
(2) Open-loop control stage (pressure isocline rise stage). Enter the open-loop control stage after rapid oil filling until the engine speed decreases.
(3) Closed-loop control stage. A stage from the engine speed decrease to the synchronization of the turbine speed and impeller speed.
(4) Total engagement stage. When the synchronization of the turbine speed and impeller speed is monitored, TCU issues a solenoid valve full open instruction and the lockup clutch is fully engaged to complete lockup.

The rapid oil filling time of the lockup clutch is affected by the factors such as oil temperature, engine speed and lockup clutch wear. The initial value set cannot completely adapt to the lockup requirements under different conditions, so an adaptive control strategy should be adopted to automatically correct the initial parameters

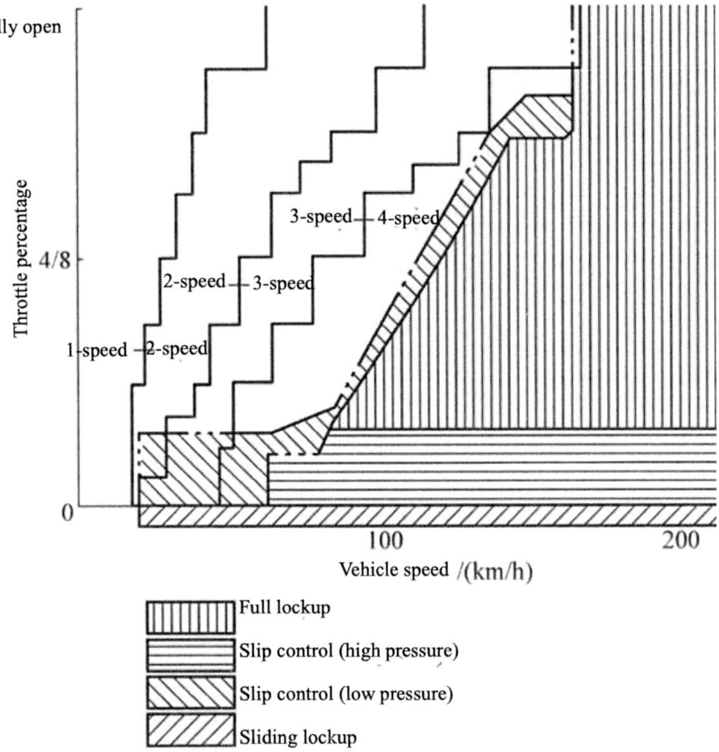

Fig. 12.27 Lockup control law

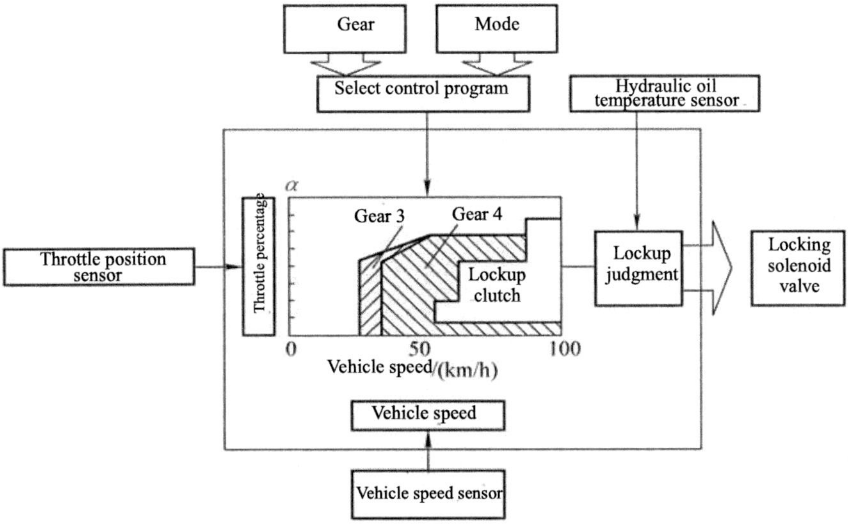

Fig. 12.28 Electronic control system of clutch lockup

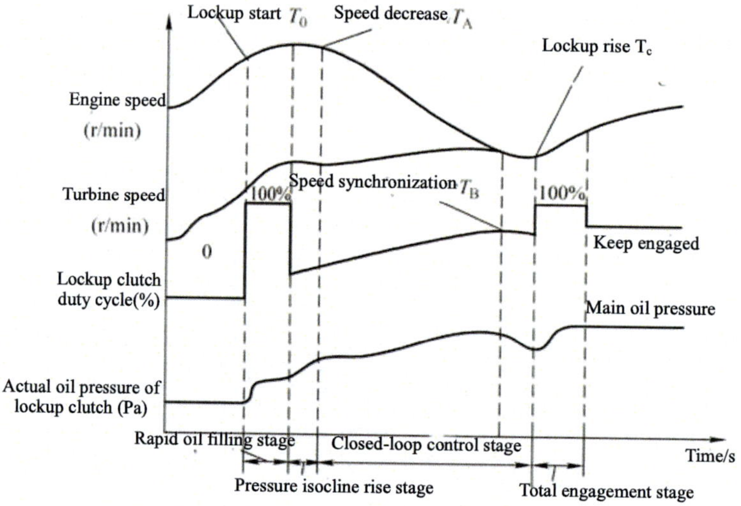

Fig. 12.29 Staged control strategies for the lockup process of the lockup clutch

according to the previous lockup parameters. Figure 12.30 shows the adaptive control principle of the rapid oil filling process of the lockup clutch. The variation Δn_{eg} of the engine speed detected before and after lockup is compared with the standard value Δn_{em} given by the reference model and the rapid oil filling time is corrected

$$\Delta n_{eg} = n_{e1} - n_{e2}$$

where, n_{e1}—engine speed before lockup of the lockup clutch;
n_{e2}—engine speed after lockup of the lockup clutch.

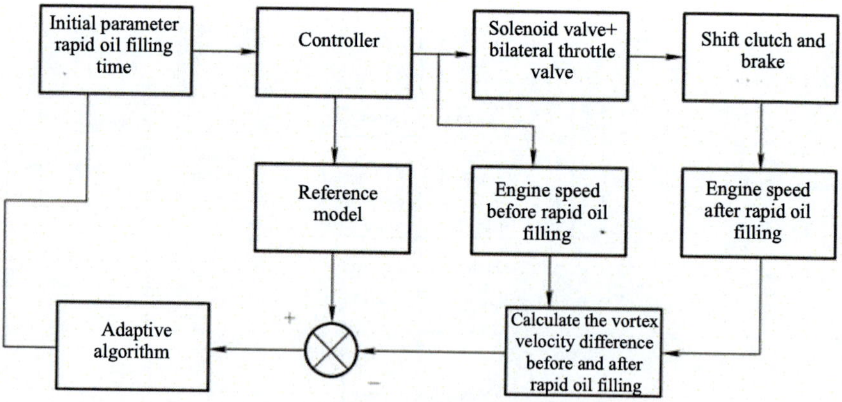

Fig. 12.30 Adaptive control principle of rapid oil filling process of lockup clutch

In the open-loop control stage, the isoclinal duty cycle control is adopted, and the size of the initial duty cycle and the growth slope of the control are determined by tests. The size of the initial duty cycle shall vary with the temperature, because the lockup oil pressure is different at the same duty cycle and different temperatures, which needs to be calibrated by test.

In the closed-loop control stage, the variation of impeller and turbine slip is taken as the control target for closed-loop control. This process is nonlinear and time-varying, so the PID iterative learning control algorithm is adopted. Formula (12.5) is P-type control algorithm formula, which is a special case of PID iterative learning control algorithm.

$$D_k(t) = D_{k-1}(t) + L_p e_k(t) \tag{12.5}$$

where, $D_k(t)$—duty cycle of electromagnetic valve PWM drive signal;

$e_k(t)$—impeller and turbine slip;

L_P—iteration factor.

Figure 12.31 shows the lockup test curve of the lockup clutch in the real vehicle road test. The lockup clutch will generate a large dynamic load during the lockup process, which will affect the life of the powertrain and the ride comfort of the vehicle, so this dynamic load shall be minimized. Figure 12.32 shows the lockup control effect of the hydraulic torque converter. It can be seen that, in the lockup process, the speed of the transmission output shaft changes steadily and the dynamic factor of the engine is about 1.85.

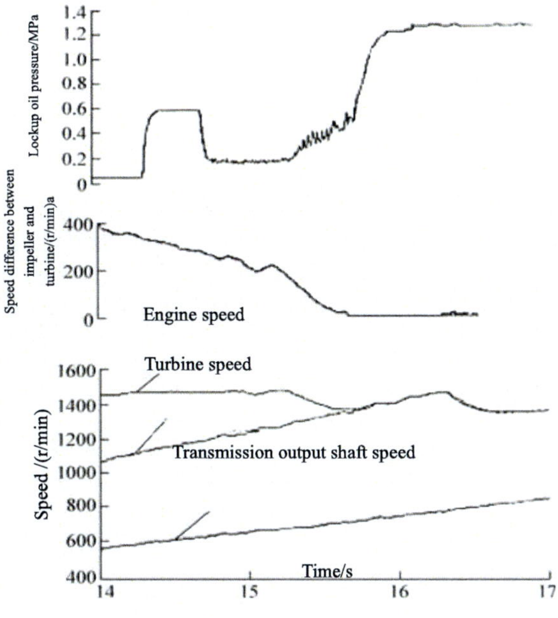

Fig. 12.31 Lockup test curve of the lockup clutch in the real vehicle road test

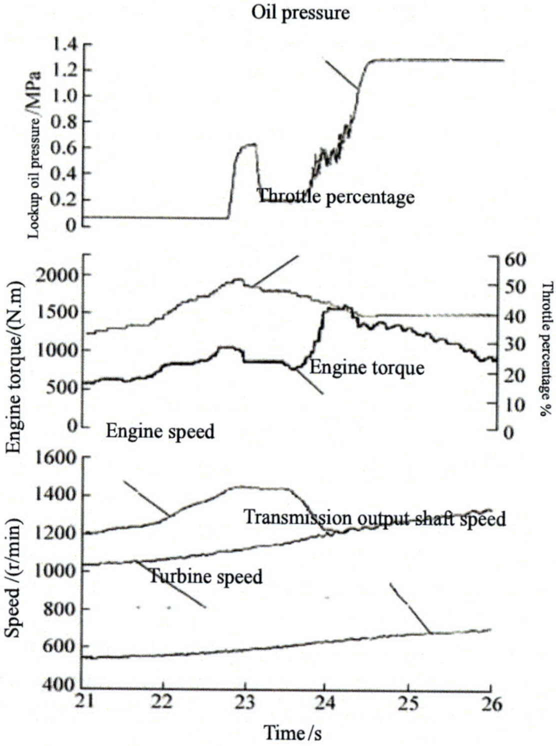

Fig. 12.32 Lockup control effect of hydraulic torque converter

The slip control can start from the torque ratio about K = 1.3 in the torque converter condition. The LA4 mode (city driving cycle mode) test shown in Fig. 12.33 shows that the clutch is mostly in slip condition and rarely in full lockup condition, which may reduce vibration and noise, improve the vehicle fuel economy and traction performance. Meanwhile, the engine braking can be used to solve the contradiction between fuel lubricity and ride comfort and improve the performance. It's important to note, however, that the full lockup must be done on the basis of an integrated torque converter; otherwise the efficiency will decrease rather than increase. In the control

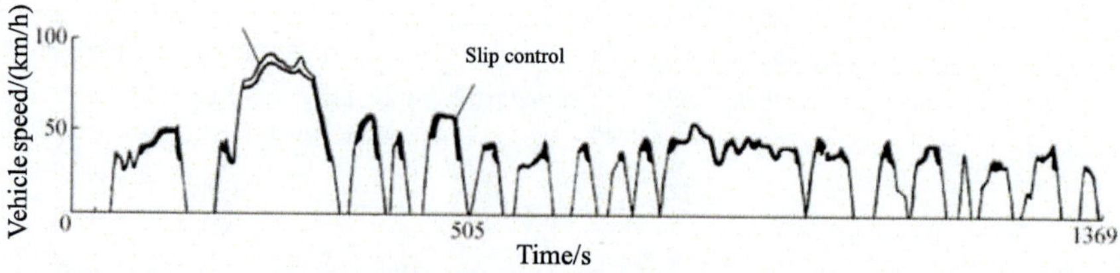

Fig. 12.33 Slip control in LA4 mode

theory, H control, instead of the traditional PID control, is used for the accurate control of slip, which is not only fast and stable, but also robust. In terms of slip control, the viscous coupling has better performance.

II. Shift law

The shift law, i.e. shift gear moment, is related to the exploitation and development of the overall potential of each assembly in the power transmission system. The shift law refers to the relationship between automatic shift gear moment and control parameters. Figure 12.34 shows an example of the control over the engine throttle percentage α and vehicle speed v, in which, the solid line represents upshift and the dashed line represents downshift. In fact, the vehicle starts and shifts in an unstable state, either accelerating or decelerating and three parameters (acceleration dv/dt, v and α) that reflect the dynamic process should be used to control the shift (Fig. 12.35). This rule, which is based on the equal acceleration of the adjacent gears before and after the gear shift, can further improve the acceleration, fuel economy and ride comfort of the vehicle, and extend the life of the components in the power transmission system, compared with the two-parameter control based on the equal steady state traction. The two-parameter control is actually just a special case of dynamic three-parameter shift at acceleration $dv/dt = 0$.

The electronic control system can store a variety of shift laws for drivers to choose, including economy law, power (kinematics) law, general (daily) law, environmental temperature law, as well as laws changing with external conditions, that is, the shift point can be set freely to conform to various laws. Fukang, for example, stores ten laws that can work separately, simultaneously or alternately, to control the automatic speed changing. In particular, the integrated control technology of fuzzy logic control and power transmission system is adopted to realize intelligent control, but the basic

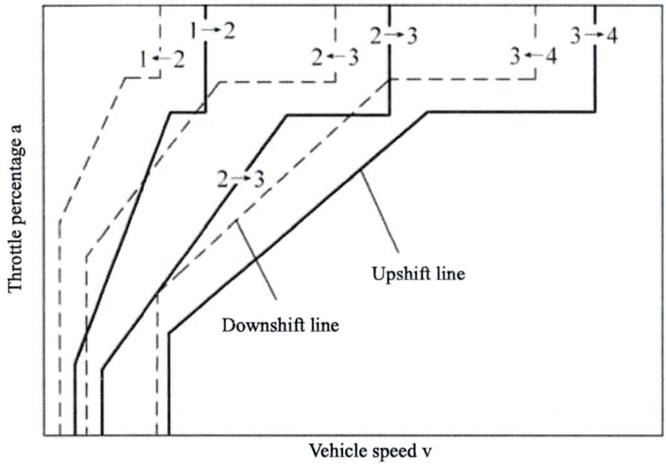

Fig. 12.34 Two-parameter control shift diagram

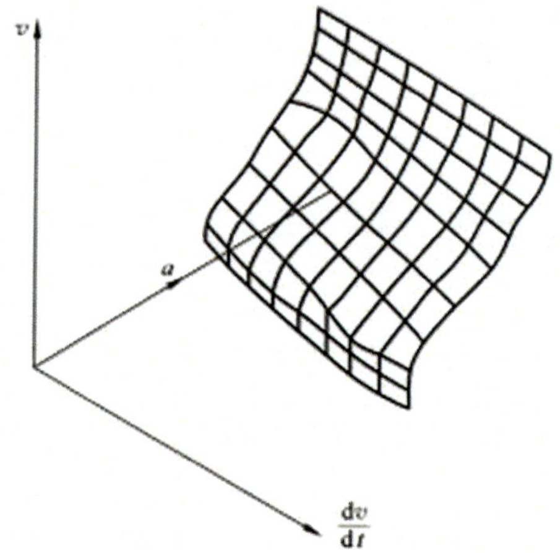

Fig. 12.35 Dynamic three-parameter optimal fuel economy shift law

control law is still two-parameter or dynamic three-parameter, and other factors are regarded as auxiliary conditions to determine the shift law. The transmission automatic control unit (TCU) calls out the corresponding law from the memory according to the location of the selector rod and compares the control parameters of dv/dt, v and α with the law. When reaching the set shift point, TCU issues an instruction to the solenoid valve to achieve automatic shift. The automatic shift control block diagram is shown in Fig. 12.36.

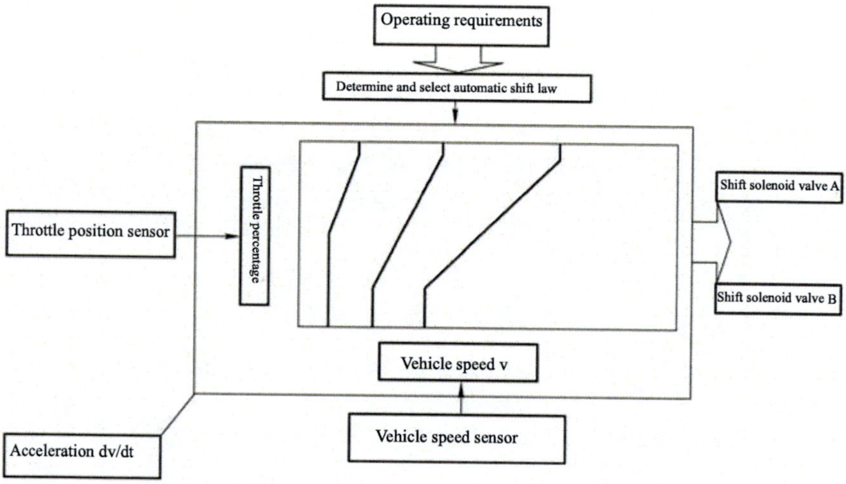

Fig. 12.36 Automatic shift control block diagram

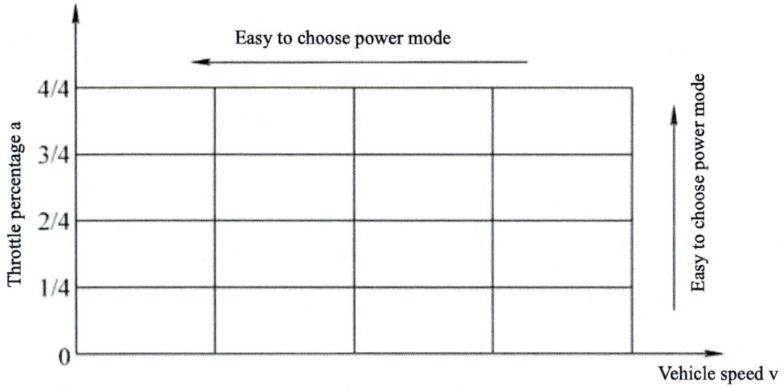

Fig. 12.37 Schematic diagram of automatic selection mode

Due to the improvement of TCU operation and control functions, when driving with general laws in daily life, the required law may be selected automatically rather than by control lever. When the acceleration pedal is pressed at a rate $d\alpha/dt$ that exceeds the rate set in the program in the drive gear (D), TCU shifts from economy to power, so the combination of the vehicle speed v and throttle percentage α is divided into a number of regions (Fig. 12.37), each with a different throttle opening rate program value. When the actual value is greater than the program value, the shift is power shift; instead, it is economic shift. The distribution law of program value is that, the lower the speed or the greater the throttle percentage is, the smaller the value is, that is, the easier it is to enter the power shift, and when the throttle percentage is less than 1/8, the power law is immediately switched the economic shift. On the other hand, the transferable manual-automatic transmissions are increasing to give drivers the pleasure of driving their own vehicles, or the desire to keep the gear through specific roads.

In order to avoid the shift cycle caused by driving resistance and speed fluctuation, two lines are used in Fig. 12.34 to represent the upshift and downshift, forming the shift speed difference Δv. Moreover, in the economic law, the adaptive control device is also introduced. In this way, when identifying large driving resistance and long duration, the shift control system will increase the shift speed difference Δv. The adaptive mode is automatically canceled in case of engine flame-out or small driving resistance identified.

The system has driver type identification, environmental condition identification, driving condition identification and other function modules. Among them, the driver type identification is to automatically select the shift law based on the inference of its characteristics of accelerator pedal stepping, braking and steering operation, thus giving intelligence to the AT; the environmental condition identification is to identify the driving resistance or the adhesion of tires to the road surface, call or correct the law; the driving condition identification is to make appropriate changes to the shift curve according to the current driving conditions (such as deceleration, downhill, curve driving or stop-go). As a result, AT can control shift by man, by time and by place. In addition, the vehicle performance will change with the increase of the use time, and this change shall also be recognized and reflected in the control, so the intelligent control technologies such as fuzzy control, adaptive control, learning control and neural network control must be adopted.

III. Shift process control

The shift quality includes the smoothness of the shift process and the component load. It is impossible for the shift actuator to alternate simultaneously as desired (e.g. brake disengagement and clutch engagement), the engine and other components will form impact due to inertia in the shift process and the dynamic friction factor and oil pressure fluctuation of the clutch in the actuator are also accompanied by torque disturbance, so the impact will occur inevitably in the shift process. For this reason, the output shaft torque disturbance shall be reduced to an acceptable degree.

(I) Clutch system modeling

Considering the process of switching between two clutches, to simplify the modeling process without affecting the implementation and validation of the estimation problem, a 2-speed ATAT is adopted in this chapter, in which, a planetary gear set is a shift gear, two clutches are actuators, and two proportional pressure control valves control two clutches respectively, as shown in Fig. 12.38.

When clutch A is engaged and clutch B is released, the vehicle is driving in gear 1 with a speed ratio of

$$i_1 = 1 + \frac{1}{\gamma} \tag{12.6}$$

where, γ—number of teeth of center gear and gear ring.

When clutch B is engaged and clutch A is released, the vehicle is driving in gear 2 with a speed ratio of

$$i_2 = 1 \tag{12.7}$$

Figure 12.39 is the schematic diagram of the PRV controlled clutch actuator system of the transmission. The system consists of the proportional solenoid valve (PSV), pressure reducing valve (PRV) and clutch assembly and can achieve automatic and high-quality shift.

The pilot pressure determines the position of the shift valve element and therefore determines the pressure and flow in the clutch piston chamber. When the pressure in the clutch piston chamber rises, the clutch piston begins to move forward until the clearance of the clutch plates is removed, and friction develops. In order to prevent overfilling and underfilling, the filling stage is controlled to end before the preset end time. To reduce the cost of the transmission sensor, no pressure sensor is installed on the clutch. Therefore, it is difficult to obtain feedback pressure of closed-loop control. In addition, because of the pressure nonlinearity and response delay of the solenoid valve, it is difficult to obtain accurate control pressure by controlling the flow. The

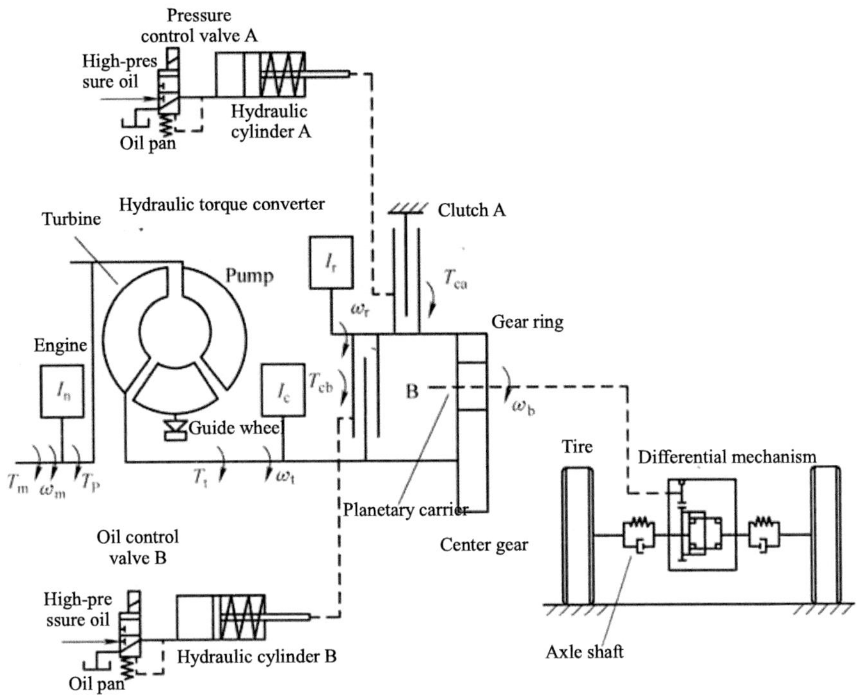

Fig. 12.38 Schematic diagram of new AT

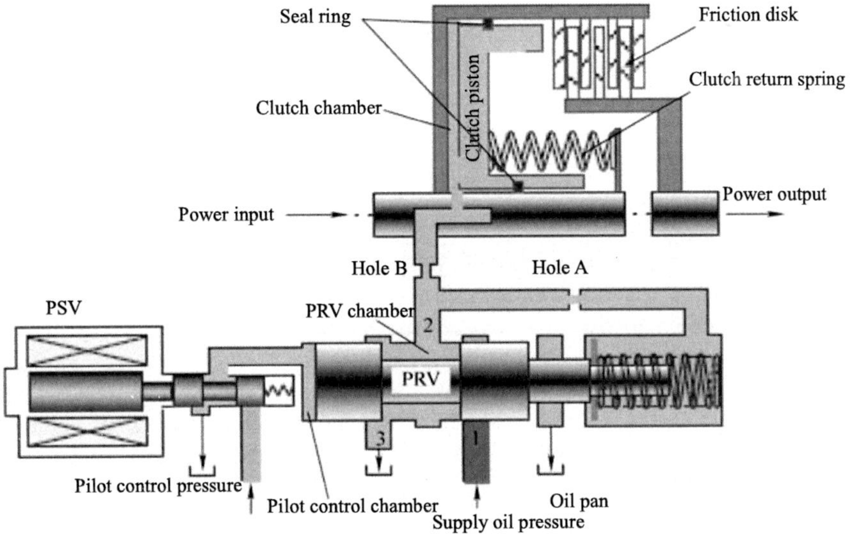

Fig. 12.39 Schematic diagram of the PRV controlled clutch actuator system

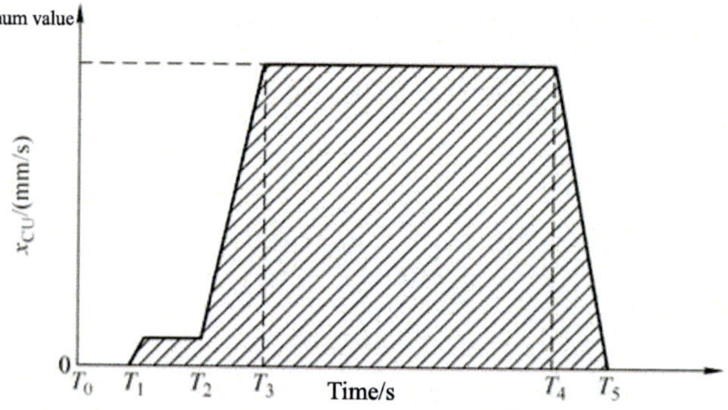

Fig. 12.40 Ideal piston speed

pressure response of the clutch is also different depending on the transmission state. Therefore, it is necessary to establish a clutch open-loop pressure control system in the clutch oil filling stage according to the electro-hydraulic clutch control system to meet the requirements of the clutch oil filling time and maximum flow. The transmission will inevitably wear out after a long period of operation. Therefore, an adaptive oil filling method based on fuzzy control is proposed, which can effectively improve the shift quality.

(II) Clutch oil prefilling control

The oil filling control objective of the clutch is to smooth out the clearance of the clutch plates in a specific time. The ideal piston speed in the clutch oil filling stage is shown in Fig. 12.40. The piston speed is close to zero, and the clutch cylinder clearance is first filled with ATF fluid in the T_0–T_2 period; the piston speed is low in the T_1–T_2 period and rapidly increases to a maximum in the T_2–T_3 period; the piston speed remains unchanged in the T_3–T_4 period; the piston speed starts to fall and the clutch starts working in the T_4–T_5 period.

Figure 12.41 shows two control methods at the clutch oil filling stage based on the piston motion in Fig. 12.40: triangular oil filling and square oil filling. In the figure, P_{pre}, P_{FP}, P_{FTP} and T represent the oil prefilling pressure, rapid oil filling pressure, stable oil filling pressure and rapid oil filling time respectively, which determines the change of the clutch oil filling pressure and the movement of the piston. In fact, based on the characteristic curves such as the pressure-current curve, the target pressure is converted into the control current of the solenoid valve. Therefore, it is necessary to analyze the target pressure and the actual pressure to optimize the clutch control. According to Figs. 12.40 and 12.41, the clutch oil filling process can be divided into three stages.

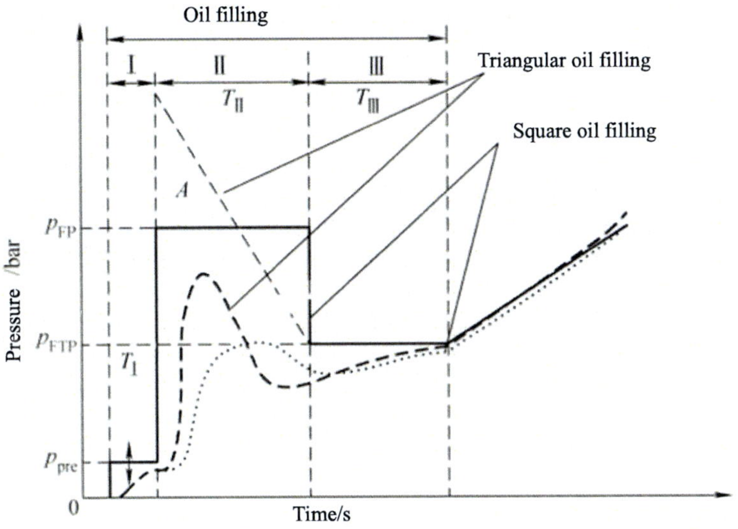

Fig. 12.41 Oil filling stage. I—oil prefilling stage, II—rapid oil filling stage, III—stable oil filling stage. *Note* 1 bar = 100 kPa

1. **Oil prefilling stage**

The oil prefilling pressure P_{pre} shall ensure that the oil filling line between the transmission shift valve and the piston is filled with ATF, generating initial pre-pressure but not exceeding the pressure required for piston movement (piston speed approaching 0). Therefore, when the clutch resistance is small, P_{pre} should not be too high to overfill. The hydraulic fluid flow must be completed within the prefilling time. In addition, the prefilling time is largely limited by the minimum shift time.

2. **Rapid oil filling stage**

The period during which the piston overcomes all resistance and begins to move rapidly is the rapid oil filling stage, which is the most important, corresponding to T_2–T_3 in Fig. 12.40. The triangular filling method can greatly increase the pressure peak and increase the piston speed. However, this method also increases the instability of the filling stage, and excessive pressure peaks require greater system flow, which affects the main pressure supply. When the square filling method is adopted, the response of pressure is relatively slow, but more stable than the triangular filling method.

The rapid filling pressure and time affect the filling speed and stable filling state. Increase of the rapid oil filling pressure can increase the reaction speed of clutch pressure and increase the initial oil filling pressure, while the increase of oil filling pressure can shorten the time for the clutch pressure to reach the target value. However, when the clearance of the clutch plates is completed, pressure fluctuations will inevitably occur in the T_3–T_4 because the piston speed is still very high and the ATF cannot absorb the residual energy immediately.

3. **Stable oil filling stage**

In the stable oil filling stage, the piston moves smoothly to eliminate the clearance of the clutch plates, corresponding to T_3–T_5 in Fig. 12.40. The control current is constant at this stage. With the movement of the piston, the elastic force of the wave spring increases gradually, so that the pressure increases gradually, and the piston does not stop moving until the clearance of the clutch plates disappears completely.

The actual oil filling pressure in the stable oil filling stage increases with the rapid oil filling time because of the increase of the filling volume and the maximum speed of the clutch piston. The clutch piston clearance then disappears and the contact point is quickly reached. Therefore, if the rapid oil filling stage takes too long, overfilling will occur and affect the shift quality.

The rapid oil filling time T_{II}, rapid oil filling pressure p_{FP}, stable rapid oil filling time T_{III} and stable oil filling pressure p_{FTP} jointly affect the clutch oil filling process. p_{FTP} affects torque transfer after engagement and takes precedence over clutch oil filling control, so p_{FTP} shall be used as an input superior to other parameters.

The rapid oil filling time T_{II} will significantly affect the clutch oil filling process, while the stable rapid oil filling time T_{III} has smaller influence. The influence of T_{III} on different transmission oil filling processes can be compensated by adjusting T_{II}. As a result, the preferred strategy is to set T_{III} as a constant and adjust T_{II}, depending on the piston speed and displacement demand.

To avoid poor shift performance, it is necessary to strictly control the proportion of the clutch oil filling time in the whole shift time, which is taken as 47% here.

Meanwhile, the rapid oil filling pressure shall be able to adapt to different environments, such as different ATF temperatures and main oil circuit pressures. Here, the rapid oil filling time is constrained to $T_{II} = 0.7T_{II,\max}$ ($T_{II,\max}$ is the maximum allowable value for rapid oil filling time).

(III) Torque phase control

The engagement and disengagement of the clutch can be considered as a process in which one clutch is engaged and the other is disengaged. This is usually divided into two phases: the inertial phase and the torque phase. In the torque phase, the engine torque is transferred from the gradually disengaged clutch to the gradually engaged clutch. Accurate timing of clutch disengagement and engagement is essential to prevent clutch torque interference and power interruption. In the inertial phase, the clutch sliding friction control has a great influence on the shift impact and shift time.

In the control of upshift, the output shaft torque will change as shown in Fig. 12.42. After the start of shift, the transmitted torque of clutch for high gear rises, resulting in the output torque change (torque phase). The engine then begins to slow down, and the release of the accumulated energy in the rotational inertia of the engine changes the output shaft speed (inertial phase). Finally, the torque recovers to the target value after the engine speed reduction. The greater the torque change of the output shaft,

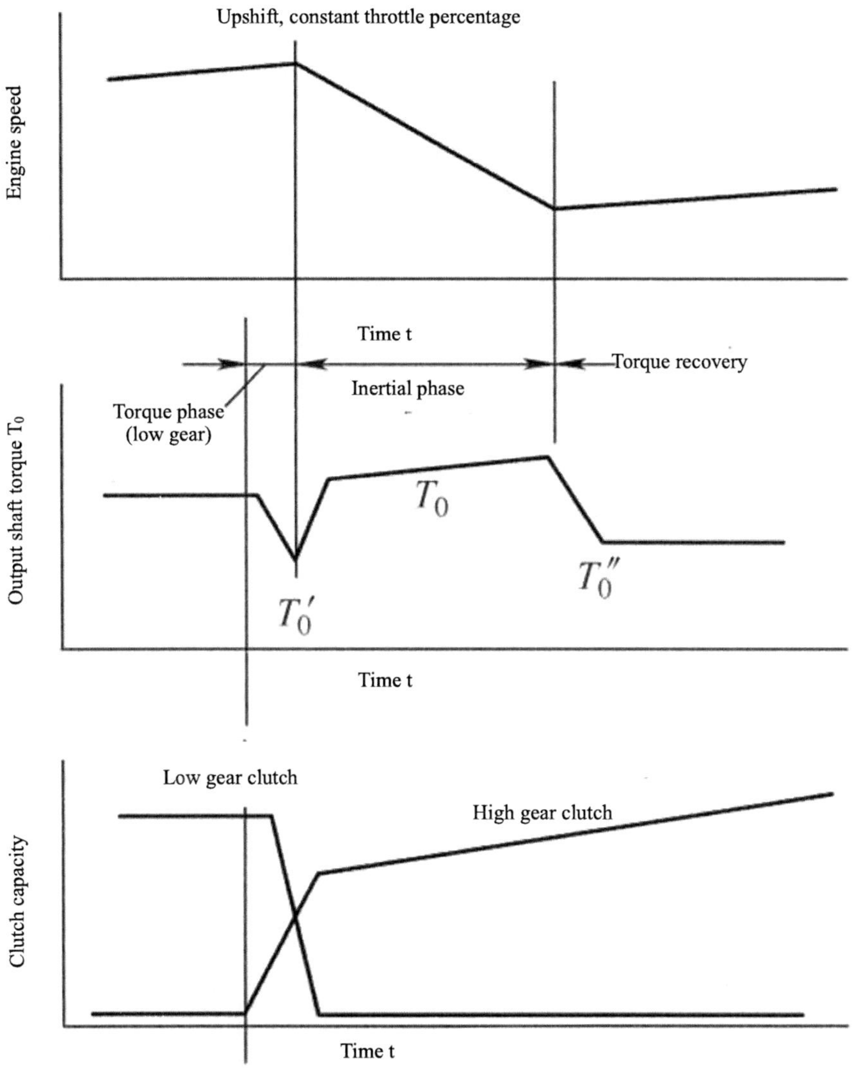

Fig. 12.42 Waveform graph of output shaft torque in shift

the greater the shift impact. The output shaft torque T_0, T'_0, T''_0 of each phase can be calculated according to the structure of the power transmission system and the connection method.

The clutch shaft speed does not change much in the torque phase. In order to achieve the smoothness of torque transfer between clutches, the clutch about to disengage needs to simulate the operation of OWC, so that it can be disengaged after switching in the direction of torque transfer.

Figure 12.43 shows the power interruption and torque interference caused by different clutch disengagement times. In the torque phase, the pressure of the clutch to be engaged increases, and in order to highlight the influence of the disengagement time of the clutch to be disengaged, three control modes are used for control. The mode shown in Fig. 12.43b provides the best control because the clutch just disengages when its torque drops to 0 at 7.93 s. The time of clutch disengagement in the mode shown in Fig. 12.43a is 0.1 s ahead of the optimal time, while the mode shown in Fig. 12.43c delays the disengagement time. The test results show that premature clutch disengagement will lead to power interruption, and the engine speed and turbine speed will increase sharply; delayed disengagement of the clutch will result in torque interference of the clutch plates, which will amplify the shift impact and increase the friction loss.

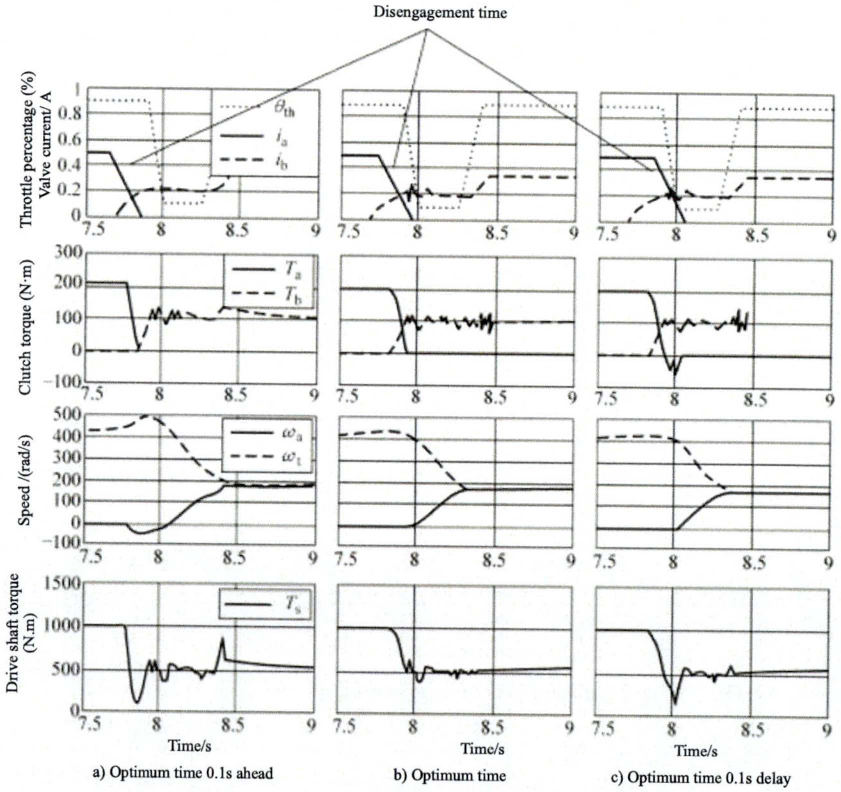

Fig. 12.43 Power interruption and torque interference caused by different clutch disengagement times. *Note* θ_{th}: throttle angle; i_a: disengaging clutch; i_b: engaging clutch; T_a: torque of disengaged clutch; T_b: torque of engaged clutch; ω_t: turbine speed; ω_a: speed difference of cut-off clutch; T_s: drive shaft torque

(IV) Inertial phase control

1. **Description of control problems**

The controlled object follows the AT shown in Fig. 12.38. The motion equation of inertial phase in the process from gear 1 to 2 is established first. This chapter discusses the use of clutch solenoid valve for clutch speed control, so the speed difference $\Delta\omega$ and pressure P_{cb} of clutch B are selected as the state variables. It is worth pointing out that although the model does not include turbine speed ω_t, both the driving moment of the engine and the driving resistance of the vehicle are constrained by the actual physical system, and their characteristics ensure the stability of the system. The state equation for the inertial phase in the process from gear 1 to 2 can be obtained

$$\dot{x}_1 = (C_{13} - C_{23})\mu(x_1)RNAx_2 + f(\omega_e, \omega_t, x_1) \tag{12.8}$$

$$x_2 = -\frac{1}{\tau_{cv}}\dot{x}_2 + \frac{K_{cv}}{\tau_{cc}}u \tag{12.9}$$

where,$x_1 = \Delta\omega x_2 = p_{cb}$

$$f(\omega_e, \omega_t, x_1) = (C_{11} - C_{21})T_t(\omega_e, \omega_t) + (C_{14} - C_{24})T_{ve}(\omega_t, x_1) - (C_{13} - C_{23})\mu(x_1)RNF_s \tag{12.10}$$

where, C_{ij}—constant coefficient determined by the inertia of the vehicle and the rotational inertia of each rotary shaft of the transmission;

u—current of valve B, $u=i_B$;

T_t—turbine torque;

T_{ve}—driving resistance at the transmission output shaft side;

K_{cv}—solenoid valve gain;

τ_{cv}—time constant of solenoid valve;

μ—static friction coefficient;

F_S—return spring force of clutch B.

In this section, the basic shift moment is used to make open-loop control of the engine torque. Therefore, the control problem is reduced to a single control input problem. The control input is the solenoid valve current i_b of the clutch B and the control target is to make the speed difference $\Delta\omega$ of the clutch B track the designed target trajectory $\Delta\omega^*$, as shown in Fig. 12.44.

The inertial phase should meet the requirements of good comfort and small friction loss simultaneously. Generally speaking, if the shift can be completed in a short time, and engine output torque can be well controlled in this process, the clutch sliding friction loss will not be too large. (That is, if the output torque is reduced in the appropriate time), the clutch friction loss will not be too large. With respect to the shift smoothness, since a large impact will be caused at the moment of the clutch synchronization, the clutch engagement process should meet the non-impact conditions. That is, at the time of synchronization, the rotational acceleration at the

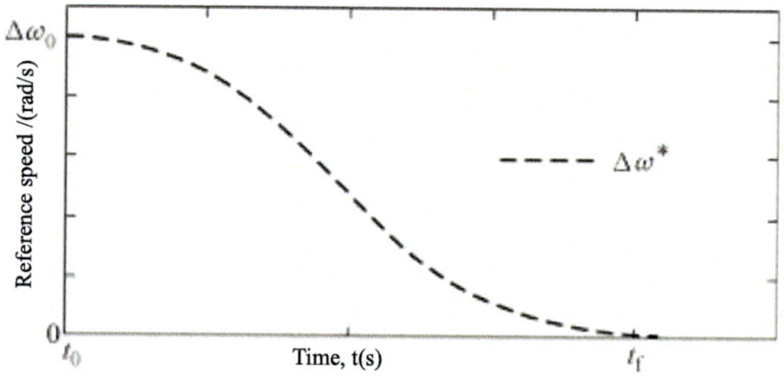

Fig. 12.44 Target control trajectory of clutch speed difference in the upshift process

input end of the clutch is equal to that at the output end. Therefore, the target trajectory needs to meet the following conditions:

(1) $t_f - t_0$ cannot exceed the expected shift time.
(2) The change rate of the clutch sliding friction speed difference at the time t_f is 0.
(3) In order to avoid saturation of the control input, the change rate of the clutch sliding friction speed at the time t_0 shall be a small value.

2. **Controller design**

According to the 92DOF controller block diagram in Fig. 12.45

$$Y(s) = P(s)\big(P^{-1}(s)M(s)R(s)\big) + K_b(s)M(s)R(s) - Y(s) \tag{12.11}$$

where, R(s) and Y(s) are Laplace conversion of r and y. If the modeling of the control object P(s) is accurate enough, the transfer function of the whole closed-loop system can be calculated

$$\frac{Y(s)}{R(s)} = M(s) \tag{12.12}$$

That is, the expected output response can be achieved by designing M(s). On the other hand, a feedback controller Kb(s) can be designed to achieve good stability

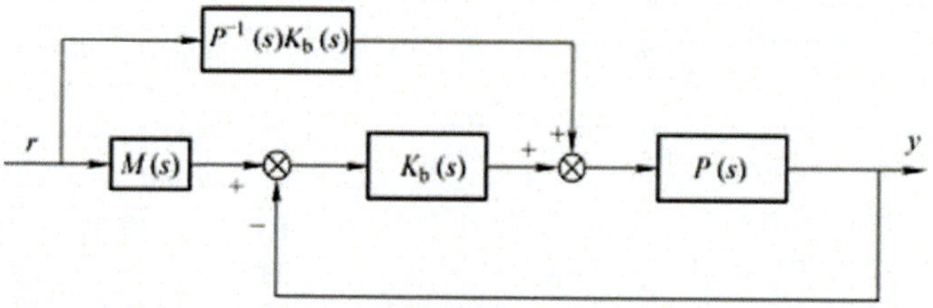

Fig. 12.45 92DOF controller

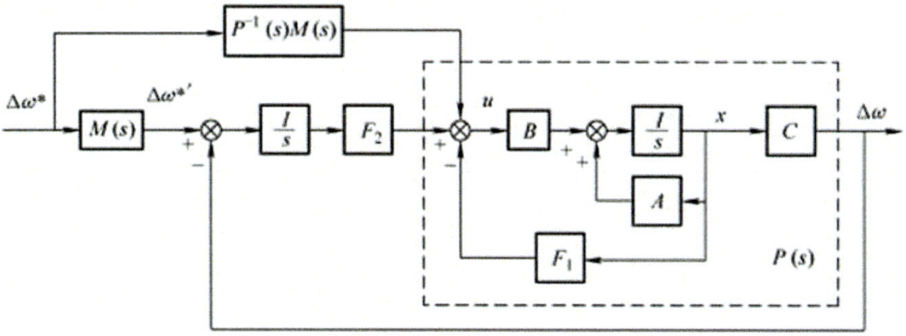

Fig. 12.46 102DOF clutch speed controller

and robustness. Ignoring the nonlinear characteristics of the friction characteristics, the dynamic friction factor is taken to be constant $\mu_0 = 0.13$, and then the kinetic equation of the clutch sliding fiction during the shift is rewritten into the following state-space equation

Based on the state equation above, a two-degree-of-freedom control system is designed to obtain the 102DOF clutch speed controller as shown in Fig. 12.46.

$$\dot{x} = Ax + Bu + Ed \tag{12.13}$$

$$y = Cx$$

$$x = [\Delta\omega P_{ch}]^T, y = \Delta\omega \tag{12.14}$$

$$A = \begin{bmatrix} 0 & (C_{13} - C_{23})\mu_0 RaNaAa \\ 0 & -\frac{1}{\tau_{cv}} \end{bmatrix}, B = \left(0 \frac{K_{cv}}{\tau_{cv}}\right)^T \tag{12.15}$$

$$E = \left(1\ 0\right)^T, C = \left(1\ 0\right) \tag{12.16}$$

$$u = i_b d = f(\omega_e, \omega_t, x_1) \tag{12.17}$$

Without feed-forward compensation, the gains F_1 and F_2 constitute a generally adopted linear servo system. The robust pole placement can be used to calculate F_1 and F_2 and the command "place" in MATLAB may be used for easy calculation.

The feed-forward compensation may be designed after the feedback gains F_1 and F_2 are obtained. First, let the part in the dotted line in Fig. 12.46 be P(s). The P(s) here is not just a control object. It also includes state feedback. Through this process, it is calculated that

$$P(s) = C\left(sI - A'\right)^{-1} B = \frac{P_n(s)}{P_d(s)} \tag{12.18}$$

where, $A' = A - BF_1$.

$P_n(s)$ is a constant and $P_d(s)$ is the quadratic function of s. $P^{-1}(s)M(s)$ must meet the regularity condition, so M(s) may be the transfer function of third order in the following form

$$M(s) = \frac{p_0^3}{(s + p_0)^3} \tag{12.19}$$

3. **Simulation result**

The controller is validated by the established AMESim simulation model. It is assumed that the matching of clutch A and clutch B can be well controlled in the torque phase. In the inertial phase, the designed clutch controller controls the sliding friction of clutch B. The feedback gain adopted in the simulation is

$$F_1 = \left[-7.8 \times 10^{-3},\ 1.9 \times 10^{-6}\right] \tag{12.20}$$

$$F_2 = [-0.081] \tag{12.21}$$

Figure 12.47a shows the simulation results when $P_0 = 100$. The matching control of the engine and transmission is not performed in the shift process. That is, the engine throttle percentage is always maintained at 90% without adjustment. The shift process starts from 5.7 s, with 5.7 to 6.1 s as the shift torque phase, and 6.1–6.5 s as the shift inertial phase. The desired inertial phase time is set at 0.4 s. The simulation results of the clutch speed difference are shown in Fig. 12.47b. The figure also gives the expected speed $\Delta\omega^*$ and $\Delta\omega^{*\prime}$, of which, $\Delta\omega^{*\prime}$ is the original expected trajectory and $\Delta\omega^*$ is expected trajectory after shaping. It can be seen that the clutch speed difference $\Delta\omega$ can well track $\Delta\omega^*$.

To observe the shift impact, the output torque T_0 of the transmission and the vehicle impact (the change rate of vehicle longitudinal acceleration, da) are also given. At the start and end of the inertial phase, there is a large output torque fluctuation which leads to a large vehicle impact, that is to say, there is a relatively obvious shift impact.

The simulation results when $P_0 = 30$ are shown in Fig. 12.48. Because the response characteristic of M(s) is slow at this time, the expected inertial phase of gear shift is set as 0.2 s to achieve the real inertial phase time of 0.4 s. It can be seen that at the beginning and end of the inertial phase, there is no sharp change in the current of the solenoid valve, which correspondingly reduces the shift impact. In particular, at 6.5 s, the end of the inertial phase, the solenoid valve current i_b has a decline process, which achieves a relatively smooth synchronization of the clutch.

Finally, combined with the comprehensive shift control of engine and transmission, the shift simulation result of engine throttle adjustment is given, as shown in Fig. 12.49. It can be seen that although the engine throttle percentage changes greatly in the shift process, the controller will still be able to work well, and the tracking error of the clutch speed difference $\Delta\omega$ can still meet the use requirements. In particular,

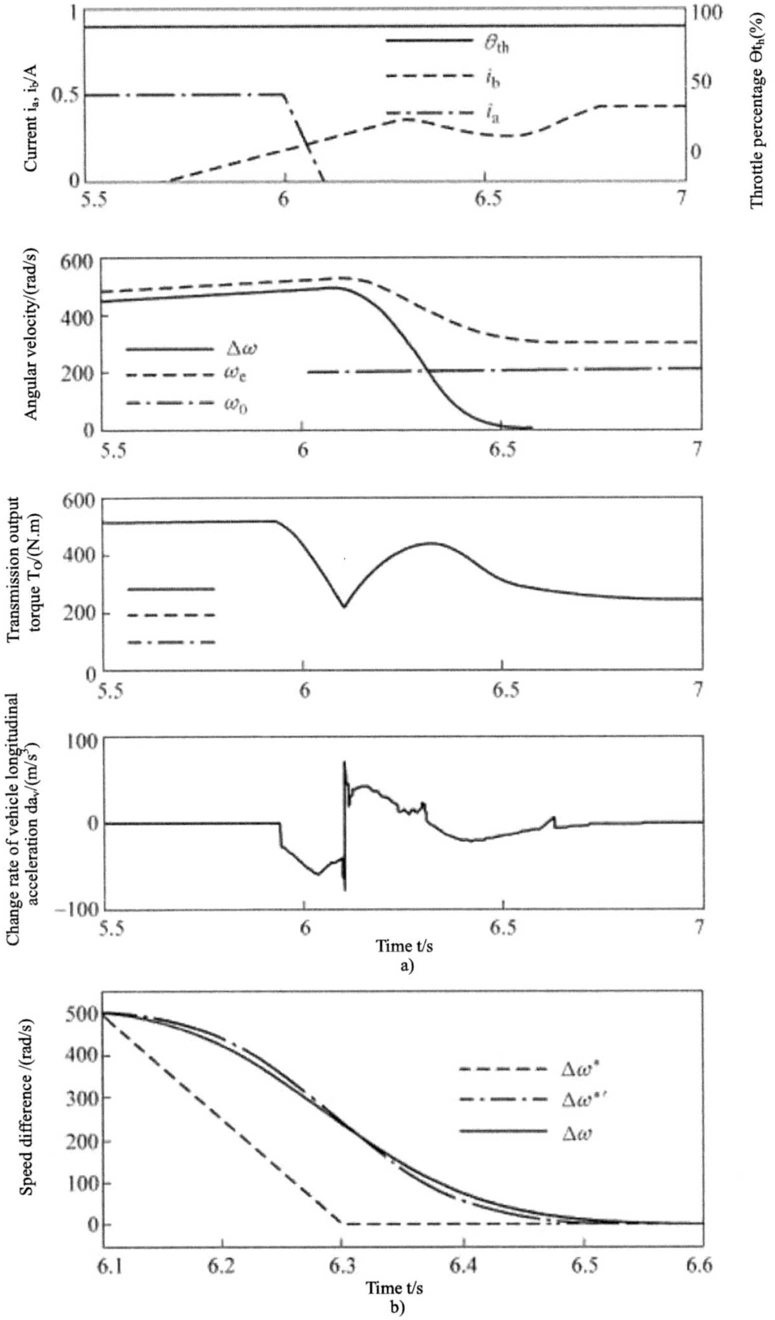

Fig. 12.47 Simulation result when $P_0 = 100$ (no engine matching)

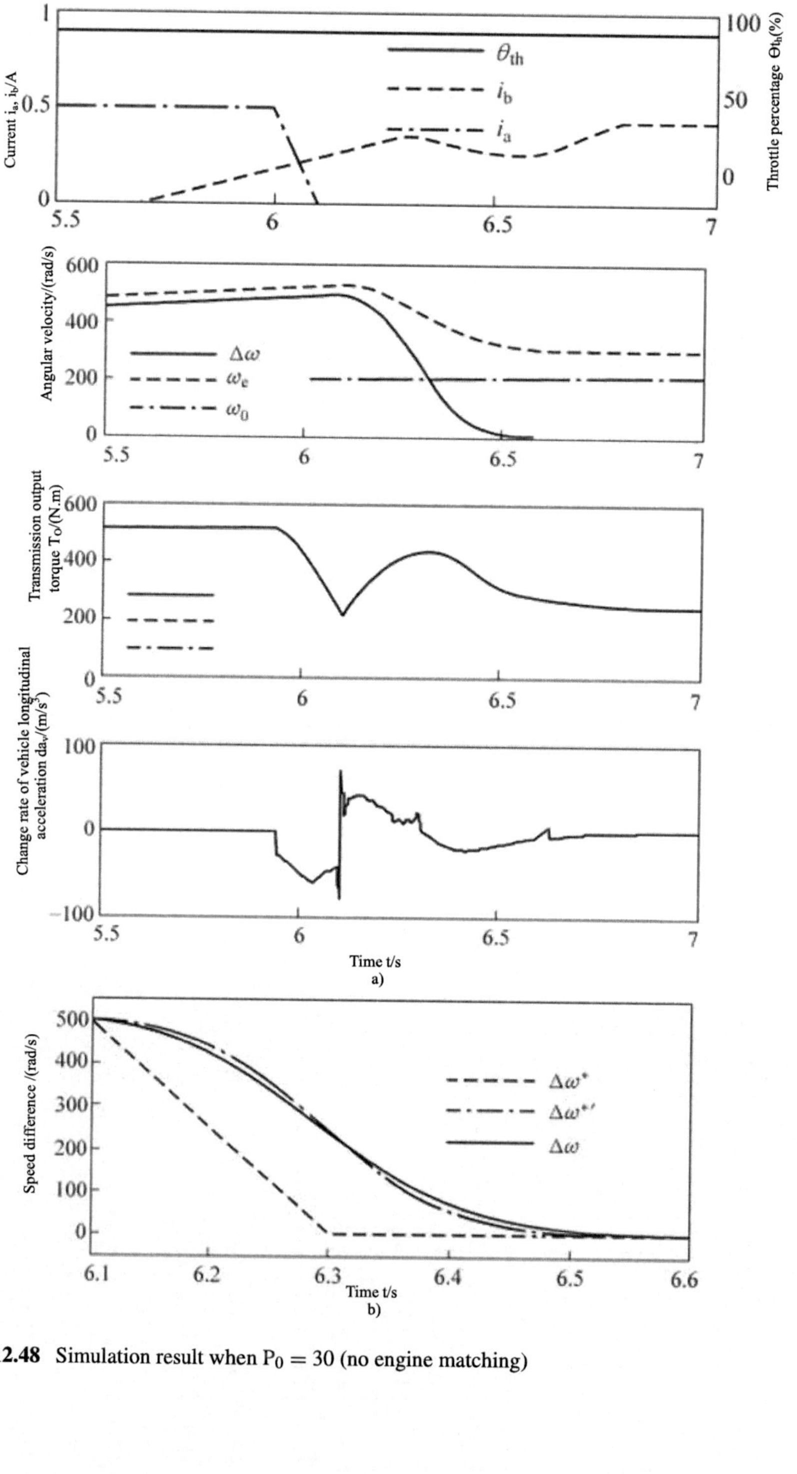

Fig. 12.48 Simulation result when $P_0 = 30$ (no engine matching)

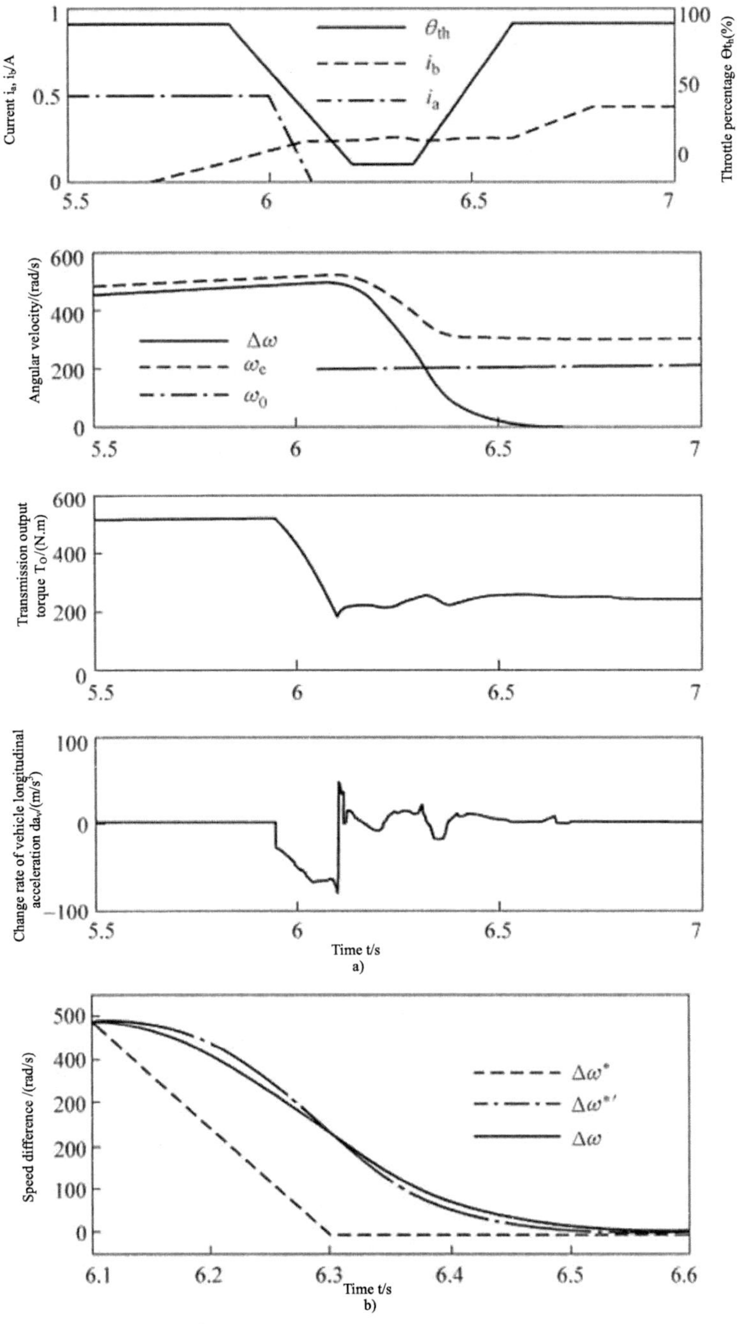

Fig. 12.49 Simulation result when $P_0 = 30$ (with engine matching)

Table 12.1 Sliding friction work W_B of clutch B (unit: J)

Shift process 1	Shift process 2	Shift process 3
27400	25400	17800

the active reduction of the engine torque makes the output torque of the transmission smoother.

The sliding friction work of the clutch B under the above three simulation conditions is shown in Table 12.1. The sliding friction work is calculated by the following formula

$$W_b = \int_{t0}^{tf} T_{cb} \Delta \omega dt \tag{12.22}$$

It can be seen that the engine active torque control during the shift can significantly reduce the load of the clutch, thus greatly reducing the sliding friction loss.

Bibliography

1. Wang Z, Li J, Mei J (2013) Development of automated transmission control unit platform. In: SAE-China congress proceedings
2. JSAE (2010) Automobile engineering manual 4: powertrain design (trans by SAE-China). Beijing Institute of Technology Press, Beijing
3. Board Automotive Handbook Editorial (2001) Automotive handbook: design. China Communications Press, Beijing
4. Chunfu Li, Guangjun Zheng, Yanqin Li et al (2010) Study on control strategy of lockup process of torque converter. Automob Technol 10:15–17
5. Wentao Sun, Huiyan Chen, Chaohua Guan et al (2009) A research on control technique for the lock-up process of hydraulic torque converter. Automot Eng 31(8):761–764
6. Gao B, Chen H (2014) Model-based control of automotive step-ratio transmissions. Int J Powertrains 3(2):197–220
7. Goetz, M, Levesley MC, Crolla, DA (2005) Dynamics and control of gearshifts on twin-clutch transmissions. Proc Inst Mech Eng, Part D: J Automob Eng 219(8):951–963
8. Gao B-Z, Chen H, Li J, Tian, et al (2012) Observer-based feedback control during torque phase of clutch-to-clutch shift process. Int J Vehicle Design 58(1):93–108
9. Bingzhao Gao (2009) Some studies on nonlinear estimation and control of automotive drivetrain. Jilin University, Changchun

Chapter 13
Automobile and Transmission Vibration and Noise

13.1 Vibration and Noise Foundation

The automotive vibration and noise is an important factor that affects the ride comfort. The automotive vibration and noise is complex and difficult to analyze, but can be represented by the basic mass spring model in most of the phenomena. It is very important to understand the automotive vibration and noise by means of mass spring model.

I. Vibration foundation

The force balance equation of the mechanical vibration system shown in Fig. 13.1 is

$$m\frac{d^2x}{dt^2} + c\frac{dx}{dt} + kx = f(t) \tag{13.1}$$

where, m—slider mass (kg);

c—damping coefficient;

k—spring stiffness coefficient.

The transfer function of the system is

$$H(\omega) = \frac{\bar{X}}{\bar{F}} = \frac{1}{k - \omega^2 m + j\omega c} \tag{13.2}$$

where, ω—circular frequency.

The amplitude-frequency characteristic is

$$|H(\omega)| = \frac{1}{\sqrt{\left(k - m\omega^2\right)^2 + (\omega c)^2}} \tag{13.3}$$

When the damping coefficient c = 0, the system is a mass-spring system. If the denominator is 0 in this case, the formula (13.3) is infinitely great and the system is

Y. Chen, *Automotive Transmissions*, Springer Tracts in Mechanical Engineering,
https://doi.org/10.1007/978-981-15-6703-2_13

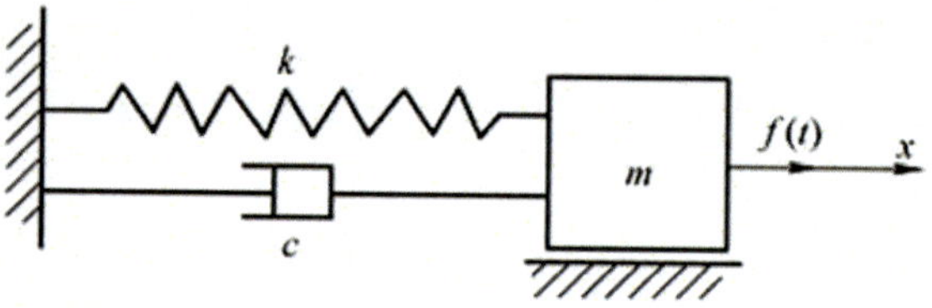

Fig. 13.1 Mechanical vibration system

in the resonance state, namely

$$\omega_0 = \sqrt{\frac{k}{m}} \tag{13.4}$$

$$f_0 = \frac{1}{2\pi}\sqrt{\frac{k}{m}} \tag{13.5}$$

where, f_0—natural frequency.

When the excitation frequency of the system is equal to or close to the natural frequency, the system resonates.

If a plurality of mechanical vibration systems shown in Fig. 13.1 is superimposed, it is called a multi-degree-of-freedom system, and the vibration of the beam belongs to a multi-degree-of-freedom system. Figure 13.2 shows the vibration mode of the beam. Among the various vibration modes, the part with the maximum amplitude is the antinode, while the part with the minimum amplitude is the node.

The key to solve the problem of vibration and noise is to understand and master the concepts of resonance, resonance frequency, vibration prevention, vibration mode,

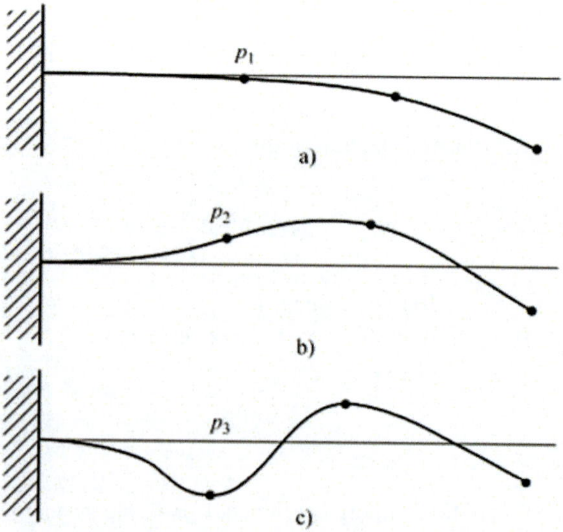

Fig. 13.2 Vibration mode of beam

node, mass-spring system, as well as dynamic vibration reduction, mass vibration reduction, natural frequency, damper, etc. The dynamic vibration reduction is to add a small mass-spring system to the main vibration system, and suppress the vibration of the main vibration system by use of the resonance frequency of the small mass-spring system; the mass vibration reduction is to add a small mass to the main vibration system. According to the Newton's laws of motion, if the mass displacement is $xt = \bar{X}\sin(\omega t)$, then $Ft = -\omega^2\bar{X}m\sin(\omega t)$, that is, under the action of the same exciting force, the larger the mass, the smaller the displacement; the natural frequency is the resonance frequency, and the number of natural frequencies of the system depends on the number of degrees of freedom; the damper is connected with the spring in parallel in the mechanical vibration system mainly to attenuate vibration.

As the object vibrates, pressure changes in the air, thus producing sound. There are solid and gas media in which vibration is transmitted, and vibration and sound are simultaneous. For example, the exhaust pipe vibration is transmitted to the vehicle body through the mount, which makes the vehicle body vibrate and produce noise.

Vibration noise is measured by a phonemeter, usually by [dB]

$$[dB] = 20\log\frac{X}{X_0} \tag{13.6}$$

where, X—measured physical quantities (force, pressure, velocity, acceleration, etc.);

X_0—reference physical quantities (force, pressure, velocity, acceleration, etc.).

II. Sound foundation

Noise is a kind of sound that causes people to be agitated or has too high volume, and then endangers human health. From the perspective of physics, noise is the sound produced when the sounding body is doing random vibration, and the waveform of noise is disorganized. From the perspective of environmental protection, any sound that affects people's normal study, work and rest, as well as people's "unwanted sound" in some situations, belong to noise.

Sound pressure level is a basic physical quantity in the field of audio frequency engineering and noise measurement to represent the sound level of the measured object. With broad physical meanings, it can be expressed by a single value (e.g. Line sound level, linear weighted sound pressure level or A weighted sound pressure level), or by the sound pressure level of each frequency band (e.g. NR curve), as well as statistical mean of the sound pressure levels measured at different points in space and total sound pressure level calculated from the sound pressure level of each frequency band.

Figure 13.3 shows an Fritz–Mosion equal loudness contour. The audible frequency range of human ear is 20–20000 Hz and the human ear is sensitive to 2–5 kHz sound and not sensitive to low frequency sound. When the sound pressure level reaches 100 dB, the equal loudness contour is horizontal, and the frequency change is not sensitive to the loudness level change.

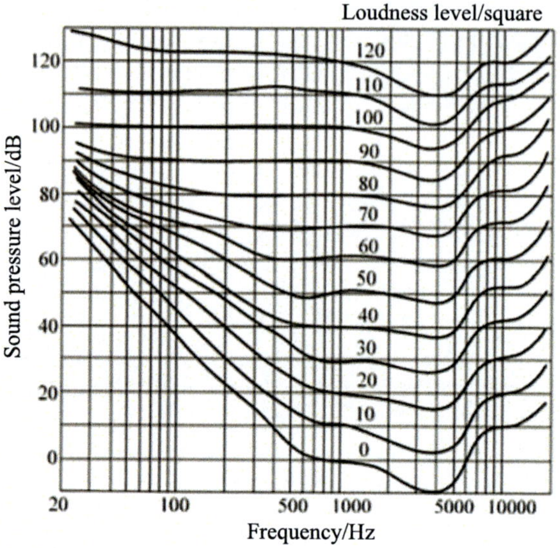

Fig. 13.3 Fritz–Mosion equal loudness contour

The frequency of the external force is near the natural frequency of the air, increasing the sound pressure. Since there are many natural frequencies in the cab of a vehicle in the range of 70–200 Hz, cavity resonant sound is formed in the cab.

The characteristics of sound mainly include resonance, beat, noise interference, sound absorption and sound isolation.

Resonance refers to the resonance phenomenon of sound, which is realized by the resonance of the sounding body and its surrounding environment. Within a certain space, there are several points of resonance in the range of 70–200 Hz, sometimes resonating, especially the intermediate speed resonance sound in the back seat in the range of 80–100 Hz.

Two sounds with similar vibration frequencies are synthesized and interfere with each other. The phenomenon of different sound levels is the beat, as shown in Fig. 13.4. The beat will occur in the driver's cab.

When you want to hear a certain sound, the sound you want to hear is not clear due to the presence of other sounds. This is the result of noise interference. In a vehicle, for example, it is also the cause of noise interference for readjustment of the radio volume suitable for normal road conditions when the vehicle enters the highway.

When sound enters the glass wool, felt and other materials, the sound wave propagates in the pores and part of the sound energy is consumed in the form of heat energy. This phenomenon is called sound absorption, which is usually not obvious in the low frequency range, but more obvious in the high frequency range. The heat shield of the engine hood reduces the noise level through the principle of sound absorption in the engine compartment.

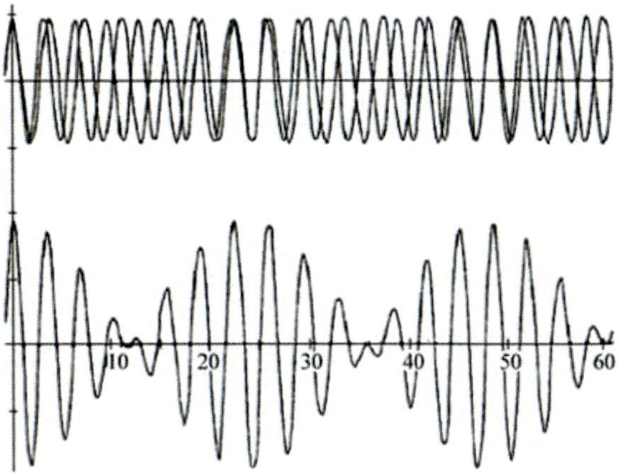

Fig. 13.4 Generation of beat

Sound insulation is the establishment of barriers between the workspace and the sound source to prevent noise propagation. For example, the dashboard between the engine compartment and the cab adopts a dual sound barrier.

13.2 Automobile Vibration and Noise

I. Exciting force of automobile vibration and noise

The vibration and noise of the automobile must be caused by the exciting force, which mainly includes the periodic exciting force generated by the engine, the periodic exciting force generated by the rotating members of the powertrain, the random exciting force generated by the rough and uneven road surface, and the exciting force generated by the step state of the transient change in the engine torque. Under the action of these exciting forces, a variety of vibrations and noises will be generated. In order to reduce the vibration and noise, it is necessary to find out the causes for the vibration and noise. Table 13.1 lists the automobile excitation sources, vibration and noise phenomena.

The exciting force generated by engine rotation has a great influence on interior noise. For example, a 4-cylinder engine generates two exciting forces per revolution and mainly consists of the following two parts:

(1) Torque change caused by combustion. For a 4-cylinder engine, the crankshaft burns once at 180° per revolution, producing two torque changes per revolution; for a 6-cylinder engine, the crankshaft burns once at 120° per revolution, producing three torque changes per revolution.
(2) Reciprocating mass inertial force. The reciprocating motion of the piston linkage mechanism produces 1, 2, 4… times of inertia force per revolution. The configuration of in-line 6-cylinder engine crankshaft can be balanced, but the secondary

Table 13.1 Automobile excitation sources, vibration and noise phenomena

Excitation source		Vibration phenomenon	Noise phenomenon
Engine	Torque (load) change Torque ripple Reciprocating mass inertial force Combustion pressure Intake and exhaust pulsation	Wobble vibration Idle vibration	Low speed resonance sound Idling Acceleration noise Resonance sound
Gear engagement			Gear noise and differential mechanism noise
Unbalanced rotating members	Transmission shaft, tire and load gear	Shaking and shimmy	Resonance sound
Powertrain coupling	Variable-speed cardan and cardan shaft thrust	Vibration in FR vehicle starting	Cavity resonance sound (beat)
Clutch friction characteristics		Clutch judder	
Tire inconsistency	Consistent tire pattern	Bouncing	Tire pattern noise
Rough and uneven road surface			Road noise
Air flow			Wind sound

components of in-line 4-cylinder engine and V6 engine cannot be balanced, which will become the exciting force.

II. Frequency range of automobile vibration and noise phenomena

The frequency range of automobile vibration and noise phenomena (except the ride comfort range) is very wide, ranging from 3 to 5 kHz. The frequencies of various phenomena are determined by the excitation sources and the powertrain frequency characteristics. In particular, the frequency range of vibration of rotating members is determined by the speed range of excitation sources.

The relationship among the engine speed n_e, engine cylinder number z and resonance frequency f is

$$f = \frac{n_e z}{130} \tag{13.7}$$

If the engine speed is $n_e = 6000$ r/min and the engine cylinder number is 4 or 6, the resonance frequency is 200 Hz or 300 Hz.

The frequency range of the automobile vibration and noise phenomena is shown in Table 13.2. The frequency range of vibration sensation is 3–300 Hz, and of noise sensation is 20–20 kHz, in which the sound with frequency of 20–40 Hz is more harsh.

Table 13.2 Frequency range of automobile vibration and noise phenomena

Name	Frequency range	Name	Frequency range
Engine noise	80–2 kHz	Wind noise	400–2000 Hz
Low speed resonance sound	25–60 Hz		
Intermediate speed resonance sound	50–100 Hz		
High speed resonance sound	90–250 Hz		
Fan noise	250–1600 Hz	Rocking vibration	3–8 Hz
Differential mechanism noise	300–800 Hz	Bouncing	8–15 Hz
MT gear noise	400–1600 Hz	Shaking	4.5–15 Hz
MT flange noise	300–2000 Hz	Clutch judder	10–16 Hz
Clutch noise	160–500 Hz	Idle vibration	5–25 Hz
Beat	80–160 Hz	Starting vibration	8–25 Hz
Road noise	80–400 Hz	Brake judder	13.5–25 Hz
Tread pattern noise	300–800 Hz	Gear shift lever vibration	50–250 Hz
Rough sound	25–100 Hz	Accelerator pedal vibration	90–250 Hz

The level of sensation noise varies from person to person. In addition, for the vibration of a certain frequency of sound, when the vibration of other frequencies becomes larger, sometimes the vibration of this frequency of sound will not be felt, which is the noise interference

III. Solutions to automobile vibration and noise

No matter what kind of automobile vibration and noise phenomenon, it can be regarded as the process that the excitation source passes through all parts of the powertrain as shown in Fig. 13.5 and the passenger finally feels the vibration and noise.

To reduce the vibration and noise, the best method is to reduce the exciting force of the excitation source and reduce the vibration transmissibility of each part of the powertrain. The improvement methods of vibration and noise are shown in Table 13.3. The application of these methods is inconsistent with mass, cost and other performance. In addition, reducing the stiffness of the engine mount can improve the idle

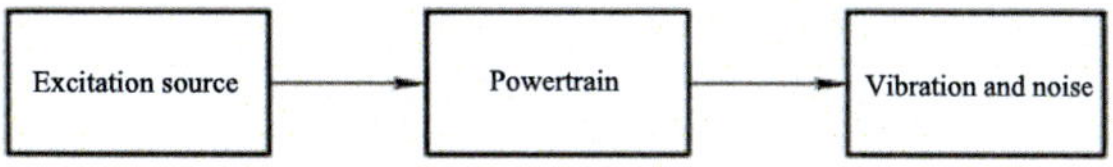

Fig. 13.5 Vibration and noise generation process

Table 13.3 Improvement methods of vibration and noise

Improvement method		Improvement performance	Adverse performance
Reduce exciting force	Tire balance Control the tire consistency Reduce and control the exciting force of gear engagement Control combustion to reduce engine vibration Improve the noise elimination performance of the muffler Add the engine balance shaft Reduce the cardan angle	Shimmy and vibration Vibration Differential mechanism and gear noise Acceleration noise Resonance sound Resonance sound Starting vibration	Equipment and cost Equipment and cost Equipment Power performance Power performance and cost Cost and mass Layout
Resonance frequency separation	Increase the steering resonance frequency Reduce the vibration resonance frequency of the mount wheel Adopt hollow drive shaft to increase resonance frequency	Idle vibration and shimmy Low speed resonance sound Resonance sound	Cost Starting vibration Cost
Adjust the resonance frequency of the components	Separate the point of resonance of the mount link from the powertrain resonance frequency, such as adding mass blocks	Differential mechanism noise	Cost and mass
Reduce the stiffness of bushing and mount	Reduce the stiffness of engine support Reduce the stiffness of mount link	Resonance sound Road noise	Shimmy Shimmy and stability
Dual vibration-proof structure	Dual vibration prevention of rear engine support Vibration prevention of mount and auxiliary frame	Gear noise Road noise	Resonance sound and cost Cost and mass
Increase the mass of vibrating parts and decrease the amplitude	Increase the inertia moment of the steering wheel	Shimmy	Stability and mass

(continued)

Table 13.3 (continued)

Improvement method		Improvement performance	Adverse performance
Vibration reduction based on dynamic vibration absorber	Dynamic vibration absorber for radiator Add front mount part Add transmission shaft	Idle vibration Resonance sound Resonance sound	Cost Cost and mass Cost and mass
Installation at the vibration modal node	Selection of installation site of exhaust pipe	Resonance sound	Layout
Vibration modal control	Engine erection adjustment and coupling of up and down vibration resonance	Vibration	Layout
Improvement of vehicle body radiation characteristics	Affix damping materials	All noises	Cost and mass
Increase sound insulation to reduce sound propagation through the air	Increase the surface density of the sound insulating materials Processing of through hole and clearance of components	Engine noise	Cost and mass
Other	Gear ratio selection (separation beat)	Beat	Power performance

vibration and interior noise, but if the shaking is intensified, the balance of other vibration and noise shall be considered.

1. **Setting of target value**

When developing a new model, the target value of vibration and noise shall be set. The target value must be appropriate, because dealing with vibration and noise is mostly related to mass and cost. Setting a too high target value will increase the mass and cost; setting a too low target value will result in product competitiveness and claims. Therefore, when setting the target value, we must fully grasp the level of market demand and the strength of competitive models, and we must consider the increase of market demand level year by year.

2. **Determination of transmission route (master input contributions)**

In case of new abnormality, the excitation source and transmission route of the phenomenon must be determined first, and the countermeasures should be studied according to the results. For an abnormality generated in a known usually transmission route, the solution must be found from multiple transmission routes. That is, the rough with high input contribution rate is usually effective. For example, the road noise is analyzed from the input of front and rear wheels and the improvement countermeasures are implemented by the route with high contribution rate.

3. **Excitation source dispersion and vehicle sensitivity**

Inconsistencies in production will cause inconsistencies in tires and wheels, which will have a great influence on the exciting force. The vibration and noise generated by the exciting force (e.g. shimmy at this time) cannot be used to judge whether the vehicle has a good shimmy performance only by evaluating the vehicle shimmy. At this point, the vehicle needs to be evaluated by increasing the level of the unbalance shimmy, and the less sensitive to the unbalance sensitivity, the better the shimmy performance.

The influence of inconsistencies in production on all performance needs to be taken into account, but the following phenomena, including the above, are particularly important in evaluating the vehicle sensitivity: vibration and shimmy caused by unbalance of tires and wheels; shimmy caused by tire inconsistency; resonance sound caused by imbalance of transmission shaft; gear noise and differential mechanism noise caused by gear engagement of transmission and differential mechanism.

4. **Relationship between vibration and noise processing and other performance**

As mentioned above, vibration and noise processing mostly has adverse effects on other performance, so it is usually necessary to compromise to solve the contradiction between each other and find an acceptable balance point. The idle vibration is illustrated as an example below.

The idle vibration refers to the phenomenon of vehicle body and steering wheel vibration during idle parking. Figure 13.6 shows the transmission route map of idle vibration.

In order to reduce the idle vibration, it is necessary to reduce the engine exciting force, reduce the transmission efficiency of the engine exciting force to the vehicle body and take corresponding measures to achieve the purpose that the vehicle body will not vibrate even if the engine generates exciting force. The following is an example of the vehicle body transmission efficiency. As shown in Fig. 13.7, the engine exciting force is the main vibration source, which is passed into the vehicle body through the engine mount, and its resonance frequency is determined by the mount spring coefficient and engine mass. If the resonance frequency is lower than the engine idle speed frequency, it has vibration prevention effect. If the mount spring coefficient is reduced, the resonance frequency is reduced and the idle vibration decreases.

On the other hand, the engine mount has the function of supporting the engine through the vehicle body and supporting the engine driving moment. At the same time, the engine wobble reduces the comfort in the driving. According to these requirements, it is best to make the engine mount hardened. The above two are contradictory, determining the need to compromise the mount spring coefficient.

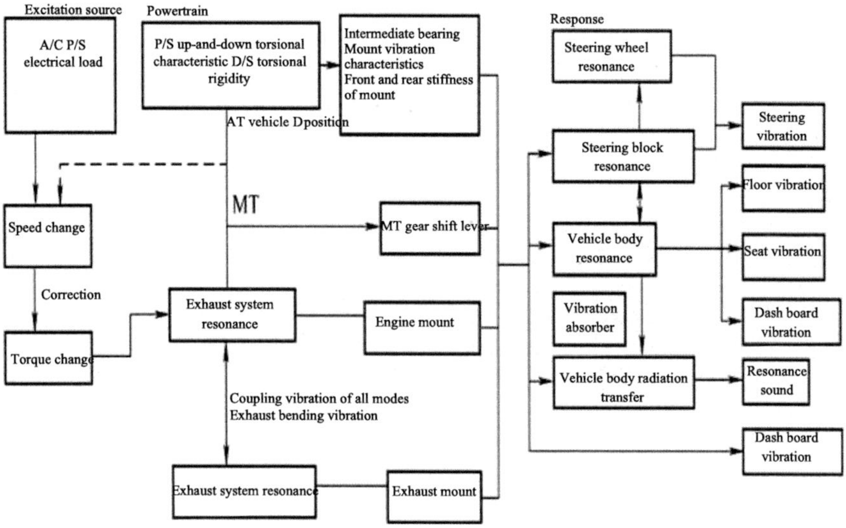

Fig. 13.6 Transmission route map of idle vibration

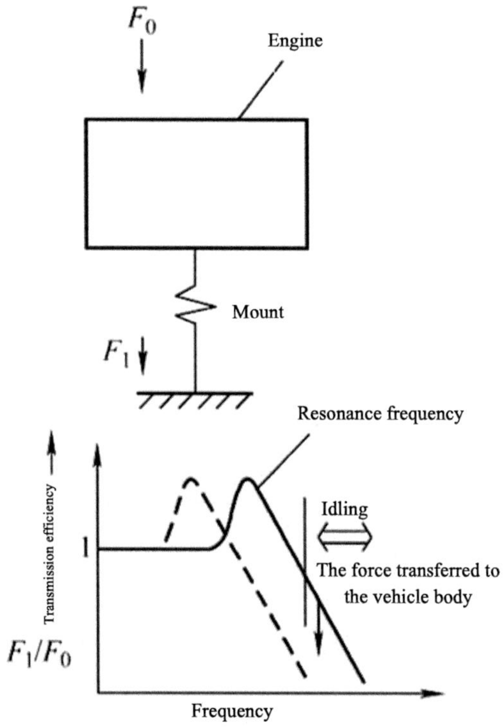

Fig. 13.7 Engine vibration

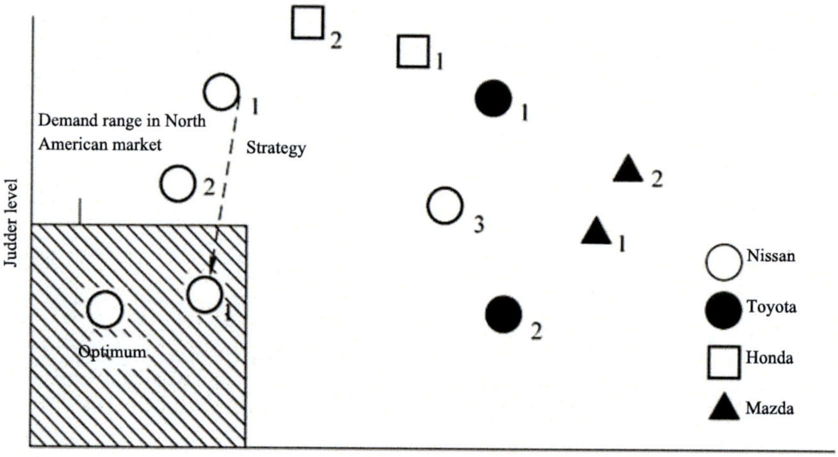

Fig. 13.8 Relationship between idle vibration level and judder level of FF vehicle

Take the level of demand in North America for example, where the restrictions on the low frequency vibration are very strict. Figure 13.8 shows the relationship between idle vibration level and judder level of FF vehicle. As can be seen from the figure, there are no models compatible with all levels of demand. In order to meet the target requirements, the parameters of the engine mount are adjusted reversely. By reducing the engine exciting force, corresponding measures are taken to ensure that the vehicle body will not vibrate even if the engine generates exciting force, so as to achieve the target value.

In this way, the vehicle can be compatible with the two conflicting performance indexes of idle vibration and engine vibration. Figure 13.9 shows the interrelations among the various performances.

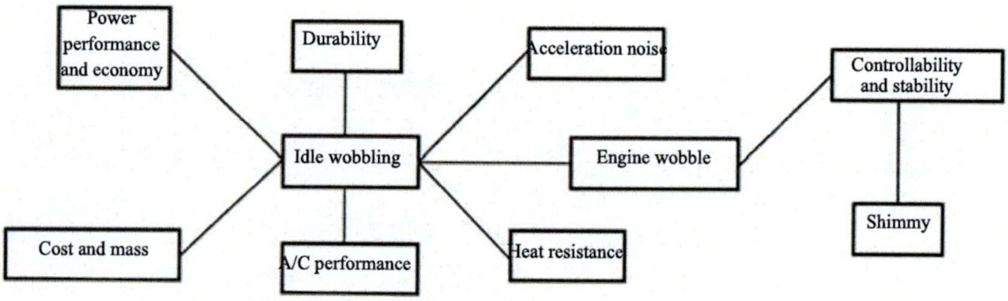

Fig. 13.9 Interrelations among various performances

13.3 Typical Automobile Vibration and Noise

Before studying the vibration and noise of automobile transmission, it is necessary to understand the vibration and noise of automobile, the causes for which include: air noise, produced by the gas vibration. If the pressure of the gas changes suddenly, the swirl disturbance will generate, thus causing the noise. The noise of the electric fan belongs to this kind of noise; mechanical noise, produced by solid vibration. When the vehicle works, the vibration is produced by the impact friction of the gears and bearings as well as the role of various sudden mechanical forces and transmitted through the air to form noise; liquid flow noise. When the liquid flows, the internal friction of the liquid, the friction between the liquid and the wall of the container or the impact of the fluid will cause the vibration of the fluid and the wall, thus causing noise; electromagnetic noise. The noise caused by the vibration of the cores and winding coils of various kinds of electrical equipment under the action of the alternating electromagnetic force is usually called hum; combustion noise. When the fuel burns, it transfers heat to the surrounding air medium, causing its temperature and pressure to change, forming eddy and vibration, thus producing noise.

I. Idle vibration

Phenomenon: vehicle body or steering wheel vibration when idling; change with the engine speed and load (air-conditioner on and off, AT gear); ear-pressing sound sometimes (idle resonance sound).

Excitation sources: torque ripple caused by engine combustion, the engine produces vibration of half of its cylinders per revolution; the engine idle speed stability is poor, producing the intermittent vibration input, which is called poor idling.

Propagation path: engine mount, suspension and exhaust system. Engine vibration is the resonance determined by the engine and engine mount, but the exhaust system resonance increases the input of the vehicle body in the idle range.

Vibration and noise position: dash board, steering wheel, seat and vehicle body. The sensitivity of the vehicle body depends on the flexural resonance of the vehicle body and the resonance of the steering system at idle speed. In order to improve the vibration characteristics, the radiator and the battery are used as dynamic vibration absorbers. Figure 13.10 shows the test results of the idle vibration at the steering wheel of a vehicle.

Target value: through the vehicle evaluation, the body (floor and seat) acceleration (RMS) is less than 0.1 m/s^2, and the steering wheel vibration acceleration (RMS) is less than 0.1 m/s^2; through the engine evaluation, the variation of rotating amplitude of the crankshaft flywheel is 30 r/min, and the variation of the average rotation speed is 60 r/min.

II. Clutch judder

Phenomenon: front and rear vehicle body judder when the clutch is engaged (half-engaged) at the vehicle starting; the dash board makes a sound that causes vibration

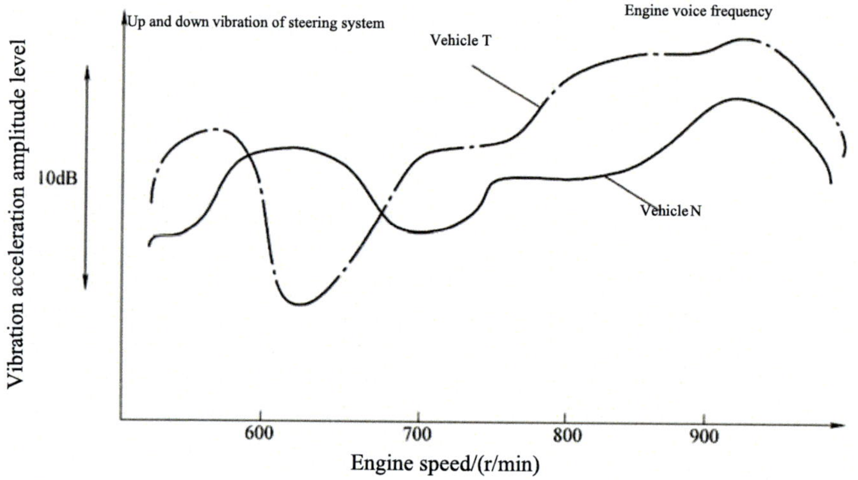

Fig. 13.10 Test results of idle vibration

in half-engagement of the clutch; the judder disappears in the total engagement of the clutch.

Excitation sources: clutch plate break and sticking oil, uneven friction characteristics of the clutch plate surface, leading to self-excited vibration; the rope in the rope-operated mechanism is too tight or damaged, and the control of the engine is sometimes interrupted at will.

Propagation path: uneven drive torque due to torsional resonance of the drive system; vehicle body vibration caused by mount input.

Vibration and noise position: floor vibration (the steering system wobbles when the floor vibrates); vehicle body shaking. Figure 13.11 shows the test results of clutch judder.

When the clutch judder is generated by the self-excited vibration, the friction coefficient of the friction plate decreases with the increase of the sliding speed, and the phenomenon of stick slip occurs between the friction surfaces, which causes the transmitted torque to fluctuate, causing the torsional vibration of the powertrain and

Fig. 13.11 Test results of clutch judder

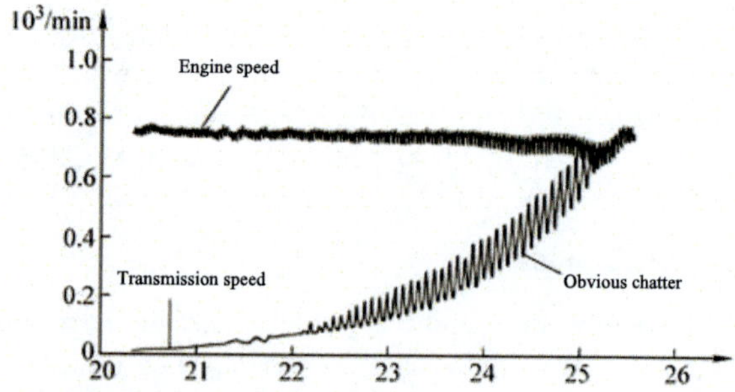

the vibration of the front and rear direction of the vehicle; the uneven clutch surface, uneven cover plate force and other factors result in the single surface friction, torque ripple, the torsional vibration of the powertrain and the vibration of the front and rear direction of the vehicle; with respect to the rope clutch control mechanism, when the clutch is engaged, the relative displacement of the body and the transmission occurs, which causes the pressure plate thrust ripple, the torsional vibration of the powertrain and the vibration of the front and rear direction of the vehicle.

III. Brake judder

Phenomenon: up and down vibration of dash board and steering wheel in braking; vibration and the same interval of pulsation from the brake pedal.

Excitation sources: uneven friction surface caused by brake hub roundness variance, eccentricity, rocking and deformation; Uneven friction surface caused by brake disc scratch, corrosion, thickness change and rocking. Figure 13.12 shows the judder caused by the brake disc thickness difference, axial circular runout, surface waviness

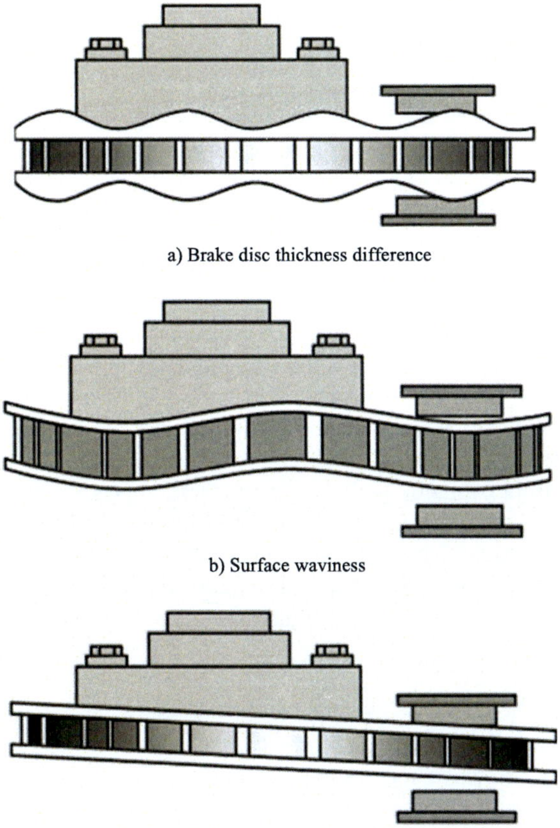

Fig. 13.12 Causes for judder caused by different geometric shapes

a) Brake disc thickness difference

b) Surface waviness

c) Axial circular runout of brake disc

and different geometric shapes. In addition, the temperature of the brake disc rises after being heated, and the expansion of the brake disc varies from place to place, resulting in the brake thickness difference and brake judder.

Propagation path: resonance of the steering tie rod and mount increases vibration; input from the mount is passed to the body; vibration from the body and wheels is transmitted to the steering system; pulsation caused by the change in brake pressure is transmitted to the brake pedal.

Vibration and noise position: up and down vibration of steering wheel; floor vibration; brake pedal pulsation.

The evaluation index of brake judder is the acceleration ratio, with the formula of

$$\text{Acceleration ration} = \frac{\text{maximum acceleration at which judder occurs}}{\text{average acceleration of a vehicle as it moves}} \tag{13.8}$$

Acceleration ratio = maximum acceleration at which judder occurs/average acceleration of a vehicle as it moves (13.8).

At the front wheel, the acceleration ratio of front and rear, left and right shall be below 2.

IV. Accelerator pedal vibration

Phenomenon: slight accelerator pedal vibration can be sensed due to different engine speed; the vibration increases when the engine speed is high.

Excitation sources: fundamental frequency vibration of the engine in the mechanical accelerator pedal cable mounting position stimulates the cable; the electronic accelerator pedal is usually fixed to the firewall by means of a pedal bracket and the pedal vibration is mainly stimulated by the two-stage reciprocating inertia force of the engine.

Propagation path: as for the mechanical accelerator pedal, the accelerator pedal cable is a powertrain and the vibration transmissibility of the accelerator pedal cable depends on the cable length, material and bending radius; as for the electronic accelerator pedal, the vibration is transmitted from the powertrain mount to the auxiliary frame, vehicle body and finally to the firewall and accelerator pedal, resulting in strong vibration in the pedal acceleration process.

Vibration and noise position: the vibration of the accelerator pedal caused by the vibration of the accelerator pedal cable; the vibration frequency of the engine sometimes overlaps with the natural frequency of the accelerator pedal cable. Figure 13.13 shows the accelerator pedal vibration test results.

V. Gear shift lever vibration

Phenomenon: gear shift lever vibration during driving, accompanied by high frequency vibration noise.

Excitation sources: the reciprocating inertial force of the engine causes the drive system to vibrate; long-distance control of gear shift lever causes the resonance in the installation site.

Fig. 13.13 Accelerator pedal vibration test results

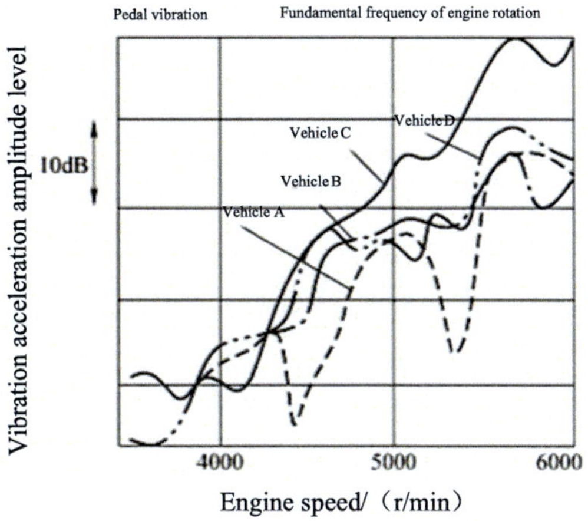

Propagation path: vibration growth rate from extension due to gear shift lever clearance and gear shift lever bending vibration; remote control of the flexural resonance of control lever and servo lever in the gear shift lever; vibration growth rate determined by the mass of the lever and spring coefficient of the bracket rubber.

Vibration and noise position: gear shift lever and shift knob. Figure 13.14 shows the gear shift lever vibration test results.

VI. Engine fan noise

Phenomenon: when a certain engine speed is reached, "cluck" and other sounds can be heard in the engine (fan working).

Excitation sources: the sound produced by the rotation of the fan is the excitation source, mainly due to the air cut by the fan blade or the turbulence generated by

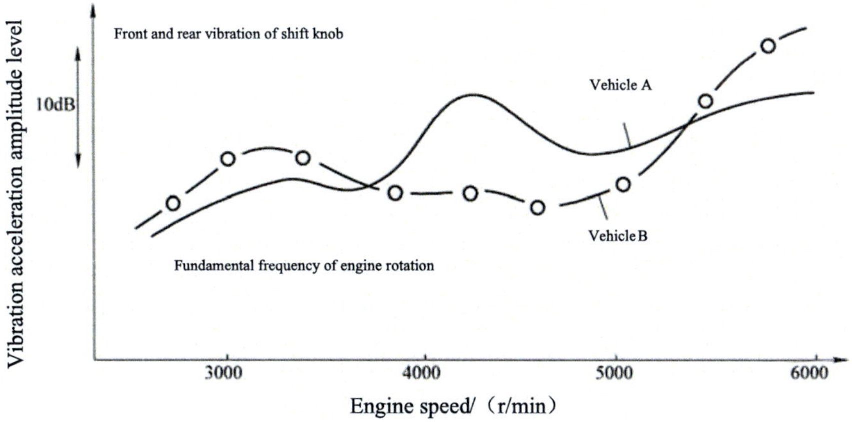

Fig. 13.14 Gear shift lever vibration test results

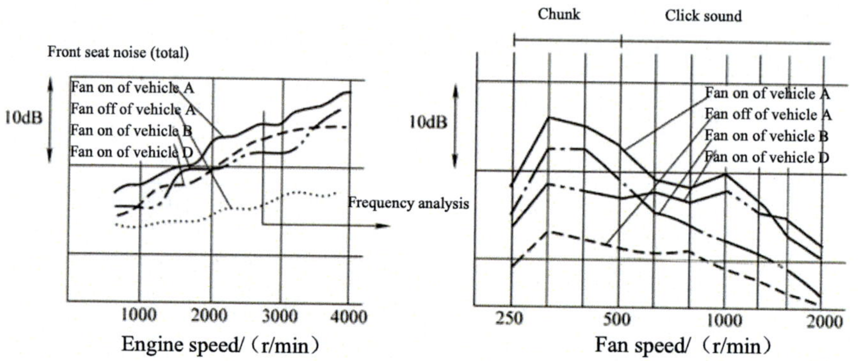

Fig. 13.15 Engine fan noise test results

the components behind the fan; when the fan is not working, the noise decreases obviously.

Propagation path: through the air.

Vibration and noise position: vehicle body. Figure 13.15 shows the engine fan noise test results.

VII. Transmission gear noise of FR vehicle

Phenomenon: when the gear shift lever is in a position, it makes a loud noise directly under the front seat.

Excitation sources: the gear meshing transmission produces vibration, which is affected by gear accuracy and gear tooth contact.

Propagation path: from the rear engine suspension to the vehicle body; from the transmission shaft and rear mount to the vehicle body; the radiation sound generated by the transmission is transmitted by air to the interior from the penetration part of the gear shift lever.

Vibration and noise position: sound produced by body shell vibration. Figure 13.16 shows the FR transmission gear noise test results.

VIII. Road noise

Phenomenon: large noise can be heard when driving on rough roads, and noise disappears on good roads.

Excitation sources: the continuous small convex and concave on the road surface become the tire exciting force. The input frequency characteristics vary with the vehicle speed and road surface.

Propagation path: vibration of the tire caused by uneven road surface is transmitted to the mount and the amplitude increases near the tire resonance frequency; the vibration amplitude increases near the tire resonance and mount resonance frequency, and the vibration is transmitted to the vehicle body through the mount push rod. Generally, the vibration input of the part with a high spring coefficient of the mount push rod is also large.

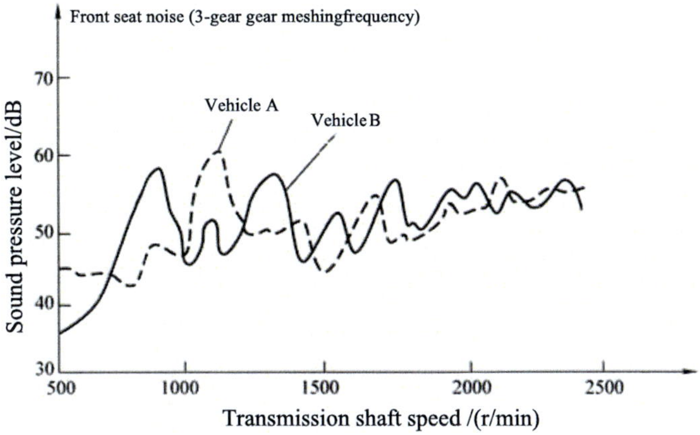

Fig. 13.16 FR transmission gear noise test results

Vibration and noise position: noise in the dash board caused by the vibration to the vehicle body through the mount. Figure 13.17 shows the road noise test results.

IX. Tread pattern noise

Phenomenon: powerful tires such as snow tires make noise when driving.

Excitation sources: the air in the tread groove is closed and expanded to produce noise; the main frequency depends on the product of the tire revolution and the number of rubber blocks on the tire. The tread condition of the tire determines the noise amplitude.

Propagation path: through the air.

Vibration and noise position: vehicle body. Figure 13.18 shows the tire noise test results.

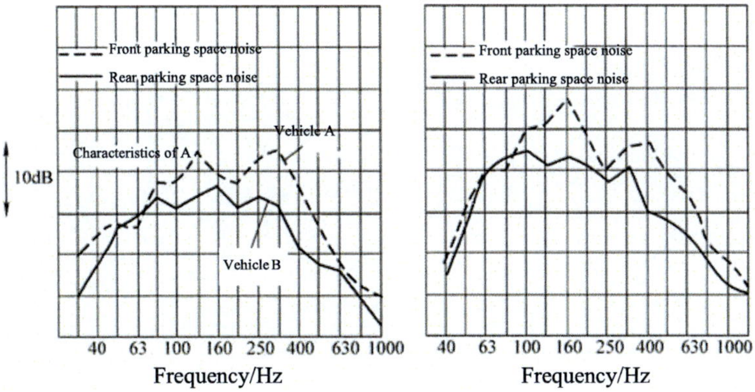

Fig. 13.17 Road noise test results

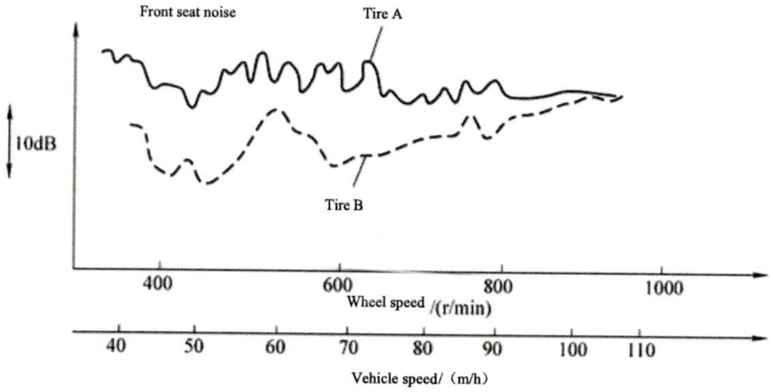

Fig. 13.18 Tire noise test results

X. Control system noise

Phenomenon: the ear-pressing low frequency sound is made in the action of the clutch, accelerator pedal and gear shift lever. The same phenomenon sometimes occurs in the AT vehicle lockup, especially in FR vehicles.

Excitation sources: the action of the control system causes positive and negative changes in the drive torque; when the drive torque changes, impact torque is generated in the parts with clearance of the drive system.

Propagation path: the rear mount is a powertrain; as the drive characteristics, the shaft sleeve stiffness and the brake clearance are affected.

Vibration and noise position: from mount to vehicle body. Figure 13.19 shows the control system noise test results.

Fig. 13.19 Control system noise test results

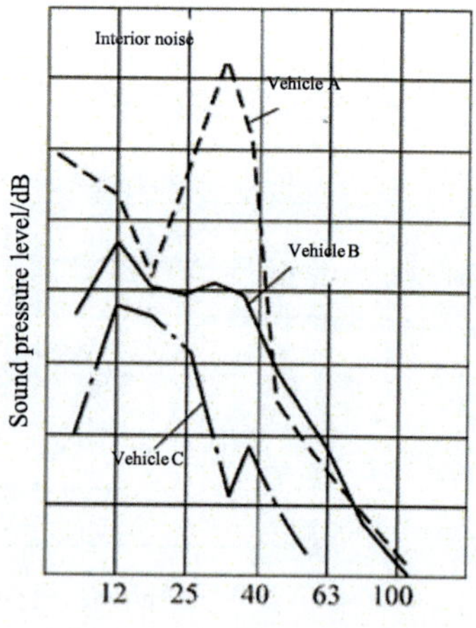

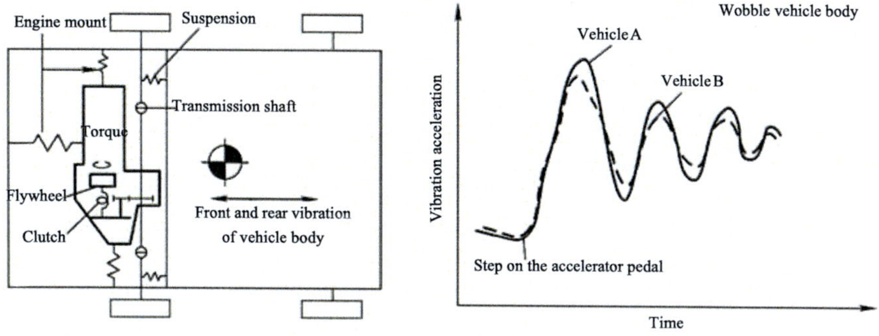

Fig. 13.20 Wobble vibration test results

XI. Wobble vibration

Phenomenon: when the accelerator pedal is pressed at the low and medium driving speed, the vehicle body has large front and rear vibration (clutch engagement).

Excitation sources: the engine torque changes into exciting force when the accelerator pedal is pressed; the torque varies in steps.

Propagation path: a drive system with the engine flywheel as mass and the clutch, transmission shaft and tire as springs causes reciprocating vibration.

Vibration and noise position: the coupling between the engine mount system and the drive system determines the front and rear vibration characteristics of the vehicle. Figure 13.20 shows the wobble vibration test results.

13.4 Analysis of Transmission Vibration and Noise

The vibration and noise is an important evaluation index in the development of automobile transmission. In the design stage of the transmission, the modal analysis is needed for the transmission case to obtain the natural frequency and the natural mode of vibration of the transmission case; the gear modification analysis is needed to guide the gear machining deviation design and avoid the transmission vibration and noise; the competitive product analysis for transmission NVH is needed to understand the NVH level of the same type of transmission. The NVH test is conducted on a designed transmission to obtain the basic NVH test data.

I. Modal analysis

The transmission modal analysis includes computational modal analysis and experimental modal analysis. The computational modal is obtained by the finite element method, that is, the characteristic parameters of the transmission such as the natural frequency and the natural mode of vibration are obtained by using the simulation software, which is conducive to the optimization of its structural performance, and the structural design defects are exposed at the early stage of the design, so as to optimize the structure. The experimental modal analysis means that the system input and output signals are identified by parameters through tests to obtain characteristic parameters of the transmission such as the natural frequency.

1. **Modal analysis of transmission case**

The free modal, constraint modal and radiation noise analysis of the case may be made for the finite element model of a designed transmission case. In the free modal analysis, RIGIDS unit is used to simulate the bolted connection between the front and rear cases of the transmission; in the constraint modal analysis, in addition to the RIGIDS simulation of the bolted connection between the front and rear cases, the bolt hole at the junction between the transmission and the engine and the bolt hole at the mount are also constrained. The same method is used for the two modal analyses and only the free modal analysis is considered in this section. Figure 13.21 shows the finite element model of a transmission case assembly. The natural frequency values of the first 8 orders of free modal of the transmission case assembly are shown in Table 13.4.

It is important to focus on the formation displacement and strain energy cloud diagrams of the first 8 orders of free modal. Only the formation displacement cloud diagram (Fig. 13.22) and formation strain energy cloud diagram (Fig. 13.23) of the first order of free modal of the transmission case are given here. Through the analysis of the first 8 orders of free modal of the transmission case assembly, the position of the modal and stress that resonate with the gear engagement frequency is found.

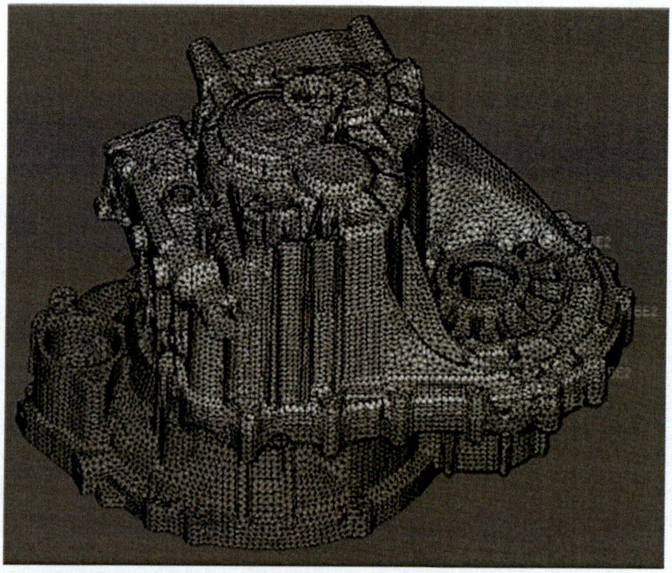

Fig. 13.21 Finite element model of a transmission case assembly

Table 13.4 Natural frequency values of the first 8 orders of free modal of the transmission case assembly

Order	1	2	3	4	5	6	7	8
Natural frequency/Hz	897.0	1153.5	1379.4	1583.9	1771.2	1806.0	1879.2	2001.6

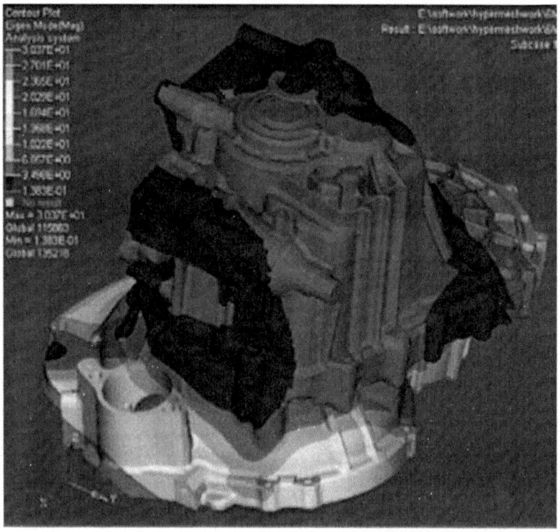

Fig. 13.22 Formation displacement cloud diagram of the first order of free modal of the transmission case (f = 897.0 Hz)

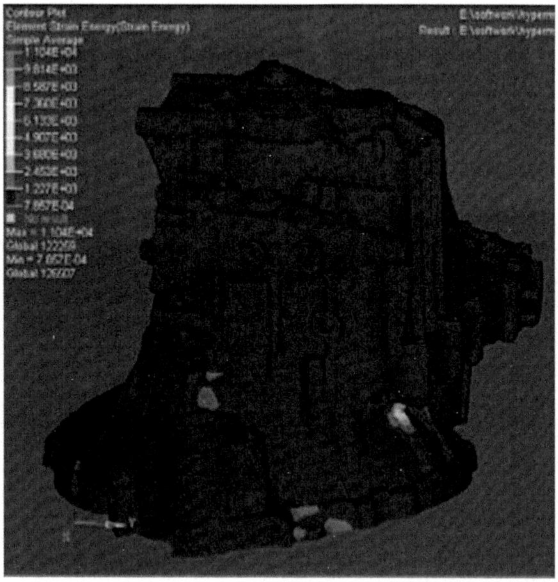

Fig. 13.23 Formation strain energy cloud diagram of the first order of free modal of the transmission case (f = 897.0 Hz)

Table 13.5 Values of modal natural frequencies when the transmission is in neutral

Order	1	2	3	4	5	6	7	8
Natural frequency/Hz	499	570	625	657	727	788	804	888

Table 13.6 Values of modal natural frequencies when the transmission is in gear 5

Order	1	2	3	4	5	6	7	8
Natural frequency/Hz	553	629	662	895	947	1113	1310	1385

The case structure is optimized by increasing the stiffeners and thickness to meet the design requirements.

2. **Modal analysis of transmission system**

Through modal analysis of all gears of the transmission system, such as neutral, 1, 2, 3, 4, 5 and 6, the natural frequencies and natural modes of vibration of the system under each gear are obtained. During the modal analysis of the system, it is necessary to constrain the fixed mount point, the joint surface between the front case and the engine, and set the bearing stiffness and gear stiffness. Table 13.5 lists the values of modal natural frequencies when the transmission is in neutral and Table 13.6 lists the values of modal natural frequencies when the transmission is in gear 5. Through modal analysis, the following conclusions are drawn: the low-order modal in neutral is mainly reflected in the bending vibration of the gear shaft, and the amplitude of the case is very small, so the noise radiated outward is not big. The main vibration of higher order of modal occurs on the case, but the excitation is weak and the influence is not great; in the high gear (gear 5), the low-order modal (below 1350 Hz) is mainly reflected in the bending vibration of the gear shaft, and the amplitude of the case is very small, so the noise radiated outward is not big. The natural frequency of the torsional modal of the gear chain in the 8th order of modal is 1385 Hz and the long-term gear engagement at this frequency should be avoided in the gear design to avoid rotation resonance.

3. **Modal analysis of transmission system test**

The transmission system is set to a resilient hanging state as shown in Fig. 13.24. Figure 13.25 shows the experimental modal test of the transmission system. The experimental modal results of the transmission system are compared with the computational modal results, and the causes for the result difference are pointed out to provide the theoretical basis for the accuracy of the computational modal analysis of the transmission.

4. **Analysis of transmission frequency response**

The frequency response analysis is to calculate the frequency response of typical point of the case to the excitation from the bearing block, which provides a reference

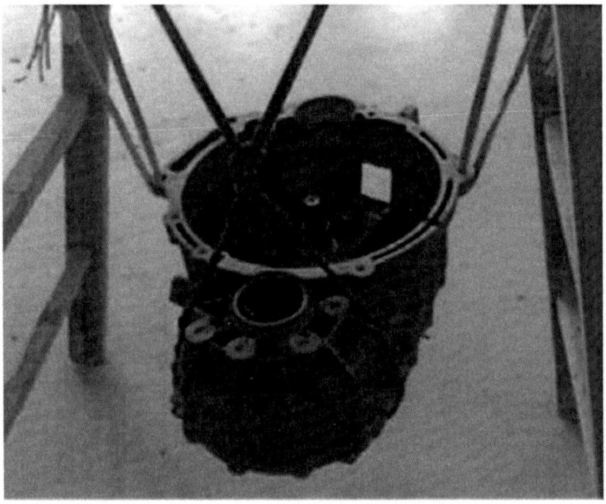

Fig. 13.24 Resilient hanging state of transmission system

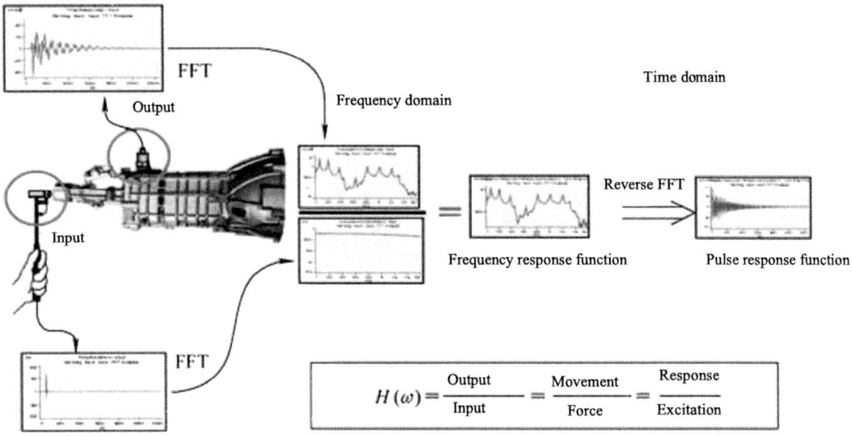

Fig. 13.25 Experimental modal test of transmission system

for the optimum structural design to improve the power performance of the case, so as to determine the position of the case where the maximum amplitude occurs during the frequency response and determine the modal that has a great influence on the frequency response.

On the front and rear cases, the bearing block is the location of the main excitation source of the case, and the excitation of each bearing block is mainly reflected in the X and Y directions (radial). We apply the sinusoidal excitation in X and Y directions of each bearing block respectively, select several response output points, determine the maximum node of the response amplitude of an excitation and the

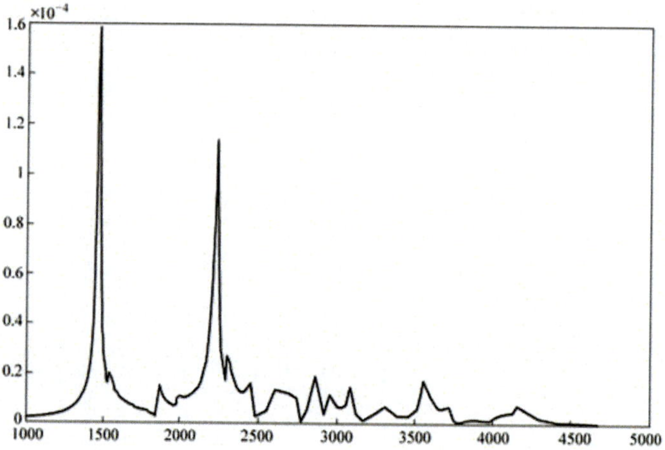

Fig. 13.26 Amplitude-frequency response results

natural frequency corresponding to the maximum response amplitude and comprehensively compare the amplitude-frequency response values corresponding to each excitation to determine the overall maximum amplitude-frequency response values and corresponding natural frequencies, so as to determine the case panel where the maximum amplitude is located and the corresponding important modal.

As there are many calculation results, only one example is listed here. Figure 13.26 shows the amplitude-frequency response results generated by a certain excitation to an output point. If the vibration amplitude of the natural frequency of the first order modal is maximum at many places under a certain excitation, the mode value at the excitation point should be suppressed so as to greatly reduce the vibration of each response point; if the vibration amplitude at a certain place is maximum under multiple excitations, the stiffness at that place shall be increased to reduce the vibration amplitude at the corresponding point.

II. Gear modification analysis

In the design stage, it is necessary to analyze the influence of different gear machining deviations on the transmission vibration and noise. Determine reasonable modification parameters by analyzing micro modification parameters, including the influence of the axial dip, profile dip, profile crowning, axial crowning and tip relief on the static transmission error of the gear, convert the modification parameters into gear machining deviation and guide the gear machining deviation design to avoid the transmission vibration and noise in the design stage. Draw up the gear teeth matching test scheme according to the simulation results, guide the teeth matching test and verify the design analysis effect.

Modify the gear micro modification parameters step by step in the following order using the MASTA simulation analysis software: analyze the impact effect of different straight axial modification on the static transmission error and determine the optimum straight axial modification parameters; analyze the modification effect of different straight profile modification and determine the optimal straight profile modification; analyze the modification effect of the profile crowning and axial crowning, evaluate the tip relief effect, finally determine the optimum gear modification scheme and draw up the teeth matching scheme according to the simulation effect.

1. **Axial modification**

Under a certain load, the influence of different axial modification on the static transmission error is analyzed. Through continuous trial calculation, taking the amplitude of static transmission error fluctuation as the target, it is only considered to calculate the optimum straight axial modification on the positive and reverse driving tooth surface of each meshing gear in the straight axial modification. The following is an example of the reverse driving condition of a gear at a certain gear to illustrate the improvement of the gear contact stress distribution before and after the design of the straight axial modification. Figures 13.27 and 13.28 show the contact stress distribution before and after the straight axial modification respectively. It can be seen from the figures that the gear contact stress is more evenly distributed axially and the concentrated stress decreases after the optimization of the straight axial modification.

2. **Straight profile modification**

On the basis of completing the straight axial modification, the influence of different straight profile modification on the static transmission error is analyzed.

3. **Gear crowning**

The optimization calculation is carried out for the gear crowning on the basis of the optimum axial modification and profile modification to obtain the optimum axial crowning and optimum profile crowning with good robustness. Figures 13.29 and

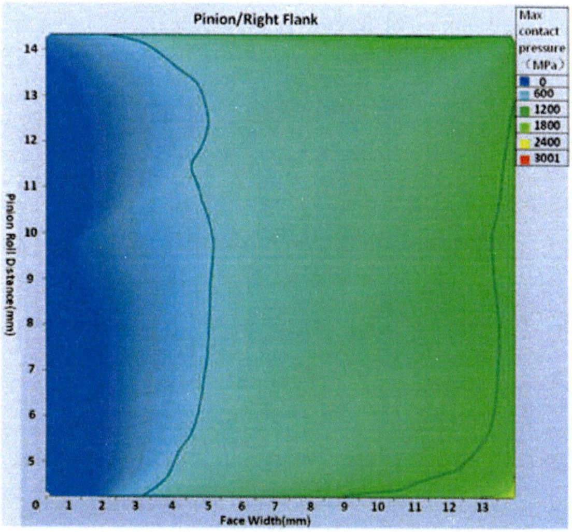

Fig. 13.27 Contact stress distribution before straight axial modification

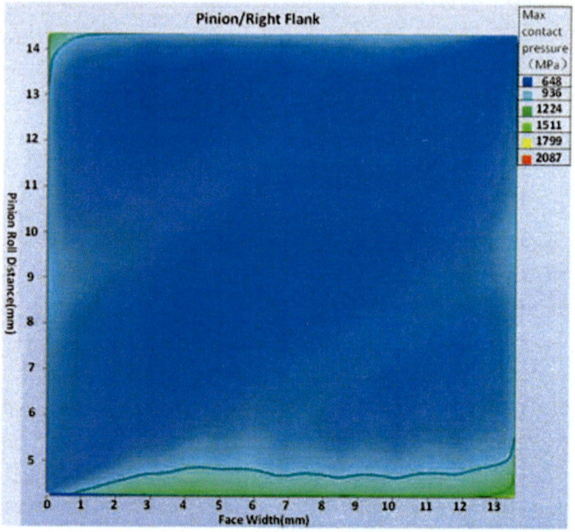

Fig. 13.28 Contact stress distribution after straight axial modification

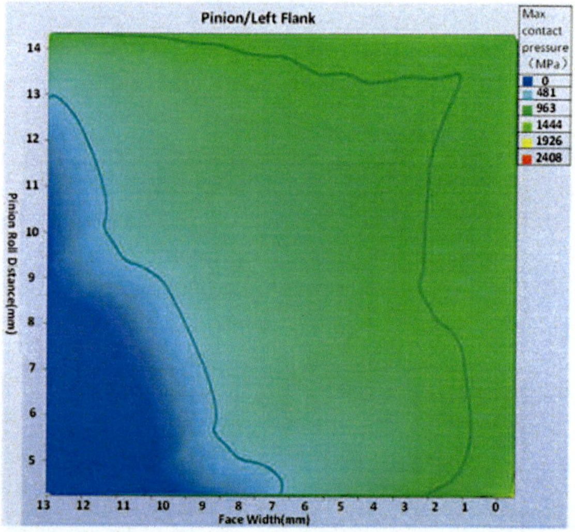

Fig. 13.29 Contact stress distribution before crowning design

13.30 show the contact stress distribution before and after the optimization of the gear crowning respectively.

4. **Tip relief**

The tip relief is designed after optimization of the straight axial modification, straight profile modification and profile crowning and then the static transmission error before and after the tip relief is compared. If the transmission error is significantly improved, it is recommended to carry out the tip relief; otherwise the tip relief is not required.

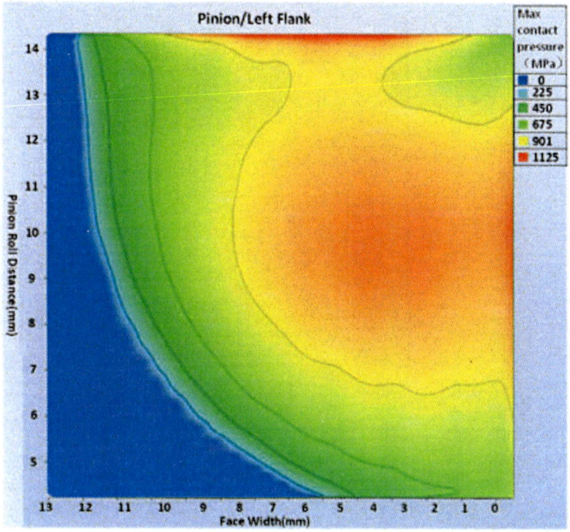

Fig. 13.30 Contact stress distribution after crowning design

Finally, the modification result is converted into gear manufacturing deviation. The straight axial modification is converted into axial dip deviation and the straight profile modification converted into straight profile dip deviation. The crowning and tip relief have consistent design relation with the gear deviation.

III. Competitive product analysis for transmission NVH

The competitive product analysis for transmission NVH is to understand the NVH level of the same kind of transmissions in typical working conditions such as acceleration and sliding, which can be used as a reference for determining the newly developed transmission NVH. The following mainly introduces the transmission NVH of a vehicle in gear 1, and the analysis method of other gears is the same. The microphone sensor in the engine compartment is arranged about 20 cm above the transmission, and the microphone sensor in the vehicle is arranged at the right ear of the driver's seat. The first order noise of the gear engagement is the main contribution of the gear engagement noise. The number of teeth of the drive gear in the gear 1 of the transmission is 17 and the number of teeth of the drive gear represents engagement of 1 order. Figure 13.31 shows the color image of noise in the interior driver's right ear at acceleration in gear 1. Obviously, the in the interior driver's right ear is more obvious in order 17; Fig. 13.32 shows the color image of noise above the transmission in the engine compartment at acceleration in gear 1. At this time, the noise is not obvious in order 17.

Figure 13.33 shows the color image of noise in the interior driver's right ear at sliding in gear 1. Obviously, the in the interior driver's right ear is more obvious in order 17; Fig. 13.34 shows the color image of noise above the transmission in the engine compartment at sliding in gear 1. At this time, the noise is not obvious in order 17.

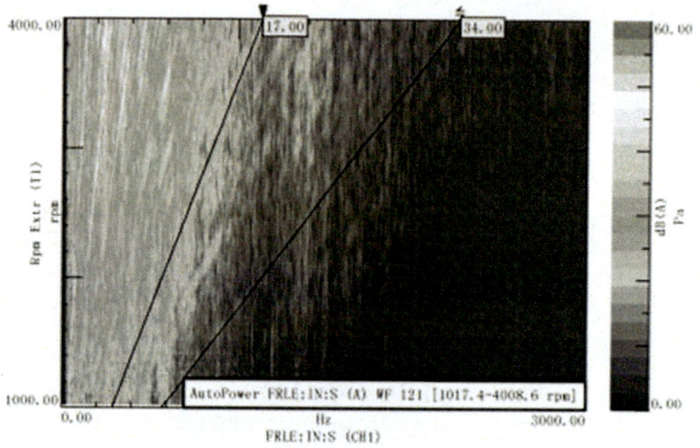

Fig. 13.31 Color image of noise in the interior driver's right ear at acceleration in gear 1

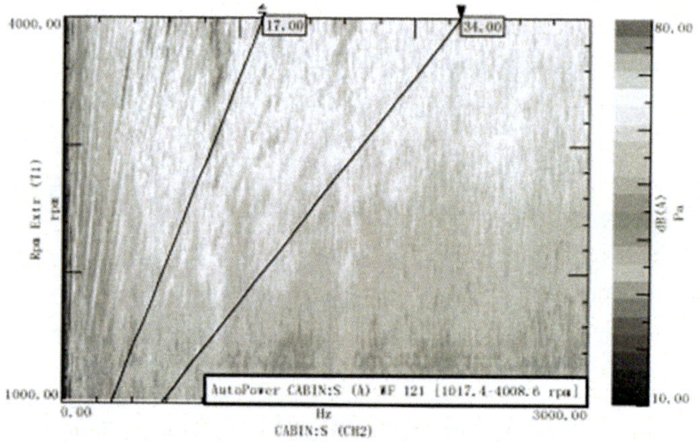

Fig. 13.32 Color image of noise above the transmission in the engine compartment at acceleration in gear 1

Figures 13.35, 13.36, 13.37, and 13.38 show the first-order noise and overall noise under typical working conditions, in which, the red line is the overall noise, and the green line is the first-order noise. Figure 13.35 shows that when the vehicle is accelerating, the noise in the interior driver's right ear has an obvious peak value of 42.01 dB (A) at the speed of 2052 r/min at the 17th order. Figure 13.35 shows that when the vehicle is accelerating, the noise above the transmission in the engine compartment has an obvious peak value of 81.89 dB (A) at the speed of 3072 r/min at the 17th order.

As shown in Fig. 13.37, the noise in the interior driver's right ear has an obvious peak value of 42.26 dB (A) at the speed of 2113 r/min at the 17th order; as shown in Fig. 13.38, the noise above the transmission in the engine compartment has an obvious peak value of 83.05 dB (A) at the speed of 3130 r/min at the 17th order.

To sum up, most of the first order noise of gear engagement in each gear of the competitive transmission at the right ear of the driver in the vehicle is less than

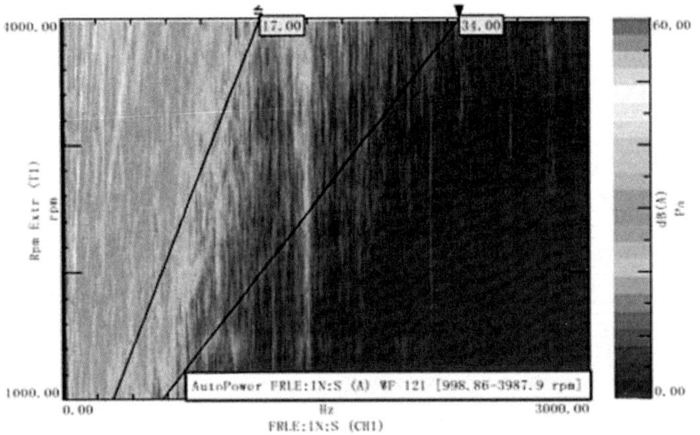

Fig. 13.33 Color image of noise in the interior driver's right ear at sliding in gear 1

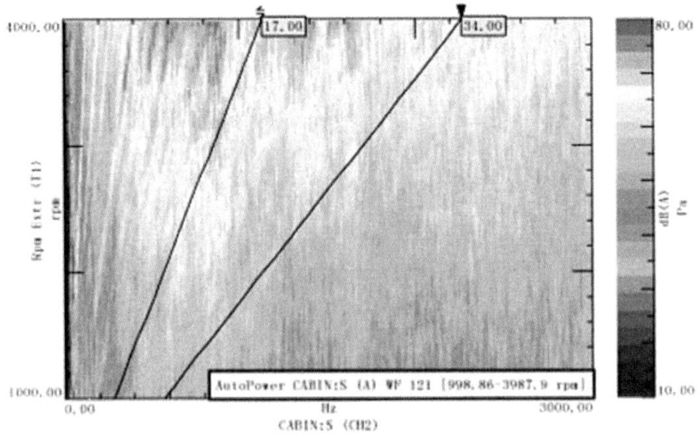

Fig. 13.34 Color image of noise above the transmission in the engine compartment at sliding in gear 1

40 dB(A), which can be used as a reference target for the development of similar new transmissions.

13.5 Typical Transmission Vibration and Noise Control

There are many kinds of transmission vibration and noise, among which gear whine and gear rattle are the most typical ones. The mechanism of production and suppression methods of the vibration and noise are introduced below.

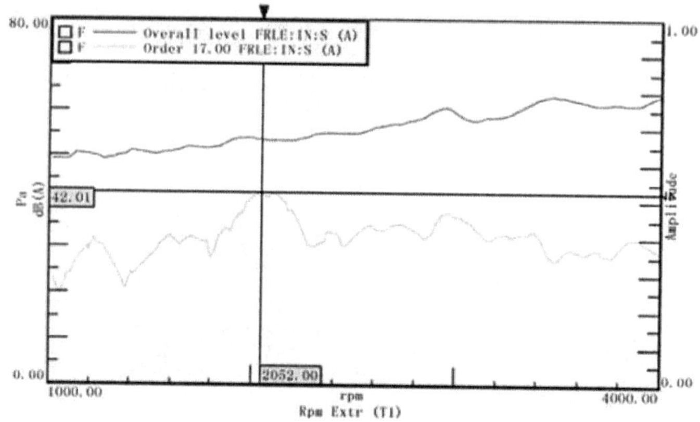

Fig. 13.35 Order diagram of noise in the interior driver's right ear at acceleration

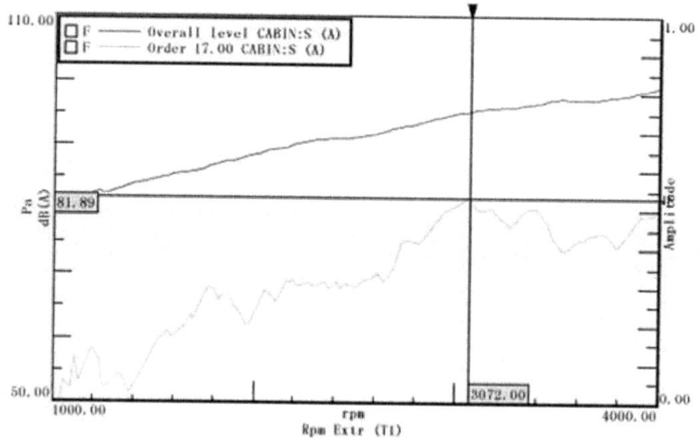

Fig. 13.36 Order diagram of noise above the transmission in the engine compartment at acceleration

I. Gear whine

The gear whine is a medium to high frequency pure tone that is easily recognizable to the ear. The primary cause for the gear whine is that the transmission error of the loaded gear pair causes the friction on the tooth surface, thus producing a periodic vibration. Resonance occurs and whine is produced when any of the harmonic frequencies of the gear is the same or very close to the natural frequency of the powertrain. The gear whine is a pain in the neck for most users and can have a serious impact on the system.

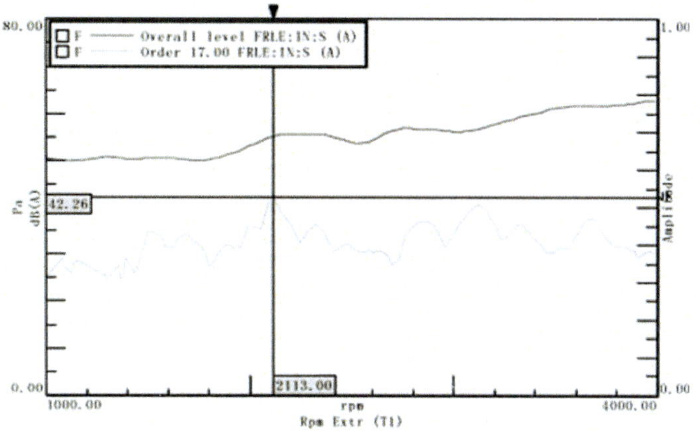

Fig. 13.37 Order diagram of noise in the interior driver's right ear

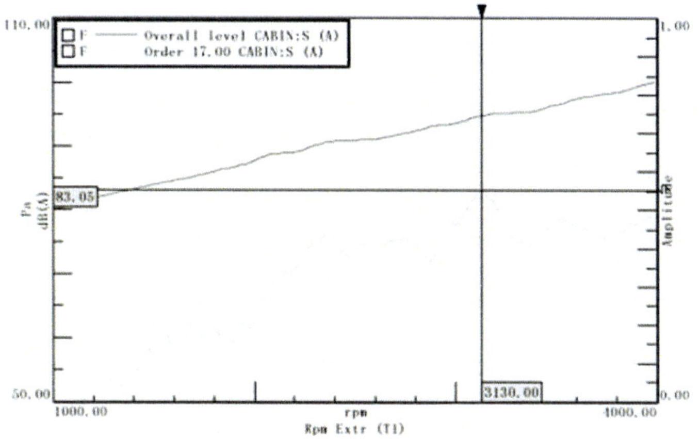

Fig. 13.38 Order diagram of noise above the transmission in the engine compartment

1. **Mechanism of production of whine**

Good gear engagement conditions are zero deflection of geometric shape, perfect centering and infinitely great gear rigidity. Figure 13.39 shows the transmission

Fig. 13.39 Actual gear engagement

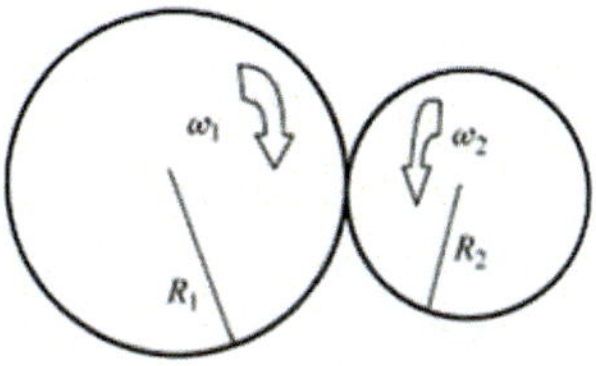

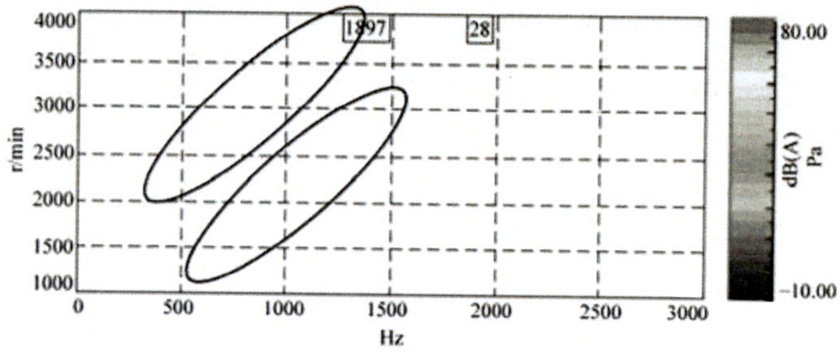

Fig. 13.40 Gear whine diagram

error in the actual gear engagement, i.e. $\omega_1 R_1 \neq \omega_2 R_2$ and the transmission error T_E is

$$T_E = \int (\omega R - \omega R) d\theta \tag{13.9}$$

The gear whine is generated by working gears, has obvious order characteristics and is related to the number of teeth and the engine speed; when the natural modal of the transmission is excited to resonance, whine becomes more apparent. Figure 13.40 is a gear whine diagram.

2. **Influence of gear parameters on whine**

The effect of transmission noise on human hearing depends on both the noise source and the transmission path. Therefore, design factors that affect both shall be sought and targeted measures shall be taken in terms of structures, parameters and materials to reduce transmission noise more effectively.

The unsteadiness of the gear engagement process causes vibration and noise. The gear is the main noise source of the transmission. In the process of transmission of power, there are many reasons and influence factors for the vibration, mainly in the following aspects: sudden change in the force on the gear; impact of gear entering and exiting the engagement; change of relative sliding and friction force between tooth surfaces during engagement; change in the tooth stiffness and elastic deformation under load conditions, resulting in the load change; uneven operation caused by the gear tooth manufacturing error.

The gear design parameters related to the above factors include modulus, number of teeth, pressure angle, helical angle, face width, addendum coefficient, modification coefficient, overlap coefficient, angle of engagement, gear modification, gear surface roughness, contact area, mounting rigidity and accuracy, load and peripheral velocity and gear pair clearance.

(1) Influence of modulus and number of teeth on noise: once the center distance is determined, the larger the number of teeth, the smaller the modulus. The noise sensitivity of gear with small modulus is lower than that of gear with large

modulus under the same overlap coefficient. Meanwhile, small modulus can increase the overlap coefficient, so that the distribution of load on the gear teeth is more uniform and the transmission is more stable.

(2) Influence of pressure angle and addendum coefficient on noise: if the gear teeth have greater flexibility and lower rigidity, the sudden change of load at the beginning and end of the gear teeth entering and exiting engagement can be reduced, and the load change caused by the gear teeth error can be compensated to some extent. The smaller the pressure angle, the lower the rigidity of the gear teeth and the stronger the adaptability. The greater the addendum coefficient, the greater the overlap coefficient, and the increase of gear height increases the flexibility of gear teeth to some extent.
(3) Influence of overlap coefficient on noise: increasing the overlap coefficient can increase the number of gear pairs engaging simultaneously, make the load distribution uniform and reduce the load change in the engagement process, which is conductive to reducing noise.
(4) Influence of helical angle, face width and contact area on noise: the relation between the noise and the helical angle & face width can be concluded as its relation with the overlap coefficient. If the face width increases, the overlap coefficient increases and the noise decreases. However, the influence of helical angle and face width on noise is limited. When the load is large, the larger the helical angle is, the lower the noise level.
(5) Influence of gear surface roughness on noise: the gear surface roughness of the normally engaged gear pair with high speed in the transmission is generally less than that of the low gear. Now many of the normally engaged gear pairs in the automobile transmission adopt the grinding process to improve the dimensional accuracy and reduce the gear surface roughness.
(6) Influence of load and peripheral velocity on noise: the greater the power (speed, load) of the same gear, the greater the noise. The sound pressure varies linearly with speed and with load.
(7) Influence of gear error and gear modification on noise: all kinds of gear errors exist objectively. Even the absolutely accurate gear will be uneven in the engagement process due to the impact of installation error and elastic deformation after load. For a gear, an ideal modification curve shall be obtained by trial and error.
(8) Influence of lubricating oil quality and injection quantity on noise: the lubricating oil has damping effect and can prevent direct contact of tooth surfaces, so the noise generally decreases with the increase of oil quantity and viscosity. However, for a transmission fully adopting helical gears, if the oil quantity is too large, the churning loss will increase and the oil temperature will rise, thus increasing the noise.

3. **Methods to suppress whine**

The methods to suppress whine include software analysis and experimental testing. The software analysis is to make the gear modification design, contact spot simulation, transfer error and vibration and noise analysis using MASTA, KISSsoft and

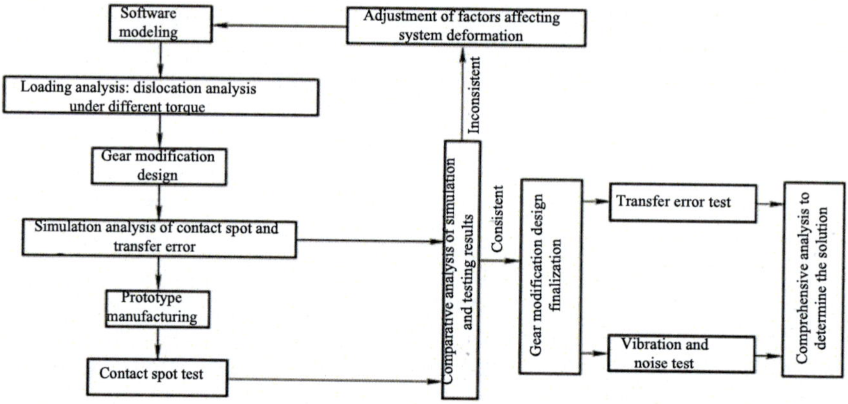

Fig. 13.41 Gear whine control method

other analysis software to optimize the gear tooth shape. The experimental testing includes contact spot test, transfer error test and vibration and noise test. The gear whine control method shown in Fig. 13.41 can optimize the contact spot, reduce the transfer error and reduce the vibration and noise.

II. Gear rattle

The gear rattle is caused by the impact between the gear meshing surfaces at drive torque due to backlash between the gear meshing surfaces and the fluctuation in the transmission input torque and speed, characterized by irregular intermittent rattle without specific frequency characteristics, and difficult to be qualitatively evaluated by frequency analysis. The gear rattle is more tolerable than the gear whine and needs to be improved in terms of reducing engine torque ripple and reducing gear backlash. Figure 13.42 is the schematic diagram of gear rattle.

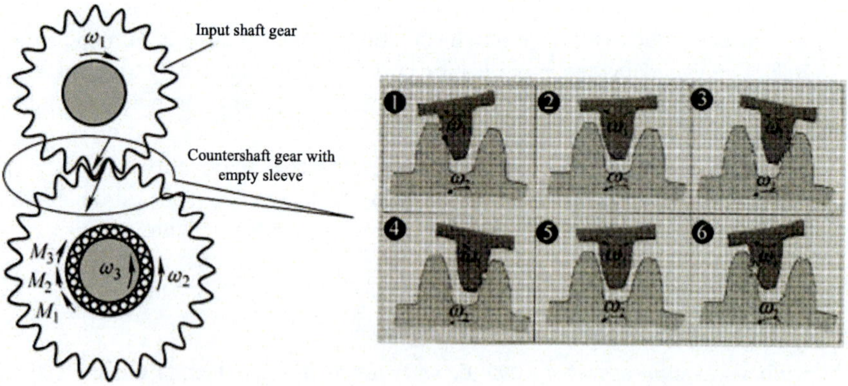

Fig. 13.42 Schematic diagram of gear rattle

The transmission gear rattle can be controlled from the following three aspects: first, reduce the engine torque ripple by increasing the rotational inertia of the flywheel, improving the engine combustion quality or accurately controlling the fuel injection quantity and time; second, reduce the rattle caused by the transmission by reducing the gear tooth clearance, increasing the shaft stiffness, reducing the rotational inertia of the gear, increasing the bearing support stiffness, increasing the transmission case stiffness or increasing the viscosity of lubricating oil; third, optimize the clutch by optimizing the torsional rigidity and damping characteristics of the clutch or reducing the clearance between the clutch driven plate spline and the transmission input shaft spline.

Bibliography

1. Jian P et al (2006) Automotive noise and vibration: principle and application. Beijing Institute of Technology Press, Beijing
2. Guan D, Su X (2004) An overview on brake vibrations and noise. Eng Mech 21(4):150–155
3. Ming Yu, Wenbin L (2003) Noise and vibration source identification in passenger car. Mach Tool Hydraul 5:117–119
4. Zhang L, Zhou H, Yu Z, et al (2001) Research on control of car interior noise due to engine vibration. J Vibr Meas Diagn 21(1):59–64
5. Yongxin C, Zuoyang Z, Maofang D (2005) Experimental studies on vehicle vibration and noise of SUVs. Automob Technol 7:25–28
6. Yong L, Xiaodong G (2004) Study on vibration and noise control of automobile transmission. Chongqing University, Chongqing
7. Jie L, Dengfeng W, Yongshun J et al (2006) The noise source identification and noise control of automobile gearbox. Noise Vibr Control 26(3):67–69
8. Yang Liang W, Xingrang ZS et al (2011) Study on idle vibration of steering system. Noise Vibr Control 31(5):80–85
9. Zhouhang S, Pengzhong L (2012) Analysis on starting Judder action mechanism of clutch self-excited vibration. Beijing Automot Eng 3:8–9
10. Zhuoping Yu, Dongxiao Y, Lijun Z et al (2005) A review on disk Brake Judder. Automot Eng 27(3):372–376
11. Changyun Z, Liting Z (2013) Analysis and optimization of accelerator pedal vibration. Automob Parts 31:48–49
12. Xuejun L, Zhi W (1999) Spectral analysis and diagnosis of gearbox noise. Vibr Shock 18(2):75–78
13. Xuanming C, Junyuan W, Weijin Ma (2011) Heavy-duty truck gear-box finite element modal analysis. Mach Des Manuf 1:22–24
14. Ye Y (2006) The study of gear yawps and gear profiling. Mech Res Appl 19(5):7–8
15. Quan S, Yuequan L, Xiaohui S et al (2010) Parameter optimization of transmission gear cluster and squeak noise control. Noise Vibr Control 6(3):46–49
16. Fujikawa T, Koike H, Oshino Y et al (2005) Definition of road roughness parameters for tire vibration noise control. Appl Acoust 66(5):501–512
17. Sheng G (2012) Vehicle noise vibration, and sound quality. SAE International, Warrendale
18. Harrison M (2004) Vehicle refinement: controlling noise and vibration in road vehicles. Amsterdam: Elsevier
19. Envia E (2002) Fan noise reduction: an overview. Int J Aeroacoust 1(1):43–64
20. Sandberg U, Ejsmont JA (2002) Tyre/road noise reference book. Springe, New York

21. Alves Filho JM, Lenzi A, Zannin PHT (2004) Effects of traffic composition on road noise: a case study. Transp Res Part D: Transp Environ 9(1):75–80
22. Liu G, Parker RG (2008) Dynamic modeling and analysis of tooth profile modification for multimesh gear vibration. J Mech Des 130(12):121402
23. Wang MY, Manoj R, Zhao W (2001) Gear rattle modelling and analysis for automotive manual transmissions. Proc Inst Mech Eng, Part D: J Automob Eng 215(2):241–258